Ecology and Ethics

Volume 8

This series is devoted to continuing research at the interfaces of ecology and ethics (embedded in the multiple fields of philosophy and ecology) to broaden our conceptual and practical frameworks in this transdisciplinary field. Confronted with global environmental change, the academic community still labors under a tradition of strong disciplinary dissociation that hinders the integration of ecological understanding and ethical values to comprehensively address the complexities of current socio-ecological problems. During the 1990s and 2000s, a transdisciplinary integration of ecology with social disciplines, especially economics, has been institutionalized via interdisciplinary societies, research programs, and mainstream journals. Work at this interface has produced novel techniques and protocols for assessing monetary values of biodiversity and ecosystem services, as illustrated by the Millennium Ecosystem Assessment. At the beginning of the 2010s, however, an equivalent integration between ecology and philosophy still remains elusive. This series undertakes the task to develop crucial theoretical and practical linkages between ecology and ethics through interdisciplinary, international, collaborative teamwork. It aims to establish a new forum and research platform to work on this vital, but until now insufficiently researched intersection between the descriptive and normative domains. The scope of this series is to facilitate the exploration of sustainable and just ways of co-inhabitation among diverse humans, and among humans and other-than-human co-inhabitants with whom we share our heterogeneous planet. It will address topics integrating the multiple fields of philosophy and ecology such as biocultural homogenization, Planetary or Earth Stewardship.

Danqiong Zhu • Ricardo Rozzi • Rika Tsuji
Benn Johnson • Terrance Caviness
Carolina Castro-Jorquera • Li Xu • Li Zhang
Editors

Linking Arts with Biocultural Conservation, Restoration, and Communication

Springer

Editors
Danqiong Zhu
Sub-Antarctic Biocultural Conservation Program
Department of Biological Sciences
University of North Texas
Denton, TX, USA

Cape Horn International Center (CHIC)
Universidad de Magallanes
Puerto Williams, Chile

Ricardo Rozzi
Sub-Antarctic Biocultural Conservation Program, Department of Philosophy and Religion and Department of Biological Sciences
University of North Texas
Denton, TX, USA

Cape Horn International Center (CHIC)
Universidad de Magallanes
Puerto Williams, Chile

Rika Tsuji
Department of Philosophy
College of Arts and Sciences
University of Rhode Island
Kingston, RI, USA

Benn Johnson
Department of Philosophy
College of Arts and Sciences
University of Rhode Island
Kingston, RI, USA

Terrance Caviness
Department of Environmental Biology
State University of New York College of Environmental Science and Forestry
Syracuse, NY, USA

Carolina Castro-Jorquera
Cape Horn International Center (CHIC)
Universidad de Magallanes
Puerto Williams, Chile

Li Xu
Department of Art Education
University of North Texas
Denton, TX, USA

Li Zhang
Laboratory of Southern Subtropical Plant Diversity, Fairy Lake Botanical Garden
Shenzhen & Chinese Academy of Sciences
Shenzhen, Guangdong, China

ISSN 2198-9729 ISSN 2198-9737 (electronic)
Ecology and Ethics
ISBN 978-3-032-04829-5 ISBN 978-3-032-04830-1 (eBook)
https://doi.org/10.1007/978-3-032-04830-1

This Springer imprint is published by the registered company Springer Nature Switzerland AG
The registered company address is: Gewerbestrasse 11, 6330 Cham, Switzerland

Foreword

Despite the unremitting efforts of philosophers such as Alasdair C. MacIntyre (1929–2005) and Hilary W. Putnam (1926–2016), the dichotomy between fact and value has remained a major challenge to the course of Western philosophy. In the field of environmental ethics, H. Rolston III argued that laws of ecology could be transformed into moral obligations only through the combination of these laws and the technical obligation to promote the circulation of nature and safeguard human survival. Unfortunately, his view seems not to be recognized by the mainstream of philosophers. In traditional Chinese moral philosophy, the dichotomy between facts and values does not seem to be a problem. The ancient Chinese philosopher Lao Tzu says, "the Dao that can be trodden is not the enduring and unchanging Dao. The name that can be named is not the enduring and unchanging name. (Conceived of as) having no name, it is the Originator of heaven and earth; (conceived of as) having a name, it is the Mother of all things" (James Legge translation). Dao, as the supreme and everlasting being, is both the cosmic thing-in-itself and the law of nature, steering and governing the movements of nature and society. He exhorted rulers to cultivate human virtues under the guidance of the Dao and to rule by *wu wei* (inaction). In other words, he made it a human virtue to follow the workings of the Dao, normalized it, and called upon people to emulate it. Only when people followed the Dao could we live in harmony with others and with heaven.

In ecology, the dichotomy between fact and value is not as severe as it might seem. Although most ecologists believe that science is value-neutral, even avoid discussing value issues in ecology and remain distant themselves from philosophy, ecology has, since its inception, served as an "economics of nature" in the course of natural resources development and utilization. In the 1970s, ecology began to provide metaphysical and theoretical support for environmental ethics, giving birth to theories and schools of thoughts such as biocentrism, ecological holism, and ecofeminism. This attribute of ecology determines that it is an interdisciplinary subject that falls in the halfway zone among the natural sciences, humanities, arts, and social sciences, and that it is more involved in the formulation and application of social value norms than any other natural sciences. This situation clearly demonstrates that ecology and philosophy are intrinsically inseparable, descriptions of different facets of the same matter, providing each other with a rich source of intellectual nourishment. For this reason, in 1997, ecologist Eugene P. Odum published the book *Ecology: A Bridge Between Science and Society*, which attempted to rectify the situation of ecology, arguing that ecology is not a traditional natural science in the strictest sense of the term, but is characterized by its scientific and humanistic aspects.

I have always believed that ecology, at least at this stage, is value-embedded, which does not prevent it from pursuing a genuine understanding of the laws of ecosystems. Leaving aside its etymological meaning by Ernst Haeckel (1834–1919), the history and current status of ecology also shows that, constrained by the existing knowledge and means of humans, ecology has not yet been able to fully attain a true representation of ecological phenomena and processes. Some core concepts and theories are inevitably constructed with a subjective tendency and marked with the values of the times.

The integration of ecology and ethics is therefore logical and not at all incongruous. The opponents of ecological value, if they were to give us a convincing reason for denying the value implicit in ecology, the only thing they could do would be to eliminate, for example, metaphorical and teleological language from ecological theories altogether. No one, at least so far, has had the nerve or the capacity to do that, and even if they did, it would be inconsistent with the way science is done.

Fortunately, the research and practical achievements of groups of scholars in the fields of philosophy of ecology, environmental conservation, eco-aesthetics, and the conservation of natural-cultural heritage, both at the theoretical level and practical level, have resulted in the fusion of ecology and ethics, having yielding remarkable achievements and giving birth to a number of novel research areas, such as biocultural ethics, environmental humanities, ecopsychology, and so on. In particular, the concepts of biocultural ethics and biocultural conservation proposed by Ricardo Rozzi have created a breakthrough for the integration of ecology and ethics.

According to Rozzi, overwhelming globalization has led to the introduction or encroachment of an increasing number of non-native species, which, while having some positive effects on local ecosystems and cultures, have resulted in the erosion of biological and cultural diversity, leading to biocultural homogenization. In order to counteract these risks, it is imperative to explore the ecological wisdom and values embedded in local biocultural systems, including the complex interrelationships among local people, other-than-human organisms in the local region, their habitat, and vital ecological relations. In this book, Rozzi reviews the career of American environmental philosopher Eugene C. Hargrove, and he credits Hargrove's pioneering analysis of the history of ideas and interdisciplinary approach to environmental ethics and aesthetics for making a significant contribution to the search for the root causes of the global environmental crisis. The impact and destruction of globalization on the cultural, educational, and value dimensions of local cultures is so severe that it directly threatens the sustainability of life and the conservation of biocultural diversity. Rozzi believes that the ethic of biocultural co-inhabitation, by promoting the well-being of co-inhabitants, respecting their way of life, and caring for the habitats they share, can counteract the increasing global biocultural homogenization.

Guided by the theories of Hargrove and Rozzi, the authors of *Linking Arts with Biocultural Conservation, Restoration, and Communication* address how to cope with the impacts of biocultural homogenization at both theoretical and practical levels, and how to develop ethical relationships between humans and their local environment through arts.

In this volume, Danqiong Zhu, Ricardo Rozzi, Rika Tsuji, Benn Johnson, Terrance Caviness, Carolina Castro-Jorquera, Li Xu, and Li Zhang offer multiple perspectives and case studies about how to promote the conservation of biological diversity and traditional culture, while ensuring the sustainable development of local economies as well as safeguarding the rights and interests of socially disadvantaged groups. I believe that these methods and measures have three main features:

(a) They emphasize the role played by arts and education in the conservation of biocultural heritage.
(b) They demonstrate the involvement of interdisciplinary and international scholars in biocultural conservation.
(c) They emphasize the role of local practices and cultures in the interaction between human beings and the natural environment, which essentially contributes to the conservation of biocultural diversity.

First of all, arts and education can provide us with a deeper understanding of the co-inhabitation of human and other-than-human beings, to experience first-hand the wonder and value of this relationship, and to enable people to uncover and safeguard biocultural diversity. For example, Tsuji and Johnson propose a relational biocultural ontology of human and other-than-human co-inhabitation. They argue that art derives from a dynamic relationship with the local land and responds to the biocultural characteristics of the place, so the goal of biocultural conservation should aim to understand and support the dynamic expression of these ontologically significant local relationships. Carolina Castro-Jorquera analyzes Cecilia Vicuña's work through a biocultural lens, arguing that art can be used as a form of practice and cognition to help repair the relationship between human beings and the environment, and to re-establish the kind of connection that our ancestors had with the land. McIntosh's embodied approach is a reversal of the traditional humanist model of education and can allow us to dialog with the biocultural community and deeply perceive the meaning of coexistence. According to Chiarini, scientific illustration, as a biocultural tool, conveys multicultural perspectives and knowledge on the same subject. It incorporates the illustrator's ethical vision and carries the motive of promoting biocultural ethics; it also allows the viewer to experience the world through the eyes or ears of others, thus breaking down the barriers between human beings and the natural world to allow for a better biocultural experience. Vezzani also expresses that art should assume an ethical role in generating works that pose questions about our relationship with cultural and biological diversity, especially at the current times of planetary crisis. Zhu examines the relationship between arts and the environment after the emergence of Artificial Intelligence. She argues that despite its ability to simulate and replicate, generating AI technology is incapable of authentically communicating biocultural connection between human beings and the natural world, resulting in detachment and alienation from the environment. The value of multiple evaluative behaviors in the symbolic, cultural, and socio-ecological systems embedded in artists' interaction with the biophysical environment cannot be conveyed in AI generated images.

Linking Arts with Biocultural Conservation, Restoration, and Communication is also an example of interdisciplinary research. Readers will learn about the theoretical deliberations and practical experiences of scholars from various disciplinary backgrounds on biocultural conservation and communication in their respective fields. For example, the implementation of cultural-landscape conservation can accomplish not only the conservation of the traditional agricultural wisdom but also promote rural revitalization, so that more disadvantaged communities can benefit from the conservation of local cultural landscape. The *Wangshan Life* model in Wuyuan County, Jiangxi, China, the nature-based solutions at Xunjinsi village combine the multiple disciplines such as landscape ecology, heritage conservation, environmental education, tourism, and arts to develop a unique conservation model that parallels biocultural conservation. The bryophyte team at the Shenzhen Fairy Lake Botanical Garden combines bryology, ecology, environmental education, popularization of science, and arts to demonstrate how to introduce small and overlooked organisms to Chinese and international citizens. Since 2007, they have systematically introduced the diversity and beauty of bryophytes through various means such as publications, paintings, videos, public talks, and exhibitions. In Franquesa-Soler and Mesa-Jurado's evaluation of practical experiences in primate conservation education in schools in southeastern Mexico, they also synthesized knowledge from multiple disciplines to promote the role of Photovoice in guiding the search for solutions that benefit both wildlife conservation and human communities.

Native and local cultures, such as traditional Chinese culture, can be integrated with biocultural ethics to consolidate the ontology of biocultural conservation, guide the restoration of a sound co-inhabitation relationship between humans and other-than-human beings, and better conserve the diversity of local biodiversity and cultures. Song argues that traditional Chinese culture lays the foundation for understanding interspecies kinship through Taoism *wu hua* (the transformation of things), Confucianism concept of *jiao hua* (inculcate), and Buddhism "redeemable properties" of all living beings. They have laid the foundation for communicating interspecies affinities and resonate with biocultural ethic in the concept of co-inhabitation. In practice, Jin's Healing Garden project integrates arts, science, ecological restoration, eco-pedagogy, activism, and community engagement to promote place-based experiential learning and explore an educational model that reimagines human-nature interactions by adhering to holistic thinking and permaculture. Drawing on the Four Gentlemen (plum, orchid, bamboo, and chrysanthemum) from the history of Chinese painting and arts, Hu and Yang adopt biocultural lens to reveal how art education can be a vehicle for repairing ethical human-plant co-inhabitation and reimagining place-based future. Zhang and Xu conducted field studies on Chinese black cardamom (*Lanxangia tsao-ko*) and explored the potential of applying watercolors for biocultural communication. These efforts transform issues of biocultural homogenization into artistic opportunities and educational assets that integrate critical and constructive activism, hope, symbiosis, and community empowerment. The traditional healing system of the Lamkaang people and the *Loh Kchet* of collective interdependence also provide us with good examples of interspecies co-inhabitation in northwest India and local cultural traditions.

If you read this book to the end, I believe you will come to the following conclusion: there is no contradiction between the promotion of biocultural conservation and the good life that humanity has been seeking. What we need to know is that the desire for a good life is equally present in beings other than us human beings. A good life should include an appreciation of the ethical value and moral status of other-than-human beings. We should learn how to live with them and to value this beautiful planet as a habitat for all living beings. Embracing such a vision might enlighten those ecologists and citizens who are caught up in the mire of a "narrow minded scientism," thereby helping us to develop a broad ecological worldview that comprises a deep philosophy of environmental values. What is vital to know is that, in our moral life, if we overlook the existence of other-than-human animals and plants, human beings will hold no value either. If nature is devoid of moral significance, human beings will cease to have moral significance either. Human moral values will ultimately be illusory.

South China Normal University
Guangzhou, Guangdong Province
China

Yonglin Ge

Acknowledgments

This volume conveys the work of a community of philosophers, scientists, artists, designers, scholars, students, and people from various regions of the world. We are grateful to our institutions in the USA, Chile, and China, which have supported our editorial work at the Cape Horn International Center (CHIC), University of Magallanes (UMAG) in Chile and at the Department of Biological Sciences, the Department of Philosophy and Religion, University of North Texas (UNT), and the Center for Environmental Philosophy (CEP) in the USA. We also thank the Turenscape Academy and College of Architecture and Landscape, Peking University, Beijing, and the research team on bryophytes at the Shenzhen Fairy Lake Botanical Garden, Shenzhen, China, whose work inspired several chapters of this book.

We appreciate the encouragement and editorial guidance provided by Springer for this eighth volume of the Ecology and Ethics book series. At UNT, we are especially thankful for Eugene Hargrove at the Center for Environmental Philosophy and Michael Thompson at the UNT office of the Sub-Antarctic Biocultural Conservation Program (SBCP), who provided invaluable assistance for the production of this volume.

Part of the inspiration for developing *Linking Arts with Biocultural Conservation, Restoration, and Communication* came from volume 5 in the Ecology and Ethics book series, "Field Environmental Philosophy: Education for Biocultural Conservation," edited by Ricardo Rozzi, Alejandro Tauro, T Wright, Noa Avriel-Avni, and Roy May, and published in 2023. Several chapters in our new volume continue the themes of field environmental philosophy, including reflections from ontological and epistemological perspectives, as well as practice in other places, particularly in Asia.

We are very grateful to those who sent their manuscripts in the first instance, even though this book contains only chapters that have passed peer review after several rounds of revisions. We are grateful to all contributors of this book, who have enthusiastically entrusted this book with their thoughts, practices, and experiences in linking place-based, locally rooted, and resilience-oriented arts with biocultural conservation, restoration, and communication.

The Chilean National Agency for Research and Development through the Cape Horn International Center, CHIC (ANID/BASAL FB210018), has supported both research and editorial work for the preparation of the volume *Linking Arts with Biocultural Conservation, Restoration, and Communication*. This book is a contribution to the Cape Horn International Center (CHIC) coordinated by UNT, Denton, Texas, and UMAG, Puerto Williams, Chile.

Contents

Editors and Contributors

About the Editors

Danqiong Zhu Sub-Antarctic Biocultural Conservation Program, Department of Biological Sciences, University of North Texas, Denton, TX, USA

Cape Horn International Center (CHIC), Universidad de Magallanes, Puerto Williams, Chile

Ricardo Rozzi Sub-Antarctic Biocultural Conservation Program, Department of Philosophy and Religion and Department of Biological Sciences, University of North Texas, Denton, TX, USA

Cape Horn International Center (CHIC), Universidad de Magallanes, Puerto Williams, Chile

Rika Tsuji Department of Philosophy, College of Arts and Sciences, University of Rhode Island, Kingston, RI, USA

Benn Johnson Department of Philosophy, College of Arts and Sciences, University of Rhode Island, Kingston, RI, USA

Terrance Caviness Department of Environmental Biology, State University of New York College of Environmental Science and Forestry, Syracuse, NY, USA

Carolina Castro-Jorquera Cape Horn International Center (CHIC), Universidad de Magallanes, Puerto Williams, Chile

Li Xu Department of Art Education, University of North Texas, Denton, TX, USA

Li Zhang Laboratory of Southern Subtropical Plant Diversity, Fairy Lake Botanical Garden, Shenzhen & Chinese Academy of Sciences, Shenzhen, Guangdong, China

Contributors

Gonzalo Arriagada Professional Photographer, Puerto Williams, Chile

Carolina Castro-Jorquera Cape Horn International Center (CHIC), Universidad de Magallanes, Puerto Williams, Chile

Terrance Caviness Department of Environmental Biology, State University of New York College of Environmental Science and Forestry, Syracuse, NY, USA

Valeria Serena Chiarini Science Writer and Illustrator, Milan, Italy

Montserrat Franquesa-Soler Instituto de Neuroetología, Universidad Veracruzana, Xalapa, México

Miku Conservación, AC, Xalapa, México

Jun Hu Fine Arts School, Hangzhou Normal University, Hangzhou, Zhejiang, China

Yifan Hu College of Architecture and Landscape, Peking University, Beijing, China

Meng Jia Department of Philosophy, Xi'an Jiaotong University, Xi'an, Shaanxi, China

Lipeng Jin School of Experimental Arts, Sichuan Institute of Fine Art, Chongqing, China

Benn Johnson Department of Philosophy, College of Arts and Sciences, University of Rhode Island, Kingston, RI, USA

Sumshot Khular Department of Philosophy and Religion, University of North Texas, Denton, TX, USA

Jianing Li College of Urban and Environmental Sciences, Peking University, Beijing, China

Zihan Li School of Chinese Language and Literature, Catholic University of Korea, Bucheon, Gyeonggi Province, South Korea

Roy Mackenzie Cape Horn International Center (CHIC), Centro Universitario Cabo de Hornos, Universidad de Magallanes, Puerto Williams, Chile

Instituto de Ciencias Aplicadas, Universidad Autónoma de Chile, Santiago, Chile

Millenium Institute Biodiversity of Antarctic and Subantarctic Ecosystems (IBASE), Santiago, Chile

Kun Ma Hangzhou Huacheng Inc., Hangzhou, Zhejiang, China

Shoshana McIntosh Department of Philosophy and Religion, University of North Texas, Denton, TX, USA

Shu Meng Kiang Wu Nursing College of Macau, Macau, China

M. Azahara Mesa-Jurado El Colegio de la Frontera Sur, Unidad Villahermosa, Sección, México

Ricardo Rozzi Sub-Antarctic Biocultural Conservation Program, Department of Philosophy and Religion and Department of Biological Sciences, University of North Texas, Denton, TX, USA

Cape Horn International Center (CHIC), Universidad de Magallanes, Puerto Williams, Chile

Yubo Song Institute of Chinese Culture and Thoughts, Northwest University, Xi'an, Shaanxi, China

Center for Environmental Philosophy and Comparative Philosophical Research, Northwest University, Xi'an, Shaanxi, China

Michael Thompson Sub-Antarctic Biocultural Conservation Program, Department of Philosophy and Religion, University of North Texas, Denton, TX, USA

Rika Tsuji Department of Philosophy, College of Arts and Sciences, University of Rhode Island, Kingston, RI, USA

Cristian Valle-Celedón Audiovisual Producer in Wolf Productions, Punta Arenas, Chile

Paola Vezzani Cape Horn International Center (CHIC), Puerto Williams, Chile

Juan Wang Department of Arts, School of Humanities and Arts, Xidian University, Xi'an, Shaanxi, China

Mengyun Xie School of Urban Design, Wuhan University, Wuhan, Hubei, China

Li Xu Department of Art Education, University of North Texas, Denton, TX, USA

Tusha Yakovleva Center for Native Peoples and Environment, The State University of New York College of Environmental Science and Forestry, Syracuse, NY

Panpan Yang Department of Art Education, University of North Texas, Denton, TX, USA

Kongjian Yu College of Architecture and Landscape, Peking University, Beijing, China

Turenscape Academy, Xixinan, Anhui, China

Hong Yun South China University of Technology School of Design, Guangzhou, Guangdong, China

Li Zhang Laboratory of Southern Subtropical Plant Diversity, Fairy Lake Botanical Garden, Shenzhen & Chinese Academy of Sciences, Shenzhen, Guangdong, China

Stephen Zhang Department of Design, College of Visual Arts and Design, University of North Texas, Denton, TX, USA

Changhui Zheng College of Humanities and Development Studies, China Agricultural University, Beijing, China

Yan Zhou Systemic Restoration Department, Land Consolidation and Rehabilitation Center, Ministry of Natural Resources, Beijing, China

Danqiong Zhu Sub-Antarctic Biocultural Conservation Program, Department of Biological Sciences, University of North Texas, Denton, TX, USA

Cape Horn International Center (CHIC), Universidad de Magallanes, Puerto Williams, Chile

Qin Zuo Laboratory of Southern Subtropical Plant Diversity, Fairy Lake Botanical Garden, Shenzhen & Chinese Academy of Sciences, Shenzhen, Guangdong, China

Chapter 1
Biocultural Communication: Arts and Sciences for Conserving Biological and Cultural Diversity

Danqiong Zhu and Ricardo Rozzi

Abstract In this introductory chapter, we propose a re-evaluation of the arts' role in biological conservation and ecological restoration by introducing the concept of *biocultural communication*. Traditionally dominated by scientific methodologies, conservation and restoration have often sidelined the epistemic, ethical, and communicative potential inherent in the arts and humanities. In response to this limitation, we develop a transdisciplinary framework that integrates science, art, and cultural diversity to promote more inclusive and context-sensitive conservation and restoration efforts. Biocultural communication enhances and transforms scientific and ecological communication by embracing transdisciplinary, intercultural, and interspecies dialogues. In this way, it fosters pluralistic engagements among both human and other-than-human communities. In contrast to the deficit model of science communication—which assumes public ignorance—the biocultural perspective acknowledges the significance of local knowledge, symbolic connections to the environment, and the agency of other-than-human beings. Through case studies from Latin America, Asia, and Europe, included in *Linking Arts with Biocultural Conservation, Restoration, and Communication*, we present artistic expressions—from installations and exhibitions to traditional practices and oral narratives—that contribute to processes of biocultural communication as well as of conservation and restoration. These processes stimulate cultural revitalization and environmental education. The arts offer a threefold contribution: (i) they enable us to visualize

D. Zhu
Sub-Antarctic Biocultural Conservation Program, Department of Biological Sciences, University of North Texas, Denton, TX, USA

Cape Horn International Center (CHIC), Universidad de Magallanes, Puerto Williams, Chile
e-mail: Danqiong.Zhu@unt.edu

R. Rozzi (✉)
Sub-Antarctic Biocultural Conservation Program, Department of Philosophy and Religion and Department of Biological Sciences, University of North Texas, Denton, TX, USA

Cape Horn International Center (CHIC), Universidad de Magallanes, Puerto Williams, Chile
e-mail: Ricardo.Rozzi@unt.edu

D. Zhu et al. (eds.), *Linking Arts with Biocultural Conservation, Restoration, and Communication*, Ecology and Ethics 8,
https://doi.org/10.1007/978-3-032-04830-1_1

relationships often overlooked by global society; (ii) they recover collective memories; and (iii) they foster a biocultural ethic of co-inhabitation in biocultural diversity. We assert that systematically incorporating the arts into conservation and restoration not only broadens its scope but also provides conceptual and practical tools to address disciplinary fragmentation and counter biocultural homogenization. This chapter sets forth the central objective of the book: to encourage us to understand and envision models of biocultural conservation and conservation that stimulate the regeneration of complex networks of meanings, affections, and both symbolic and material relationships linking human communities with the diversity of co-inhabitants with whom we share our habitats, territories, and the whole biosphere.

Keywords Science communication · Transdisciplinarity · Intercultural dialogues · Interspecies communication · Biocultural ethics

1.1 Introduction

Like practical trades such as agriculture or medicine, biological conservation and ecological restoration are both science and art. However, in the latter context, approaches have often exhibited significant bias toward scientific disciplines. In this eighth volume of the Ecology and Ethics series, we address this scientific bias by highlighting historical and contemporary perspectives that underscore the essential role the arts play in these transdisciplinary fields of conservation and restoration.

The integration of the arts reveals that the journey of conservation encompasses not only biological and ecological aspects, but also cultural dimensions. In doing so, the arts foster the development of biocultural approaches. These "art-informed" concepts and practices broaden the spectrum of biocultural approaches that have previously emerged from the medical and anthropological sciences (Armelagos et al. 1992).

In *Linking Arts with Biocultural Conservation, Restoration, and Communication*, we illustrate how the arts contribute valuable concepts and methodologies, and how diverse cultural traditions enrich them. Consequently, by emphasizing the significance of the arts, our book offers three key contributions to the theory and practice of biocultural conservation.

1. The arts play a pivotal role in shaping the social fabric that underpins biocultural conservation practices, serving as a catalyst for communication—specifically, biocultural communication—within community movements.
2. By integrating the humanities and philosophy, the arts weave together biocultural approaches that illuminate the epistemological, ontological, political, and ethical dimensions necessary for understanding and appreciating both biological and cultural diversity, along with their interconnections.
3. The arts further enhance biocultural conservation by incorporating a wide array of cultural traditions. Different chapters in our book explore the role of various artistic practices in promoting biocultural conservation within diverse cultural

and geographical contexts. Through a biocultural comparative lens, the research investigates artistic expressions grounded in diverse philosophical traditions—such as Buddhism, Taoism, Confucianism, Indigenous philosophies, and Latin American thought—and how these traditions are integrated into conservation efforts across Asia, Europe, and the Americas. This inclusion of contrasting philosophical perspectives is crucial, given the predominance of Anglo-Saxon viewpoints in environmental philosophy. By embracing non-Eurocentric perspectives, we enrich environmental philosophy and cultivate more inclusive conservation strategies. Highlighting culturally diverse artistic practices showcases their potential to enhance community engagement and support sustainable environmental stewardship, as evidenced by biocultural conservation initiatives spearheaded by artists, philosophers, biologists, educators, and community leaders.

The arts serve as catalysts of biocultural communication. At the early stage of Western civilization, Plato defined art as mimesis, or imitation, of the physical world. For Plato, artworks are copies of copies and imitations of imitations. Ontologically, artworks are inferior to the physical world, and the physical world and objects in it are inferior to the realm of forms and every form in the realm. Later, the etymological origin of the word "art" and its plural form "arts" leads us to the old Latin word *ars*, which indicates practical knowledge relevant to a specific "skill," "craft," or "technique." This Western reconceptualization of art brings the concept closer to its understanding in Chinese classical thought. In Chinese classical tradition, the term Six Arts (六艺) refers to *Li* (禮, rites), *Yue* (樂, music), *She* (射, archery), *Yu* (御, chariot riding), *Shu* (書, calligraphy), and *Shu* (數, mathematics), as well as practical knowledge. These classical arts demand intellectual, physical, aesthetic, and communicational skills.

Ultimately, *Linking Arts with Biocultural Conservation, Restoration, and Communication* emphasizes the critical role of artistic and philosophical diversity in shaping innovative biocultural conservation, restoration, and communication initiatives. Below, we begin by examining our proposed new concept, biocultural communication, and its interrelationship with biocultural restoration and conservation.

1.2 From Ecological to Biocultural Communication

Biocultural communication integrates transdisciplinary, intercultural, and interspecies communication. Consequently, it extends and transforms the Eurocentric, but globally influential, concept of ecological communication.

The conceptual framework of ecological communication was forged by the German sociologist Niklas Luhmann (1927–1998), who developed a complex analysis of the relationship between society and the environment. In his book *Ecological Communication* (1989), Luhmann examined the spheres of economics, law, science, politics, religion, and education to illustrate how these areas interact with environmental issues. This book is a foundational work in which Luhmann critically examines the notions of modernity and rationality. However, from a biocultural perspective, Luhmann's approach has two limitations that we must address. To illustrate them concisely, we will quote two excerpts from the opening of his book:

> On 15 May 1985, at the invitation of the Rhenish-Westfallian Academy of Science, I addressed their yearly assembly on the theme, Can Modern Society Adjust Itself to the Exposure to Ecological Dangers? … The main argument of the address, namely that modern society creates too little as well as too much resonance because of its structural differentiation into different function systems, was presented only in its main outline. Only from this insight does it follow that the solution to this problem can be found in new ideas about values, a new morality or an academic elaboration of an environmental ethics. (Luhman 1989, p. xvii)

Later, in Chap. 1, "Sociological Abstinence," Luhmann states that:

> Originally, sociology had been concerned with the internal aspects of society… Nature, on the other hand, could and indeed had to be left to the natural sciences. What the new discipline called sociology could discover and claim as its own field of study was either society or, if this concept was unsatisfactory, social facts. … Thus the delimitation of the discipline had to be interpreted as a demarcation of a section of reality… The problematic is reduced to structures of the social system or its subsystems. … The external sources of the problems are not even considered. And although every problem of the system is ultimately reducible to the difference between system and environment, this is not even considered. Even for the earlier theory of *societas civilis* this was no different, and the same is true for practical philosophy: what is social was viewed as civitas, as communitas perfecta or as political society, even if this included all of humanity. (Luhmann 1989, pp. 1–2)

These two quotes from the book *Ecological Communication* allow us to address both the contribution of Luhmann's proposal and its limitations. On the one hand, these two passages from his book announce a significant contribution Luhmann made by extending the concept of "ecology" to refer to the connections between social systems and the natural environment. To this end, Luhmann criticized the schizophrenia between disciplines and social functions. Furthermore, he traces the historical development of the notion of "environment," culminating in the modern definition that ultimately separated social systems from the external environment.

On the other hand, in the two previous quotes, we can detect the two significant limitations of Luhmann's analysis. First, he limits himself to examining European culture, which fails to account for the great diversity of symbolic, linguistic, and material relationships between other cultures and natural systems. Consequently, Luhmann overlooks records and analyses of intercultural communication. Second, at no point does it open the agency of other-than-human beings. Consequently, Luhmann dismisses records and analyses of interspecies communication. The chapters in our book present artistic and philosophical concepts and practices embedded in case studies of biocultural restoration and conservation, which include forms of intercultural communication and interspecies communication, thus laying the empirical groundwork for the development of a new concept: *biocultural communication*.

An examination of the essential elements of science communication indicates that many global systems operate under a "deficit model" (Simis et al. 2016). This model is based on the premise that enhancing communication can increase public support for modern science, technology, and relevant policies (Dickson 2005). Consequently, enhanced science communication would reduce skepticism, which is primarily due to a lack of adequate scientific or environmental knowledge (Schiller et al. 2001). Discourses advocating for improving environmental literacy (Simmons 1995), ecological literacy (Risser 1986; Odum 2013; Jordan et al. 2009), and eco-literacy (Orr 1992; Wooltorton 2006) are also expressions of an inherent belief in the deficit model.

By framing the public as ignorant, the deficit model overlooks the competencies and experiences that diverse communities bring to discussions about ecological issues. This limitation highlights the need for a more comprehensive communication framework that values and incorporates diverse perspectives, fostering richer conservation and restoration practices. To address this limitation, biocultural communication incorporates intercultural dialogues, which reveal a rich diversity of knowledge forms. Hence, the assumption that the public is ignorant is not generally applicable. Indigenous and other local communities, minorities, and ethnic groups contribute a diversity of knowledge forms, which do not undermine their credibility in understanding and articulating the socio-ecological issues that are vital to them. Through the integration of transdisciplinary and intercultural dialogues, biocultural communication presents itself as an invitation, a gesture, and a welcoming of plurilateral sharing, exchanging, and imparting of information relevant to nurturing the conservation and restoration of local/regional/global biocultural diversity.

Biocultural communication also encompasses interspecies communication, which is the exchange of information among different species of animals, plants, fungi, or microorganisms (Gavin 2024). This emerging field of research uncovers structural and semantic principles that challenge traditional notions of communication as a uniquely human trait (Oyewole et al. 2025). Studies on primates, cetaceans, birds, and insects have revealed underlying structural and semantic principles of animal communication (Zuberbuhler 2000; Fichtel 2004). Additionally, interspecies communication embraces interactions between humans and other-than-human species (Meijer 2019).

Biocultural communication is based on three fundamental principles:

- Promote collaborative work and the exchange of transdisciplinary knowledge about the complex interconnections between co-inhabitants, their life habits, and their shared habitats. This includes the inclusion of disciplines less considered than the natural sciences and economics, such as the arts, Indigenous and other forms of local knowledge, ethnobiology, ethology, psychology, ecotheology, formal and non-formal education, the humanities, environmental humanities, anthropology, biocultural diversity studies, philosophy, and biocultural ethics.
- Catalyze intercultural dialogues, particularly the exchange of different forms of ecological knowledge, practices, and wisdom. A key focus of these intercultural dialogues is to forge visions and actions that contribute to the well-being of humans and other-than-human communities. Inspired by the conceptual

framework of the biocultural ethic, intercultural dialogues and biocultural communication take into account the biophysical, cultural (symbolic, linguistic, and material), and institutional (social, political, economic, and technological) particularities of each biogeographical and cultural region.

- Orient forms of interspecies co-inhabitation, with special attention to the worldviews, rituals, and behavioral coordination that emerge from forms of communication between humans and other-than-human co-inhabitants. This communication explores nonverbal forms of communication, focusing on information between individuals and communities, including both human and other-than-human beings. These include, among others, behavioral coordination and information exchanges through visual, auditory, chemical, and tactile signals. Forms of interspecies communication are explored, among others, through artistic practices, ecopsychology, trans-species psychology, ethology, Indigenous and local knowledge, more-than-human anthropology, and biocultural ethics.

These transdisciplinary, intercultural, and interspecies fundamental elements underscore the need for profound philosophical, political-ecological, and practical responses to crucial themes like anthropocentrism and human exceptionalism (Haraway 2008), cognitive and interspecies injustice (Barrett et al. 2021), and ecological democracy (Romero and Dryzek 2021). Additionally, issues such as speciesism (Bradshaw and Watkins 2006; Adams 2025), interspecies justice (Healy and Pepper 2021), socioenvironmental justice (Fioret 2023; Hanazaki 2024), and biocultural ethics (Rozzi 2013) must also be considered.

Notable examples of biocultural communication that align with its principles include a multi-ethnic bird guide that integrates both Indigenous and scientific ornithology knowledge in southern South America (Rozzi 2010, 2014), the study of vegetal, animal and cultural semiosis (Kull 2009), and the reinterpretation of hydrological infrastructure and natural landscape feature (Klaver 2013). Furthermore, ethnographic and liberation theological analyses of relationships of co-inhabitation among Aymara communities, llamas, potatoes, and mountains (Mamani-Bernabé 2015; May 2015, 2017), and the interaction between honeyguides and honey hunters in Tanzania and Mozambique (Spottiswoode and Wood 2023) exemplify these interspecies connections.

Several chapters in our book further illustrate the multifaceted nature of biocultural communication through artistic interactions between humans and other species (Chiarini 2026; Hu and Yang 2026; McIntosh 2026; Meng 2026; Zhang and Xu 2026; Zuo and Zhang 2026) as well as with the land (Khular 2026; Wang 2026). These explorations present a rich tapestry of how biocultural communication can lead to deeper understandings and actions regarding the intercultural and interspecies co-inhabitation.

1.3 From Ecological to Biocultural Restoration

Biocultural communication enhances biocultural restoration. To understand the concept of biocultural restoration, it is essential to trace the origins of ecological restoration. The founding member of the Society of Ecological Restoration (SER) William Jordan and environmental historian George Lubick define ecological restoration as:

> the literal re-creation of a previously existing ecosystem, including not just some but all its parts and processes. This entails everything we do to an ecosystem or a landscape in an ongoing attempt to compensate for novel or "outside" influences on it in such a way that it continues to behave or can resume behaving as if these influences were not present. (Jordan and Lubick 2011, p. 2)

The definition of ecological restoration by Jordan and Lubick (2011) assumes that the unique, distinctive, and environmental value of an ecosystem is attached to a particular historical state. Consequently, restoration focuses on measures such as reforestation, removal of invasive species, erosion control, and reintroduction of native flora and fauna. A broader perspective was previously stated by Aldo Leopold (1887–1948), a pioneer of ecological restoration. Leopold (1949) integrated ecology, aesthetics, and ethics in restoration approaches, as expressed in his well-known maxim:

> A thing is right when it tends to preserve the integrity, stability, and beauty of the biotic community. It is wrong when it tends otherwise. (Leopold 1949, p. 211)

Over the years, ecological restoration has been defined and delineated by the community of ecologists, philosophers, and other scholars (Higgs 1994; Higgs et al. 2014; Davis and Slobokin 2004; United Nations Convention on Biological Diversity 2016; Martin 2017). Through these various efforts, the community of ecological restoration gradually crystallizes four principles regarding ecological integrity, long-term sustainability, historical knowledge, and social well-being (Suding et al. 2015).

Compared to ecological restoration, which focuses on the intrinsic value of ecosystems, biocultural restoration encompasses a broader spectrum of intrinsic, relational, and instrumental values of ecosystems. Ecologist Daniel Janzen was among the first to theorize and practice biocultural restoration. In the 1980s, given the increasingly fragmented state of the world's wilderness areas, especially in tropical regions, he suggested a distinction between ecological restoration and biocultural restoration. Based on his work in the dry tropical forests of Costa Rica, he stated that:

> Restriction of conservation to the few remaining relatively intact habitat patches automatically excludes more than 90% of tropical humanity from its direct benefits; restoration is most needed where people live. Intact habitats within successful agroecosystems are usually fragments of a small size (10 to 10,000 hectares) associated with idiosyncratic ownership, local conflicts, political wars, and displaced cultures. Restoration at the margins of such habitats is essential for their long-term survival. (Janzen 1988, p. 243)

His observations and practice in tropical regions made him realize that "[t]ropical humans are experiencing nearly total loss of this integral part of their mental lives," which is "their predecessors' contract with the natural world or the cultural offerings of the large cities that are supported by their parents' agriculture" (Janzen 1988, p. 244). In this sense, targeted restoration areas both contain wilderness or remote ecosystems, as ecological restoration primarily focuses on human-inhabited areas. Furthermore, Janzen considers it an obligation that local people be intellectually and practically involved in the restoration and management process. Biocultural restoration integrates the scientific principles that guide the restoration process with humans' cultural, mental, and spiritual relationships to their place. This biocultural integration revitalizes the contract they have signed with the local natural environment and habitat from generation to generation. Janzen's pioneering effort on biocultural restoration in Costa Rica's Guanacaste National Park integrates ecological restoration with local cultural recovery, transforming the park into a living classroom for local schools, universities, and visitors (Allen 1988).

Scientists and other professionals working on ecological restoration have increasingly recognized the importance of local cultural practices (Wehi and Lord 2017), the contribution of Indigenous and other local communities (Armesto et al. (2001) Reyes-García et al. 2019), traditional ecological knowledge (Kimmerer 2002; Uprety et al. 2012; Robinson et al. 2021), place-based values (Wickham et al. 2022), cultural landscapes (Toma and Buisson 2022), and revitalization of language and cultures (Kimmerer 2011). This recognition has led to a shift from ecological restoration practices toward biocultural restoration ones that are more sustainable in the long term, by accounting for cultural diversity, place-based knowledge, and local involvement.

1.4 From Biological to Biocultural Conservation

Biocultural communication also enhances biocultural conservation. The term conservation biology (or biological conservation) and its conception as a new professional field originated with the convening of the "First International Conference on Conservation Biology Research," held at the University of California, San Diego, California, in 1978. This conference, led by American biologists Bruce A. Wilcox and Michael E. Soulé, arose from concerns about tropical deforestation, species disappearance, and the erosion of genetic diversity within species (Soulé 1985). In an influential article, Soulé (1985) announced the emergence of a new biological science, conservation biology, which differed in two essential aspects from other applied biological sciences such as forestry, fisheries, and wildlife management. First, these traditional disciplines address problems associated with individual species and the use of particular natural resources. In contrast, conservation biology "tends to be holistic" (Soulé 1985, p. 728). Second, conservation biology embraces normative postulates "that make up the basis of an ethic of appropriate attitudes toward other forms of life" (Soulé 1985, p. 730). These postulates state:

(i) Diversity of organisms is good;
(ii) Ecological complexity is good;
(iii) Evolution is good;
(iv) Biotic diversity has intrinsic value, irrespective of its instrumental or utilitarian value. (Soulé 1985, p. 731)

However, conservation biology has been closely linked to ecology, as it investigates the population ecology (dispersal, migration, demography, effective population size, inbreeding depression, and minimum population viability) of rare or endangered species and addresses the phenomena that affect the maintenance, loss, and restoration of biodiversity, as well as the science underlying the evolutionary processes that generate genetic, population, species, and ecosystem diversity. Although conservation biologists work in the field, in natural areas, and in offices, in government, universities, nonprofit organizations, and industry, their research and education focus are on the ecological sciences, which has limited the consideration of cultural diversity and its interrelationships with biological diversity. This necessitated a new approach: the biocultural approach.

The biocultural approach emerged at the intersection of nature (human biology) and culture. In the 1950s, biocultural anthropology emerged, interested in unraveling the role that culture plays in shaping human biology and vice versa (Livingston 1958). In the 1960s, the biocultural concept was increasingly adopted by the sciences of human evolution (Bowles 1966; Baker 1969; Fischler 1979; Katz 1980), psychology, and health anthropology (Moore et al. 1980; Pepitone 1976).

In the 1970s and 1980s, scientists in the fields of ecology (Bennett et al. 1975), ethnobiology (Posey 1985), and restoration ecology (Allen 1988) explicitly adopted the biocultural concept. In addition, at the institutional level, in 1972, the World Heritage Convention (UNESCO 1972) highlighted transgenerational connections with nature, based on a cultural landscape approach. In the 1990s, the interrelationships between biological and cultural diversity were adopted by international political agendas (Bridgewater and Rotherham 2019). During the Convention on Biological Diversity (CBD) in 1992, traditional ecological knowledge (TEK) and the role of Indigenous peoples were integrated into biodiversity conservation. During the 1990s, several authors contributed new perspectives to these advances, highlighting the importance of traditional environmental knowledge types and values rooted in traditional ecological knowledge (Berkes et al. 1998). In addition, anthropologists drew attention to the role of linguistic diversity in the expression of biocultural diversity (Maffi 2001, 2005) and the crisis of language loss (Krauss 1992). To counter the extinction of native languages, it must be emphasized that it is crucial to incorporate the biocultural memory of ancestral wisdom from Indigenous peasant peoples (Toledo and Barrera-Bassols 2008). Biocultural diversity is a conceptual and methodological tool for gaining a deeper understanding of the interrelationships between cultural dynamics and ecological processes, drawing on different types of knowledge and worldviews (Nemogá 2016).

In the early 2000s, Chilean ecologist and philosopher Ricardo Rozzi coined the term “biocultural conservation” to emphasize the importance of valuing and protecting the connections between biological and cultural diversity, as well as their implications for human well-being and socio-environmental justice (Rozzi 2001, 2003). Thus, the concept of biocultural diversity became integrated into practice. Subsequently, a methodological approach to implementing biocultural conservation was developed (Rozzi et al. 2006). Biocultural diversity offered a path to sustainability through the conservation of biodiversity, culture, and languages (Caillon et al. 2017; Gavin et al. 2015; Sterling et al. 2017).

Uniting biological conservation with biocultural conservation reinforces the struggle for diversity as a source of heterogeneity in a world that tends to become more homogenous, even in remote regions (Rozzi et al. 2008a). Specifically, biocultural homogenization entails interrelated losses of native biological and cultural diversity at local, regional, and global scales. Biocultural homogenization is both a driver and a product of complex and widespread losses of biological and cultural diversity; however, its full magnitude is not yet widely recognized (Rozzi 2013, 2018). A particular and rapidly expanding problem is biotic homogenization, where non-native species can also become part of biocultures and even replace native ones to form new biocultures (Simberloff 2013). As alternatives to homogenizing practices, many resilient communities are promoting initiatives to restore or conserve biocultural relationships in rural and urban areas (Albó 2018; Hedge and James 2018; Klaver 2018; Kono 2018; Montoya-Greenheck 2018; Tang and Gavin 2018; Taylor 2018; Toyoda 2018; Zanotti 2018; Prescott et al. 2022).

1.5 About the Book

1.5.1 Part I. Theoretical Inquiries

The theoretical framework and methodology used to initiate this book are based on the framework of biocultural ethics and the method of field environmental philosophy, as proposed by Ricardo Rozzi and his collaborators (2006, 2008a, 2023a; Tauro et al. 2021; Contador et al. 2023; Crego et al. 2023; Méndez-Herranz et al. 2023; Ojeda et al. 2023). For field environmental philosophy, the arts have been essential components in the theoretical and practical endeavors in biocultural conservation and education (Rozzi 2023). In the first chapter of Part I, environmental philosopher Eugene Hargrove considers that arts and aesthetics were primary motivations for nature conservation in the United States in the nineteenth century (Hargrove 1989). Rozzi (2026) presents Hargrove's contributions to launching environmental philosophy as a new interdisciplinary academic field in the 1970s and discusses his ideas from a biocultural perspective. It is noteworthy that, in Chile and China, field environmental philosophy methods have incorporated the arts to enhance appreciation for often-overlooked organisms, such as mosses (Rozzi et al. 2008b; Zhu 2023).

Arts in our book encompass a wide range of artificial and human practices. In Part I, "theoretical inquiries," various authors examine philosophically the complex interrelationships between the arts and biocultural conservation, restoration, and communication (biocultural CRC). These theoretical inquiries introduce readers to philosophical examinations of how the arts are integrated into biocultural CRC and present novel concepts about the arts and environmental aesthetics (Zeng 2019; Zhu et al. 2026).

Philosophical inquiries comprehend a variety of perspectives, including history of ideas and environmental philosophy (Rozzi 2026), ontology (Tsuji and Johnson 2026), cross-cultural studies and intercultural dialogues (Song 2026; Hu and Yang 2026; Chiarini 2026), phenomenology and interspecies interactions (McIntosh 2026) to introduce core concepts and practices of linking arts with biocultural CRC. Tsuji and Johnson (2026) raise ontological questions to argue for the co-constitutive relationality as an alternative to fixed-object conceptions of the arts. Chapters in Part I, as well as throughout the book, address the arts as biocultural expressions grounded in localities (e.g., Yu et al. 2026a, b; Zuo and Zhang 2026; Li et al. 2026; Franquesa-Soler and Mesa-Jurado 2026).

Locality is vital to biocultural CRC because it recognizes the inextricable link among particular ecosystems or habitats, unique cultural practices or life habits, knowledge, and language that have co-evolved within them in close interactions with diverse human and other-than-human co-inhabitants. Several chapters refer to the concept of locality developed by Native American philosopher Brian Burkhart (2019) that introduces a decolonizing framework that emphasizes cultural resurgence and resistance to colonialism, particularly through his concepts of being-from-the-land and knowing-from-the-land. Complementarily, various chapters adopt the 3Hs model (Habits, co-in-Habitants, Habitats) of the biocultural ethic developed by Chilean environmental philosopher and ecologist Ricardo Rozzi. The 3Hs model has been increasingly used (Simion 2023; Prajapati and Nath 2025), since it tackles biocultural conservation as not just about conserving species or habitats in isolation, but as supporting the place-based interconnectedness among human and other-than-human co-inhabitants in their shared habitats (Rozzi 2012). The 3Hs model offers a valuable framework for guiding transdisciplinary collaborations and policy development aimed at sustaining biocultural diversity. By examining *habitats* (ecological and cultural environments), *co-inhabitants* (individuals and entities), and their *habits* (lifeways), a holistic understanding of local contexts is disclosed, which is essential for effective biocultural conservation and restoration (Tauro and Rozzi 2025). Communities of residents and Indigenous people often possess valuable and nuanced understandings of their environments that have been accumulated over generations and are deeply rooted in their languages, stories, and life habits, which involve close communication and interactions with other-than-human beings (Rozzi 2013; Kohn 2015). Biocultural communication is, therefore, contextual and sensitive to cultural differences, and involves interspecies communication that empowers local voices in pursuit of the well-being of human and other-than-human co-inhabitants.

1.5.2 Part II. Practical Knowledge Embedded in Biocultural Diversity

Practical knowledge rooted in biocultural diversity offers valuable pathways to sustainability, grounded in the skills possessed by human communities in particular biogeographic regions. This practical knowledge is formed from experiences of living sustainably, cultivating local habits and rituals, developing a linguistic approach in close relationship with other-than-human beings, establishing reciprocal relationships with the land, rivers, and sea, and co-inhabiting daily with other species in the realms of symbolic and material culture (Mbilinyi et al. 2005; Hwang and Huang 2019; Rozzi et al. 2023b). In these practical knowledge systems, the arts play a central role in their creation and transmission, often integrated into local cultural ceremonies, vernacular languages, seasonal practices, techniques, and spiritual beliefs. For example, Kuhlar (2026) presents her experience in her home community in northeast India. Kuhlar's case echoes other studies, such as the research on traditional weaving techniques practiced by the Seediq people of Taiwan, which are rooted in the cultural ecosystem and provide valuable practical knowledge about ancestral beliefs, rituals, and behaviors that support social interactions. These techniques convert ramie plants into fiber, as well as skills related to dyeing (Hwang and Huang 2019).

Biocultural homogenization, however, poses a threat to biocultural systems. Contributors in Part II explore the potential of the arts in restoring and transmitting practical knowledge embedded in various local biocultural systems to address biocultural homogenization and enhance biocultural resilience in specific regions. The artistic endeavors of independent artist Meng (2026), through her moss-themed glass artworks, demonstrate how interactions among artists, scientists, and environmental philosophers can yield artworks rich in biocultural messages and foster artistic creativity. Meng's artistic creations draw on field environmental philosophy approaches developed in southern Chile (Zhu 2023) and today inspire novel artistic practices in China, conveying the idea of biocultural co-inhabitation with other species (*sensu* Rozzi 2012).

Readers will notice that Part II exhibits a strong tendency to explore interspecies relationships, particularly those between humans and overlooked small organisms, such as bryophytes (Mackenzie et al. 2026, Chap. 17; Meng 2026; Zhu and Caviness 2026; Zuo and Zhang 2026). To achieve this aim, the concepts of cultural and biocultural keystone species are powerful "biocultural messengers" to stimulate conservation and restoration practices (Garibaldi and Turner 2004; Rozzi et al. 2023b). Zuo and Zhang (2026) explain how these concepts are integrated into a bryophyte-centered initiative that combines sciences, traditional aesthetics, field expeditions, artistic creation, science communication, and nature education. Through a novel biocultural approach, Zuo and Zhang (2026) lead *ex-situ* and *in-situ* bryophyte conservation programs at the Shenzhen Fairy Lake Botanical Garden.

In Chap. 13, Chilean sculptor Paola Vezzani (2026) calls attention to the fragility of biological and cultural diversity as well as of intercultural and interspecies encounters." Vezzani affirms that art can help us cultivate integral ways of life based on socially and environmentally just forms of biocultural co-inhabitation. Chinese educator and artist Lipeng Jin (2026, Chap. 14) uses "gardeners" as a powerful metaphor for both human and other-than-human participants that together build urban biodiversity, community cohesion, and networks of healthy multispecies relations. Based at Sichuan Fine Art Institute, Jin and collaborators have developed a transdisciplinary project that integrates art, science, ecological restoration, eco-pedagogy, activism, community engagement, and embraces the notions of systems thinking and permaculture. This eco-art and pedagogical project addresses complex issues, including climate change, biocultural homogenization, and the loss of biocultural diversity. To cope with them, Jin (2026, Chap. 14) creates a new imagery of reciprocal and cooperative partnerships through holistic "gardening" and a variety of eco-art activities that inspire caring relationships among humans and other-than-human co-inhabitants. American environmental philosopher Michael Thompson (2026) integrates field environmental philosophy methods with Kantian concepts to develop an epistemological and aesthetic approach to value and protect the miniature forests of mosses, insects, and other small organisms. Kant's description of sublimity can be visualized in the underrepresented diversity and beauty of bryophytes and lichens and helps to overcome global society's blindness and taxonomic chauvinism that exclude and oppress them (Rozzi 2019).

1.5.3 Part III. Biocultural Communication and Conservation Across Urban, Rural, and Remote Ecosystems

Part III focuses on innovative biocultural communication and conservation initiatives conducted in urban, rural, and remote ecosystems. These initiatives encompass a range of diverse practices, including the use of painting, art exhibitions, experimental dance, rural revitalization, and Indigenous botanical healing, to maintain local socio-ecological connections and conserve biocultural diversity. Chapters in Part III emphasize the role of the arts as an active agent in integrating multiple disciplines, involving local empirical knowledge and practices, bridging cross-cultural communication, and interspecies relationships. On the one hand, these roles indicate the great potential of artistic activities rooted in a specific territory for biocultural education and communication. On the other hand, they exemplify renewed meanings given to arts themselves in practices related to biocultural conservation, restoration, and communication.

Today, the arts are integrated into local biocultural contexts where they play key roles in biocultural conservation and communication initiatives by presenting biocultural connectedness in novel and appealing ways. For example, eco-themed art exhibitions (Li et al. 2026) highlight the complex interconnectedness between local communities and endemic ecosystems, biomes, or keystone cultural species. In this way, in massive urban settings, eco-themed art exhibitions reconnect people to their regional biocultural diversity and inspire them to take conservation actions across metropolitan, rural, and remote areas. Zhang and Xu (2026) describe a transdisciplinary approach to investigating and conserving biodiversity in Tengchong, Yunnan, China. Zhang and Xu employ Rozzi's field environmental philosophy methodology, combining the concept of cultural keystone species with watercolor techniques to create artworks. These artworks convey the perception of "one root of medicine and food," encapsulated in the concept of local biocultural diversity, and foster future biocultural education and communication.

Rural areas are ideal venues for fostering socio-ecological connectedness and conserving biocultural diversity. *Wangshan Life* in eastern China is a good example. In the face of multiple pressures, including rural decline and demographic loss, the Wangshan model offers a holistic framework for rural revitalization, where the arts play a crucial role. Nature-based solutions to the conservation of rural natural-cultural landscapes involve the transmission of ancestral practical knowledge, which also gives rise to an aesthetic rooted in local culture for future communication (Yu et al. 2026a, b).

The integration of arts and sciences is also critical for biocultural conservation and communication in rural and remote areas in Mexico and India. Mexican ecologists Montserrat Franquesa-Soler and Azahara Mesa-Jurado (2026) describe a participatory arts-based method, "photovoice," which they have effectively adopted and adapted to motivate young people to engage in biocultural conservation. In a remote region of northeast India, based on her personal experience and ethnographic studies, linguist Sumshot Kuhlar (2026) examines the knowledge and practices associated with the use of medicinal plants and the cultivation of land by the Indigenous Lamkaang people. Lamkaang Naga traditional healing systems are deeply intertwined with the land. Before anything can be done on the land, the Lamkaangs seek the blessing and permission from the land. *Ardii Kthung*, a land blessing ritual is performed by a priestess or priest, who require the community to listen to the land, and act of utmost importance as this can mean a life-or-death situation for villagers, who may suffer dire consequences if the listening goes wrong or the communication is not successful. Listening to dreams and interpreting these ritual signs requires a special connection to the land.

This Indigenous ritual embodies a worldview centered on collective interdependence, intricately weaving an expansive network of relationships that encompass the self, fellow humans and neighbors, ancestors, future generations, their habitat, co-inhabitants, and the cosmos. Kuhlar (2026) offers a biocultural perspective on the *Lamkaang Naga* traditional healing system aiding in the understanding and appreciation of the depth and complexities inherent to the ecological knowledge systems of Indigenous communities.

1.6 Concluding Remarks

In summary, the emphasis on transdisciplinary on-site practices involving a community of philosophers, scientists, artists, educators, and students has allowed us to explore both innovative and established sustainable and equitable forms of intercultural and interspecific co-inhabitation.

The chapters in this book present both theoretical and empirical insights regarding the significant contributions of the arts to biocultural conservation, restoration, and communication. Through the exploration of various artworks, art exhibitions, artistic initiatives, and art education, we develop an understanding of how the arts enhance appreciation for local and regional biocultural diversity. Multiple chapters illustrate how the arts act as a transformative force, reconnecting individuals with their environment, revitalizing ties to local biocultural richness, reshaping the unique interconnections of life across different biogeographic areas, and boosting regional resilience. Moreover, the arts articulate a locally rooted ethical response to the challenges of biocultural homogenization and global climate change.

It is our hope that this book will amplify the importance of local perspectives in shaping biocultural conservation, restoration, and communication efforts, while inspiring optimism for a better future through the pursuit of respectful and peaceful intercultural and interspecies co-inhabitation.

Acknowledgments We are very grateful to all contributors to the book, who enthusiastically and generously participated in forging a basis for biocultural communication, contributing with their thoughts, practices, and experiences in the arts as well as in biocultural conservation and restoration. We acknowledge the valuable comments by Roy May and Alejandra Tauro on earlier versions of this introductory chapter. The Chilean National Agency for Research and Development provided valuable support for our research and editorial work through the Cape Horn International Center, CHIC (ANID/BASAL FB210018).

References

Adams M (2025) Qualitative methods in psychology after the animal turn: human-animal and multi-species relations. Qual Res Psychol 22(1):1–14. https://doi.org/10.1080/14780887.2024.2380878

Albó X (2018) *Suma qamaña* or living well together: a contribution to biocultural conservation. In: Rozzi R, May RH Jr, Chapin FS III, Massardo F, Gavin M, Klaver I, Pauchard A, Nuñez MA, Simberloff D (eds) From biocultural homogenization to biocultural conservation, Ecology and ethics, vol 3. Springer, Dordrecht, pp 333–342. https://doi.org/10.1007/978-3-319-99513-7_21

Allen WH (1988) Biocultural restoration of a tropical forest. Bioscience 38(3):156–161. https://doi.org/10.2307/1310447

Armelagos GJ, Leatherman T, Ryan M, Sibley L (1992) Biocultural synthesis in medical anthropology. Medical Anthropology, 14(1):35-52. https://doi.org/10.1080/01459740.1992.9966065

Armesto JJ, Smith-Ramirez C, Rozzi R (2001) Conservation strategies for biodiversity and indigenous people in Chilean forest ecosystems. J R Soc N Z 31(4):865–877. https://doi.org/10.1080/03014223.2001.9517681

Baker PT (1969) Human adaptation to high altitude. Science 163:1149–1156

Barrett MJ, Hinz V, Wijngaarden V, Lovrod M (2021) Speaking with other animals through intuitive interspecies communication: towards cognitive and interspecies justice. In:

Hovorka A, McCubbin S, Van Patter L (eds) A research agenda for animal geographies. Edward Elgar Publishing, Northampton, pp 149–165. https://doi.org/10.4337/9781788979993.00018
Bennett KA, Osborne RH, Miller RJ (1975) Biocultural ecology. Annu Rev Anthropol 4:163–181
Berkes F, Kislalioglu M, Folke C, Gadgil M (1998) Minireviews: exploring the basic ecological unit: ecosystem-like concepts in traditional societies. Ecosystems 1(5):409–415. https://doi.org/10.1007/s100219900034
Bowles GT (1966) Identifying races: geography and genetics. Science 154:628–629
Bradshaw GA, Watkins M (2006) Trans-species psychology: theory and praxis. Spring J Archetype Cult 75(1):1–26
Bridgewater P, Rotherham ID (2019) A critical perspective on the concept of biocultural diversity and its emerging role in nature and heritage conservation. People Nat 1(3):291–304. https://doi.org/10.1002/pan3.10040
Burkhart B (2019) Indigenizing philosophy through the land: a trickster methodology for decolonizing environmental ethics and Indigenous futures. Michigan State University Press, East Lansing
Caillon S, Cullman G, Verschuuren B, Sterling EJ (2017) Moving beyond the human–nature dichotomy through biocultural approaches. Ecol Soc 22(4). https://doi.org/10.5751/ES-09746-220427
Chiarini VS (2026) How the invisible becomes visible: situating scientific illustration in biocultural communication. In: Zhu D, Rozzi R, Tsuji R, Johnson B, Caviness T, Castro-Jorquera C, Xu L, Zhang L (eds) Linking arts with biocultural conservation, restoration, and communication, Ecology and ethics, vol 8. Springer, Cham, pp 149–169. https://doi.org/10.1007/978-3-032-04830-1_9
Contador TA, Rendoll-Cárcamo J, Gañan M, Ojeda J, Kennedy JH, Convey P, Rozzi R (2023) Underwater with a hand lens: ecological sciences and environmental ethics to value freshwater biodiversity. In: Rozzi R, Tauro A, Wright T, Avriel-Avni N, May RH Jr (eds) Field environmental philosophy: education for biocultural conservation, Ecology and ethics series, vol 5. Springer, Cham, pp 53–69. https://doi.org/10.1007/978-3-031-23368-5_4
Crego RD, Ward N, Rozzi R (2023) The eyes of the tree: applying field environmental philosophy to tackle conservation problems at long term socio-ecological research sites. In: Rozzi R, Tauro A, Wright T, Avriel-Avni N, May RH Jr (eds) Field environmental philosophy: education for biocultural conservation, Ecology and ethics series, vol 5. Springer, Cham, pp 101–111. https://doi.org/10.1007/978-3-031-23368-5_7
Davis MA, Slobodkin LB (2004) The science and values of restoration ecology. Restor Ecol 12(1): 1–3. https://doi.org/10.1111/J.1061-2971.2004.0351.X
Dickson D (2005) The case for a 'deficit model' of science communication. Sci Dev Net 27:1–6
Fichtel C (2004) Reciprocal recognition of sifaka (*Propithecus verreauxi verreauxi*) and redfronted lemur (*Eulemur fulvus rufus*) alarm calls. Anim Cogn 7:45–52. https://doi.org/10.1007/s10071-003-0180-0
Fioret C (2023) Water justice as socioenvironmental justice. Ethics Policy Environ 26(3):406–421. https://doi.org/10.1080/21550085.2022.2090211
Fischler C (1979) La cuisine et l'esprit du temps. In: Nostrand HL (ed) La France en mutation. Newbury House, Rowley
Franquesa-Soler M, Mesa-Jurado MA (2026) Photovoice: a community-based research tool for wildlife conservation and co-existence. In: Zhu D, Rozzi R, Tsuji R, Johnson B, Caviness T, Castro-Jorquera C, Xu L, Zhang L (eds) Linking arts with biocultural conservation, restoration, and communication, Ecology and ethics, vol 8. Springer, Cham, pp 397–415. https://doi.org/10.1007/978-3-032-04830-1_25
Garibaldi A, Turner N (2004) Cultural keystone species: implications for ecological conservation and restoration. Ecol Soc 9(3). http://www.ecologyandsociety.org/vol9/iss3/art1/
Gavin S (2024) Interspecies communication: sound and music beyond humanity. University of Chicago Press, Chicago
Gavin MC, McCarter J, Mead A, Berkes F, Stepp JR, Peterson D, Tang R (2015) Defining biocultural approaches to conservation. Trends Ecol Evol 30(3):140–145. https://doi.org/10.1016/j.tree.2014.12.005

Gruber S (1972) Convention concerning the protection of the world cultural and natural heritage 1972. In: Gruber S (ed) Convention concerning the Protection of the World Cultural and Natural Heritage. UNESCO, pp 60–66

Hanazaki N (2024) Local and traditional knowledge systems, resistance, and socioenvironmental justice. J Ethnobiol Ethnomed 20(1):5. https://doi.org/10.1186/s13002-023-00641-0

Haraway DJ (2008) When species meet. University of Minnesota Press, Minneapolis

Hargrove E (1989) Foundations of environmental ethics. Prentice Hall, Englewood Cliffs/Hoboken

Healey R, Pepper A (2021) Interspecies justice: agency, self-determination, and assent. Philos Stud 178:1223–1243. https://doi.org/10.1007/s11098-020-01472-5

Hedge P, James G (2018) Challenging biocultural homogenization: experiences of the Chipko and Appiko movements in India. In: Rozzi R, May RH Jr, Chapin FS III, Massardo F, Gavin M, Klaver I, Pauchard A, Nuñez MA, Simberloff D (eds) From biocultural homogenization to biocultural conservation, Ecology and ethics, vol 3. Springer, Dordrecht, pp 427–442. https://doi.org/10.1007/978-3-319-99513-7_27

Higgs E (1994) Expanding the scope of restoration ecology. Restor Ecol 2(3):137–146

Higgs E, Falk DA, Guerrini A, Hall M, Harris J et al (2014) The changing role of history in restoration ecology. Front Ecol Environ 12(9):499–506. https://doi.org/10.1890/110267

Hu J, Yang P (2026) A biocultural encounter with the four gentlemen: art education in Chinese ecological and aesthetic traditions. In: Zhu D, Rozzi R, Tsuji R, Johnson B, Caviness T, Castro-Jorquera C, Xu L, Zhang L (eds) Linking arts with biocultural conservation, restoration, and communication. Ecology and ethics, vol 8. Springer, Cham, pp 111–129. https://doi.org/10.1007/978-3-032-04830-1_7

Hwang SH, Huang HM (2019) Cultural ecosystem of the seediq's traditional weaving techniques—a comparison of the learning differences between urban and indigenous communities. Sustainability 11(6):1519. https://doi.org/10.3390/su11061519

Janzen DH (1988) Tropical ecological and biocultural restoration. Science 239(4837):243–244. https://doi.org/10.1126/science.239.4837.243

Jin L (2026) Healing garden: eco-art actions at Sichuan Fine Art Institute. In: Zhu D, Rozzi R, Tsuji R, Johnson B, Caviness T, Castro-Jorquera C, Xu L, Zhang L (eds) Linking arts with biocultural conservation, restoration, and communication, Ecology and ethics, vol 8. Springer, Cham, pp 223–242. https://doi.org/10.1007/978-3-032-04830-1_14

Jordan W, Lubick G (2011) Making nature: whole a history of ecological restoration. Island Press, Washington, DC

Jordan R, Singer F, Vaughan J, Berkowitz A (2009) What should every citizen know about ecology? Front Ecol Environ 7(9):495–500. https://doi.org/10.1890/070113

Katz SH (1980) Biocultural evolution and the is/ought. Zygon 15:155–168. https://doi.org/10.1111/j.1467-9744.1980.tb00383.x

Khular S (2026) Lamkaang Naga Loh Kchet: collective interdependence. In: Zhu D, Rozzi R, Tsuji R, Johnson B, Caviness T, Castro-Jorquera C, Xu L, Zhang L (eds) Linking arts with biocultural conservation, restoration, and communication, Ecology and ethics, vol 8. Springer, Cham, pp 469–488. https://doi.org/10.1007/978-3-032-04830-1_28

Kimmerer RW (2002) Weaving traditional ecological knowledge into biological education: a call to action. Bioscience 52(5):432–438. https://doi.org/10.1641/0006-3568(2002)052[0432:WTEKIB]2.0.CO;2

Kimmerer RW (2011) Restoration and reciprocity: the contributions of traditional ecological knowledge. In: Egan D, Hjerpe EE, Abrams J (eds) Human dimensions of ecological restoration: integrating science, nature, and culture. Island Press/Center for Resource Economics, Washington, DC, pp 257–276

Klaver IJ (2013) Environment imagination situation. In: Rozzi R, Pickett S, Palmer C, Armesto J, Callicott J (eds) Linking ecology and ethics for a changing world, Ecology and ethics, vol 1. Springer, Dordrecht. https://doi.org/10.1007/978-94-007-7470-4_7

Klaver I (2018) Re-claiming rivers from homogenization: meandering and riverspheres. In: Rozzi R, May RH Jr, Chapin FS III, Massardo F, Gavin M, Klaver I, Pauchard A, Nuñez MA, Simberloff D (eds) From biocultural homogenization to biocultural conservation, Ecology and ethics, vol 3. Springer, Dordrecht, pp 49–69. https://doi.org/10.1007/978-3-319-99513-7_3

Kohn E (2015) Anthropology of ontologies. Annu Rev Anthropol 44(1):311–327. https://doi.org/10.1146/annurev-anthro-102214-014127

Kono T (2018) The garden as a representation of nature: a space to overcome biocultural homogenization? In: Rozzi R, May RH Jr, Chapin FS III, Massardo F, Gavin M, Klaver I, Pauchard A, Nuñez MA, Simberloff D (eds) From biocultural homogenization to biocultural conservation, Ecology and ethics, vol 3. Springer, Dordrecht, pp 459–474. https://doi.org/10.1007/978-3-319-99513-7_29

Krauss M (1992) The world's languages in crisis. Language 68(1):4–10. https://doi.org/10.1353/lan.1992.0074

Kull K (2009) Vegetative, animal, and cultural semiosis: the semiotic threshold zones. Cog Sem 4 (Suppl):8–27. https://doi.org/10.1515/cogsem.2009.4.spring2009.8

Leopold A (1949) A Sand County almanac and sketches here and there. Oxford University Press, New York

Li Z, Jia M, Ma K (2026) Immersion and interactiveness: eco-themed art exhibitions and biocultural communication in East Asia. In: Zhu D, Rozzi R, Tsuji R, Johnson B, Caviness T, Castro-Jorquera C, Xu L, Zhang L (eds) Linking arts with biocultural conservation, restoration, and communication, Ecology and ethics, vol 8. Springer, Cham, pp 367–382. https://doi.org/10.1007/978-3-032-04830-1_23

Livingstone FB (1958) Anthropological implications of sickle cell gene distribution in West Africa. Am Anthropol 60:533–562. https://doi.org/10.1525/aa.1958.60.3.02a00110

Luhmann N (1989) Ecological communication. University of Chicago Press, Chicago

Mackenzie R, Valle-Celedón C, Arriagada G (2026) Collecting Photos in the Miniature Forest: A Biocultural Approach for Bryophyte Conservation in the Cape Horn Biosphere Reserve. In: Zhu D, Rozzi R, Tsuji R, Johnson B, Caviness T, Castro-Jorquera C, Xu L, Zhang L (eds) Linking arts with biocultural conservation, restoration, and communication, Ecology and ethics, vol 8. Springer, Cham, pp 277–293. https://doi.org/10.1007/978-3-032-04830-1_17

Maffi L (ed) (2001) On biocultural diversity. Linking language, knowledge, and the environment. Smithsonian Institution Press, Washington, DC

Maffi L (2005) Linguistic, cultural, and biological diversity. Annu Rev Anthropol 34:599–617. https://doi.org/10.1146/annurev.anthro.34.081804.120437

Mamani-Bernabé V (2015) Spirituality and the Pachamama in the Andean Aymara worldview. In: Rozzi R, Chapin FS III, Callicott JB et al (eds) Earth stewardship: linking ecology and ethics in theory and practice, Ecology and ethics, vol 2. Springer, Dordrecht, pp 65–76. https://doi.org/10.1007/978-3-319-12133-8_6

Martin DM (2017) Ecological restoration should be redefined for the twenty-first century. Restor Ecol 25(5):668–673. https://doi.org/10.1111/rec.12554

May RH Jr (2015) Andean Llamas and Earth stewardship. In: Rozzi R, Chapin FS III, Callicott JB, Pickett STA, Power ME, Armesto JJ, May RH Jr (eds) Earth stewardship, Ecology and ethics, vol 2. Springer, Cham. https://doi.org/10.1007/978-3-319-12133-8_7

May RH Jr (2017) Pachasophy: landscape ethics in the Central Andes Mountains of South America. Environ Ethics 39:301–331. https://doi.org/10.5840/enviroethics201739322

Mbilinyi BP, Tumbo SD, Mahoo HF, Senkondo EM, Hatibu N (2005) Indigenous knowledge as decision support tool in rainwater harvesting. Phys Chem Earth (Pt A/B/C) 30(11–16):792–798. https://doi.org/10.1016/j.pce.2005.08.022

McIntosh S (2026) Embodied encounters: biocultural dialogues and pedagogies beyond the human. In: Zhu D, Rozzi R, Tsuji R, Johnson B, Caviness T, Castro-Jorquera C, Xu L, Zhang L (eds) Linking arts with biocultural conservation, restoration, and communication, Ecology and ethics, vol 8. Springer, Cham, pp 131–148. https://doi.org/10.1007/978-3-032-04830-1_8

Meijer E (2019) When animals speak: toward an interspecies democracy, vol 1. NYU Press, New York

Méndez-Herranz M, Marini G, Rozzi R (2023) Sub-Antarctic High Andean "Gardeners:" cultivating caring relationships. In: Rozzi R, Tauro A, Wright T, Avriel-Avni N, May RH Jr (eds) Field environmental philosophy: education for biocultural conservation, Ecology and ethics series, vol 5. Springer, Cham, pp 71–86. https://doi.org/10.1007/978-3-031-23368-5_5

Meng S (2026) The birth of "moss: the trailblazer on ruins". In: Zhu D, Rozzi R, Tsuji R, Johnson B, Caviness T, Castro-Jorquera C, Xu L, Zhang L (eds) Linking arts with biocultural conservation, restoration, and communication, Ecology and ethics, vol 8. Springer, Cham, pp 295–310. https://doi.org/10.1007/978-3-032-04830-1_18

Montoya-Greenheck F (2018) Biocultural diversity and Ngöbe people in the South Pacific of Costa Rica. In: Rozzi R, May RH Jr, Chapin FS III, Massardo F, Gavin M, Klaver I, Pauchard A, Nuñez MA, Simberloff D (eds) From biocultural homogenization to biocultural conservation, Ecology and ethics, vol 3. Springer, Dordrecht, pp 361–378. https://doi.org/10.1007/978-3-319-99513-7_23

Moore LG, Van Arsdale W, Glittenberg JE, Aldrich RA (1980) The biocultural basis of health: expanding views of medical anthropology. Waveland Press, Lake Zurich

Nemogá GR (2016) Diversidad biocultural: innovando en investigación para la conservación. Acta biol Colomb 21(1 Supl):S311–S319. https://doi.org/10.15446/abc.v21n1sup.50920

Odum EP (2013) Great ideas in ecology for the 1990s. In: Samson FB, Knopf FL (eds) Ecosystem management: selected readings. Springer, New York, pp 279–284. https://doi.org/10.2307/1311885

Ojeda J, Tauro A, Contador T, Mansilla A, Rosenfeld S, Ban N, Rozzi R (2023) "Pay attention, dive with wide-open eyes." A field environmental philosophy activity to foster reciprocity between people and the sea. In: Rozzi R, Tauro A, Wright T, Avriel-Avni N, May RH Jr (eds) Field environmental philosophy: education for biocultural conservation, Ecology and ethics series, vol 5. Springer, Cham, pp 87–100. https://doi.org/10.1007/978-3-031-23368-5_6

Orr DW (1992) Ecological literacy: education and the transition to a postmodern world. Suny Press, Albany

Oyewole JA, Awolaju BA, Ayomide PAO (2025) Exploring interspecies communication: a socio-linguistic analysis of animal signals and their human interpretations. Biol Sci 5(2):898–906. https://doi.org/10.55006/biolsciences.2025.5202

Pepitone A (1976) Toward a normative and comparative biocultural social psychology. J Pers Soc Psychol 34:641–653. https://doi.org/10.1037/0022-3514.34.4.641

Posey P (1985) Indigenous management of tropical forest ecosystems: the case of the Kayapó Indians of the Brazilian Amazon. Agrofor Syst 3:139–158. https://doi.org/10.1007/BF00122640

Prajapati A, Nath R (2025) Looking at biocultural ethics through the lens of transcultural dialogue. Ethics Sci Environ Polit 25:71–85. https://doi.org/10.3354/esep

Prescott SL, Logan AC, Bristow J, Rozzi R, Moodie R, Redvers N, Haahtela T, Warber S, Poland B, Hancock T, Berman B (2022) Exiting the Anthropocene: achieving personal and planetary health in the 21st century. Allergy 77(12):3498–3512. https://doi.org/10.1111/all.15419

Reyes-García V, Fernández-Llamazares Á, McElwee P, Molnár Z, Öllerer K et al (2019) The contributions of Indigenous Peoples and local communities to ecological restoration. Restor Ecol 27(1):3–8. https://doi.org/10.1111/rec.12894

Risser PG (1986) Address of the past president: Syracuse, New York; August 1986: ecological literacy. Bull Ecol Soc Am 67(4):264–270. https://doi.org/10.2307/2016653

Robinson JM, Gellie N, MacCarthy D, Mills JG, O'Donnell K, Redvers N (2021) Traditional ecological knowledge in restoration ecology: a call to listen deeply, to engage with, and respect Indigenous voices. Restor Ecol 29(4):e13381. https://doi.org/10.1111/rec.13381

Romero J, Dryzek JS (2021) Grounding ecological democracy: semiotics and the communicative networks of nature. Environ Value 30(4):407–429. https://doi.org/10.3197/096327120X16076972519089

Rozzi R (2001) Éticas ambientales latinoamericanas: raíces y ramas. In: Primack, R., R. Rozzi, P. Feinsinger, R. Dirzo, F. Massardo y colaboradores, Fundamentos de Conservación Biológica: Perspectivas Latinoamericanas. Fondo de Cultura Económica, México, pp 311–362

Rozzi R (2003) Biodiversity and social wellbeing in South America. In: Encyclopedia of life support systems (EOLSS). UNESCO-EOLSS. http://www.eolss.net/. https://www.eolss.net/sample-chapters/c14/E1-37-04-10.pdf

Rozzi R (2010) Introduction. In: Rozzi R et al (eds) Multi-ethnic bird guide of the sub-Antarctic forests of South America. University of North Texas Press, Denton, pp 3–32

Rozzi R (2012) Biocultural ethics: recovering the vital links between the inhabitants, their habits, and habitats. Environ Ethics 34(1):27–50. https://doi.org/10.5840/enviroethics20123414

Rozzi R (2013) Biocultural ethics: from biocultural homogenization toward biocultural conservation. In: Rozzi R, Pickett S, Palmer C, Armesto J, Callicott J (eds) Linking ecology and ethics for a changing world: values, philosophy, and action, Ecology and ethics, vol 1. Springer, Dordrecht, pp 9–32. https://doi.org/10.1007/978-94-007-7470-4_2
Rozzi R (2018) Biocultural homogenization: a wicked problem in the anthropocene. In: From biocultural homogenization to biocultural conservation, Ecology and ethics, vol 3. Springer, Cham, pp 21–48. https://doi.org/10.1007/978-3-319-99513-7_2
Rozzi R (2019) Taxonomic chauvinism, no more! Antidotes from Hume, Darwin, and biocultural ethics. Environ Ethics 41(3):249–282. https://doi.org/10.5840/enviroethics201941325
Rozzi R (2026) The contributions of Eugene Hargrove to environmental ethics and aesthetics: a biocultural tribute. In: Zhu D, Rozzi R, Tsuji R, Johnson B, Caviness T, Castro-Jorquera C, Xu L, Zhang L (eds) Linking arts with biocultural conservation, restoration, and communication, Ecology and ethics, vol 8. Springer, Cham, pp 35–59. https://doi.org/10.1007/978-3-032-04830-1_3
Rozzi R, Jiménez JE (eds) (2014) Magellanic sub-Antarctic ornithology: first decade of long-term bird studies at the Omora ethnobotanical park, Cape Horn Biosphere Reserve, Chile. University of North Texas Press, Denton, Texas, USA, pp. 345–351
Rozzi R, Massardo F, Anderson CB, Heidinger K, Silander JA Jr (2006) Ten principles for biocultural conservation at the southern tip of the Americas: the approach of the Omora Ethnobotanical Park. Ecol Soc 11(1):43. https://doi.org/10.5751/ES-01709-110143
Rozzi R, Armesto JJ, Goffinet B, Buck W, Massardo F et al (2008a) Changing lenses to assess biodiversity: patterns of species richness in sub-Antarctic plants and implications for global conservation. Front Ecol Environ 6(3):131–137. https://doi.org/10.1890/070020
Rozzi R, Arango X, Massardo F, Anderson C, Heidinger K, Moses K (2008b) Field environmental philosophy and biocultural conservation: the Omora Ethnobotanical Park educational program. Environ Ethics 30(3):325–336. https://doi.org/10.5840/enviroethics200830336
Rozzi R, Pickett ST, Palmer C, Armesto JJ, Callicott JB (eds) (2013) Linking ecology and ethics for a changing world: values, philosophy, and action, Ecology and ethics, vol 1. Springer, Dordrecht. https://doi.org/10.1007/978-94-007-7470-4
Rozzi R (2014) Ethical implications of Yahgan and Mapuche narratives about birds of the temperate forests of southern South America. In Rozzi R, Jiménez JE (eds.), Magellanic Sub-Antarctic Ornithology: First Decade of Long-Term Bird Studies at the Omora Ethnobotanical Park, Cape Horn Biosphere Reserve. University of North Texas Press, Denton, Texas, USA pp. 345–351
Rozzi R (2023) Inter-species and inter-cultural encounters: The Biocultural Education Program of the Omora Ethnobotanical Park. In: Rozzi R, Tauro A, Avriel-Avni N, Wright T, May Jr. RH (eds) Field Environmental Philosophy. Ecology and Ethics, vol 5. Springer, Cham, pp. 153–174. https://doi.org/10.1007/978-3-031-23368-5-10
Rozzi R, Tauro A, Avriel-Avni N, Wright T, May RH Jr (eds) (2023a) Field environmental philosophy: education for biocultural conservation, Ecology and ethics, vol, vol 5. Springer, Cham. https://doi.org/10.1007/978-3-031-23368-5
Rozzi R, Álvarez R, Castro V, Núñez D, Ojeda J, Tauro A, Massardo F (2023b) Biocultural calendars across four ethnolinguistic communities in southwestern South America. GeoHealth 7(4):e2022GH000623. https://doi.org/10.1029/2022GH000623
Schiller A, Hunsaker CT, Kane MA, Wolfe AK, Dale VH et al (2001) Communicating ecological indicators to decision makers and the public. Conserv Ecol 5(1). http://www.consecol.org/vol5/iss1/art19/
Simberloff D (2013) Introduced species, homogenizing biotas and cultures. In: Rozzi R, Pickett S, Palmer C, Armesto J, Callicott J (eds) Linking ecology and ethics for a changing world: values, philosophy, and action, Ecology and ethics, vol 1. Springer, Dordrecht, pp 33–48. https://doi.org/10.1007/978-94-007-7470-4_3
Simion R (2023) Considering geoengineering in an ethical biocultural model. Stud Univ Babeş-Bolyai Philos 68(2):15–32. https://doi.org/10.24193/subbphil.2023.2.02
Simis MJ, Madden H, Cacciatore MA, Yeo SK (2016) The lure of rationality: why does the deficit model persist in science communication? Public Underst Sci 25(4):400–414. https://doi.org/10.1177/096366251662974

Simmons D (1995) Environmental education, social studies, and education reform. Soc Stud Young Learner 8(1):9–11

Song Y (2026) Interspecies co-inhabitation: a dialogue between Chinese philosophical traditions and biocultural ethics. In: Zhu D, Rozzi R, Tsuji R, Johnson B, Caviness T, Castro-Jorquera C, Xu L, Zhang L (eds) Linking arts with biocultural conservation, restoration, and communication, Ecology and ethics, vol 8. Springer, Cham, pp 95–109. https://doi.org/10.1007/978-3-032-04830-1_6

Soulé ME (1985) What is conservation biology? Bioscience 35(11):727–734. https://doi.org/10.2307/1310054

Spottiswoode CN, Wood BM (2023) Culturally determined interspecies communication between humans and honeyguides. Science 382(6675):1155–1158. https://doi.org/10.1126/science.adh4129

Sterling EJ, Filardi C, Toomey A, Sigouin A, Betley E et al (2017) Biocultural approaches to well-being and sustainability indicators across scales. Nat Ecol Evol 1(12):1798–1806. https://doi.org/10.1038/s41559-017-0349-6

Suding K, Higgs E, Palmer M, Callicott JB, Anderson CB et al (2015) Committing to ecological restoration. Science 348(6235):638–640. https://doi.org/10.1126/science.aaa4216

Tang R, Gavin MC (2018) The dynamics of biocultural approaches to conservation in Inner Mongolia, China. In: Rozzi R, May RH Jr, Chapin FS III, Massardo F, Gavin M, Klaver I, Pauchard A, Nuñez MA, Simberloff D (eds) From biocultural homogenization to biocultural conservation, Ecology and ethics, vol 3. Springer, Dordrecht, pp 405–425

Tauro A, Rozzi R (2025) Biocultural ethics and Earth stewardship: a novel integration to revitalize multiple values of nature. Ecol Soc 30. https://doi.org/10.5751/ES-16362-300335

Tauro A, Ojeda J, Caviness T, Moses KP, Moreno-Terrazas R et al (2021) Field environmental philosophy: a biocultural ethic approach to education and ecotourism for sustainability. Sustainability 13(8):4526. https://doi.org/10.3390/su13084526

Taylor B (2018) Biostitutes and biocultural conservation: empire and irony in the motion picture Avatar. In: Rozzi R, May RH Jr, Chapin FS III, Massardo F, Gavin M, Klaver I, Pauchard A, Nuñez MA, Simberloff D (eds) From biocultural homogenization to biocultural conservation, Ecology and ethics, vol 3. Springer, Dordrecht, pp 71–82. https://doi.org/10.1007/978-3-319-99513-7_4

Thompson M (2026) Sublimity in miniature: aesthetic viewing and the miniature forests of Cape Horn. In: Zhu D, Rozzi R, Tsuji R, Johnson B, Caviness T, Castro-Jorquera C, Xu L, Zhang L (eds) Linking arts with biocultural conservation, restoration, and communication, Ecology and ethics, vol 8. Springer, Cham, pp 265–276. https://doi.org/10.1007/978-3-032-04830-1_16

Toledo VM, Barrera-Bassols N (2008) La memoria biocultural: la importancia ecológica de las sabidurías tradicionales, vol 3. Icaria editorial

Toma TSP, Buisson E (2022) Taking cultural landscapes into account: implications for scaling up ecological restoration. Land Use Policy 120:106233. https://doi.org/10.1016/j.landusepol.2022.106233

Toyoda M (2018) Revitalizing local commons: a democratic approach to collective management. In: Rozzi R, May RH Jr, Chapin FS III, Massardo F, Gavin M, Klaver I, Pauchard A, Nuñez MA, Simberloff D (eds) From biocultural homogenization to biocultural conservation, Ecology and ethics, vol 3. Springer, Dordrecht, pp 443–457. https://doi.org/10.1007/978-3-319-99513-7_28

Tsuji R, Johnson B (2026) Biocultural ontologies, art, and conservation. In: Zhu D, Rozzi R, Tsuji R, Johnson B, Caviness T, Castro-Jorquera C, Xu L, Zhang L (eds) Linking arts with biocultural conservation, restoration, and communication, Ecology and ethics, vol 8. Springer, Cham, pp 61 –77. https://doi.org/10.1007/978-3-032-04830-1_4

UNCBD (United Nations Convention on Biological Diversity) (2016) Ecosystem restoration: short-term action plan. CBD/COP/DEC/XIII/5, 10 December 2016

Uprety Y, Asselin H, Bergeron Y, Doyon F, Boucher JF (2012) Contribution of traditional knowledge to ecological restoration: practices and applications. Ecoscience 19(3):225–237. https://doi.org/10.2980/19-3-3530

Vezzani P (2026) "Fragility of Encounters" in Sub-Antarctic Biocultural Diversity. In: Zhu D, Rozzi R, Tsuji R, Johnson B, Caviness T, Castro-Jorquera C, Xu L, Zhang L (eds) Linking arts with biocultural conservation, restoration, and communication, Ecology and ethics, vol 8. Springer, Cham, pp 207– 222. https://doi.org/10.1007/978-3-032-04830-1_13

Wang J (2026) A mere touch of green: reconnecting people and Shanshui through poetic dance drama. In: Zhu D, Rozzi R, Tsuji R, Johnson B, Caviness T, Castro-Jorquera C, Xu L, Zhang L (eds) Linking arts with biocultural conservation, restoration, and communication, Ecology and ethics, vol 8. Springer, Cham, pp 383–395. https://doi.org/10.1007/978-3-032-04830-1_24

Wehi PM, Lord JM (2017) Importance of including cultural practices in ecological restoration. Conserv Biol 31(5):1109–1118. https://doi.org/10.1111/cobi.12915

Wickham SB, Augustine S, Forney A, Mathews DL, Shackelford N et al (2022) Incorporating place-based values into ecological restoration. Ecol Soc 27(3):32. https://doi.org/10.5751/ES-13370-270332

Wooltorton S (2006) Ecological literacy: an Australian perspective. Soc Educ 24(2):26–28

Yu K, Li J, Zhou Y, Xie M, Hu Y (2026a) Nature-based solutions to the revitalization of Wuyuan County. In: Zhu D, Rozzi R, Tsuji R, Johnson B, Caviness T, Castro-Jorquera C, Xu L, Zhang L (eds) Linking arts with biocultural conservation, restoration, and communication, Ecology and ethics, vol 8. Springer, Cham, pp 417–432. https://doi.org/10.1007/978-3-032-04830-1_26

Yu K, Li J, Zheng C, Yun H (2026b) The Wangshan life model for the revitalization of rural villages: the Xixinan case. In: Zhu D, Rozzi R, Tsuji R, Johnson B, Caviness T, Castro-Jorquera C, Xu L, Zhang L (eds) Linking arts with biocultural conservation, restoration, and communication, Ecology and ethics, vol 8. Springer, Cham, pp 433–468. https://doi.org/10.1007/978-3-032-04830-1_27

Zanotti L (2018) Biocultural approaches to conservation: water sovereignty in the Kayapó Lands. In: Rozzi R, May RH Jr, Chapin FS III, Massardo F, Gavin M, Klaver I, Pauchard A, Nuñez MA, Simberloff D (eds) From biocultural homogenization to biocultural conservation, Ecology and ethics, vol 3. Springer, Dordrecht, pp 343–359. https://doi.org/10.1007/978-3-319-99513-7_22

Zeng F (ed) (2019) Introduction to ecological aesthetics. Springer, Singapore. https://doi.org/10.1007/978-981-13-8984-9

Zhang S, Xu L (2026) A biocultural sketch of the Chinese black cardamom, Cao-guo (Lanxangia tsao-ko). In: Zhu D, Rozzi R, Tsuji R, Johnson B, Caviness T, Castro-Jorquera C, Xu L, Zhang L (eds) Linking arts with biocultural conservation, restoration, and communication, Ecology and ethics, vol 8. Springer, Cham, pp 349–365. https://doi.org/10.1007/978-3-032-04830-1_22

Zhu D (2023) Bridge the channel, enhance the inclusivity: a comparison between flagship species-centered and moss-centered conservation in Chile and China. In: Rozzi R, Tauro A, Wright T, Avriel-Avni N, May RH Jr (eds) Field environmental philosophy: education for biocultural conservation, Ecology and ethics, vol 5. Springer, Dordrecht, pp 457–482. https://doi.org/10.1007/978-3-031-23368-5_28

Zhu D, Caviness T (2026) Biocultural co-inhabitancy: an examination on scientific and indigenous nomenclature of bryophytes. In: Zhu D, Rozzi R, Tsuji R, Johnson B, Caviness T, Castro-Jorquera C, Xu L, Zhang L (eds) Linking arts with biocultural conservation, restoration, and communication, Ecology and ethics, vol 8. Springer, Cham, pp 311–328. https://doi.org/10.1007/978-3-032-04830-1_19

Zhu D, Caviness T, Rozzi R (2026) Introduction to part III: biocultural communication across urban, rural, and remote ecosystems. In: Zhu D, Rozzi R, Tsuji R, Johnson B, Caviness T, Castro-Jorquera C, Xu L, Zhang L (eds) Linking arts with biocultural conservation, restoration, and communication, Ecology and ethics, vol 8. Springer, Cham, pp 335–347. https://doi.org/10.1007/978-3-032-04830-1_21

Zuberbuhler K (2000) Interspecies semantic communication in two forest primates. Proc R Soc Lond Ser B Biol Sci 267:713–718. https://doi.org/10.1098/rspb.2000.1061

Zuo Q, Zhang L (2026) Small but essential: how to introduce the public to easily overlooked organisms? In: Zhu D, Rozzi R, Tsuji R, Johnson B, Caviness T, Castro-Jorquera C, Xu L, Zhang L (eds) Linking arts with biocultural conservation, restoration, and communication, Ecology and ethics, vol 8. Springer, Cham, pp 243–263. https://doi.org/10.1007/978-3-032-04830-1_15

Part I
Theoretical Inquiries: Linking Arts with Biocultural Conservation, Restoration, and Communication

Chapter 2
Introduction to Part I. Theoretical Inquiries: Linking Arts with Biocultural CRC

Rika Tsuji, Carolina Castro-Jorquera, Benn Johnson, and Ricardo Rozzi

Abstract Faced with the current social-environmental crisis, there is a pressing need to undertake ethical responsibilities toward diverse humans and other-than-humans. We address this ethical imperative by examining how to better understand, value, and respect biological and cultural diversity, to achieve just and sustainable forms of co-inhabitation. This first part of the book draws from the biocultural ethic framework proposed by Ricardo Rozzi, enhancing it by incorporating aesthetic dimensions that link art to biocultural conservation, restoration, and communication (CRC). Challenging traditional notions of art, we invite readers to a deeper understanding of its role beyond mere utility, and we address three key themes: (i) an ontological shift towards a biocultural understanding of art, (ii) reflection on philosophical frameworks for linking art and conservation, and (iii) embodied experiences of artistic practices that foster interspecies relationships. These themes underscore the importance of recognizing the multifaceted nature of artistic expression and its implications for biocultural communities, ultimately aiming to motivate readers towards active engagement in local biocultural conservation initiatives. By bridging the disciplinary gap between cultural and biological diversity conservation, biocultural CRC contributes to socio-environmental justice in the Anthropocene. Biocultural CRC involves collective and participatory actions that respect and

R. Tsuji (✉) · B. Johnson
Department of Philosophy, College of Arts and Sciences, University of Rhode Island, Kingston, RI, USA
e-mail: rika.tsuji@uri.edu; benn.johnson@uri.edu

C. Castro-Jorquera
Cape Horn International Center (CHIC), Universidad de Magallanes, Puerto Williams, Chile

R. Rozzi
Sub-Antarctic Biocultural Conservation Program, Department of Philosophy and Religion and Department of Biological Sciences, University of North Texas, Denton, TX, USA

Cape Horn International Center (CHIC), Universidad de Magallanes, Puerto Williams, Chile
e-mail: Ricardo.Rozzi@unt.edu

D. Zhu et al. (eds.), *Linking Arts with Biocultural Conservation, Restoration, and Communication*, Ecology and Ethics 8,
https://doi.org/10.1007/978-3-032-04830-1_2

integrate multiple knowledge systems. In these actions, art can serve not just as a tool for biocultural CRC, but can also be reinterpreted as a transformative biocultural practice that enhances interspecies and intercultural relationships of co-inhabitation.

Keywords Biocultural conservation · Biocultural ethics · Decolonial · Environmental philosophy · Locality

2.1 Introduction

A pressing need and concern in the realm of dominant Western ethics is the recognition and cultivation of interconnected relationships between human cultures and the other-than-human world (Zhu 2023; Andreozzi 2025; Prajapati and Nath 2025). To address this necessity, we explore ethical frameworks that harmoniously couple biological diversity and cultural diversity to orient just and sustainable forms of co-inhabitation. In the first part of our book, we draw from the conceptual framework of the biocultural ethic proposed by Ricardo Rozzi (2010, 2013) to broaden it through the integration of aesthetic and artistic dimensions.

More specifically, we seek to understand, recover, and shine a light on the links between art and *biocultural conservation*, *restoration*, and *communication* (CRC). Toward this aim, we introduce historical and contemporary concepts that present novel theoretical and practical reconfigurations to explain and illustrate how art and biocultural conservation are, and can be, linked. Chapters in Part I explore the connections between the arts and biocultural conservation through aesthetic, ethical, historical, political, ontological, spiritual, and phenomenological perspectives. The essays include a diversity of views rooted in contrasting places and traditions that not only challenge dominant conceptions of art but also philosophically question the relationships among artists, art practices, and biocultural communities. The conceptual framework of the biocultural ethic recognizes and values the interconnected link between cultural/linguistic and biophysical diversity (Rozzi 2012). This recognition is a critical response to Eurocentric and colonial frameworks that separate culture from nature. To recover and enhance intercultural and interspecies relationships and counter the force of biocultural homogenization, this ethic has guided a biocultural conservation approach that incorporates diverse forms of knowledge, including scientific and artistic practices (Rozzi 2013).

Biocultural conservation programs have seen an increasing interest in art as an educational, scientific, and community engagement tool (Polfus et al. 2017; Markienwicz et al. 2023; Lerski 2025). In consonance with these approaches, many chapters throughout this book offer examples of arts being incorporated into biocultural conservation. In Part I, however, we offer a theoretical and philosophical focus that helps us to question from a biocultural standpoint what arts *are,* what it means to *practice* and *experience* art, and how arts are situated within traditions, place, and community. Consequently, its exploration is not limited to questions of how art can be used as a means to an end in conservation programs.

In other words, in Part I we invite readers to think beyond art as a mere instrument. We aspire to broaden our understanding to a much wider sense of what arts may mean from biocultural standpoints. In this way, each of the chapters contributes to an expansion of our conceptions of arts, biocultural conservation, and their links, revealing paths to more reciprocal, transformative, and caring co-inhabitation with biocultural communities.

One persistent problem that must be contended with throughout the entirety of this book is the philosophical and epistemological limits of the English terms we use in our discussion. The Cleveland Museum of Art in the U.S. provides visitors with a placard for the Native North American section that notes the following:

> There is no word for "art" in Indigenous American languages, including those of North America. Though this European concept and term are now hard to avoid, they skew Native Americans' views about the aesthetically remarkable objects that Indigenous people create. The finished work and its outward appearance are important. But technical excellence, usefulness, and the process followed during creation are often at least as crucial, the latter because it is shaped by the beliefs, history, practice, and values of the maker's community. As museum director W. Richard West Jr. (Southern Cheyenne) says, Native North American arts are grounded in the communal and cultural.[1]

This placard highlights the theoretical limits of discussing non-European, in this case North American Indigenous practices using the term "art," which inevitably comes with specific metaphysical and epistemological assumptions. Nancy Marie Mithlo (2012) complexifies this issue, noting a tension among Indigenous artists and scholars between the "no word for art" approach, which comes as a part of a political rejection of coloniality, and the "artist first, Indian second" approach among many contemporary Indigenous artists, who reject identity claims in response to racist colonial practices wihin art institutions. However, Mithlo notes that growing dialogue among scholars, artists, and curators has contributed to a more nuanced approach to the discussion of Indigenous arts and artists.

In recognition of this ongoing discourse, we note the twofold risk of colonizing definitions of "art" which are deployed to perpetuate unjust power relationships through exclusion, as well as universalizing approaches that reduce bioculturally significant multidimensional practices and products to *merely aesthetic art objects*. With this risk in mind, this section begins from a pluralistic outlook, recognizing that even in the English language, for example, the term art encompasses multiple and ever-expanding meanings. Without committing to one exclusionary view of art, we share the chapters in this section as potentially complementary approaches that can help the readers develop an appreciation of the multifaceted and diverse ways that arts and biocultural CRC can be connected.

[1] Observed on May 28, 2025 in the section on Native North American art.

2.2 Three Core Concepts Linking Art with Biocultural CRC

From the chapters in Part I, we garner three key takeaway points, although the topics of the following chapters are not limited to them.

- First, there is an ontological shift toward a biocultural understanding of art. This can be seen in Tsuji and Johnson's chapter, which proposes a turn away from object-oriented "art," and in Castro Jorquera's chapter, which guides the reader's attention toward the precarious and ephemeral, as a response to biocultural homogenization.
- Second, there is recognition and reflection on foundational philosophical concepts and approaches needed for linking arts and biocultural conservation. This can be seen in Rozzi's tribute to Eugene C. Hargrove, who laid the groundwork for much of Western environmental philosophy. Consistent with Hargrove's multicultural and Rozzi's biocultural proposals, the chapters by Song as well as by Hu and Yang focus on the aesthetic roots of Chinese traditions of thought and art to explore intercultural dialogues with Western traditions.
- Third is a focus on the embodied experience of artistic practice that informs interspecies encounters. McIntosh's chapter illustrates the multispecies expansion of our embodied modes of understanding. The chapter by Chiarini reminds us that although scientific illustrations intend to be objective in the sense of accuracy, such art is no exception from the aesthetic and cultural influences of illustrators and audiences. Finally, Zhu's chapter introduces a critical discussion of the biocultural labor of artists in the advent of generative artificial intelligence (AI).

Each of these themes is also reflected in Johnson's final chapter, which is an artistic first-person narrative that draws from his face-to-face encounters with other-than-human co-inhabitants. Through these discussions, we also see that the concepts, aesthetic understandings, and practices of the arts are not homogeneous, as they reflect different interspecies and biocultural relationships.

2.3 A Concise Overview of the Chapters in Part I

Part I begins with a tribute to Eugene Carroll Hargrove, who founded the journal *Environmental Ethics* in 1979 and established a solid academic platform for the new transdisciplinary field of environmental philosophy. In Chap. 3, Ricardo Rozzi (2026) underscores two attributes in Hargrove's foundational work: its philosophical analysis of historical ideas, and (2) its transdisciplinary focus. Rozzi concludes that Hargrove's work inspires our book *Linking Arts with Biocultural Conservation, Restoration, and Communication* by stimulating interdisciplinary and intercultural approaches to environmental ethics and aesthetics. In his words, with our book we aspire to move "from the individual work in environmental ethics initiated by

Hargrove toward a collective work for biocultural conservation, in dialogue and collaboration with multiple cultural traditions that co-inhabit the heterogeneous regions of the planet" (Rozzi 2026).

In Chap. 4, Rika Tsuji and Benn Johnson (2026) expand the discussion of art and biocultural conservation beyond the scope of aesthetics to revisit biocultural art conservation through the lens of biocultural ontologies. Drawing from biocultural and critical environmental literature, such as the 3 Hs (co-inhabitants, habits, habitats) model of the biocultural ethic developed by Ricardo Rozzi (2013) and the concept of locality elaborated by Brian Burkhart (2019), the chapter offers an ontology which is relational, co-constitutive, and rooted in locality. Thus, rather than seeing art merely as a set of detachable and discrete "works" to be displayed in museums, the chapter challenges readers to reinterpret art as ongoing relational expressions of biocultural communities. In the process, Tsuji and Johnson provide examples from Japan and North America that invites us to consider the ontology of art from a biocultural perspective and imagine how this might affect efforts to do biocultural *art* conservation. Some of this philosophical approach can be found in the following chapter as well.

Focusing on the work of the Chilean artist, poet, and filmmaker Cecilia Vicuña, in Chap. 5, Carolina Castro-Jorquera (2026) examines how this artist explores themes such as language, memory, ecological destruction, and extinction, grounded in feminist and biocultural perspectives. Since the 1960s, Vicuña has developed the concept of "precarious art" (lo precario), using ephemeral materials to express the sacred and ecological value of the small and discarded, notably through works like *quipus* and *PALABRARmas*. Her early actions, such as creating ephemeral installations on the shores of the Aconcagua River in Chile, reflect a lifelong commitment to environmental and cultural resistance. Revisiting her work through Rozzi's (2018) biocultural lens, Castro-Jorquera reveals how Vicuña's art resists biocultural homogenization and reclaims interspecies and intercultural relationships, offering art as both a practice of knowledge and a prayer for the Earth.

The examination of biocultural dimensions in art and communication in Chinese traditions of thoughts is introduced in Chaps. 6 and 7. In Chap. 6, Yubo Song (2026) offers traditional Chinese understandings of interspecies relationships by focusing on Taoism, Confucianism, and Chinese Buddhism, and demonstrates how these traditions align with biocultural ethics, particularly with Rozzi's (2013) concepts of co-inhabitants and co-inhabitation with other-than-human species. Song introduces concepts from each tradition that convey interspecies and interconnected worldviews, showing that they are intricately connected with aesthetic values — expressed and practiced in various forms of arts, such as writings, paintings, music, and artifacts. In other words, Song's chapter indicates that these Taoist, Confucianist, and Chinese Buddhist worldviews, despite their differences, share an aesthetic practice and appreciation toward the interconnected attunement with self, co-inhabitants, and the world. In this way, Song's chapter demonstrates the vital link between aesthetics and biocultural ethics.

In Chap. 7, Jun Hu and Panpan Yang (2026) state that in Chinese traditions such as Confucianism and Taoism, the natural environment is considered the source of truth, symbolizing the unity of Heaven and humankind. Based on these traditions they examine Chinese aesthetics and worldviews through plants-inspired art education to reconnect students with their natural environment. Toward this aim they introduce the Chinese art concept of the "Four Gentlemen" as well as contemporary education initiatives examined through the 3Hs model of biocultural ethics (Rozzi 2013). Hu and Yang (2026) show how art classes in primary and secondary schools can play a vital role for the conservation, restoration, and communication of biocultural heritage. They conclude that creative arts education can rewire the interactions among students and citizens with other co-inhabitants, their life habits, and habitats, and restore vital connections for ethical dwelling.

Finding a link between biocultural ethics and the work of artist Miriam Simun, in Chap. 8, Shoshana McIntosh (2026) challenges limited human perceptions of other-than-human animals. McIntosh essays an embodied, cephalopod-inspired approach to understanding and coexisting with them. She explores how engaging with the octopus as a sentient, autonomous being can decenter human perspectives and foster deeper, more respectful interspecies relationships. McIntosh (2026) advocates for a transformative educational method that expands human awareness and supports multispecies flourishing by integrating phenomenology, art, and embodied pedagogy.

In Chap. 9, Valeria Serena Chiarini (2026) offers a rich historical and cross-cultural survey of scientific illustration, showing us the vital role of illustrators in sciences and biocultural conservation. Although illustration intends to be scientifically accurate, Chiarini reveals that illustrators' cultural and aesthetic perspectives influence their art execution and illustrations; illustrators both intentionally and unintentionally have taken an active role of biocultural interpretation and communication of scientific understandings. Given this, Chiarini focuses on examples of how illustrators, including herself, can visually represent otherwise invisible things, such as connections between the 3Hs (co-inhabitants, habits, and habitats) and other-than-human animal perspectives (Rozzi 2012), which are essential for biocultural conservation and communication.

Danqiong Zhu (2026) undertakes a different approach by critically examining how artistic practices of knowledge might change, given the emergence of generative artificial intelligence (AI). In Chap. 10, Zhu (2026) challenges us to revisit the biophysical links between artists and their biocultural landscapes. Drawing from a biocultural account of labor, in which biophysical and cultural-symbolic practices are linked as artists produce their artwork, the chapter argues that artwork produced by generative AI conceals the connections between artists and their environment (as well as the labor that connects them), and may even discourage specific, direct biocultural connections for viewers and artists. Zhu's chapter both explicitly and implicitly provokes the readers to question what it means to *do* art at all as a human.

Part I closes with Benn Johnson's first-person narrative about a journey to the Cape Horn Biosphere Reserve to reflect on some of the philosophical themes within the chapter. Inspired by the different aesthetic experiences of Magellanic Woodpecker, Black-faced Ibis, skua, and Mirror-winged Gulls *in situ*, in Chap. 11, Johnson (2026) explores what it might mean to begin to think about co-inhabitants not as mere specimens, but as related, precarious, and active parts of biocultural expression. Throughout Johnson's essay we see how experiencing other-than-human beings can shift one's perspective, occasionally lending to objectification of the other or the self, and at other times lending to a deeper understanding of the relationality and ethical weight of being a part of biocultural expression. Various species, including birds, mosses, and lichens of the Cape Horn region of Chile, resist becoming mere things through their unique expressions.

In an era marked by trans-disciplinarity and trans-culturalism (Prajapati and Nath 2025), Part I introduces harmonious reconciliations between cultural and biological diversity. We acknowledge that there are already manifold personal and collective initiatives committed to better understanding, valuing, and defending life and its biocultural diversity. These initiatives are led by courageous leaders of Indigenous and other local communities as well as artists and scientists that transcend their specialties to embrace a genuine sense of socio-environmental justice.

These biocultural conservation, restoration, and communication (CRC) initiatives have been largely underrepresented by the media, formal education, and the prevailing culture of global society (Rozzi 2019; Georgiadou et al. 2020; Tauro and Rozzi 2025). We hope that chapters in Part I motivate readers to search for biocultural CRC initiatives in their own localities and regions of the world. In this way, we can contribute to transforming a uniform Anthropocene into a multifaceted era that nurtures multiple aspects of biocultural ethics, including collaborations among artists, scientists, philosophers, and local people. In short, our book *Linking Arts with Biocultural Conservation, Restoration, and Communication* aims to stimulate collective work for biocultural conservation that is fostered by multiple cultural traditions that co-inhabit the heterogeneous regions of the planet.

Acknowledgments We would like to thank Danqiong Zhu for her valuable comments on draft versions of this chapter, and to acknowledge the support from the Cape Horn International Center (CHIC)-ANID PIA/BASAL (PFB210018).

References

Andreozzi M (2025) Reimagining ethics. Non-anthropocentric perspectives on morality, Ecology and ethics, vol 7. Springer, Cham

Burkhart B (2019) Indigenizing philosophy through the land: a trickster methodology for decolonizing environmental ethics and indigenous futures. Michigan State University Press, East Lansing

Castro-Jorquera C (2026) The precarious art of Cecilia vicuña: finding the ecological and sacred value of the small and the discarded. In: Zhu D, Rozzi R, Tsuji R, Johnson B, Caviness T, Castro-Jorquera C, Xu L, Zhang L (eds) Linking arts with biocultural conservation, restoration, and communication, Ecology and ethics, vol 8. Springer, Cham, pp 25–33. https://doi.org/10.1007/978-3-032-04830-1_2

Chiarini VS (2026) How the invisible becomes visible: situating scientific illustration in biocultural communication. In: Zhu D, Rozzi R, Tsuji R, Johnson B, Caviness T, Castro-Jorquera C, Xu L, Zhang L (eds) Linking arts with biocultural conservation, restoration, and communication, Ecology and ethics, vol 8. Springer, Cham, pp 149–169. https://doi.org/10.1007/978-3-032-04830-1_9

Georgiadou Z, Florou V, Ilia-Georgiadou I (2020) Damnatio memoriae: contemporary condemnation of memory. Design/Arts/Culture 1. https://doi.org/10.12681/dac.25913
Hu J, Yang P (2026) A biocultural encounter with the four gentlemen: art education in Chinese ecological and aesthetic traditions. In: Zhu D, Rozzi R, Tsuji R, Johnson B, Caviness T, Castro-Jorquera C, Xu L, Zhang L (eds) Linking arts with biocultural conservation, restoration, and communication, Ecology and ethics, vol 8. Springer, Cham, pp 111–129. https://doi.org/10.1007/978-3-032-04830-1_7
Johnson B (2026) A bird in the hand: experiencing birds in the subantarctic. In: Zhu D, Rozzi R, Tsuji R, Johnson B, Caviness T, Castro-Jorquera C, Xu L, Zhang L (eds) Linking arts with biocultural conservation, restoration, and communication, Ecology and ethics, vol 8. Springer, Cham, pp 189–193. https://doi.org/10.1007/978-3-032-04830-1_11
Lerski MB (2025) Identifying intangible and biocultural heritage elements toward environmental understanding: engaging stakeholders through art. Ecol Soc 30(1):5. https://doi.org/10.5751/ES-15631-300105
Markiewicz J, Goralnik L, O'Connell K (2023) Screens on trails: digital environmental science, arts, and humanities learning for biocultural conservation. In: Rozzi R, Tauro A, Avriel-Avni N, Wright T, May RH Jr (eds) Field environmental philosophy: education for biocultural conservation, Ecology and ethics, vol 5. Springer, Cham, pp 555–576. https://doi.org/10.1007/978-3-031-23368-5_33
McIntosh S (2026) Embodied encounters: biocultural dialogues and pedagogies beyond the human. In: Zhu D, Rozzi R, Tsuji R, Johnson B, Caviness T, Castro-Jorquera C, Xu L, Zhang L (eds) Linking arts with biocultural conservation, restoration, and communication, Ecology and ethics, vol 8. Springer, Cham, pp 131–148. https://doi.org/10.1007/978-3-032-04830-1_8
Mithlo NM (2012) No word for art in our language?: old questions, new paradigms. Wicazo Sa Rev 27(1):111–126. https://doi.org/10.1353/wic.2012.0005
Polfus JL, Simmons D, Neyelle M, Bayha W, Andrew F, Andrew L, Merkle BG, Rice K, Manseau M (2017) Creative convergence: exploring biocultural diversity through art. Ecol Soc 22(2):4. https://doi.org/10.5751/ES-08711-220204
Prajapati A, Nath R (2025) Looking at biocultural ethics through the lens of transcultural dialogue. Ethics Sci Environ Polit 25:71–85. https://doi.org/10.3354/esep00221
Rozzi R (2010) Field environmental philosophy: regaining an understanding of the vital links between the inhabitants, their habits and the regional habitats. Dialogue Universalism 1(2):85–109. https://doi.org/10.5840/du20102011/127
Rozzi R (2012) Biocultural ethics: recovering the vital links between the inhabitants, their habits, and habitats. Environ Ethics 34(1):27–50. https://doi.org/10.5840/enviroethics20123414
Rozzi R (2013) Biocultural ethics: from biocultural homogenization toward biocultural conservation. In: Rozzi R, Palmer C, Armesto JJ, Callicott JB, Pickett S (eds) Linking ecology and ethics for a changing world: values, philosophy, and action, Ecology and ethics, vol 1. Springer, Dordrecht, pp 9–32. https://doi.org/10.1007/978-94-007-7470-4_2
Rozzi R (2018) Biocultural homogenization: a wicked problem in the Anthropocene. In: Rozzi R, May RH Jr, Chapin FS III, Massardo F, Gavin M, Klaver I, Pauchard A, Nuñez MA, Simberloff D (eds) From biocultural homogenization to biocultural conservation, Ecology and ethics, vol 3. Springer, Cham, pp 21–48. https://doi.org/10.1007/978-3-319-99513-7_2
Rozzi R (2019) Taxonomic chauvinism, no more! Antidotes from Hume, Darwin, and biocultural ethics. Environ Ethics 41(3):249–282. https://doi.org/10.5840/enviroethics201941325
Rozzi R (2026) The contributions of Eugene Hargrove to environmental ethics and aesthetics: a biocultural tribute. In: Zhu D, Rozzi R, Tsuji R, Johnson B, Caviness T, Castro-Jorquera C, Xu L, Zhang L (eds) Linking arts with biocultural conservation, restoration, and communication, Ecology and ethics, vol 8. Springer, Cham, pp 35–59. https://doi.org/10.1007/978-3-032-04830-1_3

Song Y (2026) Interspecies co-inhabitation: a dialogue between Chinese philosophical traditions and biocultural ethics. In: Zhu D, Rozzi R, Tsuji R, Johnson B, Caviness T, Castro-Jorquera C, Xu L, Zhang L (eds) Linking arts with biocultural conservation, restoration, and communication, Ecology and ethics, vol 8. Springer, Cham, pp 95–109. https://doi.org/10.1007/978-3-032-04830-1_6

Tauro A, Rozzi R (2025) Biocultural ethics and earth stewardship: a novel integration to revitalize multiple values of nature. Ecol Soc. 30. https://doi.org/10.5751/ES-16362-300335

Tsuji R, Johnson B (2026) Biocultural ontologies, art, and conservation. In: Zhu D, Rozzi R, Tsuji R, Johnson B, Caviness T, Castro-Jorquera C, Xu L, Zhang L (eds) Linking arts with biocultural conservation, restoration, and communication, Ecology and ethics, vol 8. Springer, Cham, pp 61–77. https://doi.org/10.1007/978-3-032-04830-1_4

Zhu D (2023) Bridge the channel, enhance the inclusivity: a comparison between flagship species-centered and moss-centered conservation in Chile and China. In: Rozzi R, Tauro A, Wright T, Avriel-Avni N, May RH Jr (eds) Field environmental philosophy: education for biocultural conservation, Ecology and ethics, vol 5. Springer, Dordrecht, pp 457–482. https://doi.org/10.1007/978-3-031-23368-5_28

Zhu D (2026) Arts and environment after the emergence of artificial intelligence. In: Zhu D, Rozzi R, Tsuji R, Johnson B, Caviness T, Castro-Jorquera C, Xu L, Zhang L (eds) Linking arts with biocultural conservation, restoration, and communication, Ecology and ethics, vol 8. Springer, Cham, pp 171–187. https://doi.org/10.1007/978-3-032-04830-1_10

Chapter 3
The Contribution of Eugene Hargrove to Environmental Ethics and Aesthetics: A Biocultural Tribute

Ricardo Rozzi

Abstract Biocultural co-inhabitation implies a triple ethical duty: to foster the well-being of the co-inhabitants, to respect their lifestyles, and to care for the habitats shared with them. To address this imperative of the biocultural ethic, I examine the work of American environmental philosopher Eugene C. Hargrove, who founded the journal *Environ*mental Ethics (EE) and established a solid academic platform for the new transdisciplinary field of environmental philosophy at the end of the twentieth century. I argue that for twenty-first century environmental ethics, his pioneering work is relevant due to two fundamental attributes: his philosophical analysis of the history of ideas and his transdisciplinary approach. These attributes are vital to address the ultimate causes of the global socio-environmental crisis, which constitutes one of the greatest threats to the sustainability of life and the conservation of its biological and cultural diversity. The ultimate causes of this crisis are cultural, educational, and value based. I highlight how Hargrove uses his transdisciplinary approach rooted in the history of ideas to focus on aesthetic values. In this way, he expands the spectrum of the values of nature that should be considered in environmental policy and education. Toward this aim, he documents a broad diversity of thinkers, artists, and other citizens who have treasured nature since the dawn of Western civilization. Importantly, Hargrove emphasizes that members of different societies and ethnic groups should cultivate forms of environmental ethics rooted in the values of their own culture. In summary, he proposes that environmental ethics should be transdisciplinary, international, and intercultural in order to be effective and just. For these reasons, I argue that Hargrove's contribution to environmental ethics and aesthetics takes on special relevance in countering pervasive processes of biocultural homogenization that are amplified by global society, and in inspiring, instead, contemporary forms of biocultural conservation.

R. Rozzi (✉)
Sub-Antarctic Biocultural Conservation Program, Department of Philosophy and Religion and Department of Biological Sciences, University of North Texas, Denton, TX, USA

Cape Horn International Center (CHIC), Universidad de Magallanes, Puerto Williams, Chile
e-mail: Ricardo.Rozzi@unt.edu

D. Zhu et al. (eds.), *Linking Arts with Biocultural Conservation, Restoration, and Communication*, Ecology and Ethics 8,
https://doi.org/10.1007/978-3-032-04830-1_3

Keywords Biocultural conservation · Biocultural ethics · Environmental philosophy · History of ideas · Transdisciplinary

3.1 Introduction

Are we willing to co-inhabit the planet by valuing, respecting, and caring for biological and cultural diversity? This is the fundamental question we ask ourselves from the perspective of biocultural ethics (Rozzi 2012a). To answer affirmatively, I offer this chapter inspired by the work of American environmental philosopher Eugene C. Hargrove, better known to his students and friends as "Gene." In 1979, Hargrove created a platform for the new transdisciplinary field of environmental ethics by establishing the eponymous journal *Environmental Ethics* (EE), and by publishing the book *Foundations of Environmental Ethics* a decade later. More than 30 years after its publication, this pioneering work of environmental ethics takes on special relevance due to two attributes: (1) its philosophical analysis of historical ideas (Hargrove 1979a, 1980, 1989) and (2) its transdisciplinary approach (Hargrove 1992a, 2008).

These two core attributes of Hargrove's work are vital for understanding (and hopefully resolving) the ultimate causes of climate change and the global socio-environmental crisis. Today, this crisis is one of the greatest threats to the sustainability of life, and its ultimate causes are cultural, educational, and value based. In my discussion of Hargrove's lasting contribution to the field of environmental ethics, I highlight how, based on the two former attributes, he focuses on aesthetics to broaden the spectrum of nature values as well as the diversity of thinkers, artists, and other citizens who have treasured the natural world throughout the history of Western civilization. Finally, by underscoring that members of different societies and ethnic groups cultivate environmental ethics rooted in the values of their own culture, Hargrove's proposal takes on special relevance to counteract processes of biocultural homogenization that are pervasive in the twenty-first century. In summary, for environmental ethics to be effective and just, Hargrove identifies three essential characteristics: transdisciplinary, international, and intercultural.

3.1.1 *History of Ideas*

The first attribute of Hargrove's foundational book (1989) is its philosophical approach based on the history of ideas. This approach is especially relevant because throughout history, diverse cultures around the world have marveled at and developed relations of care for nature. In ancient China, more than 2500 years ago, the garden symbolized the best place for rest, meditation, and reconnection with nature. More than 2000 years ago, Epicurus taught his ethics in a garden on the outskirts of Athens. Metaphorically, I imagine that for Epicurus, the ethical life would be analogous to that of gardeners who take care of themselves as much as they do for plants. This ethic emanates from cultivating relationships of reciprocity between human and other-than-human beings (see Box 3.1). For the twenty-first century, we can draw from Epicurus and many other schools of thought on seminal values and life habits to foster a biocultural ethic that guides global society toward more just and sustainable ways of co-inhabiting the planet.

Box 3.1 The Concept of "Other-Than-Humans"

I have used the term "other-than-humans" (Rozzi 2018) instead of the more common term "non-humans" for five main reasons.

First, to avoid dichotomous thinking between "humans" and "non-humans," since the latter term creates a fracture or abysmal separation between humans and other beings. This dichotomy is not innocent because it establishes a dualistic distinction that separates one species (humans) from myriad species that are grouped together as the "rest" ("non-humans").

Second, the term "non-humans" is hierarchical and oppressive, since it defines these beings by negation; by the lack of "exceptional qualities" that only humans possess. This definition of imperfection, of lack, represents them in a situation of deprivation that provides a justification (actual or potential) for the exploitation of "non-humans" by humans.

Third, the expression "other-than-human" helps us understand that these beings inhabit not only biophysical nature but also the images, symbols, and values of our cultures. Therefore, they are co-inhabitants of our biocultural communities that encompass biophysical and symbolic-linguistic dimensions of reality. Both dimensions exist in both our waking and dream phases of life. Consequently, we regain attention to the importance of the dream world.

Fourth, the expression "other-than-human" helps us avoid dualism between "biotic" and "abiotic" beings, which were even included in Arthur Tansley's definition of ecosystem (Johnson 2020). In contrast, it invites us to understand that individuals are intertwined in webs of existence in which matter and energy flow between different levels of organization and interactions of beings that make up both small ecosystems (for example, a lagoon located in a city square or on a mountaintop) and the entire biosphere, where human beings co-inhabit with the entirety of beings (who live "inside" and "outside" the illusory limits of our bodies defined by the outer layer of epidermal cells).

Fifth, political dimensions of reality are also incorporated, since these concepts are understood not only from a Western perspective (philosophical or scientific), but also from ecological worldviews that belong to multiple cultures. For many Indigenous peoples, such as the Aymara and Quechua in South America, the sun, the moon, lightning, frost, hail, mountains, lakes, springs, plants, fungi, animals, and humans are subjects that can be understood as other-than-human co-inhabitants. Through worldviews, rituals, and care practices, the interconnections between humans and the set of beings with

(continued)

Box 3.1 (continued)
whom we co-inhabit the lakes, oceans, land, air, and skies are recognized. This understanding has been incorporated into recent legislative documents that aim to overcome the dichotomies between humans and nature introduced by colonialist concepts and practices responsible for some of the most negative anthropogenic impacts driving the Anthropocene today (Álvarez et al. 2023).

Hargrove focuses on the historical drift of environmental ideas and practices in Western civilization and uncovers a specially inspiring precedent for supporting an environmental ethic that springs from the collaborative work of naturalists and artists. In particular, he carefully examines how the Hudson River School of painters in New York motivated a cultural and political shift toward the appreciation of nature. Throughout his work, Hargrove documents facts, values, and practices that can inspire cultural and political transformations today, but which, however, have remained latent because they are little known. During the eighteenth and nineteenth centuries, naturalist expeditions as well as landscape gardening acquired special relevance for the scientific and aesthetic appreciation of nature. In *Foundations of Environmental Ethics*, he states that:

> In the nonformal garden …, the natural qualities and characteristics of the plants became matters of great interest, and the plants were elevated to the status of self-contained and self-organizing entities worthy of admiration and study for their own sake. This transformation was paralleled by a similar development then taking place in botany and in the other natural history sciences… Not surprisingly, these scientists came to focus some of their attention on the aesthetic aspects of the natural objects and other phenomena they studied. When writing in their field notebooks, most of them … jotted down their aesthetic judgments alongside their factual descriptions. In this way, they began to attend to values as well as to facts… both botany and gardening provided the foundations for a broader association of science and art. (Hargrove 1989, pp. 84–85)

3.1.2 A Transdisciplinary Approach

The second attribute of Hargrove's work is his stubborn transdisciplinary approach, which marks a turning point in the history of environmental philosophy. He has always encouraged his students and colleagues to incorporate ethical, aesthetic, ecological, ontological, epistemological, political, and educational perspectives into their philosophical analyses (Hargrove 2008, 2015). In this way, he encouraged a new generation of environmental philosophers to integrate theory and practice. Thanks to this approach, since the 1990s, philosophers and other thinkers have become interested in multiple ways of knowing and have become involved in environmental decision-making and policymaking, in educational reforms, and in actions promoting biodiversity conservation and socio-environmental justice. In his approach to education, he writes that:

> A good noncontroversial starting point [to teach environmental ethics] may be to begin with the values that are often listed in the purpose statements of environmental laws. Teachers could teach the history of ideas behind those values and their relationship to environmental concern... Comparative value discussion can be used to clarify traditional values and in countries with indigenous populations with values originating in different histories of ideas, such as the values of the First Nation peoples in Canada and the Mapuche in Chile, which can be used to promote better understanding between major social groups. (Hargrove 2008, p. 263)

In this way, his transdisciplinary approach is contextual and multicultural. And he also incorporates methodological guidelines. For example, he writes that,

> In deciding which values to teach, I recommend that educators work backward from what is needed to facilitate environmental decisions by the citizen and the environmental professional to what is needed to prepare them educationally for these adult decisions. Look first at your environmental laws and your environmental policies to see what specific values might be most appropriate for environmental decision making. (Hargrove 2008, p. 264)

Finally, Hargrove points out that

> Environmental education should not be an imposition of new values, but a strengthening of existing values. In most Western countries, for example, there is a three-century-old nature tradition in landscape painting, in nature poetry, in landscape gardening, and in natural history science... However, because of their educational experiences, in which they have been taught that their sense of beauty is an arbitrary, subjective, irrational expression of emotion, rather than an evolved social ideal. (Hargrove 2008, pp. 266–267)

In summary, Hargrove combines his two essential approaches, the history of ideas and transdisciplinary practices, to forge his proposal for environmental ethics. Both approaches acquire a quadruple relevance today.

- First, Hargrove's approach contributes to correcting the lack of philosophical memory in global society. This loss of memory is driven by the hegemonic notion of progress that has severed the historical roots of the ethical and aesthetical that have been essential to the development of Western civilization.
- Second, Hargrove provides an inter- and trans-disciplinary approach to environmental thought and action. His approach overcomes the ultra-specialization and atomization into discrete disciplines that were forged by twentieth-century academia.
- Third, to confront the complex socio-environmental scenarios of the twenty-first century, Hargrove introduces examples of real, historical cases that serve as models for integrating concepts (theory) and actions (practice).
- Fourth, Hargrove proposes a transdisciplinary approach that draws on multiple disciplines and cultures with their traditions of environmental thought.

Along with this quadruple relevance of Hargrove's work, four of the editors (Danqiong Zhu, Rika Tsuji, Benn Johnson, and I) of our book entitled *Linking Arts with Biocultural Conservation, Restoration, and Communication* have been his students and later colleagues. His friendship and wisdom have encouraged and guided us along a path of environmental ethics that considers aesthetics as a central dimension of the field. Furthermore, Hargrove established fruitful and long-standing collaborations with thinkers from around the globe. We have sought to incorporate all these attributes into our own book, including ethical and aesthetic perspectives from different disciplines and heterogeneous regions of the world. We offer this chapter as a tribute to our mentor, Gene Hargrove.

3.2 Hargrove: A Creative and Generous Philosopher

In 1992, amid the expectations generated by the Earth Summit in Rio de Janeiro, I learned that the Department of Philosophy at the University of North Texas (UNT) was creating a master's program in environmental ethics, a new field of philosophy that had been incorporated into academia in the 1970s. This was going to become the first graduate program in environmental ethics worldwide. And Hargrove created it with an approach that challenged the disciplinary specialization that characterized most philosophy departments. Excited by this news, which opened a horizon for investigating the values of nature beyond utilitarian approaches based on economic cost-benefit analysis for society (which were widely prevalent in 1992, exalting only the instrumental value of biodiversity), I wrote to the creator of this new graduate program, "Professor and Chair of the Department of Philosophy at UNT, Dr. Eugene Carroll Hargrove."

3.2.1 My Meeting with Gene Hargrove Upon Arriving from Chile

Hargrove's response to my letter was immediate. He kindly invited me to visit his Center for Environmental Philosophy (CEP) and to attend some courses in the new master's program. This welcome encouraged me to apply from Chile for a Fulbright scholarship that would allow me to pursue graduate studies in the U.S. The scholarship required me to apply to a doctoral program, which also excited me because it would allow me to achieve my idea of integrating philosophy and ecology into conservation biology. For a couple of years, I corresponded with Gene, in conjunction with ecologist John Silander at the University of Connecticut (UConn), which offered a doctoral program in biological conservation with an interdisciplinary approach. I began my graduate studies in ecology at UConn in 1996, and in 1997, I was able to travel to Texas to visit CEP and take master's courses in environmental ethics.

Hargrove himself met me at the airport. It was unexpected for me that the head of the Philosophy Department would personally pick up a young Chilean graduate student who was coming to Texas for the first time. On the way, he told me that he had never liked his given names and preferred to be called "Gene." He then invited me to lunch to learn about Tex-Mex cuisine. His kindness went hand-in-hand with a genuine interest in promoting the field of environmental ethics in Latin America (in addition to his genuine enjoyment of the culinary arts of the Mexican-American population residing in Texas).

Since that time, I have continuously collaborated with Gene on projects that seek to incorporate environmental ethics into environmental policy design, educational programs, and work on the conservation of biological and cultural diversity (Rozzi et al. 1998; Hargrove et al. 2008). So much so that in 2000, I applied for an assistant professor position in the Philosophy Department at UNT. When I was selected for this position, with my heart torn by my desire to continue living and working in biological conservation in Chile, in the final interview I said, "Gene, you know that I have a desire and commitment to continue working in conservation in Chile." His laconic response was, "We will help you." And so, it has been to this day.

In 2000, with the University of Magallanes (UMAG) and the Omora Foundation in Chile, I led the creation of Omora Park, conceived as a space for research, education, and biocultural conservation in the Cape Horn Biosphere Reserve at the southernmost tip of South America (Rozzi et al. 2006). With Gene's help, in 2004 we consolidated this initiative through the creation of the International Subantarctic Biocultural Conservation Program, co-managed by UMAG and the Omora Foundation in Chile and by UNT and CEP in the U.S. (Hargrove et al. 2008). With this program, we strengthened a series of international courses in field environmental philosophy, which have been annually offered by UMAG and UNT since 2005 (Rozzi et al. 2006, 2008). With the support of Gene and other collaborators, we have run these courses conducted at Omora Park, Cape Horn, Chile, for over 20 years. Moreover, in 2010, we established at Omora Park a network of "climate change sentinels," investigating value-based and educational dimensions in conjunction with biophysical dimensions of global change (Rozzi et al. 2012). In collaboration with CEP and UNT, we also inaugurated two bilingual publication series in Spanish and English. The first included special issues of the EE journal on Latin American environmental ethics. The second included books on subantarctic biocultural conservation published by UNT Press in collaboration with UMAG Press.

The initial encounter with Hargrove linked to the range of workshops, research, publications, and courses far exceed a purely academic sphere. Together with colleagues and collaborators, Gene broke the academic "ivory tower" in which philosophy programs are typically developed. Instead, he established at UNT a school of transdisciplinary thought and action committed to the sustainability of life. Before continuing, it is important to briefly recount some events that marked Gene's life, which I consider inseparable from the ideas that permeate his work.

3.2.2 Hargrove: Philosopher, Caver, and Conservationist

Gene was born on October 22, 1944, in Detroit, Michigan, in the Great Lakes region located on the US-Canadian border where his mother's mother was working in a munitions factory during World War II. His father, Oren Keith Hargrove, was an

Army Air Force mechanical engineer working in Laredo, Texas, and his mother, Eleanor Mae Lade, worked in commerce. After Gene was born, Gene and his mother went back to Laredo, Texas. Much to his parents' surprise, the first word Gene ever spoke was "agua," water in Spanish, causing the babysitter to be fired.

Then, at the age of two, his family settled in Saint Louis, Missouri, where Gene spent his childhood and youth. During his later school years, Gene planned to become a journalist and worked for a newspaper for almost a year before going to college. The newspaper editors and reporters recommended that he study English at the University of Missouri. Gene welcomed the suggestion, but during his first years of college, he took several required philosophy courses and discovered he enjoyed the subject much more, so he switched majors.

In 1966, he graduated with a bachelor's degree in philosophy from the University of Missouri. He then received his master's degree from the same university in 1967. After a brilliant academic career, he became an army officer having finished the Reserve Officers' Training Corps (ROTC) program in college, and served two years in Germany during the Vietnam War. He stayed in Germany for another one and half years there as an English teacher. Upon his return to the U.S., he resumed his philosophy studies at the University of Missouri, and in December 1974, he completed his doctorate with the defense of his thesis, "Wittgenstein and Ethics." He was able to continue his research on the relationship between Ludwig Wittgenstein's philosophy and ethics through a postdoctoral fellowship at the University of Vienna, Austria, during 1976 and 1977. He proposed that Wittgenstein's examination of logical necessity and rule-governed justification in science bears similarities to ethical reasoning. To this end, he analogized the distinction between the stages of discovery and justification in science with the stages of decision and judgment in ethics. His research on ethical decision-making processes had, from the outset, a practical application in the field of conservation.

Gene never limited his life to academic activity. In 1966, in his student days, he became involved in cave exploration. Throughout his life, speleology has held a lifelong attraction for him, fueled by the fascinating nature of its geological labyrinths and the varied life forms that sustain these natural caves. While working on his PhD, he undertook a lead role in exploring a large cave near the university campus, located in Rock Bridge Memorial State Park in Columbia, Missouri, which was established in 1967 (Hargrove 1968). This cave is called the Devil's Icebox because cool air gushes out of it year-round, maintaining a stable ambience temperature of 13 °C, even in midsummer. Residents of the county were amazed to hear that the cave was the largest object in the county when Gene lectured on it at the county library, and its complex network of galleries, cracks, fissures, and underground passages (some of which are flooded) are home to animals adapted to the unique conditions of light, silence, relative humidity, pressure, and temperature. In fact, this cave is home to several threatened cave species in Missouri, including the endangered Indiana bat (*Myotis sodalis*) and the pink planarian (*Kenkia glandulosa*, a species endemic to the cave).

Upon returning from Germany in 1970, Gene was shocked to learn that the cave was likely to be seriously threatened by the influx of contaminated sewage from a proposed housing development. He immediately committed himself to action and devoted much of his time to protecting the cave while developing his doctorate. He wrote extensively about conservation issues in caving journals and his activities were covered daily in two local newspapers (Hargrove 1973a, b). He examined problems associated with land use (and abuse) and proposed solutions at multiple scales, from the community to the national level (Hargrove et al. 1973). The governor followed his efforts closely. Gene warned authorities that inappropriate land use by landowners or managers would lead to the destruction of cave life, and even the cave itself, if the problems of water pollution and quarrying were not remedied. He proposed four types of solutions: (a) actions led by the National Speleological Society, (b) legal action through state laws to regulate water pollution and land use, (c) zoning and territorial planning in each commune or county, and (d) a large-scale educational program aimed at changing public attitudes toward the preservation of natural phenomena, both living and nonliving (Hargrove 1973c). The cave was saved in part by becoming part of Rock Bridge Memorial State Park. As a result of these actions and proposals, Gene was identified in the early 1970s as an "activist environmentalist specializing in professional ethics." This is how he became involved in environmental ethics or environmental philosophy.

To incorporate ethical values into political arguments, Hargrove chose to focus on historical research into the origins of philosophical ideas. In a Wittgensteinian sense, his research helps elucidate philosophical confusion, because it brings to light assumptions that are often not explicitly stated at a conscious level (Hargrove 1985). For example, those responsible for trying to damage the cave by creating a residential development on the sinkhole plain over the cave argued from the point of view of private land ownership as if they could freely destroy their property, without considering the consequences of such actions for society as a whole or for the environment.

To elucidate the philosophical confusion underlying this assumption, Hargrove traced the historical origin of the concept of private property (Hargrove 1980). He identified ancient Germanic land-use practices that existed as early as 100 B.C. Such practices later spread throughout Northern Europe due to the Germanic conquest and were brought to North America by the English in the seventeenth century, where they were enshrined in the U.S. Constitution, a document inspired by John Locke's property theory and the writings of Thomas Jefferson. A key problem with this view is that it fails to consider the consequences of micro-scale practices, defined at the individual level by the owners of private lands, on the macro-scale, which includes the entire set of private and public lands contained within a single watershed or other ecosystem.

During the value and political debate over the Devil's Icebox, Hargrove offered an alternative vision to individualism and private property. As Aristotle put it, ethics is understood from the perspective of the individual, and politics from the perspective of the group. Similarly, Aldo Leopold (considered the father of twentieth-century American environmental ethics) concluded, in his famous essay "The Land Ethic" (Leopold 1949), that the moral characteristics that make people moral individuals are the same characteristics that make them good citizens (Hargrove 2015). Leopold

(1949) also believed there were shortcomings to what the government could and should do, and that an ethic on the part of individuals was required to fill the gap between their individual behavior and government regulations.

Hargrove has made a decisive contribution to understanding the communitarian meaning of environmental ethics and to incorporating the "intrinsic value" of other-than-human beings into political and cultural debates. He has thus broadened the conceptual and value spectrum of discourses focused solely on the "instrumental" or economic value of nature, popularized in the 1980s and 1990s with concepts such as "natural capital" and "ecosystem services," which, while necessary, are not sufficient for a genuine cultural transformation toward the sustainability of life.

For our book, *Linking Arts with Biocultural Conservation, Restoration, and Communication*, it is highly relevant that Hargrove has identified a plurality of values and has always been categorical in affirming that living beings, ecosystems, and the entire biosphere have intrinsic value. It is increasingly clear that one of the main indirect drivers of climate change and, more broadly, rapid global socio-environmental change is the fact that political decisions focus almost exclusively on the instrumental value of nature, excluding other values from its spectrum (Pascual et al. 2023). Biodiversity has both instrumental and intrinsic values, which are not mutually exclusive (Rozzi 1999). Furthermore, instrumental values are not necessarily anthropocentric. For example, a wide variety of insects are instrumentally valuable to birds, but not to humans. We can therefore affirm that these insects are instrumentally valuable, even when they are not directly useful to humans, but rather are valuable to other types of organisms. Therefore, these insects have an instrumental value that is not anthropocentric; that is, they have eco- or bio-centric values (Osorio-García and Roberto-Alban 2023). Furthermore, the fact that they have instrumental value does not deny that insects (the most diverse group of organisms on a global scale) have intrinsic, aesthetic, and other values.

3.2.3 The Importance of Aesthetics in Conservation

In identifying multiple values of nature, Hargrove captured the cultural relevance of the beauty of a landscape as a value that appeals to government officials and environmental decision-makers. This perspective converges with the vision of Leopold, who in his Land Ethic concluded that "a thing is right because it tends to preserve the integrity, stability, and beauty of the biotic community." The three attributes (integrity, stability, and beauty) are complementary and mutually reinforcing.

Hargrove became interested in how beauty acquired great relevance for nature preservation in the United States since the mid-nineteenth century. Most of those places painted by a sensitive and influential artist or recorded by a prominent photographer in the nineteenth century, went on to become national parks or national monuments, beginning with Yosemite and Yellowstone. Hargrove promptly realized that the aesthetic appreciation of nature is by no means limited to its expression in painting or photography, since scientific information about nature can also stimulate an aesthetic experience that promotes its preservation. As we learn more about nature in terms of its natural beauty and its scientific interest, we will come to value it more holistically and become more determinedly involved in its protection.

Gene addressed these issues during his student years. Later, in 1978, he obtained a Rockefeller Foundation Fellowship in Environmental Affairs to investigate the history of the ideas behind the arguments involved in nature conservation. He conducted part of his research at the Smithsonian Institution in Washington, D.C., where he met administrators of the National Endowment for the Humanities (NEH). At the time, NEH was initiating programs to fund institutes in engineering and ethics, technology, and the humanities. Hargrove suggested to one of the NEH administrators that an institute on ethics and the environment would be a good idea. After admitting that there was almost nothing on ethics and environmental issues, these administrators encouraged Gene to write an interdisciplinary book.

3.2.4 The World's First Environmental Ethics Journal

Hargrove transformed the idea presented to the NEH; he created instead the journal *Environmental Ethics*. He published his first issue when he joined as an assistant professor of philosophy the University of New Mexico in Albuquerque (Hargrove 1979b). Gene was not only the journal's founder and editor, but he was its editor-in-chief from 1979 to 2020 (for over four decades!). The journal initially focused on topics related to the history of ideas in environmental thought, the aesthetic appreciation of nature and the landscape, and the relevance of art to the appreciation and care of wildlife.

Worldwide, *Environmental Ethics* (EE) was the first journal entirely dedicated to environmental philosophy, and it contributed to the consolidation of this field as a new subdiscipline of philosophy. For the long-term sustainability of the journal, Hargrove created a Center for Environmental Philosophy (CEP), administered by a nonprofit organization, Environmental Philosophy, Inc. The CEP assumed responsibility for the quarterly publication of the journal EE as Gene, and his nonprofit organization, moved to the Institute of Ecology at the University of Georgia in 1981. Finally, in 1989, Hargrove, and with him the CEP and the journal, relocated to UNT, where he served as chair of the philosophy department. As noted earlier, at UNT, he created the world's first graduate program in environmental philosophy and did so in close collaboration with ecology and other environmental programs. As part of his work at the CEP, in the 1990s, Hargrove also created the Environmental Ethics Books series to reprint foundational works in environmental ethics that were out of print.

Along with his concern for higher education, Gene has shown a special interest in teaching environmental ethics to children. He has made innovative proposals for introducing it across the board in public primary and secondary schools and in graduate programs that train environmental educators and other professionals (Hargrove 2008; Poole et al. 2013).

Throughout his career, Gene Hargrove has also played a fundamental role in graduate education. He not only founded the first master's program in environmental ethics in 1994, but also the first doctorate in this field in 2005. I was fortunate as a student to meet Gene in the first of these programs, and later, as a colleague, to collaborate on the development of this doctorate at UNT. The CEP has strengthened these programs because, in addition to publishing the journal EE, it has organized numerous conferences and provided a space for visiting scholars from around the world to conduct research on environmental ethics in partnership with the UNT Philosophy Department. Thus, visitors can attend classes free of charge and interact with colleagues and graduate students. It is noteworthy that to be a CEP visitor, it is not necessary to be a philosopher; they can be scientists or scholars in the humanities and social sciences relevant to environmental ethics.

In short, between 1978 and 2024, Gene conducted remarkably multifaceted activities. Along with being editor of EE, he was president of the CEP, a scholar at the University of Georgia, and later director of the Department of Philosophy and Religion at UNT, while also being an amateur speleologist and manager of multiple nature conservation initiatives. He has even made proposals to protect the moon and all the planets in the solar system! Since childhood, Hargrove has had a strong attraction to science fiction literature and space exploration. During the 1980s, he became interested in protecting natural satellites such as the Moon (from the threats of its appropriation as private property and, therefore, its unregulated exploitation) and planets such as Mars (from the threats of its terraforming and, therefore, its degradation for human use and abuse) (Hargrove 1986). In the twenty-first century, as we note with growing concern these possibilities of extraterrestrial environmental degradation, Hargrove's perspective on environmental ethics takes on increasing relevance and validity.

In conclusion, based on multiple complementary disciplines and practices, Hargrove has played a crucial role in the development of environmental ethics as a new branch of philosophy imbued with inter- and trans-disciplinary approaches to solving complex socio-environmental problems, which acquired global dimensions during the second half of the twentieth century and continue into the twenty-first century.

3.3 Ethics and Aesthetics in Theory and Practice

Hargrove has the virtue of combining theory and practice. Philosophy and philosophers are often characterized as abstract and opaque. The attraction to purely theoretical knowledge, generated by the mere curiosity to know, carries two risks. The first risk is that theoretical knowledge driven by pure curiosity has no limits. Consequently, it can consume one's entire academic life in a constant "snooping," without addressing the urgent needs of society, much less of the broader community of living beings with whom we share the biosphere. The second risk is that even if it assumes social action, it does not pay attention to the historical, cultural, political, economic, and biophysical context. As criticized by Argentine thinkers Juan Carlos

Scannone (1973) and Rodolfo Kusch (2010), modern science and philosophy have maintained an epistemic hegemony linked to a colonialist agenda that dismisses the wisdom of the people and the rooting of thought in everyday praxis that takes place in particular social, cultural, and ecological contexts (Rozzi 2012b).

Hargrove has overcome both risks. He has avoided the first because he has made both theoretical and practical contributions. On the one hand, he has formulated a solid theoretical proposal from the history of ideas in philosophy, and on the other, he has made fundamental practical contributions. Through the creation of a journal and a study center, he masterfully convened and orchestrated an epistemic community that fostered the development and discussion of ideas on environmental philosophy. In his work as editor, he has been scrupulously disciplined in avoiding the primacy of one theoretical approach at the expense of another. The journal EE has published dozens of articles with theoretical and methodological approaches on topics as diverse as animal liberation, biocentrism, biocultural ethics, deep ecology, ecocentrism, ecofeminism, environmental pragmatism, liberation philosophy, political ecology, process philosophy, sustainability under strong and weak anthropocentrism paradigms, non-anthropocentric holism, and the intersections of environmental ethics with various religions, including Christianity, Buddhism, Judaism, Islam, and Taoism.

Regarding the second risk (philosophical approaches that ignore social, cultural, historical, and biogeographical contexts), Gene has consistently asserted that environmental ethics must be adapted to each culture and region (Hargrove 2008). In his publications and teaching, he has integrated this contextualist approach with particular attention to the fields of education and aesthetics. Regarding education, for example, he recommends teaching and examining the values included in the environmental laws of each region, since values are understood not as purely subjective inclinations, but as social expressions recorded in the governance, culture, and history of each society.

Hargrove's work has great didactic value because it offers guidance and encouragement to investigate our own histories and cultures, identifying within them values and ideas that guide our ways of living and relating to our natural environment. This flexibilization of the structures of our social models and forms of education can be key to appropriately resolving complex problems of the current global socio-environmental crisis.

In his integration of environmental education and environmental ethics, as noted earlier, Hargrove has emphasized that this integration should not occur through the imposition of new values, but rather through the reinforcement of existing ones. Among these values, some come from naturalist traditions that, since the dawn of humanity, have been captured in petroglyphs, drawings, and paintings through the representation of plants, animals, fungi, mountains, seas, and other landscapes that took on heterogeneous forms among indigenous peoples of Africa, America, Asia, Oceania, and Europe (Hargrove 1989, 1992b). In the final section of this chapter, I will examine how Hargrove pays attention to the collaborative work of artists and scientists in modern natural history who forged conservationist attitudes toward nature for aesthetic reasons.

3.4 Aesthetics and Naturalists: A Biocultural Analysis of Hargrove's Approach

Eugene Hargrove has investigated how landscape painting has inspired naturalists and the public to conserve landscapes and beautiful entities of nature. This inspiration sprang from a virtuoso collaboration between scientists and artists. Scientists studied the properties of nature and generated discoveries that were of particular interest to painters in their appreciation of landscapes and human and other-than-human co-inhabitants with their varied biocultural interactions. Reciprocally, artists focused on the qualitative properties of nature and inspired the habits of aesthetic appreciation in scientists. With this feedback loop, both naturalist accounts and artistic works combine qualitative and quantitative properties of nature. In Hargrove's words,

> In landscape painting in the nineteenth century, there is … [a] close working relationship between science and art … The need for visual documentation in the natural history sciences encouraged and speeded the transition from a purely scientific point of view to a more aesthetic one in the sciences, and vice versa in art. Lacking cameras, scientists had either to develop artistic abilities themselves or to bring artists with them into the field. The resulting interplay between naturalists, geologists, and artists led to a common perception of the natural world, shared by both groups, that placed emphasis on both fact and value. (Hargrove 1989, p. 84)

3.4.1 The Hudson River School

From a biocultural perspective, we can expand Hargrove's analysis. As he has pointed out, at the beginning of the nineteenth century, European and American painters combined scientific and aesthetic values in their appreciation of diverse, beautiful, and ecologically essential beings. We can affirm that these painters, among them who founded the Hudson River School in the United States, also contributed to making visible the delicate and complex webs of biocultural interactions that occur in landscapes.

The founder of the Hudson River School, the painter Thomas Cole, held a particular fascination for Hargrove. Cole was born in England in 1801 and immigrated to the United States as a youth. At the age of 22, he embarked on a steamboat along the Hudson River, landing in the town of Catskill. He soon headed west, entering the eastern Catskill Mountains of New York, where he began to paint his first landscapes. Cole died prematurely in 1848, but on his painting expeditions he was accompanied by other painters, such as Asher Brown Durand, and poets, such as William Cullen Bryant. Durand was born in 1796 in New England, where he painted one of his most famous paintings, Kindred Spirits, in 1849. In it, he portrays Thomas Cole and William Bryant, thus illustrating the close relationship that existed between writers and painters in the genesis of the Hudson River School. This posthumous tribute to Cole has left us a visual memory of the great importance that the arts had in the appreciation of nature in the United States in the nineteenth century.

In the twentieth century, Cole's painting deeply moved Hargrove (Box 3.2). The title Kindred Spirits refers to the last line of the poem "O Solitude," written in 1816 by the English Romantic poet John Keats:

Is my soul's pleasure; and it sure must be
Almost the highest bliss of human-kind,
When to thy haunts two kindred spirits flee.

Keats begins his poem by proclaiming that:

O Solitude! if I must with thee dwell,
Let it not be among the jumbled heap
Of murky buildings: climb with me the steep,—
Nature's observatory—whence the dell,
In flowery slopes, its river's crystal swell

Just as the twentieth-century philosopher and speleologist Hargrove was fascinated by the beauty of nature and committed to defending it, in the nineteenth century poets like Keats and Bryant were appalled by the indolent expansion of industrial society and cities. Alongside poets and thinkers, painters like Cole and Durand delved into nature, where they found a more spiritual experience and discovered a profound sense of the sublime in life in its astonishing diversity.

Box 3.2 Hargrove's Kindred Spirits

Eugene Hargrove was fascinated by the Hudson River School, and was particularly moved by Asher Brown Durand's 1849 painting, Kindred Spirit, in homage to his mentor and friend Thomas Cole (Fig. 3.1). Cole was one of the first and most important mentors in the formation of the Hudson River School. By portraying him alongside his friend, the poet William Cullen Bryant, Durand expresses the close friendship and collaboration that existed between painters and writers in capturing the sublime in nature.

Given the profound influence the Hudson River School had on Hargrove, with him and Alexandria Poole, who was a research assistant in the Sub-Antarctic Biocultural Conservation Program at UNT and later Associate Director for CEP, I co-organized a trip to the Catskills Mountains in May 2011 (Fig. 3.2). We made the trip with a group of philosophers and ecologists participating in the 14th Cary Conference held at the Cary Institute for Ecosystem Studies in Millbrook, New York. This conference motivated me to create the Springer Ecology and Ethics book series, of which our current volume, *Linking Arts with Biocultural Conservation, Restoration, and Communication*, is a part.

On this trip, we visited Olana, the historic village that was the home of Thomas Cole's greatest disciple, Frederic Edwin Church. This landscape painter was well known for his travels to South America, where he created oil paintings of the Andes and other environments. His paintings spread the beauty and exuberant diversity of the Neotropics in the United States. Church expanded the Hudson River School's conception beyond his country, and his art was a primary inspiration for Hargrove's argument for appreciating biodiversity with scientific meticulousness, emotions, and aesthetic and ethical values.

(continued)

Box 3.2 (continued)

Fig. 3.1 A split image showing a painting and a photograph. (Left) *Kindred Spirits*, a painting by Asher Brown Durand in the Catskill Mountains, in the Hudson River Valley, in the state of New York (United States). Durand portrays the painter Thomas Cole (right) and his friend the poet William Cullen Bryant (left) standing on an immense rock covered in moss at its edges, overlooking a ravine where the vegetation and ecosystems stand out for their realism. The play of light in the painting expresses the grandiose and sublime in nature, a value that fascinated the painters of the Hudson River School. (Image licensed under the Creative Commons Attribution-Share Alike 4.0 International license: https://commons.wikimedia.org/wiki/File:Asher_Durand_Kindred_Spirits.jpg). (Right) Eugene Hargrove (right) and Alexandria Poole (left) pose on a large rock covered in moss at a location near where the painter Thomas Cole and his friend the poet William Cullen Bryant were depicted by Asher Brown Durand in his painting Kindred Spirits. Poole is an environmental philosopher who studied with Ricardo Rozzi and has collaborated with Hargrove since 2007. This photograph expresses both the deep friendship and appreciation that Hargrove has generated in other thinkers and the impact of his thought on the twenty-first century. (Photograph taken by ecologist Eric Keeling on May 17, 2011)

Fig. 3.2 (Left) Photograph of Olana, the historic home designed by Frederic Edwin Church with a view of the Hudson River Valley and the Catskill Mountains. (Photograph by Peter Aaron Otto; licensed under the Creative Commons Attribution-Share Alike 4.0 International license. https://commons.wikimedia.org/wiki/File:OlanaExterior.jpg). (Right) Photograph of Ricardo Rozzi (left) and Eugene Hargrove (right) walking through the Olana Village grounds. (Photographs taken by Alexandria Poole on May 17, 2011)

Under Cole's leadership, the Hudson River School was formed in the 1820s by a group of artists who forged an aesthetic vision influenced by Romanticism. Hudson River School landscapes are characterized by their realistic, detailed, but often idealized depiction of nature. They criticized the historical process of conquest of American territories, which was driven by a development model insensitive to the beauty of natural landscapes. In their paintings, the Hudson River School painters illustrate how peaceful agricultural landscapes, and their villages, were rapidly disappearing, along with remnants of wild nature. They lamented that this was happening just as these landscapes were beginning to be appreciated for their qualities of sublime beauty.

Looking back at the history of European art that preceded and inspired the Hudson River School, Hargrove affirms that the tradition of aesthetic appreciation of nature

> began in Europe in the late 1600s, first with an appreciation of the sublime in the Alps and then more generally an appreciation of nature as picturesque, "pretty enough to be a picture" (or more specifically a picture by Claude Lorrain). This picturesque travel tradition is now called tourism. Because of this tradition, natural beauty is one of the most important reasons given at public hearing for the protection of natural places. (Hargrove 2008, p. 268)

He concludes that his central thesis is that

> For the environment to be properly protected, value perspectives other than economic value perspectives must play a leading role. By promoting understanding among all the voices speaking for the environment politically from so-called non-economic perspectives, the chance that the management of a society's environment will eventually be in terms of a balanced value system reflecting the views of society as a whole, and not just the business and industrial communities, will be greatly improved. The alternative is simply an environment managed for private economic gain. (Hargrove 2008, p. 271)

3.4.2 A Decolonial Critique of the Naturalist Tradition

Regarding the naturalist tradition narrated by Hargrove, a biocultural analysis cannot avoid a historical and political critique: the colonialist nature of European expeditions between the fifteenth and twentieth centuries. So-called "natural history" has generated narratives dominated by a Eurocentric, hegemonic, and homogenizing vision of the biological and cultural diversity of the colonies (Pratt 1992). The language of these naturalist texts is seemingly neutral and innocent, "as if" their accounts did not seek to transform this natural environment or establish any apparatus of domination. However, their accounts have served as a means of intellectual appropriation of these colonized worlds, forging a discourse that has catalyzed processes of biocultural homogenization (Rozzi 2018).

Linguist Mary Louise Pratt has investigated "contact zones," geographical and social spaces where two or more cultures meet, clash, and struggle with each other, often in contexts of highly asymmetrical power relations, such as colonialism. Pratt (1992) critically analyzes European colonialism built on and in contact with the "rest of the world" through the accounts of naturalists and commanders of expeditionary forces to Africa and the Americas between the eighteenth and twentieth centuries.

The homogenization of biotas and cultures has caused profound suffering and violent oppression among the original inhabitants (human and other-than-human) and the destruction of their unique lifestyles that were deployed in different habitats. These were later replaced by urban and rural lifestyles and re-created habitats according to European origin concepts (Paredes-Castellanos and Rozzi 2018). These Europeanized naturalist narratives, presented as avant-garde, were used as aesthetic tools by Spanish but American born "Creole" intellectual elites to "re-invent America" during the nineteenth century (Pratt and Pratt 2007). These elites, which included statesmen such as Andrés Bello (Venezuela) and Domingo Faustino Sarmiento (Argentina), drew on writings such as those of Alexander von Humboldt and Charles Darwin to legitimize their ideological projects that founded new social orders in Latin America.

Humboldt espoused an "anti-conquest" stance but also claimed to have "reinvented nature" with his travel accounts of South America. The concept of "the invention of nature" has acquired a new presence with the eponymous book written by German historian and writer Andrea Wulf (2016), who recounts the expeditions and discoveries of Alexander von Humboldt, exalting the German naturalist's vision. From a critical perspective, it cannot be ignored that his travels through the Americas and his writings coincided with European capitalist expansion. Humboldt was undoubtedly an intrepid explorer and scientist who understood ecological interconnections and distinguished climatic patterns that influence science to this day. As is well known, in his life and work, Humboldt chose to dedicate himself to scientific observation and poetic narrative. He resided in Europe between 1807 and 1834, publishing about his travels and ideas, which played a key role in the discursive and ideological reinvention of Latin America. At the beginning of the nineteenth century, the colonies were gaining independence and Spanish hegemony was crumbling. Consequently, both the European commercial classes and the Creole elites grasped the need for a "reinvention" that would legitimize their aspirations for economic and political expansion. For the post-independence Spanish colonies, Humboldt's writings represented an influential voice in this transatlantic dialogue.

Less known are the life choices made by naturalists who prioritized biodiversity conservation and solidarity with Indigenous communities oppressed by colonialism. An exemplary case is Humboldt's traveling companion, the French physician and botanist Aimé Jacques Alexandre Goujaud. He was better known as Bonpland (Bon-Plant = Good Plant), due to the nickname his paternal grandfather gave to Aimé's father, having been born in the vineyards of Saint-Maurice, France. In Latin America, Aimé permanently adopted the pseudonym Aimé Bonpland. He even signed the names of new plant species he discovered in the "New World" with the botanical abbreviation Bonpl.

In 1799, Humboldt and Bonpland obtained permission from the Spanish court to travel freely within the Spanish colonies. Their voyage extended from 1799 to 1804, extensively exploring South and North America. In contrast to Humboldt, Bonpland returned to South America in 1814 and remained in the La Plata River basin until his death. He dedicated most of his life to community work in health, agriculture, ethnopharmacology, and environmental conservation in Argentina, Brazil, and Paraguay (Nye 2019). In fact, in recognition of his work, he obtained Paraguayan nationality. Bonpland is one of many naturalists who chose to co-inhabit with the human and other-than-human communities he initially came to study, but later became part of, adopting lifestyle habits and vital commitments, along with writing stories, that today can inspire genuine biocultural conservation actions.

3.4.3 From Naturalists to Biocultural Conservationists

From my perspective of biocultural conservation, it is essential to appreciate the diversity of stories and life stories left by naturalists like Bonpland, who, abandoning their Eurocentrism, empathized and stood in solidarity with "otherness." Not with an abstract otherness, but with concrete beings with their autonomous intentions. These beings (human and other-than-human) have moved many artists, naturalists, and scientists who have embarked on explorations through remote territories. I want to emphasize that these cases are not isolated; on the contrary, they are abundant and continue to occur on all continents. For example, recently in the Argentine-Paraguayan Chaco, the English anthropologist John Palmer established a relationship of co-inhabitation and defense with the Wichí ethnic group; he considered solidarity with "otherness" to be the noblest moral virtue (Box 3.3). At the beginning of his testimonial book, *Wichí Good Will: An Indigenous Spirituality* (Palmer 2005), he alludes to the American Declaration of the Rights and Duties of Man and quotes its preamble:

> It is the duty of man [*sic*] to serve the spirit with all his powers and resources, because the spirit is the supreme purpose of human existence and its highest category.
>
> It is the duty of man to exercise, maintain, and stimulate culture by all means at his disposal, because culture is the highest social and historical expression of the spirit.
>
> And since morality and good manners constitute the noblest flowering of culture, it is the duty of every man to always respect them. (OEA 1948)

The life stories of naturalists like Bonpland, ethnographers like John Palmer, missionaries like Bartolomé de las Casas (Mignolo 2011; Restrepo 2018), and environmentalists like Berta Cáceres (May 2018) help us understand the diversity of values and habits in violent processes such as the conquest of territories and their communities of life. In this diversity, among those committed to the defense of indigenous peoples and their habitats, we find options and inspiration to undertake participatory and co-creative actions of biocultural conservation amidst the complex scenarios of the Anthropocene.

Box 3.3 John Palmer: Towards Biocultural Conservation

With the aim of completing his doctoral thesis for Oxford University, the English anthropologist and ethnographer John Palmer (1953–2023) arrived in the city of Salta, northern Argentina, in the mid-1970s. This motivated him to contact the indigenous Wichi Lapacho Mocho (*Hoktek t'oi*) community, which lives 18 kilometers from the remote village of Tartagal, located 365 km north of Salta. After several trips to Europe and nearly three decades of living with the Wichi community, Palmer settled in this community in 1999. He had been accepted as part of the community and married the young woman Tojueia, with whom he formed a family of six children, leaving an indelible mark on the Chaco region of Salta.

From the moment he arrived, Palmer witnessed the violent contradictions between the Wichí community's way of life and institutional education (which boys and girls are legally required to attend). This conflict between this Indigenous people's communal relationship with nature and private property ownership was linked to constant abuses by Salta society for the exploitation of natural resources. These abuses even occurred in territories that the Province of Salta had ceded to the Wichí community as a form of historical restitution.

Noticing these contradictions, Palmer put his academic interests aside and chose to become part of the Wichí community. He shared their ways of life, in their ancestral habitats, with co-inhabitants not only human but also arboreal and many other species of the Chaco savannah. He married a young Wichí woman and took on the defense of their territories as the community's legal representative against the constant abuses suffered by them, their habitats, and their co-inhabitants. He expressed his empathy and solidarity with the "otherness." This otherness is not abstract, but it is made up of real and distinct living beings in his Wichí community, including palm trees, quebracho trees, and other trees, native honeybees and other invertebrates, capybaras and other mammals, tinamous and other birds, *tariras* and other freshwater fish, as well as the streams, rivers, and savannas of their territories (Martínez 2013). Palmer's story was captured in the documentary *El Etnógrafo* (The Ethnographer), directed by Argentine artist Ulises Rosell (2012), which conveys cultural contrasts as well as intercultural encounters and fusions.

In an era marked by multiculturalism, Palmer's family seems to emerge as a harmonious reconciliation of everything that sometimes seems like a distant or impossible romanticism. Palmer's life established genuine contact with an Indigenous community. The Lapacho Mocho community continues to resist encroachment on their lands and culture. Palmer passed away at the end of May 2023, but his children, educated in the Wichi languages, Spanish, and English, possess Indigenous wisdom and are familiar with Western culture and society. Perhaps Palmer and his family show us a path toward more sustainable and just biocultural futures in the twenty-first century.

I have discussed at length naturalist, ethnographic, artistic, and political testimonies embedded in encounters with communities of living beings (human and other-than-human) for two fundamental reasons.

First, these intercultural encounters in ancestral habitats transform values, life habits, and commitments to the defense of life in its biological and cultural diversity. Hargrove (1989) presents numerous examples of naturalists and painters such as Frederick Edwin Church, who furthered the Hudson River School of painting by revealing astonishing mountain, jungle, and river formations in North and South America, paying special attention to realistic details and dramatic luminosity in landscapes that should be admired and cared for. In this way, Hargrove helps us recover the memory of environmental and biocultural values with testimonies of life and values that have been abolished in global culture.

In this sense, Hargrove's work subverts the *damnatio memoriae* or "damnation of memory" erected by the governance of global society with new strategies oriented by a notion of "progress" and "uniformity" that erases cultural traditions (Quesada-García 2004; Georgiadou et al. 2020). In this regard, Hargrove had a special interest in the novel *1984*, published by George Orwell in 1949 in its original English version as *Nineteen Eighty-Four*. This novel describes a fictional society where the enemy is subjected to "vaporization," which consists of physically and symbolically eliminating them, erasing all records of them in monuments, books, newspapers, or other documents. The concept of an Orwellian society is currently used to describe a type of society where historical memory—especially the memory of Indigenous people and of non-economic values that since the dawn of Western civilization have been considered essential for justice and the good life—has been condemned to oblivion.

From a biocultural ethics perspective, I propose adopting and adapting Gene's approach to activate the recovery of memory about cultural traditions and socio-environmental values that have existed and/or continue to exist in our own communities. For example, John Palmer and Berta Cáceres offer current accounts of enduring commitments to defending values and the life of communities of co-inhabitants among the Wichí indigenous peoples in South America and the Lenca in Central America, respectively (Roselo 2012; May 2018).

Secondly, Hargrove draws attention to the abundant personal and collective initiatives committed to defending life, examining biocultural diversity in different eras and regions of the world. We thus understand a courageous and creative ethical commitment among artists, naturalists, and leaders that transcends their "technical specialties" and brings them closer to philosophical lives committed to a genuine sense of socio-environmental justice. These life stories have been largely underrepresented by the media, formal education, and the prevailing culture of global society (Rozzi 2019; Tauro et al. 2021; Tauro and Rozzi 2025).

Gene's didactic description of the multifaceted work of artists and naturalists (with their life commitments embedded in ethical and aesthetic values) could motivate readers to seek inspiring testimonies in the histories, biogeographies, and cultures of diverse regions of the world. In this way, Hargrove's work inspires and exhorts us to transform a uniform Anthropocene into a multifaceted one that nurtures multiple aspects of environmental ethics. Indeed, Hargrove's appreciation for the work of naturalists and their aesthetic approach endure in initiatives such as the collaborative work developed by UNT and CEP at Omora Park (Contador et al. 2023; Crego et al. 2023; Ojeda et al. 2023; Thompson 2026; Vezzani 2026) and other regions of the world (Schüttler et al. 2023; Tauro 2023; Zhu 2023), which utilizes the pedagogy of Field Environmental Philosophy (FEP). This educational and research methodology merges the works of artists (such as Chilean visual artist Paola Vezzani), scientists, philosophers, and local people.

This type of collaboration, like the one that inspires our book *Linking Arts with Biocultural Conservation, Restoration, and Communication*, generates a triple complementarity: interdisciplinary, intergeneric, and interhemispheric. The result is a biocultural metamorphosis: from the individual work in environmental ethics initiated by Hargrove toward a collective work for biocultural conservation, in dialogue and collaboration with multiple cultural traditions that co-inhabit the heterogeneous regions of the planet.

Acknowledgments I would like to thank Francisca Massardo, Paloma González, Roy May, Rika Tsuji, and Benn Johnson for their valuable comments on draft versions of this chapter, and to acknowledge the support from the Cape Horn International Center (CHIC)-ANID PIA/BASAL (PFB210018).

References

Álvarez R, Araos F, Núñez D et al (2023) Otros-que-humanos: tensiones ontológicas en la implementación de la ley Lafkenche. Cultura-Hombre-Sociedad CUHSO (Temuco) 33(1). Available via http://www.scielo.cl/scielo.php?script=sci_arttext&pid=S2452-610X2023005000101&lng=es&nrm=iso

Contador TA, Rendoll-Cárcamo J, Gañan M, Ojeda-Villarroel J, Kennedy JH, Convey P, Rozzi R (2023) Underwater with a hand lens: ecological sciences and environmental ethics to value freshwater biodiversity. In: Rozzi R, Tauro A, Wright T, Avriel-Avni N, May RH Jr (eds) Field environmental philosophy: education for biocultural conservation, Ecology and ethics, vol 5. Springer, Dordrecht, pp 53–70. https://doi.org/10.1007/978-3-031-23368-5_4

Crego RD, Ward N, Rozzi R (2023) The eyes of the tree: applying field environmental philosophy to tackle conservation problems at long term socio-ecological research sites. In: Rozzi R, Tauro A, Wright T, Avriel-Avni N, May RH Jr (eds) Field environmental philosophy: education for biocultural conservation, Ecology and ethics, vol 5. Springer, Dordrecht, pp 101–112. https://doi.org/10.1007/978-3-031-23368-5_7

Georgiadou Z, Florou V, Ilia-Georgiadou I (2020) Damnatio memoriae: contemporary condemnation of memory. Design/Arts/Culture 1. https://doi.org/10.12681/dac.25913

Hargrove E (1968) Relation of a chert zone to development of Devils Icebox, Boone County, Missouri. Missouri Speleol 10(1):15–21

Hargrove E (1973a) The development of devils icebox. Missouri Speleol 13(3):89–98

Hargrove E (1973b) Report of the devils icebox-rockbridge park conservation task force. Missouri Speleol 13(3):99–106

Hargrove E (1973c) Alternatives to the letter writing approach in conservation. NSS Bull Q J Nat Speleol Soc 39(1):13–16

Hargrove EC (1979a) The historical foundations of American environmental attitudes. Environ Ethics 1(3):209–240. https://doi.org/10.5840/enviroethics19791314

Hargrove E (1979b) From the editor: how, when, where, and why. Environ Ethics 1(1):1. https://doi.org/10.5840/enviroethics1979111

Hargrove E (1980) Anglo-American land use attitudes. Environ Ethics 2(2):121–148. https://doi.org/10.5840/enviroethics19802228

Hargrove E (1985) The role of rules in ethical decision making. Inquiry 28(1–4):3–42. https://doi.org/10.1080/00201748508602046

Hargrove E (ed) (1986) Beyond spaceship earth: environmental ethics and the solar system. Sierra Club Books, San Francisco

Hargrove E (1989) Foundations of environmental ethics. Prentice Hall, Englewood Cliffs

Hargrove E (1992a) Weak anthropocentric intrinsic value. Monist 75(2):183–207

Hargrove E (ed) (1992b) The animal rights/environmental ethics debate: the environmental perspective. Suny Press, New York

Hargrove E (2008) A traditional and multicultural approach to environmental ethics and primary and secondary school levels. Environ Ethics 30(3):263–271. https://doi.org/10.5840/enviroethics200830332

Hargrove E (2015) Stewardship versus citizenship. In: Rozzi R, Chapin FS III, Callicott JB, Pickett STA, Power ME, Armesto JJ, May RH Jr (eds) Earth stewardship: linking ecology and ethics in theory and practice, Ecology and ethics, vol 2. Springer, Cham, pp 315–323. https://doi.org/10.1007/978-3-319-12133-8_20

Hargrove E, Nicholson D, Vineyard JD et al (1973) Report of the devils icebox. Rock Bridge Park Conservation Task Force of the National Speleological Society, Alabama

Hargrove E, Arroyo MTK, Raven PH, Mooney H (2008) Omora Ethnobotanical Park and the UNESCO Cape Horn Biosphere Reserve. Ecol Soc 13(2):49. Available via http://www.ecologyandsociety.org/vol13/iss2/art49/

Johnson B (2020) Abiotic ecosystems? A critical examination of Arthur Tansley's ecosystem definition. Environ Ethics 42(1):39–53. https://doi.org/10.5840/enviroethics20204215

Kusch R (2010) Indigenous and popular thinking in América. Duke University Press, Durham

Leopold A (1949) A Sand County almanac, with essays on conservation from Round River. Ballantine Books, 1970, New York. Originally published as A Sand County almanac and sketches here and there. Oxford University Press, New York

Martínez G (2013) Use of fauna in the traditional medicine of native Toba (qom) from the Argentine Gran Chaco region: an ethnozoological and conservationist approach. Ethnobiol Conserv 2(2): 1–43. https://doi.org/10.15451/ec2013-8-2.2-1-43

May RH Jr (2018) Land grabbing and violence against environmentalists. In: Rozzi R, May RH Jr, Chapin FS III et al (eds) From biocultural homogenization to biocultural conservation, Ecology and ethics, vol 3. Springer, Dordrecht, pp 109–124. https://doi.org/10.1007/978-3-319-99513-7_7

Mignolo W (2011) Epistemic disobedience and the decolonial option: a manifesto. Transmodernity 1(2). https://doi.org/10.5070/T412011807

Nye A (2019) Aimé Bonpland: a land ethic in the La Plata. Environ Ethics 41(4):361–379. https://doi.org/10.5840/enviroethics201941442

OEA (1948) Declaración Americana de los Derechos y Deberes del Hombre. Organización de los Estados Americanos (OEA), Bogotá. Disponible en documentos básicos de la Comisión Interamericana de Derechos Humanos (CIDH) https://www.oas.org/es/cidh/mandato/basicos/declaracion.asp

Ojeda J, Tauro A, Contador T, Mansilla A, Rosenfeld S, Ban N, Rozzi R (2023) "Pay attention, dive with your eyes wide open:" a field environmental philosophy activity to foster reciprocity between people and nature. In: Rozzi R, Tauro A, Wright T, Avriel-Avni N, May RH Jr (eds) Field environmental philosophy: education for biocultural conservation, Ecology and ethics, vol 5. Springer, Dordrecht, pp 87–100. https://doi.org/10.1007/978-3-031-23368-5_6

Osorio-García SN, Roberto-Alban NF (2023) Ética, ecología y ecosofía: perspectivas divergentes para refundamentar la bioética global. Rev Latinoam Bioet 23(1):121–136. Available via https://www.redalyc.org/journal/1270/127075971008/html/
Palmer J (2005) La Buena Voluntad Wichí: Una Espiritualidad Indígena. APCD/CECAZO/EPRAZOL, Grupo de Trabajo Ruta, Formosa, p 81
Paredes-Castellanos A, Rozzi R (2018) Biocultural exoticism in the feminine landscape of Latin America. In: Rozzi R, May RH Jr, Chapin FS III et al (eds) From biocultural homogenization to biocultural conservation, Ecology and ethics, vol 3. Springer, Dordrecht, pp 167–183. https://doi.org/10.1007/978-3-319-99513-7_10
Pascual U, Balvanera P, Anderson CB et al (2023) Diverse values of nature for sustainability. Nature 620:813–823. https://doi.org/10.1038/s41586-023-06406-9
Poole AK, Hargrove E, Day P et al (2013) A call for ethics literacy in environmental education. In: Rozzi R, Pickett STA, Palmer C et al (eds) Linking ecology and ethics for a changing world: values, philosophy, and action, ecology and ethics, Ecology and ethics, vol 1. Springer, Dordrecht, pp 349–372. https://doi.org/10.1007/978-94-007-7470-4_28
Pratt ML (1992) Imperial eyes: travel writing and transculturation. Routledge, London
Pratt ML, Pratt IE (2007) Alexander von Humboldt and the reinvention of América, 2nd edn. Routledge, London
Quesada-García S (2004) La *damnatio memoriae* o la negación de la memoria. Aldaba 16:104–115
Restrepo LF (2018) The colonial face-to-face and the human condition: writing and subjectivity in Bartolomé de las Casas. In: Orique DT, Roldán-Figueroa R (eds) Bartolomé de las casas, O.P. history, philosophy, and theology in the age of European expansion, Studies in the history of Christian traditions, vol 189. Brill, Leiden, pp 260–279. https://doi.org/10.1163/9789004387669_012
Rosel U (2012) El etnógrafo. Revista Cine Documental 7, http://revista.cinedocumental.com.ar/numeros/index.php/criticas/9-numero7/26-el-etnografo-ulises-rosell-2012
Rozzi R, Hargrove E, Armesto JJ, Pickett STA, Silander J (1998) "Natural drift" as a post-modern metaphor. Revista Chilena de Historia Natural 71:9-21
Rozzi R (1999) The reciprocal links between evolutionary-ecological sciences and environmental ethics. Bioscience 49(11):911–921. https://doi.org/10.2307/1313650
Rozzi R (2012a) Biocultural ethics: the vital links between the inhabitants, their habits and regional habitats. Environ Ethics 34:27–50. https://doi.org/10.5840/enviroethics20123414
Rozzi R (2012b) South American environmental philosophy: ancestral Amerindian roots and emergent academic branches. Environ Ethics 34:343–366. https://doi.org/10.5840/enviroethics201234436
Rozzi R (2018) Biocultural homogenization: a wicked problem in the Anthropocene. In: Rozzi R, May RH Jr, Chapin FS III et al (eds) From biocultural homogenization to biocultural conservation, Ecology and ethics, vol 3. Springer, Dordrecht, pp 21–47. https://doi.org/10.1007/978-3-319-99513-7_2
Rozzi R (2019) Taxonomic Chauvinism, no more! Antidotes from Hume, Darwin, and stewbiocultural ethics. Environ Ethics 41(3):249–282. https://doi.org/10.5840/enviroethics201941325
Rozzi R, Massardo F, Anderson C et al (2006) Ten principles for biocultural conservation at the Southern Tip of the Americas: the approach of the Omora Ethnobotanical Park. Ecol Soc 11(1): 43. Available via http://www.ecologyandsociety.org/vol11/iss1/art43/
Rozzi R, Arango X, Massardo F et al (2008) Field environmental philosophy and biocultural conservation: the Omora Ethnobotanical Park educational program. Environ Ethics 30(3): 325–336. https://doi.org/10.5840/enviroethics200830336
Rozzi R, Armesto JJ, Gutiérrez JR et al (2012) Integrating ecology and environmental ethics: Earth stewardship in the southern end of the Americas. Bioscience 62(3):226–236. https://doi.org/10.1525/bio.2012.62.3.4

Scannone JC (1973) Trascendencia, praxis liberadora y lenguaje. Hacia una filosofía de la religión posmoderna y latinoamericanamente situada. In: Ardiles O (ed) Hacia una filosofía de la liberación latinoamericana. Bonum, Buenos Aires, pp 245–269

Schüttler E, Mackenzie R, Muñoz-Petersen L (2023) Biocultural conservation in biosphere reserves in temperate regions of Chile, Estonia, Germany, and Sweden. In: Rozzi R, Tauro A, Wright T, Avriel-Avni N, May RH Jr (eds) Field environmental philosophy: education for biocultural conservation, Ecology and ethics, vol 5. Springer, Dordrecht, pp 483–502. https://doi.org/10.1007/978-3-031-23368-5_29

Tauro A (2023) Biocultural reconnection. Recovering the sense of community through education. In: Rozzi R, Tauro A, Wright T, Avriel-Avni N, May RH Jr (eds) Field environmental philosophy: education for biocultural conservation, Ecology and ethics, vol 5. Springer, Dordrecht, pp 329–338. https://doi.org/10.1007/978-3-031-23368-5_20

Tauro A, Rozzi R (2025) Biocultural ethics and Earth stewardship: a novel integration to revitalize multiple values of nature. Ecol Soc. 30. https://doi.org/10.5751/ES-16362-300335

Tauro A, Ojeda J, Caviness T, Moses K, Moreno R, Wright T, Zhu D, Poole A, Massardo F, Rozzi R (2021) Field environmental philosophy: a biocultural ethic approach to education and ecotourism for sustainability. Sustainability 13:4526. https://doi.org/10.3390/su13084526

Thompson M (2026) Sublimity in miniature: aesthetic viewing and the miniature forests of Cape Horn. In: Zhu D, Rozzi R, Tsuji R, Johnson B, Caviness T, Castro-Jorquera C, Xu L, Zhang L (eds) Linking arts with biocultural conservation, restoration, and communication, Ecology and ethics, vol 8. Springer, Cham, pp 265–276. https://doi.org/10.1007/978-3-032-04830-1_16

Vezzani P (2026) "Fragility of encounters" in subantarctic biocultural diversity. In: Zhu D, Rozzi R, Tsuji R, Johnson B, Caviness T, Castro-Jorquera C, Xu L, Zhang L (eds) Linking arts with biocultural conservation, restoration, and communication, Ecology and ethics, vol 8. Springer, Cham, pp 207–222. https://doi.org/10.1007/978-3-032-04830-1_13

Wulf A (2016) The invention of nature: Alexander von Humboldt's new world. Vintage Books, New York

Zhu D (2023) Bridge the channel, enhance the inclusivity: a comparison between flagship species-centered and Moss-centered conservation in Chile and China. In: Rozzi R, Tauro A, Wright T, Avriel-Avni N, May RH Jr (eds) Field environmental philosophy: education for biocultural conservation, Ecology and ethics, vol 5. Springer, Dordrecht, pp 457–482. https://doi.org/10.1007/978-3-031-23368-5_28

Chapter 4
Biocultural Ontologies, Art, and Conservation

Rika Tsuji and Benn Johnson

Abstract Biocultural conservation focuses on the restoration and conservation of relationships between human inhabitants, other than human co-inhabitants, and the land. Such conservation is rooted in relational ontologies situated from locality. This chapter draws from biocultural literature to offer an account of what such ontological positions might mean for biocultural conservation of art. Both conservation and art exhibition rooted in colonial ontologies tend to reduce Indigenous and local communities' living relationships with the land to mere historical or art objects. We argue that biocultural conservation presents ontologies of co-constitutive relationality that challenge fixed-object conceptions of art, bringing us to a more nuanced view of art as biocultural expression from locality.

Keywords Art · Biocultural conservation · Biocultural ontology · Locality · Relationality

4.1 Introduction

Biocultural diversity enhances the sustainability of life, both human and other-than-human (Rozzi 2012a; Bridgewater and Rotherham 2019; Martin et al. 2022; Nemogá et al. 2022; Cavaliere et al. 2024). However, biocultural diversity, including much Indigenous, local biocultural knowledge, practice, and relationship with the land, is at great risk of loss or has already been lost through the ongoing sociopolitical and cultural hegemony of neoliberal globalization, an extension of colonialism (Rozzi 2013). Chilean ecologist and philosopher Ricardo Rozzi (2012b, p. 29) links this suppression of diverse forms of knowledge to a broader process that he calls *biocultural homogenization*, which "entails simultaneous and interdigitated losses of native biological and cultural diversity at local, regional, and global scales," and

R. Tsuji (✉) · B. Johnson
Department of Philosophy, College of Arts and Sciences, University of Rhode Island, Kingston, RI, USA
e-mail: rika.tsuji@uri.edu; benn.johnson@uri.edu

D. Zhu et al. (eds.), *Linking Arts with Biocultural Conservation, Restoration, and Communication*, Ecology and Ethics 8,
https://doi.org/10.1007/978-3-032-04830-1_4

"leads to the disruption of the interrelationships between cultures and their land." As Rozzi (2010) explains, this process of biocultural homogenization involves deep injustices to marginalized communities, and thus it is important for biocultural conservation to respect the diverse worldviews and ways of living of Indigenous and local people. Through such an understanding, biocultural conservation aims to overcome the currently prevailing worldview of human-nature dualism by recalling our vital links to our environment for the sustenance of diverse livelihoods and practices.

Art has gained increasing attention as a communication tool for biocultural conservation (Polfus et al. 2017; Brennan 2018; López et al. 2020; Martin et al. 2022; Rozzi and Tauro 2023). Visual and musical expressions have proven to be effective means of sensitizing people to cultures, nature and their conservation (Swanson 2023; Markiewicz et al. 2023). We invite readers and colleagues to understand and value art in broader and more complex ways than as a mere *tool* to communicate science and conservation. A biocultural approach should also consider art as the expression and embodiment of diverse biocultural relationships. Thus, such artistic expressions, which can include both tangible and intangible dimensions of biocultural relationships, are worthy in themselves of conservation or restoration (e.g. Khular 2026).

Such a conception of biocultural art conservation contrasts with the widespread colonial and homogenizing structures of art institutions and relationships. For example, many art institutions like museums and galleries carry troubled histories of colonial practices, as gatekeepers of a civilized/uncivilized dichotomy (Nichols 2023). Exhibition of present Indigenous art as "historical artifacts" is a prevailing way of erasing the voices and livelihoods and struggles of existing communities (Martin 2017; Nixon 2020; Hays-Gilpin et al. 2020; Nichols 2023). Furthermore, frequently the acquisition of Indigenous cultural products, including human remains, involves theft or otherwise unjust displacement (La Follette 2017; Nixon 2020; Barg and Palus 2021). These forms of exhibition turn "Indigenous art" into mere objects to be seen. Instead, it is urgent to conserve such art as the expression of life practices that are vital to Indigenous and other local communities' relationships with the land (Rozzi 2023). The conservation of Indigenous art should not mean protection of important objects in sterile glass cases, separated from historical context and means of acquisition.

In face of these unjust and biased ways of portraying Indigenous art at museums and other settings, we must rethink the question: what *is* conserved through biocultural conservation? To philosophically engage with the "is" within this question, we will focus on ontology, i.e., we will examine what exists and how beings are interrelated. Although we discuss the question of what art is from a biocultural framework, we have intentionally left vague any aesthetic conception or definition of art, as that is beyond the purview of our paper (and as mentioned in the previous paragraph, is often bound with colonial gatekeeping). In other words, we are not interested in a universalizing inquiry into what counts as art versus not-art, but are rather concerned with the things and relationships that are said to be conserved within the scope of art and biocultural conservation. Ultimately, we will argue that biocultural conservation presents ontologies of co-constitutive relationality that challenge fixed-object conceptions of art, bringing us to a more full view of art as biocultural expression from locality.

4.2 Ontologies and Critiques in Biocultural Literature

Present in much of the biocultural literature involving ontology is a critique of the predominant ontological framework, influenced by Western dualistic and colonial thinking, which perceives a stark hierarchical split between humans and nature (Rozzi 2013, 2018; Caillon et al. 2017; Klaver 2018; González-Morales 2022; Crego et al. 2023). In this ontological understanding, humans are the only active agents, and nature is composed of passive objects. Rozzi, for instance, problematizes this ontological separation between subject and object because it fails to recognize "the intentionality and interests" of other-than-humans (2018, p. 38). He points out that this subject/object split is also found in conceptions like ecosystem services where natural objects serve the interests of human subjects. As Rozzi (2018) points out, Irene Klaver's analysis is helpful here to show that this objectification of the other-than-human environment is tied to "a value attribution with an implied sense of superiority (culture, agency) versus inferiority (nature, passivity) and hence an implied legitimation for use, domination, and exploitation" (Klaver 2013, p. 93). This predominant ontological framework justifies and perpetuates an anthropocentric relationship with other-than-humans which is manifested in extractive and dominative practices including economic, artistic, and other cultural expressions.

The dualistic ontological framework is tied to colonial practices that privilege specific sociopolitical and cultural values that are used to justify the erasure of Indigenous and traditional communities and their relationships with other-than-humans (Depuy et al. 2021). For example, settler colonial ontologies of land replace Indigenous ontologies of land by viewing Indigenous territory as *terra nullius* (or empty land), thereby "justifying" land grabbing and displacement of people for commodification (Burow et al. 2018). Although not explicitly discussed using the term ontology, Fouad Makki (2018) and Roy May Jr. (2018) also describe how this reduction of land to objectified, relationless commodity in Ethiopia and Latin America, respectively, are strategically employed to grab land and justify violence against local people. Sophie Caillon and collaborators note that the dualist notion of humans and nature is "ahistorical, a denial of the often generations-long relationships between people and their environments" (Caillon et al. 2017, para 4). The previous examples illustrate how the ontological framework employed in such violence of colonialism is both historically decontextualized and unjust toward Indigenous and traditional communities and their relationships with other-than-humans.

As a response to the critique of dualism, scholars have explored and drawn from Indigenous worldviews to outline ontological positions that reflect biocultural relationships and that foster dialogue among plural worldviews. For example, René Moreno-Terrazas-Troyo and collaborators (2023) draw from the Maya-Quiché and

Aztec-Mexica cultural views of the sea and the sky to foster a biocultural ecotourism rooted in a relational ontology, as opposed to colonial and commodifying dualisms. They emphasize the recovery of these worldviews and the exploration of relational ontologies that "will reconnect/entangle/root us with the whole" (Moreno-Terrazas-Troyo et al. 2023, p. 126). Laura Zanotti (2018) has demonstrated how biocultural conservation's relational approach can provide discursive space to recognize the Kayapó people's relationship with land and water sovereignty amidst socio-political struggle.

Moreno-Terrazas-Troyo (2023) as well as Zanotti (2018) use the ontological framework of Rozzi (2012b, 2013) who proposes a biocultural ethic that values the co-constitutive links among (1) habitats, (2) human and other-than-human co-inhabitants, and (3) their habits, which he calls the 3Hs. In the creation of his framework, Rozzi (2012b) draws from parallels among early Western philosophy, contemporary ecological sciences, and the philosophy of the Mapuche-Pewenche people (pp. 38–44). Although in his 2012 work on biocultural conservation, Rozzi only mentions the ethical and epistemological nature of this work, he later reflects more directly on the ontological component that this framework entails (2013, p. 10, 2018, pp. 36–39). He claims that the ontological framework embedded within the 3Hs consists of three major aspects:

> (i) Co-inhabitants are subjects not objects.
> (ii) Co-inhabitants co-constitute their identities and their well-being by dwelling with other human and other-than-human beings.
> (iii) Co-inhabitants share habitats that they co-structure through co-inhabitation relationships. They establish ecological relationships of complementarity and reciprocity that occur through exchanges of matter and energy. Therefore, the care and conservation of habitats is the condition of possibility for the existence and well-being of the diverse co-inhabitants. (Rozzi 2018, p. 39)

Rozzi's ontological description recognizes the inseparable links among the 3 Hs, which help to overcome nature-culture dualism by resituating humans within habitats and amongst other-than humans with whom our ontological identities are co-constructed. In the same way, it recognizes habitats and other-than-humans as subjects in the creation of a common world. In other words, by describing the 3Hs as relational and co-constitutive, he articulates an image of beings and ways of being in the habitat which cannot be ontologically isolated from each other. Importantly, Rozzi (2018) argues that understanding these vital ontological links among habitats, humans, and other-than-human co-inhabitants can help us to reorient ourselves ethically in biocultural conservation (pp. 10, 14). Rozzi's co-constitutive ontological framework is developed specifically for the implementation of *in-situ* biocultural conservation programs; therefore, it lays out a broad, barebones ontology that is sensitive and adaptable to local worldviews but also communicable with multiple stakeholders, such as policymakers, scientists, Indigenous communities, artists, and students.

Other scholars within biocultural conservation literature have taken a different, though often complementary approach, specifically arguing for a turn to the conservation and promotion of Indigenous and local ontologies themselves (Caillon et al.

2017; Campion et al. 2023). For example, Caillon and collaborators (2017) problematize conservation approaches that are rooted in the dualistic ontology of humans and nature and turn to the inextricable links between the well-being of Indigenous people and local communities and their environments, which are expressed through their knowledge, practices, and ontologies. Thus, these authors emphasize the acceptance of plural ontologies for just biocultural conservation approaches that recognize the important feedback loop among ontology, epistemology, and practice, and take diverse ontologies themselves as worthy of biocultural conservation.

Sarah-Lan Mathez-Stiefel and her collaborators (2007) have elaborated a nuanced account of the inextricable link between biological and cultural diversity. Based on their work in the Chorojo community in the Bolivian Andes, they argue that at a local, rather than global scale, ontological, rather than linguistic diversity, is most closely aligned with biological diversity. They maintain that culture cannot be reduced to its linguistic components; consequently, they argue that the interconnected ontological, epistemological, normative, and practical aspects of culture must be incorporated into our understanding of biocultural communities.

Although Mathez-Stiefel and her collaborators (2007) discuss "ontological communities" and "culture," they do not presuppose that fixed, unchanging cultures are discrete ontological facts. They draw from the work of Cocks (2006), who argues that traditional definitions of culture as systems of thought and action within a community fail to capture the dynamic processes of transcultural hybridity, and tend to mythologize people from Indigenous communities as monolithically eco-friendly. Rather, Cocks claims that biocultural conservation must be extended to social groups that are not traditionally thought of as Indigenous or local, thereby broadening the domain of biocultural conservation. Mathez-Stiefel and her collaborators build upon this approach, recognizing that the communities in the Bolivian Andes are culturally and dynamically hybrid, sharing linguistic and cultural elements, while often maintaining unique ontologies and practices. Thus, they conclude that (although important) it is not sufficient to merely preserve diverse languages; rather, we should "promote a diversity of worldviews in order to maintain the diversity of frameworks of interactions with the natural world," and that doing so requires a recognition and redress of the unjust power dynamics between what they call ontological communities (Mathez-Stiefel et al. 2007, p. 79).

Although biocultural communities are not discrete, unchanging identities, this unfixed interpretation does not suggest an ontological view which is detached from land. In his book *Indigenizing Philosophy through the Land* (2019), Brian Burkhart resituates philosophy in locality. He explains that "locality is being-from-the-land and knowing-from-the-land. I use 'locality' as a term of art in this book as a way to reference the manner in which being, meaning, and knowing are rooted in the land" (Burkhart 2019, p. xiv). That is, the land is not a mere background of things in knowing or being but is ontologically inseparable from our being-from-the-land. In this sense, he problematizes colonial philosophy's delocalized view of the world, claiming that:

> One of the layers of delocality can be seen in the manner by which 'things' are established as foundational in Western ontologies and epistemologies…Western philosophy focuses almost exclusively on the being of things and establishing the being of things and the relationships between things as a foundational knowledge. (Burkhart 2019, p. 100)

In other words, Western philosophy's reduction of all reality to static things, which agential realist Karen Barad (2003, p. 812) calls "thingification," is a prior condition for abstracting "things" from their relationship to the land. Burkhart clarifies in his explanation of this ontological view which he calls substance metaphysics that "the primary building blocks of reality must be static things that can exist across time and space" (2019, p. 101).

Burkhart's ontological view may also speak to Rozzi's co-constitutive vision of biocultural conservation, because the reduction of the 3Hs to mere things (individual, delocalized components) is part of the process of abstraction that justifies biocultural homogenization and exploitation. In this way, as opposed to the dominant Western philosophy of delocality, Burkhart's contribution clarifies an embedded relationality with land in all interconnected aspects of our being, knowing, and acting, as "the ontological and epistemic relationship between morality and the land that arises in locality" (Burkhart 2019, p. xxx).

Through this biocultural conservation and critical environmental literature, we begin to see a clearer picture of the main ontological themes emerging within the field. First is a strong critique of nature-culture dualism present in dominant ontological worldviews, which tend to value the cultural over the natural, and which tend to erase or justify violence toward Indigenous people. Rozzi's 3Hs approach provides a co-constitutive ontology for biocultural conservation programs that centralizes relationality and de-emphasizes isolated individuals separated from their co-inhabitants and habitats.

As the literature shows, many ontological perspectives already contain deeply co-constitutive and relational realities. Furthermore, the literature also ontologically unsettles our conception of *bioculture*, since "cultures" themselves cannot be thought of as discrete, ontologically bounded objects. Through a critique of abstract identities and the "thingification" of ontology, which is part of the colonial project of delocalization, our understanding of bioculture is deepened. Through each of these, we gain a picture of ontology as relational, co-inhabitants as co-constitutive, and being as arising from locality. Each of these ontological contributions helps us to deepen our discussion on art and biocultural conservation in the next sections.

4.3 Understanding Art from Biocultural Ontologies

Some tensions and questions arise in relation to our initial question: what *is* conserved by biocultural conservation? Notice again here the focus on "is." What does it mean to exist and how do things exist within the fields, institutions, disciplines, traditions, or practices that could be considered art. We are interested in what it means to say something is a "piece" of art or an "artwork," or even what is the temporality of an art object from a biocultural ontological perspective.

There are some linguistic limitations in this kind of discussion because much of art language in English appears to be already object-oriented (thingified), especially if we are concerned with conservation or preservation of art. To speak of "artworks" and "art pieces" as if they were distinct, delocalized things already calls our attention to an ontological presupposition that may exist in our discourse surrounding art. Thus, in this section we show how reframing art through a biocultural lens may help broaden our understanding of what it means to conserve art. To do so, we illustrate with some examples of art that resists such objectification, in order to show how relational ontologies might contribute to a biocultural understanding of art.

4.3.1 Landscape Relationship as Biocultural Art: Setouchi Triennale and Teshima

The first example is from the Setouchi Triennale, an international art festival in Japan. In the Seto Inland Sea, since 2010, the festival has been held triennially mainly on 12 outlying islands. In 1934, the Seto Inland Sea region became Japan's first national park, Setonaikai National Park, and this region began to be widely known for its beautiful inland archipelago seascape. White sand and green pine 白砂青松 ("*hakusya-seisyo*") is a Japanese idiomatic term that expresses a biocultural aesthetic of the coastal landscape. Matsushima and Ferreira (2022, p. 90) describe *hakusya-seisyo* as "the beauty of a coastal landscape in Japan," yet point out that such a coastal landscape has been developed through the long-term interaction of the people with the beach. Masanori Nishida (1999) explains that during postwar development, the islands and inland sea area underwent industrialization, consequently losing the iconic seascape of white sand and green pine. Along with postwar economic development and landscape destruction, many island communities in the Seto Inland Sea also suffered from severe depopulation as many young residents left for cities to find jobs. To counteract these negative trends, the Setouchi Triennale adopted the theme of "restoration of the sea" with the aim to revitalize through art and responsible tourism these island communities that were suffering from severe depopulation (Setouchi Triennale Executive Committee n.d.).

Among the islands of the Setouchi Triennale, Teshima is the second largest, with a current population of around 760 people (Kagawa Prefecture 2023). Its name in Japanese consists of two Chinese characters, 豊島, literally meaning "rich" or "abundant" island. Like the name suggests, Teshima's rich natural environments have supported people's livelihood through agriculture, fishing, forestry, and quarries. However, Teshima was once called a "garbage island" because the island became home to one of the worst illegal toxic dumping sites in Japanese history. In response, the local residents formed a grass-roots movement and through legal action—the Teshima court battle, have fought the environmental injustice they

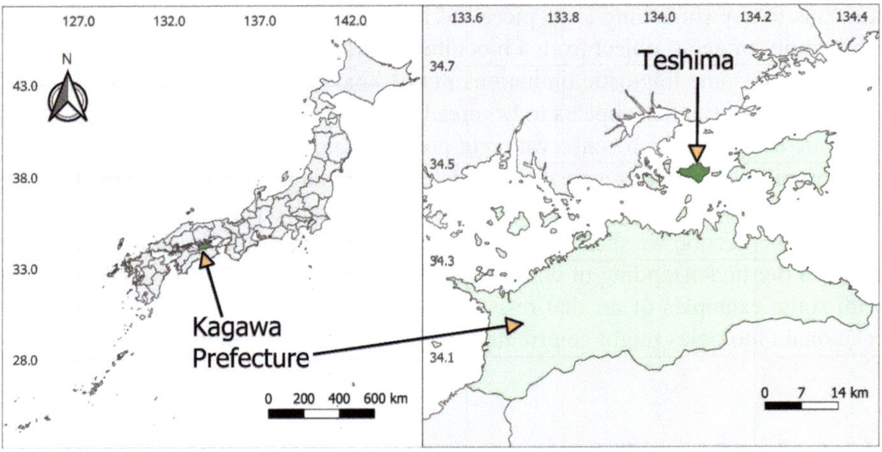

Fig. 4.1 Map of Japan (left) and Kagawa prefecture (right), showing location of Teshima. (Map prepared by Benn Johnson using QGIS and map data from GADM)

experienced since 1975 (Teshima Residents Council on Waste Management 2005; Awang et al. 2017). The site had been a beautiful white sand and green pine coastline but was turned into an environmentally hazardous illegal dumping place. Toxic industrial waste, such as shredder dust, was produced on Teshima and also brought from outside the island (Suzuki and Kin 2020). In the memoir of the vice president of the defense council in the Teshima court battle, Shinro Okawa (2004, p. 221) emphasizes that the residential group's motive was rooted in their care for the island and concern for their future descendants (Fig. 4.1).

Okawa describes the movement as a selfless act. The residents' group states that "our wish for the 'return of our rich island' is the heartfelt wish of those who grieve for their irreplaceable hometown and cherish the lives that are born and nurtured here," while pointing out that the root of this injustice is in the exploitation of a depopulated rural area within a "disposable society" (Okawa 2004, p. 76). In 2000, through ardent and persistent organization and activism, the residents' group obligated the local government to remove the toxic wastes and to restore the ecosystems and landscape of the coastline (Teshima Residents Council on Waste Management 2005; Okawa 2004). In 2024, the restoration project is still in process.

Since the arrival of the Setouchi Triennale on Teshima, however, the media and tourism industries have rebranded Teshima as an "art island" (Marett 2023). Much research on the Setouchi Triennale has focused on economic impacts to islands or on visitor perspectives of art installations. A different type of analysis has been developed by Meng Qu, who has studied the residents' perceptions of the art, not merely as aesthetic objects, but also as part of the broader material culture and landscape of Teshima (Qu 2019, 2020). Although the goal of festival organizers and curators is to create localized artworks that reflect the culture and landscape of the environment,

Qu finds that some Teshima residents have mixed feelings about the art that has been created. For example, one interviewee of Qu's research states about the art that:

> …without understanding Teshima's background, such as our industrial waste problem, you [are limited to] playful feelings, like Teshima is just some amusement park. If visitors are walking around looking at artwork like they are in a theme park [while important issues go unaddressed], it doesn't seem to have much to do with revitalization to me. (in Qu 2019, p. 32)

The analogy of the theme park demonstrates local skepticism about the way the island is used. As previously mentioned, the locals in Teshima have experienced their island being reduced to disposability. In this sense, some residents are concerned with the festival being merely "a fad to draw fickle Japanese consumers" (Qu 2019, p. 34), a new exploitation of Teshima's landscape as an art tourism product. Even though the goal of this festival is revitalization, Qu (2020, p. 258) remarks that "some business owners treat [the art] as a new type of 'garbage' from outside generated by tourism and locals worry that if one day the art becomes redundant, the entire island will be full of 'art garbage.'"

Rather, Qu notes that some locals perceive that "the natural and cultural landscape of the island was the 'real art' visitors should focus on" (Qu 2019, p. 34). For example, instead of promoting the Teshima Museum of Art, which is the major art tourism attraction or other Setouchi Triennale art sites, one of Qu's interviewees states that:

> You can see the rice terraces in the landscape at the side [of Teshima Museum of Art]… Depending on the four seasons, all the landscapes will change. What are those landscapes presenting? That is also art for me. So, separately, I would like to see the natural as art, not the art they created. (in Qu 2020, p. 258)

It seems that the biocultural landscape is itself an artistic expression to the locals. Although there is no written account of local ontological perspectives on Teshima, Rozzi's ontological framework helps to show how the Teshima landscape can be a form of biocultural art. The landscape is a living expression co-developed through the relationship among humans, other-than-humans, and the land. Consequently, the residents care for the island.

Paraphrasing Burkhart (2019, p. 128) we could say that there is a "continuous manifestation of the locality" of the island. In this sense, the living art that *is* Teshima and its constantly evolving biocultural community cannot be separated from context; it is not an art*work* that can be placed in a museum, presented on a stage, or framed. The natural coastal landscape of white sand and green pine, as well as the rice terraces that were formed through residents' multi-generational practices, motivated people's resistance to exploitation and objectification of the island (Fig. 4.2).

The example of Teshima rearticulates our understanding of art; it is not merely discrete objects or works produced by artists but is also a continuous expression of biocultural relationality in locality. Thus, what it means to conserve art bioculturally will have to be adjusted accordingly, in order to account for this ontological reinterpretation. Art conservation in this full biocultural sense must recognize the co-constitutive relationships among the co-inhabitants, their habitats, and habits for the continuity of such expression into the future.

Fig. 4.2 The rice terraces located in Karato, Teshima. Locals call this particular place Karato-oka, which means Karato hill, and the bay in the background is called Karato-hama, or Karato bay. The terraces of Karato-oka are part of an ongoing restoration collaboration between locals, the government, and the Fukutake foundation (NPO Teshima Tourism Association, n.d.). (Photo credit: Masataka Yatsuzuka)

4.3.2 Authenticity and Fixity: Mibu no Hana-Taue

In general terms, a focus on continuity of expression, however, should not be mistaken for fixity or permanence. Such a contrast is also clarified in light of Harsh Vardhan Bhati and Yaffa Epstein's (2023) work on tangible and intangible biocultural heritage. Bhati and Epstein (2023) describe the World Heritage Committee's overemphasis on tangible cultural heritage, noting in particular the importance often placed on permanence and authenticity, claiming that such an emphasis often neglects the intangible and ever-changing biocultural links within biocultural communities and leads to privileging the protection of Western cultural sites. Authenticity and permanence both serve to create an imagined bioculture, reduced to essential characteristics that cannot or should not change. This privileging of the permanent, authentic, and tangible over the intangible resonates with Burkhart's critique that Western substance metaphysics presumes a delocal reality wherein "static things can exist across time and space" (Burkhart 2019, p. 101). It is only within such a paradigm that "authenticity" is even found, whereas, echoing Cocks, we can understand that cultures are in fact hybrid and changing.

A second example from Japan, the agricultural ritual for preservation of rice transplanting called the Mibu no Hana-Taue, addresses the concept of "authenticity." Hiroyuki Hashimoto offers a valuable analysis of this ritual in two communities, Mibu and Kawahigashi, in Hiroshima prefecture, Japan. The Mibu no Hana-taue was listed as UNESCO Intangible Cultural Heritage of Humanity in 2011. This practice, which originally started as forms of singing that accompanied rice transplanting, was lost but later revived as two distinct types of performances and competitions to attract local tourists. In his article "Between Preservation and Tourism: Folk Performing Arts in Contemporary Japan," Hashimoto (2003) describes that the two current styles reflect a sort of objectified tourist culture. The first is thought of as more traditional, in which participants perform with bulls in rice fields. The second, which has continued to evolve, includes organized dance and freedom of expression, as it takes place in athletic fields. However, even in the first version, Hashimoto explains that because rice planting technologies no longer require bulls, animals are often transported from other communities to Mibu in order to perform.

Through the example of Mibu no Hana-taue, it becomes clear that it is difficult to link cultural authenticity to "fixed objects" of conservation. As Hashimoto (2003, p. 134) affirms, "[performers] continue to interpret/reinterpret Mibu no Hana-taue in order to secure their own sense of purpose and meaning, [which] requires us to decenter our perspective on authentic cultural phenomena." Ultimately, however, he concludes that we should turn to the performers' own interpretations to "discover an authentic cultural phenomenon" (Hashimoto 2003, p. 234).

From a framework of biocultural conservation ontology, however, attempting to locate authentic cultural phenomena is unnecessary. This is because a relational biocultural ontology defocuses our perspective on static objects such as individuals and cultural manifestations, and problematizes the very tendency to determine their authenticity. As criticized by Burkhart (2019), the constant attention to whether Indigenous (or in this case, local) voices are authentic is a project of coloniality. In his words, "the search for authenticity in the context of locality is an attempt to obscure locality with delocality" (Burkhart 2019, p. 199). Echoing him, we could say that a search for authenticity erases the agency of marginalized voices by putting them on trial for their "authenticity," which simultaneously reinforces the construction of Western subjectivity and agency as "delocal" and abstracted from land and relationality.

4.3.3 Specimen Logic and Abstraction from Land: Yeii *and Museum Displays*

Problems of neocolonial abstraction may also be found in interpretations of biocultural practices in locality. Burkhart's discussion of Navajo sand-painting, or *Yeii* is helpful for understanding the risk of such abstraction. In *Indigenizing Philosophy Through the Land*, Burkhart (2019, p. 128) criticizes that the delocal lens would portray the *Yeii* of the Navajo people as merely abstract, artistic expressions and representations of things, instead of seeing them as a "continual manifestation of locality." He explains that:

> The *Yeii* (spirit beings) that are formed by the colored sand in the paintings do not represent these *Yeii* as attempted pictures or symbolic representations of them; the material sand shaped into the form of the *Yeii* become the *Yeii* themselves. The forms of the *Yeii* in the sand painting are the actual *Yeii* rather than mere symbolic representations of them. The *Yeii* are present in the forms that are shaped by the painter in the ceremony, and as the *Yeii* are made present in the painting, they bring their power or medicine to heal the patient who then sits directly on the sand painting. (Burkhart 2019, p. 129)

As Burkhart describes, the portrayal of *Yeii*'s forms as merely "artworks" is problematic not simply from an aesthetic perspective. It is also an ontological failure because it categorically reduces *Yeii* to simple symbolic representations, which obscures the locality and relationality of the practices genuinely grounded in the land.

Such "delocation" is made possible under the ontological assumption that things and relationships are fixed and detachable from their co-constitutive relationality. This practice of abstracting things from their biocultural conditions (from locality) is also part of another problem. Janice Neri (2011) coined the concept of "specimen logic" to criticize the modern practice of collecting insects and other living beings isolated from their habitats. Hence, in most biological collections material relations and living context of the specimens are omitted. Consequently, these collections preserve a delocalized index of "the things themself [*sic*]." José Miguel Esteban (2018) has elaborated on the criticisms linked to the specimen logic, adding that this has been an inherent practice in colonial wildlife imagery that portrays organisms as isolated individuals against blank backgrounds. Esteban (2018) claims that such logic, propagated through colonial artwork, has constructed an image of habitats as neutral white space, setting the stage for biocultural homogenization.

With this specimen logic, essentialized objects are displayed in museums and galleries, shipped around the world for study and admiration. Moreover, this logic applies not only to representations of living organisms and habitats, but may also transform tangible or intangible biocultural expressions and relationships themselves into specimens under the name of "artworks" or "artifacts." Under this ontological paradigm, these artworks and artifacts may then be hung on the white walls of museums, seemingly stripped of their biocultural relations or meanings in their locality. Such collections of numerous "things" and transformation of them into digestible "data" should not be confused with conservation of biocultural diversity.

4.3.4 *Efforts to Decolonize Museums*

To address the former problem, in 2014, the National Museum of Ethnology of Japan launched a project named "Info-Forum Museum," thus establishing an international collaborative network involving other museums and research institutes worldwide as well as members of the source communities (people who created ethnographic objects or used them and their descendants). In 2020, in collaboration with the Hopi people from Arizona in Southwestern United States, a special issue on "Decolonizing Museum Catalogs" was edited. In this issue, Kelley Hays-Gilpin, Atsunori Ito, and Robert Breunig published "Defining and Exploring the Problem" in which they denounce that anthropological museum collections of marginalized communities' material culture are problematically rooted in colonialism and that the preservation of such collections have often constructed a false narrative that such cultures are dying off. Furthermore, when descendants of these communities want to connect with their heritage, Hays-Gilpin and collaborators (2020, para 2) explain that "they are often disappointed in the quantity and quality of information in our museum catalog records…[which has] no culturally appropriate meanings to the descendants of those who made and used the objects." With this example, we call attention to the fact that although this is the historically dominant way of institutionalization of art within a colonial framework, Indigenous communities, art theorists, and museologists have been working to decolonize museums, attempting to use them as active spaces for communication of local knowledge rather than disconnected collections of objects (Hays-Gilpin et al. 2020).

Another example that illustrates ongoing efforts for decolonizing museums is offered by Ruth Phillips. She describes the *Anishinaabeg: Art & Power* exhibit at Toronto's Royal Ontario Museum, which, curated in part by members of the Anishinaabeg community, "structured ancestral and contemporary temporalities, cosmic spatial zones, and the surface geography of the earth so that visitors experienced them as relational, co-existent, and traversable" (Phillips 2022, p. 59). The curators of the exhibition describe that their goal was to reach Anishinaabe people. According to one of the curators, Alan Corbiere, their goal has been "to show the richness, beauty and complexity of our culture" (in Phillips 2022, p. 61).

In summary, although museums are often sites of delocalization and specimen logic, they can alternatively become spaces for dialogical communication, restoration, and connection within and among biocultural communities. In such cases, the agency of biocultural communities is vital to move beyond mere objectification; such agency is not the delocal agency of Western colonial subjects, but rather arises from relationality, from the co-constitution of co-inhabitants through their habits within habitats.

4.4 Conclusion

In this chapter, we attempted to articulate ontologically what it might look like to do biocultural art conservation and to show how such conservation might differ from colonial practices, which start from the assumption of a delocalized, static ontology. Through our analysis, we show that biocultural ontologies embody co-constitutive relationality that defocuses from fixed objects. This invites us to understand biocultural communities and their art through co-constitutive relationality rooted in the land, as shown through the examples from Indigenous and local communities throughout this chapter. Thus, biocultural conservation of art should honor such relationality, and simultaneously, this view allows us to avoid reducing art to mere collections of things to be displayed. As seen in Teshima's case, local people may perceive the biocultural landscape itself as living and changing art, which is deeply connected with their care for the land.

Biocultural communities and the art expressed through them are dynamic relationships from the land, thus biocultural conservation of art should aim to understand and support the expression of these ontologically dynamic relationships. Art conservation in this sense will require more than simply keeping "works" of art safe, retouching them, or cataloguing them. Rather it should aim to conserve the human and other-than-human co-inhabitants' relationships through which art is expressed, as well as their connection to the land, which is not ontologically separate from such "works." That said, conservation of these dynamic relationships should not essentialize relationships or fixate on authenticity because biocultural communities are themselves never ontologically fixed. Conservation will entail recognition of the agency that forms through relation with the land, and biocultural conservation of art.

There remain some unaddressed questions about the relationship between biocultural ontologies and art. Many of these have distinctly ethical and political connotations. For example, what conservation norms and practices should we adopt toward long-extinct biocultural communities? Moreover, might there occasionally be room for a sort of "strategic thingification" in biocultural art conservation, akin to how biocultural conservationists use the concept of keystone species to support broader biocultural conservation efforts? Equally important are questions of conflicting ontologies; for example, given that communities and cultures are not sealed, what does it mean to navigate conservation within non-homogenous communities for whom conflicting ontologies are held? Although these questions require further exploration, we do not have the space to consider them here. However, we hope that this entry point into an examination of biocultural ontologies and art will assist in deepening further discussion and research.

Acknowledgments We would like to thank Danqiong Zhu, Ricardo Rozzi, and Roy May for their valuable comments on draft versions of this chapter.

References

Awang A, Abdullah A, Tengku Hamzah TAA, Azhari MS (2017) Japan's environmental pollution issues: 1950s to 1970s. Int J East Asian Stud 6(1):21–34

Barad K (2003) Posthumanist performativity: toward an understanding of how matter comes to matter. Signs 28(3):801–831

Barg DM, Palus ES (2021) Recovering looted artifacts and the art of deciding what to curate: the Cerberus collection. Coll 17(1):76–89

Bhati HV, Epstein Y (2023) Protection of biocultural heritage in the anthropocene: towards reconciling natural, cultural, tangible and intangible heritage. J Environ Law 35:353–375

Brennan RE (2018) Re-storying marine conservation: integrating art and science to explore and articulate ideas, visions and expressions of marine space. Ocean Coast Manag 162:110–126
Bridgewater P, Rotherham ID (2019) A critical perspective on the concept of biocultural diversity and its emerging role in nature and heritage conservation. People Nat 1(3):291–304
Burkhart B (2019) Indigenizing philosophy through the land: a trickster methodology for decolonizing environmental ethics and Indigenous futures. Michigan State University Press, East Lansing
Burow PB, Brock S, Dove MR (2018) Unsettling the land: indigeneity, ontology, and hybridity in settler colonialism. Environ Soc Adv Res 9:57–74
Caillon S, Cullman G, Verschuuren B, Sterling EJ (2017) Moving beyond the human–nature dichotomy through biocultural approaches: including ecological well-being in resilience indicators. Ecol Soc 22(4):27
Campion OB, West S, Degnian K, Djarrbal M, Ignjic E, Ramandjarri C, Malibirr GW, Guwankil M, Djigirr P, Biridjala F, O'Ryan S, Austin BJ (2023) Balpara: a practical approach to working with ontological difference in Indigenous land & sea management. Soc Nat Resour 37(5): 695–715. https://doi.org/10.1080/08941920.2023.2199690
Cavaliere CT, Branstrator JR, Cheer JM (2024) Intersectional emancipation for biocultural conservation: an exploratory neolocalism framework. J Travel Res 0(0). https://doi.org/10.1177/00472875241247315
Cocks M (2006) Biocultural diversity: moving beyond the realm of "Indigenous" and "local" people. Hum Ecol 34(2):85–200
Crego RD, Ward N, Rozzi R (2023) The eyes of the tree: applying field environmental philosophy to tackle conservation problems at long-term socio-ecological research sites. In: Rozzi R, Tauro A, Avriel-Avni N, Wright T, May RH Jr (eds) Field environmental philosophy: education for biocultural conservation, Ecology and ethics, vol 5. Springer, Cham, pp 101–111. https://doi.org/10.1007/978-3-031-23368-5_7
DePuy W, Weger J, Foster K, Bonanno AM, Kumar S, Lear K, Basilio R, German L (2021) Environmental governance: broadening ontological spaces for a more livable world. Environ Plan E Nat Space 5(2):947–975
Esteban M (2018) Dürer's Rhinoceros: biocultural homogenization of the visual construction of nature. In: Rozzi R, May RH Jr, Chapin FS III, Massardo F, Gavin M, Klaver I, Pauchard A, Nuñez MA, Simberloff D (eds) From biocultural homogenization to biocultural conservation, Ecology and ethics, vol 3. Springer, Cham, pp 137–165
González Morales V (2022) Epistemological, ontological, and ethical dimensions of biocultural rights: the case of the Atrato river, Colombia. Doctorate dissertation, University of North Texas, Denton, Texas
Hashimoto H (2003) Between preservation and tourism: folk performing arts in contemporary Japan. Asian Folklore Stud 62(2):225–236
Hays-Gilpin K, Ito A, Breunig R (2020) Decolonizing museum catalogs: defining and exploring the problem. Trajectoria 1: article no. 2. National Museum of Ethnology, Japan, Senri
Kagawa Prefecture (2023) 香川県離島振興計画 (令和5年度~令和 14 年度) [Kagawa Prefecture remote islands promotion plan (FY2023–FY2032)]. https://www.pref.kagawa.lg.jp/documents/8225/kagawakenritousinkou2023.pdf. Accessed 31 Dec 2023
Khular S (2026) Lamkaang naga loh kchet: Collective Interdependence. In: Zhu D, Rozzi R, Tsuji R, Johnson B, Caviness T, Castro-Jorquera C, Xu L, Zhang L (eds) Linking arts with biocultural conservation, restoration, and communication, Ecology and ethics, vol 8. Springer, Cham, pp 469–488. https://doi.org/10.1007/978-3-032-04830-1_28
Klaver IJ (2013) Environment imagination situation. In: Rozzi R, Palmer C, Armesto JJ, Callicott JB, Pickett S (eds) Linking ecology and ethics for a changing world: values, philosophy, and action, Ecology and ethics, vol 1. Springer, Dordrecht, pp 85–105. https://doi.org/10.1007/978-94-007-7470-4_7
Klaver I (2018) Re-claiming rivers from homogenization: meandering and riverspheres. In: Rozzi R, May RH Jr, Chapin FS III, Massardo F, Gavin M, Klaver I, Pauchard A, Nuñez

MA, Simberloff D (eds) From biocultural homogenization to biocultural conservation, Ecology and ethics, vol 3. Springer, Cham, pp 49–69. https://doi.org/10.1007/978-3-319-99513-7_3

La Follette L (2017) Looted antiquities, art museums and restitution in the United States since 1970. J Contemp Hist 52(3):669–687

López RF, Wickson F, Hausner HV (2020) Bridging different perspectives for biocultural conservation: art-based participatory research on native maize conservation in two indigenous farming communities in Oaxaca, Mexico. Environ Dev Sustain 22:7427–7451

Makki F (2018) Chapter 5: the political ecology of land grabs in Ethiopia. In: Rozzi R, May RH Jr, Chapin FS III, Massardo F, Gavin M, Klaver I, Pauchard A, Nuñez MA, Simberloff D (eds) From biocultural homogenization to biocultural conservation, Ecology and ethics, vol 3. Springer, Cham pp, pp 84–95. https://doi.org/10.1007/978-3-319-99513-7_5

Marett C (2023) A guide to Japan's art islands: Naoshima, Teshima and Inujima, Tatler, 2023, March 7. https://www.tatlerasia.com/lifestyle/travel/japan-art-islands

Markiewicz J, Goralnik L, O'Connell K (2023) Screens on trails: digital environmental science, arts, and humanities learning for biocultural conservation. In: Rozzi R, Tauro A, Avriel-Avni N, Wright T, May RH Jr (eds) Field environmental philosophy, Ecology and ethics, vol 5. Springer, Cham. https://doi.org/10.1007/978-3-031-23368-5_33

Martin LA (2017) Anger and reconciliation: a very brief history of exhibiting contemporary Indigenous art in Canada. Afterall 43:108–115

Martin A, O'Farrell P, Kumar R, Eser U, Faith DP et al (2022) The role of diverse values of nature in visioning and transforming towards just and sustainable futures. In: Christie M, Balvanera P, Pascual U, Baptiste B, González-Jiménez D (eds) Methodological assessment report on the diverse values and valuation of nature of the intergovernmental science-policy platform on biodiversity and ecosystem services. IPBES Secretariat, Bonn. https://doi.org/10.5281/zenodo.6522326

Mathez-Stiefel SL, Boillat S, Rist S (2007) Promoting the diversity of worldviews: an ontological approach to bio-cultural diversity. In: Haverkort B, Rist S (eds) Endogenous development and bio-cultural diversity. Compas/ETC, Leusden, pp 67–81

Matsushima H, Ferreira SMF (2022) Urbanization of coastal areas: loss of coastal dune ecosystems in Japan. In: Gonçalves SC, Ferreira SMF (eds) Sandy beaches as endangered ecosystems: environmental problems, possible assessment and management solutions. CRC Press, Boca Raton, pp 89–106

May RH Jr (2018) Land grabbing and violence against environmentalists. In: Rozzi R, May RH Jr, Chapin FS III, Massardo F, Gavin M, Klaver I, Pauchard A, Nuñez MA, Simberloff D (eds) From biocultural homogenization to biocultural conservation, Ecology and ethics, vol 3. Springer, Cham, pp 109–123. https://doi.org/10.1007/978-3-319-99513-7_7

Moreno-Terrazas-Troyo R, Díaz-Gómez Z, González-Galván H, Cariño-Olvera M, Monteforte-Sánche M (2023) Starfish and sky stars: field environmental philosophy education and ecotourism experiences in Baja California Sur, Mexico. In: Rozzi R, Tauro A, Avriel-Avni N, Wright T, May RH Jr (eds) Field environmental philosophy: education for biocultural conservation, Ecology and ethics, vol 5. Springer, Cham, pp 113–130. https://doi.org/10.1007/978-3-031-23368-5_8

Nemogá GR, Appasamy A, Romanow CA (2022) Protecting indigenous and local knowledge through a biocultural diversity framework. J Environ Dev 31(3):223–252

Neri J (2011) The insect and the image: visualizing nature in early Modern Europe, 1500–1700. The University of Minnesota Press, Minneapolis

Nichols K (2023) Indigenous dispossession and settler colonial art galleries: anguish at the National Gallery of Victoria. Art Hist 46(1):102–123

Nishida M (1999) 瀬戸内海の発見—意味の風景から視覚の風景へ [Discovery of the Seto Inland Sea: from a landscape of meanings to a visual landscape]. Chūō Kōron Shinsha, Tokyo

Nixon L (2020) A brief history of Indigenous art and resistance. In: King H (ed) A culture of exploitation: "reconciliation" and the Institutions of Canadian Art. Yellowhead Institute, Toronto, pp 6–9

NPO Teshima Tourism Association (n.d.) About Teshima, Teshima Tourism Navi. https://teshima-navi.jp/en/about-2/. Accessed 13 Oct 2024

Okawa S (2004) 豊島産業廃棄物不法投棄事件ー巨大な壁に挑んだ二五年のたたかい [Teshima illegal dumping of industrial waste incident: a 25-year struggle against a huge wall]. Nihon Hyoron Sha, Tokyo

Phillips RB (2022) The issue is moot: decolonizing art/artifact. J Mater Cult 27(1):48–70

Polfus JL, Simmons D, Neyelle M, Bayha W, Andrew F, Andrew L, Merkle BG, Rice K, Manseau M (2017) Creative convergence: exploring biocultural diversity through art. Ecol Soc 22(2):4

Qu M (2019) Art interventions on Japanese islands: the promise and pitfalls of artistic interpretations of community. Int J Soc Polit Commun Agendas Arts 14(3):19–38

Qu M (2020) Teshima—from island art to the art island: art on/for a previously declining Japanese Inland Sea Island. Shima 14(2):250–265

Rozzi R (2010) Field environmental philosophy: regaining an understanding of the vital links between the inhabitants, their habits and the regional habitats. Dialogue Universalism 1(2): 85–109. https://doi.org/10.5840/du20102011/127

Rozzi R (2012a) South American environmental philosophy: ancestral Amerindian roots and emergent academic branches. Environ Ethics 34(4):343–366. https://doi.org/10.5840/enviroethics201234436

Rozzi R (2012b) Biocultural ethics: recovering the vital links between the inhabitants, their habits, and habitats. Environ Ethics 34(1):38–44. https://doi.org/10.5840/enviroethics2012341

Rozzi R (2013) Biocultural ethics: from biocultural homogenization toward biocultural conservation. In: Rozzi R, Palmer C, Armesto JJ, Callicott JB, Pickett S (eds) Linking ecology and ethics for a changing world: values, philosophy, and action, Ecology and ethics, vol 1. Springer, Dordrecht, pp 9–32. https://doi.org/10.1007/978-94-007-7470-4_2

Rozzi R (2018) Biocultural homogenization: a wicked problem in the Anthropocene. In: Rozzi R, May RH Jr, Chapin FS III, Massardo F, Gavin M, Klaver I, Pauchard A, Nuñez MA, Simberloff D (eds) From biocultural homogenization to biocultural conservation, Ecology and ethics, vol 3. Springer, Cham, pp 21–48. https://doi.org/10.1007/978-3-319-99513-7_2

Rozzi R (2023) Inter-species and inter-cultural encounters: the biocultural education program of the Omora Ethnobotanical Park. In: Rozzi R, Tauro A, Wright T, Avriel-Avni N, May RH Jr (eds) Field environmental philosophy: education for biocultural conservation, ecology and ethics series, vol 5. Springer, Dordrecht, pp 153–174. https://doi.org/10.1007/978-3-031-23368-5_10

Rozzi R, Tauro A (2023) The multiple lenses of ecotourism with a hand lens: fundamental concepts and practices. In: Rozzi R, Tauro A, Avriel-Avni N, Wright T, RH M Jr (eds) Field environmental philosophy, Ecology and ethics, vol 5. Springer, Cham. https://doi.org/10.1007/978-3-031-23368-5_3

Setouchi Triennale Executive Committee (n.d.) The philosophy of the setouchi triennale, Art Setouchi. https://setouchi-artfest.jp/en/sdgs/. Accessed 26 Dec 2023

Suzuki T, Kin Y (2020) 20 years on, residents set to return 'garbage island' in western Japan to pristine state. The Mainichi (online). https://mainichi.jp/english/articles/20200606/p2a/00m/0fe/004000c

Swanson FJ (2023) The H.J. Andrews experimental forest long-term ecological research program, Oregon, USA: a historical biocultural perspective. In: Rozzi R, Tauro A, Avriel-Avni N, Wright T, May RH Jr (eds) Field environmental philosophy, Ecology and ethics, vol 5. Springer, Cham. https://doi.org/10.1007/978-3-031-23368-5_32

Teshima Residents Council on Waste Management [廃棄物対策豊島住民会議] (2005) 豊かさを問うII—調停成立5周年をむかえて: 豊島事件の記録 [Questioning Wealth II—celebrating the 5th anniversary of the establishment of mediation: records of the Teshima Incident]. Teshima Residents Council on Waste Management [廃棄物対策豊島住民会議]

Zanotti L (2018) Biocultural approaches to conservation: water sovereignty in the Kayapó Lands. In: Rozzi R, May RH Jr, Chapin FS III, Massardo F, Gavin M, Klaver I, Pauchard A, Nuñez MA, Simberloff D (eds) From biocultural homogenization to biocultural conservation, Ecology and ethics, vol 3. Springer, Cham, pp 343–359. https://doi.org/10.1007/978-3-319-99513-7_22

Chapter 5
The Precarious Art of Cecilia Vicuña: Finding the Ecological and Sacred Value of the Small and the Discarded

Carolina Castro-Jorquera

The wild vicuña is sacred to indigenous Andean cultures. 'She lives in the highest places, so she is considered not to belong to the humans, but to the Apus, the lords.' The strong fibers of her wool (with a molecular structure that provides the best defense against the cold) long provided warmth and wealth, before the Spaniards replaced ecologically balanced native herds of domesticated llamas and alpacas, wild vicuñas and guanacos, with cattle and sheep whose cloven feet and habit of uprooting vegetation eroded the terrain and made its people poor. Legend has it that vicuñas are born at the sources of springs, and the fiber made from their wool is symbolically associated with the thread of running water or the stream of life.
Lucy R. Lippard, "Spinning the common thread," in The Precarious. The Art and Poetry of Cecilia Vicuña. *(Vicuña and De Zegher 1997, p. 7)*

Abstract The artwork of Chilean contemporary artist, poet, and filmmaker Cecilia Vicuña is noted for themes of language, memory, dissolution, extinction, and exile. As a teenager, she began to point out the relevance of her work to the politics of ecological destruction, cultural homogenization, and economic disparity, particularly how such phenomena disenfranchise the already powerless. Her commitment to feminist forms and methodologies is considered a unifying theme across her diverse body of work, among which quipus, "PALABRARmas," and precarious stand out. She conceived the concept of "precarious art" (*lo precario*) in the 1960s to highlight the sacred and ecological value of the small and discarded. In 1966, on the seashore of Concon, mouth of the Aconcagua river in Chile, she collected debris, shells, feathers, seeds, and small objects with which she built the first precarious objects to be erased by sea waves. This work continues even today. In 1965, she referred to the quipu ("knot" in Quechua) for the first time in her diary with the phrase "The quipu

C. Castro-Jorquera (✉)
Cape Horn International Center (CHIC), Universidad de Magallanes, Puerto Williams, Chile

D. Zhu et al. (eds.), *Linking Arts with Biocultural Conservation, Restoration, and Communication*, Ecology and Ethics 8,
https://doi.org/10.1007/978-3-032-04830-1_5

that remembers nothing," which marks the beginning of her work around this concept and element of the Andean world that has become a constant inspiration for her work and which is also part of "lo precario." The word precarious is derived from prayer; i.e., obtained by praying, and in Vicuña's work, it means that art can become a prayer for the Earth. In this text, I propose revisiting Cecilia Vicuña's concept of precarious art through the biocultural lens proposed by Chilean ecologist and philosopher Ricardo Rozzi, emphasizing the importance of making biocultural homogenization visible. Rozzi aims to deconstruct homogenization to recover inter-cultural and inter-species relationships of co-inhabitation. An appreciation of the vast biological and cultural diversity of life is essential for cultivating relationships of complementarity and reciprocity among human and other-than-human co-inhabitants. In turn, Vicuña offers creative art habits to show how art can be a practice and a form of knowledge through which we can restore our relationships with the environment and our ancestral entanglement with the land.

Keywords Biocultural art · Chile · Contemporary art · Ecofeminism · Quipus

5.1 Introduction

Cecilia Vicuña stands as a pioneering figure in the realm of contemporary art, renowned for her profound exploration of themes such as language, memory, dissolution, extinction, and exile. Since her teenage years, she has underscored the political dimensions of her work, particularly its impacts on causing ecological destruction, cultural homogenization, and economic disparity. Vicuña has shed light on how these environmental, social, and political problems further disenfranchise the already powerless. For over 50 years, her work has exceeded all conventional art categories. Notions such as *conceptual art*, *land art*, or *arte povera* "are insufficient to understand or describe the passionate entanglement that she weaves between word and seed, between sound and thread, between quipu and blood, between body and earth, between garbage and cosmos" (López 2019, p. 15). Vicuña has devoted a significant part of her artistic practice to studying, interpreting, and reactivating the *quipus*, which were banned by the Spanish conquistadors during their colonization of South America. Drawing on her indigenous heritage, Vicuña channels this ancient, sensorial mode of communication into immersive installations and participatory performances.

> A quipu (also *khipu* or *kipu*) is a device of knotted, colored strings of vegetable or animal fiber used by the Incas and other Andean cultures to record information. Since the first international Quipu Conference in 1988, despite the significant research carried out in recent decades, the meanings of *quipus* are still debated among specialists. Manuel Medrano and Galen Brokaw have pointed out that, "scholars in the twentieth century often dismissed the quipu as a mnemonic device that, while not writing, enabled effective socioeconomic and political administration. However, colonial sources claim that the quipu was used for a variety of purposes beyond tribute and storehouse accounting, including the recording of censuses, laws, genealogy, and even history. Such assertions have led some to argue in favor of a division between numerical and narrative quipus." (Medrano and Brokaw 2023, pp. 1–25)

Fig. 5.1 Cecilia Vicuña, *Quipu Menstrual, la sangre de los glaciares* (*Quipu Menstrual, The Blood of the Glaciers*), 2006. Installation, wool, document (letter to Michelle Bachelet), video, invitation, fragment of poem. Glaciar El Plomo, Chile. (Photograph by James O'Hern. Courtesy of the Artist. © 2025 Cecilia Vicuña)

In Vicuña's early understanding of the quipu, she knew that they kept a knowledge of great relevance to our present time, such as the locations of the glacial peaks and water sources. It is known that the Incas practiced some ritual sacrifices in these places, such as the case of the "niño del Cerro El Plomo" buried on the summit of a mountain of the same name, where the Mapocho Valley, one of the southernmost points of the Inca territory, begins. This is one of the reasons that led Vicuña to establish a link between water and blood, expressed in works like *Quipu Menstrual* (Fig. 5.1).

Vicuñas' work stands for water, forests, butterflies, and Andean cats, among many *earthly beings,* including humans, but most of all, she stands for connections. She says, "Everything is falling apart because of a lack of connections. Weaving is the connection that is missing, the connection between people and themselves, people and nature" (Vicuña and De Zegher 1997, p. 12). Notably, in the 1960s, she conceived the concept of "precarious art" to emphasize the sacred value of the small and discarded, death, and transformation. She experientially grabbed this by collecting debris, shells, feathers, seeds, and small objects on the seashore of Concón, at the mouth of the Aconcagua River in central Chile, giving rise to a characteristic of her work, where garbage and the cosmos are related (Fig. 5.2). In

Fig. 5.2 Cecilia Vicuña, *Dissolution*, 2009, site-specific performance installation. Concón, Chile. (Photograph by James O'Hern. Courtesy of the Artist. © 2025 Cecilia Vicuña)

1966, she created installations destined to be erased by the sea waves. "Precarious is what is obtained by prayer," Vicuña has written. "Uncertain, exposed to hazards, insecure. From the Latin *precarius*, from *precis*; prayer". The word *oir* (to hear) was originally the same word *orar* (to pray). "Reciprocity. By praying, you reconnect" (Vicuña and De Zegher 1997, p. 10).

In her approach to the concept of precarious art, we can find solid statements for an early ecological awareness. Historically, art has been a place where the representation of the world takes place, from expressions in caves thousands of years old to the present day, the different forms of art that have emerged have been at the service of asking us about the world we inhabit, how we inhabit it, and with whom we share it. In this regard, it is worth asking ourselves how the vision of the world offered by Vicuña's precarious art can show us the ecological and sacred value of that which we have made invisible or discarded, leaving it out of our imagination, to put it back before our eyes. How can quipus or precarious objects open ways of thinking and seeing, which global society has forgotten and ignored, but are they still present and within the realms of possibilities? How can the concept of "the precarious" introduce us to a wider form of co-existence, in permanent dialogue with water, animals, insects, plants, fungi, cells, bacteria, and the diversity of other-than-human beings that shape us and transcend us?

In this essay, I embark on a journey to explore Cecilia Vicuña's concept of precarious art—"lo precario"—through the biocultural lens proposed by the Chilean ecologist and philosopher Ricardo Rozzi (1999, 2012). Through this lens, I aim to elucidate how art can make biocultural homogenization visible and help us to recover our relationships of inhabitation, complementarity, and reciprocity with both humans and other-than-human beings. In Rozzi's words,

> This biocultural (re)connection is urgent because a root cause of the current socio-environmental crisis stems from the losses of biological and cultural diversity. To achieve biocultural reconnections, we propose multisensorial educational methodologies grounded

> in analogical thinking, the integration of science, humanities, and other disciplines like art, and habits of care of the multiplicity of human and other-than-human co-inhabitants and the singularity of their shared habitats. (Rozzi 2023, p. 18)

In this context, Vicuña's sustained practice, highlighting the importance of Andean thought, the silent but massive extinction of species, the care of seeds, as well as the importance of preserving the sacredness of rivers and mountains, emerges as a compelling invitation to appreciate how art can serve as both a practice and a form of knowledge through which we can restore our relationships with the environment and thereby reimagine kinder futures. More specifically, it can guide members of global society to realize how paying attention to the most unperceived beings (things that are small and often discarded) can show us their precious ecological value for this great being we call Earth (Vicuña and De Zegher 1997; Rozzi et al. 2008).

In doing so, I have organized this chapter into three parts: (1) The Precarious, (2) The Political Ecology of Vicuña's Art, and (3) Reorienting our Vision: Precarious Art Through a Biocultural Lens.

5.2 The Precarious/"Lo Precario"

From the beginning, Vicuña's notion of precarious art challenged conventional understandings of permanence and value in artmaking. In prevailing artistic trends of the 1960s (such as *Land Art* or *Participatory Art*), the work of art happens on the land (outdoors), brings elements from the land to the gallery space, or was created through the participation of people in addition to the artist. In contrast, Vicuña's assemblages do not respond to the concerns of the art world, but rather seek to participate in a greater dialogue, in a humble act of communication with nature (Vicuña and López 2021, 2024). As if they were sacred offerings, the precarious objects were presented as a spontaneous action "as a way to honor the balance, complementary dimension and reciprocity of the natural world" (López 2019, p. 40).

Precarious objects can be understood as visual poems, metaphors in space, and small pieces of garbage, where scraps of stone, wood, feathers, shells, cloth, wool, and other human-made detritus are gently juxtaposed. In 2013, one of the first precarious things I noticed when I met Cecilia Vicuña was a tangle of wool that had a stick that went through it, along with smaller ones holding it up (Castro-Jorquera 2020). In this arrangement, plastic threads and one or two feathers at its tips held it all in perfect balance. Originally, these artworks were made to be placed on the seashore to be erased by the ocean. However, some of these precarious objects were recrated and included in art collections, which are displayed in museum galleries, often arranged on a layer of sand (Figs. 5.3 and 5.4).

Born of contemplation and made of refuse, Vicuñas' *earthworks*, as Catherine de Zegher named them, are prayers for the Earth, as most of her works are (De Zegher 2019). Understanding art as a form of awareness of our place in the world, her art has

Fig. 5.3 Cecilia Vicuña, *Pueblo de Altares*, 1990/2017, site-specific installation of precarious objects, mixed media. Installation view, "Cecilia Vicuña: About to Happen", CAC New Orleans (CACNO), 2017. (Courtesy the artist, CACNO, and Lehmann Maupin, New York, Seoul, and London. Photo: Alex Marks. © 2025 Cecilia Vicuña)

occurred on the margins, outside of artistic institutions; for example, on the street, on the beach, in her own home or studio, and in a silent way (although always noisy) cultivating forms of reciprocity with the environment and its diverse forms of life and existence, without distinguishing their categories. Curator Miguel Lopez illustrates this point by stating that:

> [O]ver the years Vicuña has insisted on paying attention to another kind of poetic event never recognized in books, recitals, or places dedicated to literature or canonical cultural expressions. In her opinion, poetry emerged spontaneously and anonymously as a result of physical connections. It emerged in the inventive and often uncontrolled dimensions of the act of living itself—that is, in the "biological-cultural-political-technological" compositions where people are just part of a broader and more complex landscape of forces and powers. (in López 2019, p. 13)

By utilizing materials that are often overlooked or discarded, Vicuña elevates the unnoticed to the realm of the sacred, inviting us to reconsider our relationship with the environment and all its co-inhabitants, in consonance with Rozzi's cornerstone concept. Through her art, Vicuña transforms the act of creation into a form of prayer, a gesture of reverence and connection to the Earth. As Julia Bryan-Wilson pointed out, "In her environmental interventions, Vicuña assiduously returned to the idea of planet Earth as a house, a conception consonant with numerous indigenous worldviews" (in López 2019, p. 45). Vicuña pays tribute to these worldviews.

Fig. 5.4 Cecilia Vicuña, *Pueblo de Altares*, 1990/2019, site-specific installation of precarious objects, mixed media, detail. Installation view, "Seehearing the Enlightened Failure", Kunstinstituut Melly (FKA Witte de With), Rotterdam, 2019. (Courtesy the artist and Lehmann Maupin, New York, Seoul, and London. Photo: Kristien Daem. © 2025 Cecilia Vicuña)

Over the years, the concept of precarious art took a broader meaning in Vicuñas art, including not just precarious objects but her entire practice. Her actions are often site-specific and ephemeral. Her art objects are conceived as "a moment of emergency," or "the future potential of the unformed." This is what Vicuña calls "lo precario," a way to remind us of our fragility and kinship with the natural world.

5.3 The Political Ecology of Vicuña's Art

At the heart of Vicuña's artistic practice lies a profound "concern for nature." French philosopher, anthropologist, and sociologist Bruno Latour (1991, 2004) stated that political ecology is, and should be, a concern for nature in politics. At a political level, for Vicuña, this means that the way in which humans and other-than-humans co-inhabit must be of political interest. Her deep critique of the politics of ecological destruction and cultural homogenization imposed by capitalism and neocolonialism can be found across her poetic and artistic body of work. She recognizes that these phenomena are not isolated but interconnected, perpetuated by systems of power that prioritize the gain of short-term profit over the well-being of people and other species co-inhabiting the planet.

Vicuña's work is a powerful indictment of the injustices derived from this narrow economic focus of politics, shedding light on how marginalized communities bear the brunt of environmental degradation and cultural erasure. One aspect that Vicuña has emphasized is the loss of languages, as well as the extinction of species, which radically affects the loss of cultural and biological diversity, two terms that, as Rozzi (2012) has pointed out in his writings, are intertwined. According to Ethnologue, Gary Simons (2019), today more than nine languages cease to be spoken every year, languages are repositories of wisdom passed down through generations, and often that wisdom is ecological.

Vicuña's insistence on using diverse, contextual, secular languages in her performances and poems opens up the submerged dimensions of a great diversity of knowledge, deeply linked to the land. The way we name things, the way they have been named by Indigenous and other local peoples in each place, is a recognition of biocultural diversity. This diversity has undergone a deep crisis with the hegemony of scientific knowledge. To reverse this trend, it is urgently necessary to explore new approaches, holistic and generous ones, as Ricardo Rozzi exemplifies with the word Earth:

> *Earth* is not only a biophysical entity, but it is also a word. A word that influences the way we understand and relate to the biophysical reality of the planet. Very often, scientists forget the gravity of words and focus their research on the biophysical reality. Conversely, philosophers often focus on examining the language of cultural reality, ignoring the biophysical realm. The biocultural ethic unites biological and physical realities with human cultural attainment. It provides Earth stewardship with a conceptual framework that integrates the biophysical and symbolic-linguistic realms of reality. (Rozzi 2015, p. 88)

When the diverse meanings of a word are lost or a complete language disappears, with them, specific ways of naming and valuing things, living beings, and territories are missing. With these losses, many worlds are disappearing too. Curator Catherine de Zegher, who has been a pioneer in researching and making visible Vicuña's work in European contexts, points out that:

> In Cecilia Vicuña's art and poetry, particularly as they relate to protests of resistance, the investigation of language and the politics of definition are always at stake, because "naming" for her is the most political act of all. Her work concerns "la batalla de los significados" (the battle of the meanings). According to Vicuña, submission and poverty begin with the acceptance of definitions that others create for you. (Vicuña and De Zegher 1997, p. 24)

In February 2024, Cecilia and I joined a conservation movement in defense of the Juncal Valley, the headwaters of the Aconcagua River. We called on people, local organizations, and artists to carry out *Quipu de Encuentros Juncal-Aconcagua* (Fig. 5.5). This collective poetic act aimed at defending the glaciers, waters, and biodiversity of the Aconcagua Valley. Its clean waters originate in the Juncal Valley, which in 2010 was designated as a Ramsar site, the Parque Andino Juncal. This place is protected under this international convention for the conservation of wetlands; however, due to Chilean laws that prioritize mining over biodiversity conservation, its glaciers and waterways have been threatened by mining. In 2024, Vicuña

Fig. 5.5 Cecilia Vicuña, *Quipu de Encuentros Juncal-Aconcagua*, 2024, collective performance with unspun wool, Parque Andino Juncal, Valparaíso, Chile. (Photograph by Nicolás Amaro. Courtesy of the Artist. © 2025 Cecilia Vicuña)

Fig. 5.6 Cecilia Vicuña, *Brain Forest Quipu*, 2022. Multimedia installation, Turbine Hall, Tate Modern, London. Hyundai Commission. (Photograph by Sonal Bakrania. Courtesy of the Artist. © 2025 Cecilia Vicuña)

joined the park team, the authorities, and local communities in an unprecedented territorial defense movement that succeeded in stopping the entry of mining companies, at least for a while.

Returning to the political ecology proposed by Latour (1991, 2004; Buraj 2023) and the aforementioned questions about language, we must ask ourselves: What were we defending when we spoke of a "glacier"? Were we defending only a body of water or a vital network, a nourishing agent entangled in a web of human and other-than-human assemblages? Is it possible to imagine that artistic actions such as *Quipu de Encuentros Juncal-Aconcagua* can contribute directly to biocultural conservation, restoration, and communication? I would like to think so. However, artistic actions would not have the impact required without close links to alliances with scientists, educators, decision-makers, and defenders of territories committed to the care of ecosystems.

In actions like this, Vicuña seeks to amplify the voices of those who are most affected by ecological devastation, offering a platform for resistance and resilience. From the mining threat to Chile's glaciers in artworks like *Quipu Menstrual* (Fig. 5.1) to the deforestation of the Amazon rainforest in her *Brain Forest Quipu* (Fig. 5.6) that was exhibited at the Tate Gallery in London (2023). Furthermore, the displacement of rural and indigenous communities is expressed in most of her poems, oral performances, and precarious objects. In this way, Vicuña's work bears witness to the profound consequences of unchecked capitalism and colonialism. By highlighting these issues, she invites viewers to confront their complicity and imagine alternative futures rooted in justice and solidarity between co-inhabitants.

5.4 Reorienting Our Vision: Precarious Art Through a Biocultural Lens

In recent years, Ricardo Rozzi has proposed a biocultural approach to understanding human-environment relationships. By identifying links among co-inhabitants, their life habits, and shared habitats ("3Hs"), Rozzi offers a hermeneutic model to understand and value the interconnectedness of biological and cultural diversity.

Rozzi points out that

> [P]urely scientific or other cognitive arguments have not been sufficient to open the "ecological eyes" of most citizens to appreciate the magnitude of the current social and environmental crisis, and the essential need to appreciate life in its cultural and biological diversity. (Rozzi 2023, p. 17)

In our book, we aim to give voice and even agency to two arenas: "other-than-human life forms" and "other-than-dominant cultures" that together comprise the rich biocultural world we live in. Hence, to protect biodiversity, we need cultural transformations, and one of the most challenging ones is to change our lenses, our ways of seeing the world (Rozzi et al 2008). In Rozzi's words:

> [T]he biocultural ethic demands overcoming multiple forms of chauvinism... [we need] not only to solve the problems derived from *anthropocentrism* but also from *ethnocentrisms*. It is necessary to leave ethnocentrisms behind to affirm the value of the diversity of cultures *whose knowledge has been made invisible*. This knowledge should complement Western science and philosophy to address global socioenvironmental change problems, such as the sixth mass extinction, that affect multiple cultures. *Comprehensive socioenvironmental justice can only be achieved with the participation of multiple cultures.* (Rozzi 2019, pp. 278–279; emphasis added)

From this perspective, Vicuña's concept of precarious art takes on new significance, as it not only invites us to pay attention to little things, little beings, and little gestures, but to challenge the dominant gaze looking at the fragility of the natural world as a major virtue, and learn and value Indigenous ways of worlding and their intimate knowledge of the land. The concept of worlding underscores the dynamic, continuous becoming of our coexistences, in terms of Ernesto Spinelli (2016). In turn, by highlighting the sacred value of the small and discarded, Vicuña's precarious art invites us to reconsider our relationship with the environment and the subjects that inhabit it. In doing so, it invites us to adopt a more holistic and reciprocal approach to conservation, regeneration, and restoration, one that recognizes the intrinsic value of both biological and cultural diversity. Through her art, Vicuña reminds us that preserving the diversity of life on Earth requires not only protecting ecosystems but also honoring the knowledge, language, and traditions of Indigenous peoples who have lived with the land for centuries.

In this context, we can ask ourselves: How can we encourage diverse ways of knowing ecosystems not only through sciences but also through the arts and humanities to develop a sense of place and responsibility toward the conservation of biocultural diversity? (Rozzi 2023).

One of the aspects that Rozzi and Vicuña insist on is solidarity, questioning how we can situate ourselves in others' positions, and our ability to exchange points of view. Art can offer a wide range of possibilities in that as a practice and as a tool it implies pushing our ways of thinking towards an analogous direction, opening us to other ways of knowing and empathizing with the great diversity of ways of making worlds together, or worlding in the Heideggerian notion reintroduced by Indian postcolonial thinker Gayatri Spivak (1985). Rozzi has stated that his deepest motivation is to make:

> [A]n open invitation to collaboratively (inter-disciplinarily and inter-culturally) cultivate an appreciation for the small invertebrates (and other under-appreciated and under-valued co-inhabitants of biodiversity). These little co-inhabitants are precious in themselves. Additionally, we share our local habitats with them, and the well-being of the global biosphere (including human societies) depends on them. (Rozzi 2019, p. 282)

To deepen the shift Rozzi proposes, we must take seriously his invitation to reorient our perception, not only to look at life differently but different forms of life. In 2000, Rozzi, with a team of philosophers, artists, and scientists at the Omora Ethnobotanical Park in Puerto Williams, capital of the Antarctic Province of Chile, created an innovative educational and special interest tourism activity that is called Ecotourism with a Hand-Lens. Notably, this "change of lenses" stimulated the creation of the UNESCO Cape Horn Biosphere Reserve, the first protected area in the world that has been designated based on the diversity of mosses and liverworts and other small organisms that have been little appreciated in international conservation (Rozzi 2019, p. 276). In his words:

> To appreciate the ecological, aesthetic, economic, and ethical values of the micro-cosmos composed by little plants (e.g., mosses, liverworts), lichens, phytoplankton, bacteria, fungi, invertebrates, and other tiny organisms, we have proposed a 'change of lenses' to assess biodiversity. (Rozzi 2019, p. 276)

This "change of lenses" is not metaphorical—it implies a transformation in epistemology, aesthetics, and ethics. It challenges us to recognize that vast portions of life's richness are invisible to the dominant gaze, which tends to prioritize large, charismatic species or dramatic environmental events. It is here that art can perform an essential operation: making visible (and even audible or tangible) that which remains overlooked, and even more, art can bring a complementary way of knowing ecosystems through attention and imagination. Precarious art, as practiced by Vicuña, embraces fragility and detail as a form of attention and resistance. The scale of the small (as well as the inanimate) becomes a deliberate refusal of extractive aesthetics that rely on spectacle, scale, and permanence.

From this perspective, *worlding* becomes a collective and situated practice of co-creation, not merely among humans but with and through other-than-human life forms or co-inhabitants. For instance, in previously mentioned works done by Vicuña, the Pacific Ocean, the Aconcagua River, and the Juncal Glacier would be co-creators of her art. These principles do not separate the social from the ecological, or the spiritual from the material. It embodies a relational ontology where rivers are ancestors, mountains are protectors, and plants are teachers. Artistic practices grounded in or inspired by such ontologies do not simply "represent" these relationships—they enact them. In this sense, art becomes a mode of *worlding*, of participating in a world made through relations of reciprocity, reverence, and coinhabitation.

In this sense, Art, together with Rozzi's ethical approach of the 3Hs, considering habitats, habits, and co-inhabitants, can also open pathways to recognize and respect more-than-human agencies. Artworks and art practices invite us to listen and recognize their expressions, working as a new avenue for them to speak, rethinking their role in sustaining ecosystems, making visible the networks of dependency we tend to ignore. The question is not just how to *speak for* these beings, but how to *co-inhabit the world with* them—how to draft aesthetic and epistemic forms that do not erase, but amplify, their presence and existence.

In this context, the "small" and "discarded" are not diminutive categories, but radical ones. It opens an ethical and political stance: to attend to that which capitalism deems negligible, to resist through care, fragility, and attentiveness. It is a commitment to multispecies solidarity, and a recognition that true biocultural conservation begins by seeing the world not from above, but from within—through the mosses, lichens, insects, rivers, mountains, and invisible flows that hold it all together.

5.5 Conclusion: Restoring Relationships Through Art and Biocultural Reciprocity

Cecilia Vicuña's precarious art offers a powerful testament to the interconnectedness of the environment and culture. Through her ephemeral installations and poetic interventions, she challenges us to reconsider our relationship with the natural world and the communities that co-inhabit it. By embracing impermanence and valuing the small and discarded, Vicuña's art serves as a prayer for the Earth, a reminder of the sacredness of all life forms.

As we revisit Vicuña's concept of precarious art through a biocultural lens, we are reminded of the importance of fostering reciprocal relationships with humans and other-than-human beings. In an era marked by ecological destruction and biocultural homogenization (Rozzi 2018), Vicuña's work offers a beacon of hope, pointing towards a future where art becomes a practice of restoration and regeneration.

Through her art, we are invited to imagine new possibilities for coexistence, where diversity is celebrated, and relationships are nurtured with care and reverence.

In conclusion, Cecilia Vicuña's precarious art is a powerful testament to not just the sacred, but the ecological value of the small and the discarded. By embracing impermanence and valuing the marginalized, Vicuña's art offers hope and resilience in the face of ecological destruction and biocultural homogenization. Through her art, she reminds us of the sacred interconnectedness of all life forms and invites us to reimagine our place within the web of life and the worlds we co-create.

Acknowledgments I thank Danqiong Zhu and Ricardo Rozzi for their valuable comments on draft versions of this chapter, and the Parque Andino Juncal team for their support. Funding was provided by the Cape Horn International Center (CHIC - ANID/BASAL FB210018) and FONDART, Ministerio de las Culturas, las Artes y el Patrimonio, línea de Investigación 2024 (Folio 707461).

References

Buraj I (2023) Bruno Latour: new challenges and inspirations in political ecology. Studia Philosophiae Christianae, 59(1): 49–68

Castro-Jorquera C (2020) El camino de la conciencia. Mira Schendel, Víctor Grippo y Cecilia Vicuña. Ediciones Universidad Finis Terrae: Santiago, Chile

De Zegher C (2019) Talk with Catherine de Zegher: Cecilia Vicuña Arte Precario. Audio available https://www.kunstinstituutmelly.nl/en/programs/4615-talk-with-catherine-de-zegher-cecilia-vicuna-arte-precario

Latour B (1991) The impact of science studies on political philosophy. Sci Technol Hum Values 16(1):3–19

Latour B (2004) Politics of nature: how to bring the sciences into democracy. Harvard University Press, Cambridge, MA

López MA (ed) (2019) Seehearing the enlightened failure. Witte de With Publishers, Rotterdam

Medrano M, Brokaw G (2023) Oxford bibliographies in Latin. Am Stud. https://doi.org/10.1093/OBO/9780199766581-0175

Rozzi R (1999) The reciprocal links between evolutionary-ecological sciences and environmental ethics. BioScience 49(11):911–921. https://doi.org/10.2307/1313650

Rozzi R (2012) Biocultural ethics. Environ Ethics 34(1):27–50. https://doi.org/10.5840/enviroethics20123414

Rozzi R (2015) Earth stewardship and the biocultural ethic: Latin American perspectives. In: Rozzi R, Chapin FS III, Callicott JB, Pickett STA, Power ME, Armesto JJ, May RH Jr (eds) Earth stewardship: linking ecology and ethics in theory and practice. Springer, Cham, pp 87–112. https://doi.org/10.1007/978-3-319-12133-8_8

Rozzi R (2018) Biocultural homogenization: a wicked problem in the anthropocene. In: Rozzi R, May RH Jr, Chapin FS III, Massardo F, Gavin M, Klaver I, Pauchard A, Nuñez MA, Simberloff D (eds) From biocultural homogenization to biocultural conservation, Ecology and ethics, vol 3. Springer, Dordrecht, pp 21–47. https://doi.org/10.1007/978-3-319-99513-7_2

Rozzi R (2019) Taxonomic chauvinism, no more! Antidotes from Hume, Darwin, and biocultural ethics. Environ Ethics 41(3):253–288. https://doi.org/10.5840/enviroethics201941325

Rozzi R (2023) Field environmental philosophy: concepts and case studies. In: Rozzi R, Tauro A, Wright T, Avriel-Avni N, May RH Jr (eds) Field environmental philosophy: education for biocultural conservation, Ecology and ethics series, vol 5. Springer, Dordrecht, pp 17–25. https://doi.org/10.1007/978-3-031-23368-5_2

Rozzi R, Armesto JJ, Goffinet B, Buck W, Massardo F et al (2008) Changing lenses to assess biodiversity: patterns of species richness in sub-Antarctic plants and implications for global conservation. Front Ecol Environ 6(3):131–137. https://doi.org/10.1890/070020

Simons GF (2019) Two centuries of spreading language loss. Proc Ling Soc Am 4:27–21

Spinelli E (2016) Relatedness—contextualising being and doing in existential therapy. Exist Anal 27(2):303–329

Spivak GC (1985) Three women's texts and a critique of imperialism. Crit Inq 12(1):243–261

Vicuña C, De Zegher C (eds) (1997) The precarious. The art and poetry of Cecilia Vicuña/ Quipoem. University Press of New England, Lebanon

Vicuña C, López M (eds) (2021) Veroir el fracaso iluminado. Turner, Nashville

Vicuña C, López M (eds) (2024) Cecilia Vicuña. Soñar el Agua. Una retrospectiva del Futuro (1964- …). Museo Nacional de Bellas Artes/MALBA, Santiago de Chile/Buenos Aires/ Pinacoteca de Sao Paulo

Chapter 6
Interspecies Co-inhabitation: A Dialogue Between Chinese Philosophical Traditions and Biocultural Ethics

Yubo Song

Abstract Interspecies relationships are of critical relevance in the field of biocultural conservation and restoration, particularly as stated by Ricardo Rozzi's biocultural ethic concept of other-than-human species as humans' co-inhabitants. Traditional Chinese culture can offer meaningful support to the concept of co-inhabitants, which will also serve as a basis for communicating biocultural ethics to Chinese society. In this chapter, the feasible routes and connotations of communicating biocultural ethics to Chinese citizens are addressed, with a focus on ways of understanding interspecies relationships through Taoism, Confucianism, and Buddhism. First, in terms of Taoism, the concept of "All under the sky there is one breath of life" (通天下一氣爾) in *Zhuangzi*. This key expression invites readers to appreciate interspecies relationships. In this way, the aesthetically appealing text of *Zhuangzi* is an expression of art that conveys biocultural ethical messages. Through the observation and contemplation of all living things, human beings not only realize the intriguing nature of the universe in which we are inevitably placed, but also use it to feel full and comfortable with all mysterious interfaces and transformation. The core concept in Taoism is Naturalness, which does not mean a metaphysical existence; in contrast, it emphasizes the emergence of each particular creature. Second, for Confucian doctrine, the metaphysics of morality is central. In the depth of Confucian thought, the aspiration to "inculcate" (教化) is constantly present, and the transcendental sense of the subject, of benevolence ultimately needs to be merged in the chorus of "the oneness of all creatures" (萬物一體). A universe without morality and other-than-human beings in it would be unacceptable for the gods. In a universe that encompass all creatures, the central virtues are fertility, nurture, and completion (化育流行). At the same time, the work of humankind is also highly valued in Confucianism. These concepts are conveyed through

Y. Song (✉)
Institute of Chinese Thoughts and Culture, Northwest University, Xi'an, China

Center for Environmental Philosophy and Comparative Philosophical Research, Northwest University, Xi'an, China
e-mail: phusis@nwu.edu.cn

D. Zhu et al. (eds.), *Linking Arts with Biocultural Conservation, Restoration, and Communication*, Ecology and Ethics 8,
https://doi.org/10.1007/978-3-032-04830-1_6

the arts of mentoring virtuous citizens. Third, Chinese Buddhism is the outgrowth of Indian Buddhism's "Chineseization," which has added the richness of diverse traditional Chinese cultures. By viewing all living beings from the perspective of Buddha-dhātu and the Dharma, one would recognize the uplifting and redeemable properties of all living beings, as recognized by the major Chinese Buddhist schools. These ethical concepts are taught through the arts of metaphorical communication in Chinese Zen. Its philosophical depth is mainly concerned with the annihilation of egoism. In sum, traditional Chinese culture has laid the foundation for communicating interspecies kinship through Taosim's "transformation of things" (物化), Confucianism's view on "fertility, nurture, and completion," and Chinese Buddhism's view on the "redeemable properties" of all living beings. These three concepts converge with the notion of co-inhabitant proposed by Rozzi's biocultural ethic.

Keywords Arts · Biocultural conservation · Buddhism · Confucianism · Taoism

6.1 Introduction

Reflections on interspecies relationships of co-inhabitation are of great significance to the discussion of the biocultural ethic proposed by Chilean ecologist and philosopher Ricardo Rozzi, particularly to his central concept of co-inhabitants. Traditional Chinese culture provides a powerful support for such concepts from both theoretical and practical perspectives. Inspired by Rozzi's framework of biocultural ethics (Rozzi 2013, 2018), serving as a programmatic approach, this chapter seeks to propose some resources for reference on this issue from the perspectives of Taoism, Confucianism, and Chinese Buddhism.

6.2 Taoism

6.2.1 Place and Xiaoyao (逍遥)

Canadian geographer Edward Relph once said that "an authentic sense of place is above all that of being inside and belonging to your place both as an individual and as a member of a community, and to know this without reflecting upon it" (Relph 1976, p. 107). His profound insight suggests a state of total commitment, which constitutes "the basis of man's [sic] relation to the world" (Relph 1976, p. 62). In the age of globalization, especially in countries that rely on automobiles for transportation and daily commutes, frequent and rapid migratory flows have diluted this sense of place, preventing it from coalescing in a profound manner. The space of mobility, like the flow of capital, does not foster deep relationships and a sense of belonging. Wherever the wheels pass through, place degenerates into placelessness.

The Chinese Taoist Zhuangzi also referred to the perception of place and customs. According to Zhuangzi, if the customs of a place cannot be stabilized, it is impossible to talk about the unity of or harmony between humans and nature. His discussion of "Qiuli" (丘里, Hamlets and Villages) is an expression of this view of the links between place and customs. In *Zhuangzi*, a conversation mentions that:

> Hamlets and Villages are formed by the union—say of ten surnames and a hundred names, and are considered to be (the source of) manners and customs. The differences between them are united to form their common character, and what is common to them is separately apportioned to form the differences (丘里者, 合十姓百名而以為風俗也。合異以為同, 散同以為異。). (Zhuangzi 1891)

According to *Zhuangzi*, the customs of a place are like a Gestalt, which is neither the sum of all the things in the habitat nor an abstract idea above them, but a kind of vitality and atmosphere embodied in a whole. Beings in such a situation, including human beings, are neither as dead as a parched tree branch, nor do they believe that there is a master over them. With the intention of the sages, it is suggested that subjectivity should be actualized, or that subjectivity should be submerged in the intention of "thoroughly comprehending the condition in which all things are intermingled as a whole" (周尽一体) (Zhuangzi 1891). Although it is not necessary, as Lao Tzu says, that one can swim with the elk or can inspect the eggs inside the bird's nest, this is where the rhythm of the place begins.

The moment of *Xiaoyao* (逍遙, free and easy wandering, also translated as the enjoyment of untroubled ease) begets when one is in the midst of the circumstances of such a place. Of course, one might point out that this place-centered insight may be as confining as a frog at the bottom of a well. What Taoism is trying to explain, however, is that it is not enough for us to wait for or rely on various existential substances to satisfy our mind. It is only when our minds are shifted, as if we are no longer spontaneous droplets in a vortex, but are intuiting the center of the vortex, the hollow hole of near nothingness, that we become a ubiquitous being that retains its core. This can be called *wú-dài* (無待, flow with the course of nature) or *wú-yòng-zhī-yòng* (無用之用, the usefulness of being useless).

6.2.2 Nature and Luring

Nature is a key concept in Taoism. However, it should be made clear that the Taoist concept of nature is not a metaphysical entity, but rather a diffuse expression of subtle intrigue. When a person is angry, he or she can be "furious," and when a person is relaxed, his or her entire face may be beamed with happiness. The same is true for plants and animals. This is the display of nature and the actual subsistence. Therefore, the concept of nature is essentially an expression of multiplicity, plurality, and heterogeneity. We cannot merely abstract a metaphysical or ontological "thing-in-itself" of nature from the multiplicity, plurality, and heterogeneity and interpret the operation of all things in terms of it. That kind of reductionist and fundamentalist approach is not in line with Zhuangzi's genuine intention.

Based on such an apprehension, we may be able to grasp the meaning of Zhuangzi's *Xiaoyao*, and we may be able to lift our philosophical vigilance and observe all things in the world, including human beings, other-than-human beings, and of course, their habits and habitats (converging with the biocultural ethic's perspective). Creatures in the natural world, such as grass, trees, birds, insects, snakes and turtles, and various craft operations in human societies, such as of boatmen and craftsmen, are all under the scrutiny or lens. Under such a lens, the world becomes full of wonders and is no longer a mirrored reflection of an objects-oriented vision. As stated by American philosopher Philip Shields,

> When…Wittgenstein shows us the strangeness of the familiar, he is trying to shift our perspective from the mundane to the religious and to recapture the special sense of wonder and awe which he felt was extinguished by the prevailing scientific Weltanschauung [worldview]. (Shields 1993, p. 111)

This view also inspires the idea that opposites are interdependent and mutually empowering. Take for example the word "*Youran*" (誘然, luring, charming, captivating) mentioned in *Zhuangzi*. When a pine tree sprouts and grows, and casts its own shadow under the sunlight, all kinds of bushes and herbs will flourish around the pine tree, as if the pine tree has lured these plants and related animals to this place, or plants and related animals were charmed. This is the lure or charm. In this case, *Youran* is nature, and there is no abstract nature beyond or above *Youran*.

The term supervenience emphasized by American scholar Jae-kwon Kim (1993) is probably an echo of the *Youran* concept of nature. If one were to explore the underlying causes of this naturalness, one would probably "not know what it is" (不知其然). In his essay, Rozzi (2013) brilliantly illustrates this theme with the life practices of the South American Indians, such as how the *Pewenche* habit of gathering *Pewen* seeds in the *Pewen* forest habitats. This is the true meaning of the concept of co-inhabitants and co-inhabitation (Rozzi 2013).

6.2.3 *The Foetid and Putrid Versus the Spirit-Like and Wonderful* (臭腐與神奇)

In their social life, human beings focus on ethics from different aesthetic orientations, and form judgments of good and evil (Hargrove 1989). To be social seems to be natural for humans, but it is also the cause of conflicts. Further, there are considerable conflicts, not only between human beings, but also between human beings and the natural world. According to *Zhuangzi*, the things that people love are praised as "the spirit-like and wonderful" and the things people loathe are denounced as "the foetid and putrid." Perhaps the various kinds of discrimination that exist among human beings are not far from this differentiation. These are things that people oppose on the surface, but perhaps they do not thoroughly examine in their deepest mind. The view put forward by *Zhuangzi* over 2000 years ago stated that:

> [T]he foetid and putrid is transformed again into the spirit-like and wonderful, and the spirit-like and wonderful is transformed again into the foetid and putrid (臭腐復化為神奇, 神奇復化為臭腐). (Zhuangzi 1891, p. n/a)

This statement is still thought provoking and communicates an important truth to our own days. The reason for this sentiment is that Zhuangzi recognized that everything in the universe is connected by one breath of life. Hence, the hierarchical thinking of human societies that stands on a bureaucratic system is at odds with this cosmic reality. Perhaps only the colonialists, both the old and the new, arrogantly retain this distorted idea.

In Zhuangzi's view, all living things, including human life and death, are also linked. The fallen leaves return to their roots and decay, and the organic matter then enters the workings of a new life. It is in this way that the quintessence of heaven and earth are transformed in all things, as God descends from heaven, and light comes forth from the earth. Only when the mind is open to this one breath of life, the body may return to its original honesty, integrity, and truthfulness.

Environmental philosophy could draw inspiration from the thoughts in *Zhuangzi*. A prevailing Western environmental philosophical discourse is based on the notion of individual rights. However, this line of thought—from human rights to animal rights, or even the rights of germs or bacteria—has its limitations.

Norwegian philosopher Arne Næss, who founded the school of deep ecology, criticized the discourse based on rights. Instead, Næss called for

> The idea of totality cannot be discounted … [as] … the entire cell unit's dynamics extend far beyond its observable boundary. … The cell walls are not independent of their surroundings—they are not walls in a commonsense way. We are dealing with an 'all-pervasive network' of forces and interactions. (Næss 1989, pp. 78–79)

Næss not only emphasized the presence of ecological normative awareness in any policy decision, but also acutely recognized the limitations of logical reasoning. He asked people to reflect on the disorganization between logical derivations and basic norms, "the road which should lead from the basic principles of democracy, freedom, justice, and well-being to concrete policies for the realization of these principles is as uncharted as a back-country track!" (Næss 1989, p. 77) This, of course, also creates more complex requirements for the future of education.

6.3 Confucianism

6.3.1 Qiequ *and* Zhiqu *(*竊曲與致曲*)*

Generally, Confucianism appears to give people a stereotypical impression because it places special emphasis on the primacy of etiquette and is almost cumbersome in its detailed provisions. As stated in the chapter of "Li Qi" in the book of *Li Ki* (also

Fig. 6.1 Qiequ Wen Ding (竊曲紋鼎, Curved Pattern Tripod), unearthed cultural artefacts of Western Zhou dynasty (1046 BC–771 BC), 23.9 cm in height and 23 cm in diameter. Inventory of National Museum of China, Open access at https://www.chnmuseum.cn/zp/zpml/csp/202008/t20200826_247388_wap.shtml

translated as *Classic of Rites*), "while the important rules are 300, and the smaller rules 3000, the result to which they all lead is one and the same" (經禮三百, 曲禮三千, 其致一也) (Confucianists 1885) extensively ranging from adult rites, weddings, and funerals to social, diplomatic etiquette. Convoluted and detailed instructions for the rites were given in different scenarios.

Confucius also stated in the "Yan Yuan" chapter in *Analects* that "to subdue one's self and return to propriety, is perfect virtue" (克己復禮為仁) (Confucius 1861). This requires people to comply with the various provisions of rites and etiquette, so that they can be perfected in the maturity of their subjectivity. This is the reason why we refer to China's Confucian-dominated cultural tradition as a tradition of ritual and music culture. In this regard, etiquette, in essence, is not merely an external guideline for behaviors or a spontaneous display of interior sentiments, but rather it runs through a major thread that leads to the Way of Heaven.

This main thread stems from the faith tradition of China since the Shang and Zhou dynasties (1600 BC–256 BC). Since the Shang and Zhou dynasties, China has worshipped divine dragons with two mouths, symbolizing the two major issues of human existence, life and death. In archaeological excavations, this belief is shown in the form of *kui* dragons (夔龙), also known as *Qiequ* patterns (竊曲紋) or curved patterns, which are found on a variety of bronzes (Fig. 6.1) and objects made of other materials. The basic feature of the *Qiequ* pattern is a horizontal S-, C-, or G-shape, which is in line with the characteristic that "the top and bottom are all curved" (Fig. 6.2).

No words can describe the *Qiequ* pattern. It is better to describe it in poetry, music, or dance, or even to interpret it in a religious/spiritual dimension. This is precisely the significance of music in the tradition of rites and music culture. How marvelous is a song involving life and death! Each individual's life journey is to contribute its own tune to this cosmic chorus, including humans and more-than-human beings. With regard to human communities and societies, the tune of the adult ceremony is the harmony of loveliness and solemnity. In the wedding

Fig. 6.2 Various *Qiequ* Patterns. (Image was adjusted based on Peng 2002)

ceremony, the tune is the transmission of responsibility and the harmony of two families. In social etiquette, it is the harmony of interactions between the self and the other. In funerals and burials, the tune is the solemn harmony between nurturing the living ones and sending off the dead.

This is what the chapter of "*Zhongyong*" (《中庸》, also translated as "Doctrine of the Mean") in the book of *Li Ki* (1885) refers to the term *Zhiqu* (致曲, literally fulfill the curved); i.e., to cultivate to the utmost the "shoots" of goodness. The maturity of every human being denotes maturity or subjectivity, which comes close to an apprehension of the sacred in the religious sense. *Qu*, the curved, as subtle matters and as twisted and curved as the *Qiequ* pattern, urges a Confucianist to humbly practice the smallest things, so that they can eventually harmonize with the spirit of heaven and earth.

6.3.2 The Subject and All Creatures 主體与萬物

From the above discussion we discover the theme of the relationship between subjectivity and the world, as revealed in Confucian doctrines. The establishment of subjectivity in traditional Confucianism is mainly achieved through the discourse of the gentleman. Unlike the Western tradition of establishing subjectivity, the Confucian teaching is based on the attainment of the rhythm and cadence of a cosmic song to experience the manifestation of this subjectivity; i.e., the attainment of the characters or virtues of a *Junzi* (君子, gentleman).

We need only to take one syntactical example to reveal this point. In the chapter of "Wei Zheng" in *Analects*, it is said, "A gentleman creates unity without taking sides; A petty person takes sides without creating unity" (君子周而不比, 小人比而不周) (Confucius 1861). It meant that a gentleman lived in harmony with the others, while a villain was good at forming cliques for personal gain. Syntactically, the word "with/out" deserves further deliberation.

That is to say, the role played by such a transitive word implies an interesting Confucian view on subjectivity. Like the rhythms in music and the transitions of bending and stretching in dance, the word "with/out" linguistically demonstrates the kind of beauty, or cosmic mystery, that is at work here. The establishment of the gentleman's character, or the question of Confucian subjectivity, is to labor and practice on this habitual point. This is where the mystery of Chinese ritual and music civilization for the cultivation of personal characters and integrity lies. In this kind of daily practice, the benevolent enjoy the mountains, the intelligent enjoy the water. From then on, the habitats, the landscape of mountains and rivers also feature a profound humanistic appeal.

This subjective awareness of the problem is neither embroiled in anthropocentrism nor in Darwinism. Rather, it is a way of presenting a beautiful movement in the choral melody of human beings and the natural world in praxis. As the French philosopher François Jullien puts it,

> By directly experiencing the fundamental 'and' of interpersonal harmony in every tiny opportunity, man [sic] stands high above the intuition of the harmony that is often active in the great operation of things; he realizes what is 'heaven'. (Jullien 2000)

This perception is right. This idea of the interconnectedness of all creatures might be expressed with words of French philosopher and activist Simone Weil, who in her book *The Need for Roots* (1955) stated that:

> Every human being needs to have multiple roots. It is necessary for him to draw well-nigh the whole of his moral, intellectual and spiritual life by way of the environment of which he forms a part. (quoted from Relph 1976, p. 38)

6.3.3 Cultivation and Evolution (教化與進化)

In Confucianism, the evolution of the world is a fact that must be confronted directly, and so is the evolution of social history. Confucius' attitude is that damage and benefit are facts that must be faced forthrightly. Moreover, in addition to respect towards evolution, human beings are actually entrusted with the duty of cultivation.

In his illuminating explanation of Aldo Leopold's land ethics, American environmental philosopher J. Baird Callicott (1980) also highlights this doctrine of evolution in a broader sense. When a gentleman follows the traditions of rites and music to improve his own characters, he is not only sublimating his own characters, but it also needs to be reminded that he is also carrying the mission of the world or the cosmos. He bears responsibility for the conservation and sublimation of the world. To use the terms from the Book of *Zhong Yong*, it is humans' mission to "assist the transforming and nourishing powers of Heaven and Earth." As described in the following quotes,

> It is only he who is possessed of the most complete sincerity that can exist under heaven, who can give its full development to his nature (唯天下至誠, 為能盡其性).
>
> Able to give its full development to his own nature, he can do the same to the nature of other men (能盡其性, 則能盡人之性).
>
> Able to give its full development to the nature of other men, he can give their full development to the natures of animals and things (能盡人之性, 則能盡物之性).
>
> Able to give their full development to the natures of creatures and things, he can assist the transforming and nourishing powers of Heaven and Earth (能盡物之性, 則可以贊天地之化育).
>
> Able to assist the transforming and nourishing powers of Heaven and Earth, he may with Heaven and Earth form a tension. 可以贊天地之化育, 則可以與天地參矣。(Confucianists 1885)

To put it simply, this mission of participation in nurturing is to intervene in the natural world in a way that is consistent with the life course of living beings, that is, from the height of genesis, in a restrained manner, and in a way that is also applicable to regulating the rhythm of the development of human societies. When spring comes to the earth, all creatures begin to revive, birds and beasts begin to multiply, and we as human beings should not interrupt this rhythm. To safeguard the integrity of the biotic habitats, it is also imperative not to over-exploit the mountains, forests and marshlands. When it comes to summer, fall, and winter, human beings should also engage in activities based on the corresponding seasonal laws. Such practices or habits, from a religious perspective, are to guard the tranquility of the God of the various habitats on our planet. This concept converges with the perspective of biocultural calendars that highlight the attunement between biological and cultural seasonality (Rozzi et al. 2023b).

As stated by German philosopher Martin Heidegger (1977), a forest has a forest's crux, and a habitat has a habitat's locus. Our countryside life has been implicitly telling us that this is the case since childhood. When village hills are occupied by monocultural economic forests, the fun of childhood trips into the mountains has completely disappeared. Gone are the old pine trees, and the child excitement of gathering circling mushrooms underneath them. The hidden places deep in the bushes under the diverse trees have vanished, and wild animals such as wolves have lost their hiding places, and with them the sense of wonder. Accompanying the decline of these things was the fall of Liaozhai Zhiyi (聊齋), or the *Strange Tales from Liaozhai*, which had been popular since the old days. Alas, tediousness began to well up in our heads.

6.4 Chinese Buddhism

6.4.1 World and Dharma

Chinese Zen Buddhism is the greatest fruit of the "Chineseization" of Indian Buddhism, and it has profoundly affected the spirituality of the people (Song 2012). There is a well-known saying, "Thriving green bamboos are all Dharmakaya,

Fig. 6.3 Full painting and partial close-up of *Geese Alighting on a Sandbank* (平沙落雁圖). By monk Mu-Xi (牧溪, act. thirteenth century C.E.), ink on paper; h. 33.1 cm, w. 110.1 cm. China. Southern Song dynasty. (Collected in Idemitsu Museum of Arts, Tokyo)

the luxuriant yellow flowers are all Prajna (wisdom)" (青青翠竹盡是法身，鬱鬱黄花無非般若) (Ding 2015). The verdant bamboo and yellow flowers are a representation of the beautiful environment in which the Zen monks reside, and the breath of the wilderness is vivid. Over the years of spiritual enquiry into the universe and human life, the Zen monk's remarkable skill lies in sublimating the concept of the world suggested by the surrounding environment into a realm of dharma, as opposed to the worldview of the common people. Every blade of grass, wood, bird, and insect in the living environment seems to be enshrouded in a dense atmosphere. This is certainly not a subjective imagination like daydreaming, but an epiphany that goes straight to the heart.

Viewing landscape paintings of the Song and Yuan dynasties is akin to viewing Impressionist paintings, because we do not try to grasp a precise representation of the landscapes. In contrast, the pictures of a landscape generated with a camera may suggest the impression that everything is clear. However, this photographic viewing may become the greatest concealment of the world's connotation. In landscape paintings, especially on the vast rivers and lakes, or in the deep valleys, the large blank spaces are where the true meaning resides (Fig. 6.3). The same is true of the visionary fantasy of oil paintings (Fig. 6.4). In these Zen and Western works of art, the world is not dispelled, but rather the charm is intensified.

Let's borrow the terminology of the Hua-yen School of Chinese Buddhism, which claims that the world is transformed into the Dharma Realm. The Zen monks appreciated this profoundly; the landscape painters of the Song and Yuan dynasties perceived this; and the Impressionist painters presumably realized this as well. There's nothing odd about this; the inability to appreciate this is strange. Technology is constantly evolving, and photographic techniques seem to be becoming more and more realistic in reproducing the phenomenal world. However, the more realistic or high resolution the images are, the more corrosive they seem to be to the world's connotations. What I mean is that the authentic metaphysical appeal and the obscure aesthetic pregnancies of nature are ruined.

This affinity towards nature, or towards the world, is not the result of deliberate behavior. Sacredness does not lie in loftiness. It may seem strange that the sacred resides within the mundane, or even that the mundane in the eyes of high intellectuals is the sacred. Why is it so bizarre? Those who are familiar with Chinese culture are often intrigued by the classical novel *A Dream of Red Mansions* (紅樓夢), which

Fig. 6.4 *Impression, Sunrise* by Claude Monet (1872), oil on canvas, h. 48 cm; w. 63 cm. Musée Marmottan Monet, Paris. (Source: Wiki Commons)

teaches us that self-exaltation is an illusion, and that the real step towards the sacred may be secularization. For the whole mystery of the salvation of people and worlds may well be contained in this.

6.4.2 *Redemption and Metaphysics*

When it comes to redemption, Chinese Buddhism presents itself as a question of *Buddha-dhātu* (佛性). Just as in the Judeo-Christian cultural tradition the question of original sin is at the core, in Buddhism the focus is on *Avidyā* (無明, ignorance). The gist of Mahayana Buddhism is that the sin of the self always exists if the other has not been redeemed. The enlightenment or liberation advocated by Buddhism is inseparable from others, and likewise, from the true awakening of the self. The redemption of Chinese Mahayana Buddhism thus attempts to address a metaphysical ambivalence of both self-related and other-related subjectivity. Therefore, it is necessary to be aware that in the case of *Buddha-dhātu*, just as in the case of original sin, we cannot talk about it in a dichotomic sense, such as existence or non-existence, presence or absence, being or nothingness, *you* (有) or *wu* (無), etc. To realize this is the birth of the wisdom of *Śūnyatā* (空性, emptiness).

Looking at the problem from the perspective of existence or non-existence presents a kind of metaphysical inertia. In the view of Vaisravana and Huayan

Buddhism, this metaphysical inertia is transformed into the insight that if one wishes to enter the Dharma realm, the central problem is to remove the obstacle of *jñeyāvaraṇa* (所知障, knowledge barriers). What is *jñeyāvaraṇa?* It is in fact manifested as an awareness of objectification in an extremely subtle sense. Human beings are accustomed to the superficial arguments of theoretical or cognitive rationality, and when people encounter the moment of antinomies, the argument must be reverted to the starting point. To terminate such Sisyphean destiny, all things need to be transformed. To be specific, everything around us, including other humans, other-than-human beings, the symbiotic habitats, serves as an interdependent and reciprocal redeemer. If humans venerate the things in our environment, these things will return the favor and become the awakener of human wisdom on the religious dimension. Everything that appears to be impermanent on the surface thus takes on the substance of true permanence, and this is the complete realization of "the Dharma Realm of One Truth Domain" (一真法界). The process of practicing this is of course prolonged and arduous and cannot be discussed in detail here.

6.5 Concluding Remarks

The human mind often holds a reductionistic latent position, a phenomenon that has consistently existed in theology and philosophy. Contemporary Chinese culture has also been affected by this phenomenon. When people, especially the ruling class, are unable to escape the impact of this reductionist mindset, they generate a brutal ethical problem because "the value of capital is placed above the value of the life of "the rest" who represents the vast majority of human and non-human beings" (Rozzi 2015a, p. 125). As we witness in contemporary global society, the roots of all problems are reduced to the problem of capital, which becomes the ultimate surplus. Rozzi's (2018) account reveals how the reduction of philosophical education and ethical examination have fostered global biocultural homogenization. He has criticized that the broad archaic meaning of the Greek word *ethos*, from which the word ethics derives, has been lost (Rozzi 2013). This loss has led to a cultural and ethical degradation of the human condition (Rozzi 2018). The innate rich human subjectivity of both individuals and global society has been lost, and the alienation of humans from nature and place has become one of its main manifestations. People have not only become place-less people, rootless sojourners, but, more frighteningly, become people perpetually unrested and flustered. This is a dreadful prospect that is leading global society into a Necrocene, an era in which death, oppression, and extinction of biological and cultural diversity prevail (Rozzi 2023, p. 315).

Fortunately, there is a thriving revival of traditional cultures in contemporary Chinese society in which many people find skeptical and criticize the superficiality of the life habits and injustice caused by the rapidly expanding cultural and economic patterns of global society. Various scholars around the world have also identified and criticized these problematic trends. Many Chinese and international scholars argue in favor of wisdom rooted in traditional knowledge. This rich cultural heritage, which has developed over centuries cannot be arbitrarily suddenly dismissed.

With an ecumenic perspective, Rozzi (2015b) examines Pope Francis' encyclical Laudato Si (2015) and other ecotheological concepts and movements to affirm that, if humans are to retain dignity, society should not trespass, or exert tyranny on the sacred places and habitats that human beings share with other-than-human beings. He underscores that the Argentine-born Pope Francis took his name from Francis of Assisi and in his Laudato Si encyclical expressed that:

> "One of the greatest challenges of our time is to convert ourselves to a type of development that knows how to respect creation …when I look at America, also my own homeland [South America], so many forests, all cut, that have become lands… that can no longer give life. *This is our sin, exploiting the Earth and not allowing her to give us what she has within her*" (Pope Francis 2014). Saint Francis is not a frozen historical figure but continues to inspire a contemporary, holistic Earth stewardship, grounded in reality rather than dogma. (Rozzi 2015b, p. 102)

Here, what I underscore is that an essential prerequisite in understanding interspecies kinship is to embrace the biocultural conditions of being human. Rozzi analyzes, through the 3Hs model (Habitats, life Habits, and co-in-Habitants) of the biocultural ethic the work by Brazilian Catholic liberation theologian and philosopher Leonardo Boff, to disclose innovative religious and cultural paths that could help transform the current era of the Necrocene (dominated by death) into a new era of a Biocene, "an age of life, encompassing its biological and cultural diversity" (Rozzi 2023, p. 315). Elaborating further on the term Necrocene, originally coined by American environmental historian Justin McBrien (2016), Rozzi's introduction of the concept of a Biocene comes with an ethical plea to co-inhabit with other people, other cultures, and other species. Rozzi's call for inter-cultural and inter-species coinhabitation is deeply anchored in the etymological root of the word human derived from the Latin word *humus*. To recover this biocultural significance of the word human, Rozzi (2023) draws not only on theological concepts and texts, but also on historical precedents from the Romans and contemporary biogeochemistry that scientifically demonstrate that human beings are composed of the same elements as all other beings with whom we share planet Earth and the cosmos. In this chapter, an examination of Chinese Taoism, Confucianism, and Buddhism thoughts suggests that ancient Chinese philosophers presented commensurate ecological wisdom in their consideration of the human condition. This concise review enables us to open a dialogue between Chinese traditions and Rozzi's biocultural ethics as well as other religious and philosophical, ancient and contemporary, core ethical concepts of Western traditions. Considering these biocultural and Western concepts in the light of Chinese philosophical discourses from Taoism ("All under the sky there is one breath of life"), Confucianism ("the oneness of all creatures"), and Chinese Buddhism ("Buddha-dhātu" of all living things), we can identify remarkable convergences. When Taoism senses that all beings share the same breath, when Confucianism connects human destiny with the destiny of other-than-human beings as one, and when Chinese Buddhism affirms that all things are redeemable, are these not the most resolute response to an interspecies kinship that stimulate an urgent call to peaceful and respectful co-inhabitation?

References

Callicott JB (1980) Animal liberation: a triangular affair. Environ Ethics 2(4):311–338

Confucianists. 《禮記》. English edition: Confucianists (1885) The Li Ki (Trans. Legge J). In: Chinese text project. Available via https://ctext.org/liji. Accessed 1 Oct 2023

Confucius. 《論語》. English edition: Confucius (1861) Analects (Trans. Legge J). In: Chinese text project. Available at https://ctext.org/analects. Accessed 1 Oct 2023

Ding F (2015) 《佛學大辭典》(A dictionary of Chinese Buddhist terms). Shanghai Bookstore, Shanghai

Francis (2014) Exploitation of nature sin of our time. The Huffington Post USA. http://www.huffingtonpost.com/2014/07/05/pope-francis-nature-environment-sin-_n_5559631.html. Accessed 5 July 2014

Francis (2015) Laudato si'. Encyclical letter on care for our common home. Libreria Editrice Vaticana, Vatican City

Hargrove EC (1989) Foundations of environmental ethics. Environmental Ethics Books, Denton

Heidegger M (1977) The end of philosophy and the task of thinking. In: Heidegger M, Heidegger M (eds) Basic writings: from Being and time (1927) to The task of thinking (1964). HarperCollins Publishers, New York, pp 427–449

Jullien F (1995) Le Détour et l'accès: Stratégies du sens en Chine, en Grèce. English edition: Jullien F (2000) Detour and access: strategies of meaning in China and Greece (Trans. Hawkes S). Zone Books, New York

Kim J (1993) Supervenience and mind: selected philosophical essays. Cambridge University Press, New York

McBrien J (2016) Accumulating extinction: planetary catastrophism in the Necrocene. In: Anthropocene or Capitalocene? Nature, history, and the crisis of capitalism. PM Press, Oakland, pp 116–137

Næss A (1989) Ecology, community and lifestyle: outline of an ecosophy. Cambridge University Press, New York

Peng Y (2002) Study on the Western Zhou bronzes with curved Qiequ patterns (西周青銅器竊曲紋研究). J Archaeol (考古學報) 4:421–436

Relph E (1976) Place and placelessness. Pion, London

Rozzi R (2013) Biocultural ethics: from biocultural homogenization toward biocultural conservation. In: Rozzi R, Pickett ST, Palmer C et al (eds) Linking ecology and ethics for a changing world: values, philosophy, and action, Ecology and ethics, vol 1. Springer, Dordrecht, pp 9–32. https://doi.org/10.1007/978-94-007-7470-4_2

Rozzi R (2015a) Earth stewardship and the biocultural ethic: Latin American perspectives. In: Rozzi R, Chapin FS, Callicott JB, Pickett STA, Power ME, Armesto JJ, May RH Jr (eds) Earth stewardship: linking ecology and ethics in theory and practice, Ecology and ethics, vol 2. Springer, Cham, pp 87–112. https://doi.org/10.1007/978-3-319-12133-8_8

Rozzi R (2015b) Implications of the biocultural ethic for earth stewardship. In: Rozzi R, Chapin FS III, Callicott JB, Pickett STA, Power ME, Armesto JJ, May RH Jr (eds) Earth stewardship: linking ecology and ethics in theory and practice, Ecology and ethics, vol 2. Springer, Cham, pp 113–136. https://doi.org/10.1007/978-3-319-12133-8_9

Rozzi R (2018) Biocultural homogenization: a wicked problem in the anthropocene. In: Rozzi R, May RH, Chapin FS, Massardo F, Gavin MC et al (eds) From biocultural homogenization to biocultural conservation, Ecology and ethics, vol 3. Springer, Cham, pp 21–48. https://doi.org/10.1007/978-3-319-99513-7_2

Rozzi R (2023a) Humans are humus: an analysis of boff's panentheistic ecotheology in the framework of the biocultural ethic. In: Luca V (ed) Pantheism and ecology: cosmological, philosophical, and theological perspectives. Springer, Cham, pp 313–329. https://doi.org/10.1007/978-3-031-40040-7_25

Rozzi R, Álvarez R, Castro V, Núñez D, Ojeda J, Tauro A, Massardo F (2023b) Biocultural calendars across four ethnolinguistic communities in southwestern South America. GeoHealth,7 (4):e2022GH000623. https://doi.org/10.1029/2022GH000623

Shields PR (1993) Logic and sin in the writings of Ludwig Wittgenstein. University of Chicago Press, New York

Song Y (2012) A study of the course of Chineseization of Buddhism. Shaanxi People's Press, Xi'an

Zhuangzi. 《莊子》. English edition: Zhuangzi (1891) Zhuangzi (Trans. Legge J). In: Chinese text project. Available via http://ctext.org/zhuangzi. Accessed 1 Oct 2023

Chapter 7
Plants, Poetry, and Painting: A Biocultural Encounter with the Chinese Four Gentlemen

Jun Hu and Panpan Yang

Abstract Current environmental degradation and the classroom-centric approach to art education raise significant pedagogical challenges. As art educators concerned with Chinese traditions, we aim to explore the connection between art and education regarding the harmonious relationship between humans and nature. Our study focuses on the integration of biocultural conservation, restoration, and communication into art education. In Chinese traditions such as Confucianism and Taoism, the natural environment is considered the source of truth. It symbolizes the unity of Heaven and humankind. Drawing from this philosophy, we aim to reconnect students with the natural environment by cultivating Chinese aesthetics and worldview through plants-inspired art education. We envision that art classes in primary and secondary schools can play a relevant role for appreciating biocultural heritage. The Chinese art concept of the "Four Gentlemen" (四君子) was adopted to convey Chinese ethical values, which could broaden and enrich current approaches to biocultural conservation, restoration, and communication.

Keywords Art education · Biocultural ethic · Confucianism · Four gentlemen · Taoism

7.1 Introduction

Since the end of the twentieth-century, rapid economic growth in China, industrialization and urbanization have led to a significant decline in rural population and an accelerated migration of younger generations to urban areas. This demographic shift poses serious challenges to the sustainable conservation, restoration, and education

J. Hu
Fine Arts School, Hangzhou Normal University, Hangzhou, China

P. Yang (✉)
Department of Art Education, University of North Texas, Denton, TX, USA
e-mail: Panpan.Yang@unt.edu

D. Zhu et al. (eds.), *Linking Arts with Biocultural Conservation, Restoration, and Communication*, Ecology and Ethics 8,
https://doi.org/10.1007/978-3-032-04830-1_7

of local and regional biocultural diversity, which has been deeply rooted in China's agrarian civilization over two millennia. Contemporarily, cultural diversity is threatened by the "hollowing out" and aging problems in China's rural regions, making it difficult for traditional values and cultural heritage to survive and to be passed down (Liu et al. 2021). At the same time, as rural schools are deserted due to severe shortage in enrollment, students gathered in urban schools are suffering serious "nature deficiency" problems due to physical distancing from nature, worsened by harsh academic competition enforced by compulsive examinations for quality schooling (Huang et al. 2020).

To raise students' awareness about the problem of biocultural homogenization (*sensu* Rozzi 2018), the classroom-centric approach of teaching is far from sufficient to significantly impact students. They need hands-on experience rather than knowledge alone to internalize the value of biocultural diversity (Rozzi 2023). They also need practice rather than subject teaching alone to prepare them to overcome narrow utilitarian attitudes toward nature and culture (Begel and Ko 2019). As Chinese art educators, we aim to explore the connection between art, nature, and people's well-being with inspirations drawn from ancient Chinese philosophy, such as Confucianism and Taoism. Chinese worldviews and aesthetics consider nature as the source of truth and wisdom, as well as the consequent pedagogy of "Nurture by Nature/Dao Fa Zi Ran" (道法自然) in art making, learning and teaching, which cherishes an aesthetics centering on the unity of humans and nature as "Human-and-Nature Being One/Tian Ren He Yi" (天人合一) (Hu 2022). We aim to reconnect urban students with other-than-human co-inhabitants and human villagers through nature-inspired art addressing the nature deficiency problem and intergenerational rural art education, respectively. To achieve this, we promote an arts-based educational research approach that intertwines nature and art, fostering contemporary art education that supports the conservation, restoration, and communication of biocultural diversity.

7.2 Traditional Chinese Aesthetics Examined Through a Biocultural Lens

Our practices over the past few years have been grounded in traditional Chinese philosophy and aesthetics. The 3Hs model of the biocultural ethic values the vital links among *co-in-Habitants* and their life *Habits* in their shared *Habitats* (Rozzi 2013). This biocultural model offers a valuable lens for reinterpreting traditional Chinese ethics in a more ecologically integrated way. For centuries, Chinese villages have embodied rich biocultural diversity. We maintain that this unconsciously reflects a harmony among the "3Hs" (co-inhabitants, habits, and habitats) as stated by biocultural ethics (Rozzi 2012). A prime example is the concept of Four Gentlemen, or the Four Noble Ones, a prominent theme in poetry and painting during the Song Dynasty (960–1279). It refers to four plants, namely, the plum blossom (梅),

Fig. 7.1 *Mei-lan-zhu-ju* (梅兰竹菊四屏, a series of paintings of the four gentlemen), 108 by 120 cm, ink on paper by Wu Changshuo (吴昌硕, 1844–1927) in 1920

orchid (蘭), bamboo (竹), and chrysanthemum (菊), which symbolizes Confucian *Junzi* (君子), or "Four Gentlemen/Treasures" (Chu 2023) (Fig. 7.1). When viewed through Rozzi's 3Hs lens, these plants offer a compelling illustration of biocultural conservation and the harmonious relationship between *Habits*, *Habitat*, and *co-in-Habitants*.

Given that the Four Gentlemen have been central subjects in traditional Chinese ink-and-wash painting for over a thousand years, and that they symbolize the moral characters, virtues, and spiritual attributes of Confucian *junzi*, the interconnectedness of human habits and these plants are evident. The Chinese plum, blooming in winter's coldest months, symbolizes resilience and inner beauty in harsh conditions (Hsü 1996). The orchid, graceful yet modest, stands for humility and nobility (Tay 2005). The bamboo, hollow yet strong, represents integrity and the moral flexibility to bend without breaking (Tay 2005). The chrysanthemum, flourishing in autumn's chill, symbolizes vitality and endurance (Welch 2013).

While these plants provide year-round natural beauty, they have become essential for gardening. For example, poet Lin Bu (林逋; 967–1028), a lifelong bachelor renowned for his verses on plum blossom, as the legend has it, treated plum trees as

Fig. 7.2 Left: *Yuanming Wearing Chrysanthemum* (渊明簪菊圖) by Chen Hongshou (1599–1652). Right: *Cai-ju* (采菊, Picking Chrysanthemum) by Chen Hongshou

his wife and cranes as his children (梅妻鶴子), for he lived in seclusion on Gushan Island in Hangzhou planting plum trees. Su Shi (苏轼; 1037–1101), a household poet and master of literature, wrote in his poem, "I would rather eat no meat than live without bamboo. Eating no meat makes one thin, but living without bamboo makes one vulgar" (《於潛僧綠筠軒》寧可食無肉, 不可居無竹。無肉令人瘦, 無竹令人俗。). Yuanming Tao (陶渊明, 365–427), a literary scholar famous for his seclusion in the countryside, wrote a renowned poem that states, "I pluck chrysanthemums under the eastern hedge, and gaze afar towards the southern mountains" (《飲酒·其五》採菊東籬下, 悠然見南山). Since then, the chrysanthemum has been regarded as the symbol of hermits and idyllic life.

The aesthetic appreciation of the four types of plants leads to the appreciation of moral values of Chinese literati and of the intraspecific variation in these four types of plants. For example, plants represented by chrysanthemums contain profound cultural connotations and they also demonstrate the richness of biodiversity in the process of long-term artificial cultivation. This integration is illustrated, for instance, by Chrysanthemum that blooms in bright colors during chilly autumn, a time when most flowers wither, and wearing these flowers in the hair became a custom and might have contributed to breeding the chrysanthemum since the Song dynasty (960–1279) (Way 2020). They also became an ideal and a common choice for celebrating festivals and meetings in autumn. During the Chongyang Festival (重陽節), also known as the Double Ninth Festival, which dedicated to the respect and care for seniors in Chinese tradition, people would climb mountains, drink Chrysanthemum wine, and wear the Chrysanthemum blossom in their hair as symbols of warding off evil and wishing for longevity. These habits and customs have not only been documented in essays and poems of that period but also illustrated in master paintings, including "Yuanming Wearing Chrysanthemum" (《渊明簪菊圖》by Chen Hongshou) (陈洪绶, 1599–1652) (Fig. 7.2). The widespread use of Chrysanthemum during the Chongyang Festival and other occasions has contributed to the conservation and growth of its biodiversity through intentional cultivation, breeding, and management. At least 3000 varieties have been cultivated in China, making annual Chrysanthemum festivals in cities, including Beijing, stunning spectacles (Zeng 2021).

Another example is based on the high diversity of bamboo species in China. The 762 bamboo species in China make up more than half of the world's bamboo species (Shi et al. 2020). Since the Song dynasty, people's aesthetic appreciation of bamboo has coincided with economic utilization. Ou-yang Xiu (欧阳修, 1007–1072) presented to the emperor an essay "Mourning on Bamboo Cutting" (戕竹记) on the necessity of the preserving bamboo forests around the capital city of Luoyang, in which he highly praised the economic, landscape, gardening, aesthetic, and emotional values bamboo can offer to people. This way of co-inhabitation between human and bamboo manifests "a unity between human and nature" (天人合一). This biocultural intimate relationship has probably set the ground for contemporary bamboo-related policy and environmental practices in China. On November 7, 2022, the Chinese government and the International Bamboo and Vine Organization officially initiated the action of "Replacing Plastics with Bamboo" (以竹代塑). In this regard, bamboo not only cultivated the moral characters of Ancient Confucian scholars but also has begun to play significant roles in the contemporary national goal of building an ecological civilization in China, to foster technological innovations for building a clean, affluent, beautiful, and sustainable home for future generations.

7.3 Two Relevant Problems After Our Biocultural Examination

Biocultural ethics offers a unique lens for understanding how distinct habits, habitats, and the co-inhabitants' interrelationships in urban and rural areas shape the ethical, cultural, and environmental dynamics within these spaces. In response to two problems in post-reformation China, namely, the problem of nature deficiency of urban children and the hollowing out problem in villages, the framework of biocultural ethics informs that these two problems are indeed caused by the dis-connection among co-inhabitants, habits, and habitats (Rozzi 2012).

Strictly speaking, human residents cannot completely detach from their habitats, and their daily lifestyles are shaped by environmental and behavioral patterns. From a biocultural ethics perspective, it is more likely that the core problem is that people do not recognize the connection between themselves and their environment (including other-than-human organisms and the ecological links among them) (Rozzi 2018). Moreover, the relationship between most people and their environment has been weakened or distorted by a surge of global biocultural homogenization (Rozzi 2018). Habits in cities tend to prioritize efficiency, productivity, and convenience, which fit in the urban habitats that are characterized by high population density, infrastructure, and concrete jungles, which has strong impacts on human and other-than-human co-inhabitants due to the scarcity of natural elements and limited green spaces, consequently having negative effects on biodiversity and ecological connections (Grimm et al. 2008). Urban children seldom have enough green space or companionship with other species within an urban setting, which in turn may weaken ecological awareness. Although concerns about the inadequate contact of children and citizens with natural settings were raised in the late twenty-century (Moore 1997), the situation has not changed in the twenty-first century.

Similarly, in the issue of rural hollowing out, the migration of young people to cities disrupts the village's social structure and cultural heritage. Losses in rural population alter people's connection with their local environment, leading to a decline in traditional agricultural and cultural activities. Villages are gradually abandoned or transformed. In rural communities, social ties are weakened while interpersonal alienation is increased. Rural habitats are typically abundant in natural landscapes, including farmlands, forests, rivers, and open spaces that support a more diverse range of flora and fauna, providing essential ecosystems services and a closer connection to nature. However, rural habitats are being deserted with a demography that has shifted toward a dominance of urban populations and new life habits (Pannell 2003).

Before globalization, residents in rural settings often engaged in agricultural activities, experienced a slower pace of life, and maintained close ties within a society structured around acquaintances and kinship. The dramatic change after globalization raises new questions. For example, can rural life, with its strong human-nature connections, evolve to incorporate the richness of urban social and technological engagement without compromising its ecological intimacy? Some scholars suggest that bridging these worlds—through sustainable rural development, digital connectivity, and eco-centered community planning—can allow for a synthesis of both lifestyles, fostering a form of modern rurality that supports human and non-human kinship while benefiting from global networks (Woods 2007).

Since the end of the twenty-century, In China, younger generations are adopting urban habits and distancing themselves from rural traditions. However, at the end of the second decade of the twenty-first century, there has been a growing interest among the young generation in China in returning to a slower-paced, rural lifestyle, driven by factors such as urban burnout, digital entrepreneurship opportunities in rural areas, and government initiatives supporting rural revitalization (Wang 2023). While it remains to be seen whether this movement is substantial enough to counterbalance urban migration, it highlights a more complex relationship between youth and rural habitats.

The causes of the above two problems are disclosed by introducing the 3Hs model of the biocultural ethic proposed by Chilean philosopher and conservationist Ricardo Rozzi (2013) (Fig. 7.3). The interior triangle would reveal that urban habits of children are detached from their habitats or more-than-human beings in their environment, causing nature deficiency. The intermediate triangle helps to demonstrate that rural habitats are marginalized by the institutional-sociopolitical-technological domain dominated by urban habits, leading to hollowing out problems in

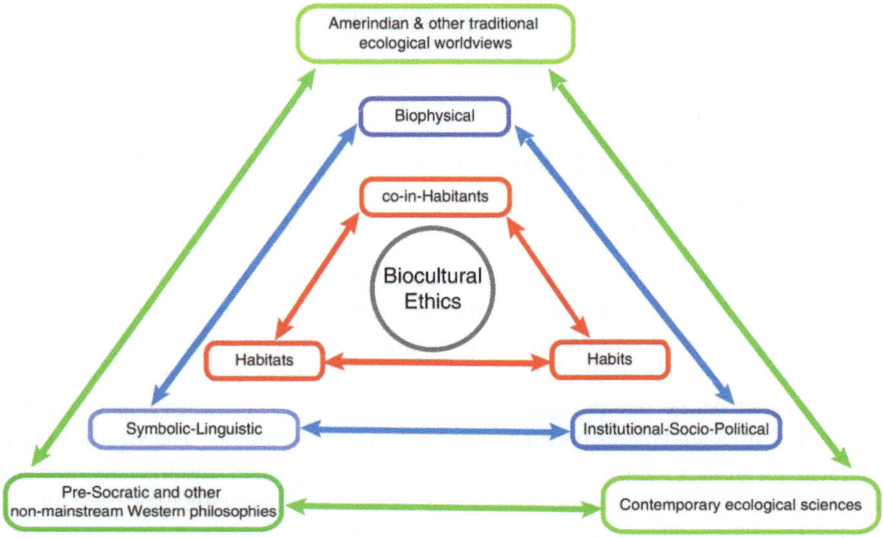

Fig. 7.3 Formal proposal of the biocultural ethic illustrated by a combination of three inclusive triangles. Interior triangle: The "3Hs": co-in-Habitants, their life Habits, and shared Habitats. Intermediate triangle: The 3Hs' interrelationships involve biophysical, symbolic-linguistic, and institutional-sociopolitical-technological domains. Outside triangle: The 3Hs' interrelationships are understood by three families of worldviews: (i) pre-Socratic and other non-mainstream Western philosophies, (ii) Native American and other non-Western ecological worldviews, and (iii) contemporary ecological-evolutionary sciences. (Adapted from Rozzi 2013)

villages. The outermost triangle explains the absence of traditional Chinese ecological worldviews, indicating that the philosophy of "the unity between humans and nature" in contemporary Chinese society needs to be brought back.

7.4 Problem-Solving Proposals Based on Biocultural Ethics

Based on our observations and analyses, we propose some solutions to address the two problems raised above. In the biocultural ethic's model (Rozzi 2013), the outermost triangle, historical precedents should be identified that may inspire applicable solutions. Moreover, important links should be established between traditional ecological views, the government's ecological imperatives, and contemporary ecological sciences, thereby providing theoretical support for solving the above problems. In the intermediate triangle, dominance of urban habits in the institutional-social-political sphere should be challenged, leveraging policy to support institutionalized framing for initiatives. In the inner triangle, art-based research should be conducted to shape new habits that facilitate the conservation, restoration, and exchange of biocultural diversity through specific projects, such as the triad of

Plant Poetry Painting in our studies, and other initiatives in China's movement of rebuilding villages through arts (艺术乡建).

7.4.1 Re-connecting Urban Children with Rural Habitats and Plants

To address the Inner Triangle level of the 3Hs model (Fig. 7.3), we conduct art-based education in Plant Poetry and Painting by exposing urban children to natural surroundings. In rural art education, urban children come to the countryside where village elders become co-teachers, sharing their vast knowledge of the local environment, and native plants co-inhabit the area. In the Plant Poetry and Painting Program, native wild plants become art subjects and art mediums. These projects aim to raise awareness among urban children about neglected biodiversity and its potential for artistic expression. Under this initiative, urban habits and their drawbacks are addressed by providing urban children access to rural habitats and face-to-face contact with other-than-human plants, in a way that is consistent with field environmental philosophy methodological approach that puts the "3Hs" model of the biocultural ethic into action (Tauro et al. 2021; Rozzi 2023).

7.4.2 Reverse Power Structure in Education

To address the intermediate triangle of the biocultural ethic's 3Hs model (Fig. 7.3), we challenge the dominance of urbanized education in the institutional-socio-political sphere by training village elders as co-teachers to reverse the power structure in education. Industrialization and urbanization have marginalized rural areas, and subject-based education has trivialized the holistic knowledge possessed by village elders. Through our project "Intergenerational Rural Arts Education," village elders played an important role in the dissemination of knowledge about biocultural diversity. They have local knowledge that has been passed down for centuries and have proven its validity in their lives. The "Intergenerational Village Arts Education" project created a bridge between rural and urban areas for exploring the possibilities of villages as a major venue for biocultural education. Through this experimental effort, we challenged the existing institutional-socio-political framework by empowering villagers to share their voices in the conservation, restoration, and communication of biocultural diversity. This effort transforms lonely and desolate village elders into confident co-teachers who share their knowledge of natural and cultural heritage with the younger generation.

7.4.3 Links Between Chinese Traditional Aesthetics and Contemporary Ecological Sciences

To address the outermost triangular level of the biocultural ethic's 3Hs model (Fig. 7.3), we combine traditional Chinese aesthetics with artistic practice, integrating them into a worldview and practice based on the construction of an ecological civilization. This practice aims to link human habits with contemporary ecological science amid artistic creation, replacing dominant consumerism and utilitarianism with the value of beauty, thereby supporting human liberation and the establishment of a human-nature relationship that is more in line with biocultural ethics and traditional Chinese worldviews. Ancient Chinese traditions of b*onsai* (盆景), gardening, and ink-and-wash painting embody the principle of "Nurture-after-Nature/Dao Fa Zi Ran" (道法自然), emphasizing the following of the natural way and the pursuit of "unity of heaven and Humankind," i.e., the harmony between human beings and nature (Liu 2008; Tuan 1971). These artistic practices represent not only artistic styles, but also a way of life that celebrates relationships of co-inhabitation among humans and more-than-human beings. This makes a unique contribution to the conservation of the world's biocultural diversity.

Over the centuries, these traditions have had a significant impact on Chinese aesthetics. This influence has also extended to other East Asian countries, culminating in the Japanese concept of wabi-sabi (侘寂), which represents the beauty of imperfection, transience, and simplicity (Koren 2008). Wabi-sabi represents a unique form of biocultural identity (Berque and Schwartz 1997). Through wabi-sabi people recognize the importance of the non-existent and overcome the limits of merely utilitarian cognition and sensation (Fang et al. 2023). Thus, from this standpoint, the purpose of artistic creation is not to represent nature as it is, but to complicate the interrelationship between humans and nature, embracing poetic ambiguity. In this context, aesthetics does not merely please the eye with formal beauty but resists the reduction of the human-nature relationship to purely utilitarian concerns. Through poetic complexity, it offers new opportunities for the emergence of artistic biocultural diversity. In other words, creativity beyond ecological knowledge can be embraced by historical precedents such as the Four Gentlemen. This poetic complexity, deeply rooted in Zen Buddhism, values transience and imperfection, revealing beauty in the natural cycles of growth and decay and reflecting a profound appreciation of humility, simplicity, and imperfection.

In addition to artistic expression in Chinese b*onsai*, gardening, architectural decoration, embroidery, poetry, ink-and-wash painting, and music, the aesthetic value of the Four Gentlemen was also documented and transmitted by traditional Chinese intellectuals. The first book on bamboo appeared as early as the Southern Dynasty (420–589), and *Zhu Pu* (竹譜), or the Bamboo Genealogy, written by Dai Kaizhi (戴凱之), became a symbol of the culmination of bamboo culture. The book documented the characteristics, cultivation and maintenance methods of more than 70 types of bamboo, among which Linglong bamboo (铃笼竹) was listed as the first, as it can be used to make musical instruments. Regarding Chinese plum, Chengda

Fan (范成大, 1126–1193), a well-known poet during the Song Dynasty, devoted himself to the collection and breeding of plum species motivated by his fondness for plum, and compiled the world's first monograph on plum, *Mei Pu* (梅譜), or the *Genealogy of Plum*, at the age of 67. The compilations on orchids are even more numerous. The *Pu of Jinzhang Orchid* (金漳蘭譜, 1233) was compiled by Zhao Shigeng (趙時庚) in the Song Dynasty, which included 32 valuable orchids. The *Wang's Pu of Orchid* (王氏蘭譜, 1247) was compiled by Wang Guixue (王貴學) a little later, which included 50 types of orchids. The earliest extant *Liu's Pu of Chrysanthemum* (劉氏菊譜, 1104), written by Liu Meng (劉蒙) of the Song Dynasty, describes in detail the morphological characteristics of 35 types of Chrysanthemums popular at that time. These precedent works show that there exists a considerable amount of biocultural heritage in the traditional Chinese aesthetic culture that can be mined and studied for contemporary biocultural conservation and education.

It is worth noting that although these aesthetic concepts have profoundly influenced people, because of China's vast size and cultural diversity, the views are consequently rich and varied. Not all Chinese can identify with these aesthetic principles. Thus, it would be more accurate to say that these traditions have shaped the identity of many Chinese rather than emphasizing a monolithic cultural identity.

7.5 Implementation of Two Art Educational Projects

Two projects, "Plant Poetry and Painting" and "Intergenerational Rural Art Education" were introduced to address contemporary problems from a biocultural standpoint. Plant Poetry and Painting was launched in 2019 and concluded with a farm workshop at the 3rd Asian Art Photography Symposium at Hangzhou Normal University in China. It aimed to explore the issue of "nature deficit" and guide urban children to create artworks inspired by plants in the fields. "Intergenerational Rural Art Education" was launched in 2021 and included in the "100 Projects for Building a Better China" of the China Academy of Arts. The results of this project were presented at the "Song of the Earth: A Chronicle of a Beautiful China 2023", held from February 28 to March 27, 2023, at the China Millennium Art Museum in Beijing. Since July 2023, it has been on permanent exhibition at the Beautiful China Learning Center of the Central Academy of Culture and Tourism Management of the Central State Administration.

7.5.1 Plant Poetry and Painting in Fenghuangxi

In the Plant Poetry and Painting project, we embraced the aesthetic tradition of the Four Gentlemen and creatively anthropomorphized plants. During a one-day field trip to the village of Fenghuangxi, 70 kilometers from downtown Hangzhou, the

Fig. 7.4 Students walking with closed eyes and touching plants. (Photograph by Jun Hu)

students were guided to explore the unique traits and beauty of plants such as rosemary, perilla, parsley, mugwort, clover, and ramie. While exploring the diversity of plants, students each wrote a riddle poem about an unknown plant and created colorful prints on canvas with pigments extracted from plants in the local habitat, based on the poem's theme. The expectation is not that the children would gain systematic knowledge about biodiversity or ecosystems during the one-day field trip, but that their awareness of biodiversity and its significance would be awakened through face-to-face contact with these plants and appreciation of their aesthetic values.

In this program, embodied cognition replaced didactic lectures. Students first walked with closed eyes in the field to unlock their way of experiencing nature through other senses, thus stimulating curiosity to further explore the unknown biodiversity. After closing their eyes, students would search for ten different plants by touch and smell in a designated area (Fig. 7.4). As they were amazed by the previously unnoticed diversity of plants, touch and smell enhanced their embodied knowledge, which prepared them to select plants as their artistic subjects for later poetic and artistic endeavors. When students had accumulated a great deal of embodied knowledge by comparing and classifying the shapes, colors, textures, and smells of unknown plants, they then searched for the names of the plants and learned the associated botanical knowledge, metaphorical meanings, and other cultural connotations through online resources. Equipped with such information,

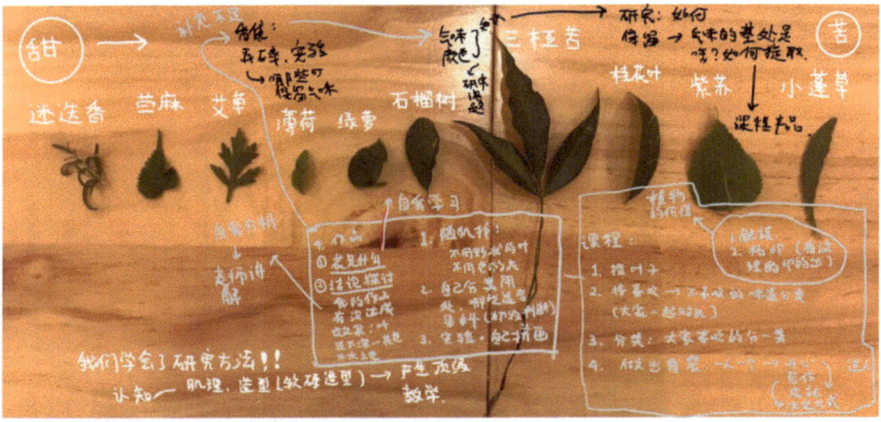

Fig. 7.5 Students classify collected plant leaves and explain what they have learned

Fig. 7.6 Interactive Art Exhibition: Students present their artworks and interpret them

they then selected the plants they were interested in, created riddle poems, and printed the riddle poems with metaphorical imagery on canvas (Fig. 7.5). The riddle poems would then be displayed in an interactive art exhibition where visitors would be invited to a riddle party (Fig. 7.6).

The Plant Poetry and Painting project was an endeavor in response to the national policy, "Opinions of State Department of Education on Promoting Field Trips for Primary and Secondary School Students," issued by 11 state departments in November 2016, which required all schools to arrange 3–5 days of extracurricular activities for each student every school year. The farm where the project was conducted had previously provided outdoor hands-on activities for school children, but these activities were mostly utility-oriented, with labor attitudes, discipline and self-control being the focus of learning. Due to the lack of wonder in the learning process, students' attitudes towards the environment had not changed much.

However, the "Plant Poetry and Painting" approach goes beyond utility to create a deep and complex interrelationship between arts, environment, habits, and learning, generating aesthetic value that exceeds mere functionality. In the field activities, artistic creativity is intertwined with plant diversity, and the essentials of the Four Gentlemen are reinterpreted. While creating riddle poems, a new bond was formed between students and plants, truly revealing "the unity of heaven and humans" (Roetz 2013). This phenomenological approach, reviving the Chinese tradition of the Four Gentlemen in a contemporary context, explored the diversity of plant life. Plants as co-inhabitants and their scientific and aesthetic value were recognized by students through embodied cognition and artistic engagement. Further, field activities of Plant Poetry and Painting also offered chances for reshaping the habits of urban children to make new artistic, cognitive, and ethical connections with their habitats and the more-than-human creatures within the habitats, such as plants.

7.5.2 Intergenerational Rural Art Education in Songxi

Songxi (嵩溪, literally "mountain stream"), a 1000-year-old village in Zhejiang province in the Southeast China, exemplifies a significant national as well as East Asian problem in recent decades known as "hollowing out," caused by aging and rapid urbanization due to fast industrialization (Pagano 2018). The village has won many national honors, such as the title of National Ecological and Cultural Village. However, the population in Songxi has declined from approximately 3000 to 500, as the labor force, together with their school-age children, migrated from rural areas to work in towns and cities, leaving behind only the elderly and the disabled (Pagano 2018). Most of the ancient buildings are vacant or crumbling (Fig. 7.7). Due to the absence of the younger generation, the village's economy and cultural heritage is in jeopardy. If no intervention is taken, the village will eventually die out within a generation as the population continues to decrease.

In 2021, the research team of the Art Education Program at Hangzhou Normal University conducted a comprehensive study of the village and proposed that intergenerational art education could be a viable solution. By creating a new interface between urban and rural areas, it would train village elders to become "co-teachers" in the arts program so that their knowledge of biocultural diversity could be passed on to urban residents, especially urban children. The rationale for the

Fig. 7.7 Old residential buildings in Songxi village, Zhejiang. (Photographs by Panpan Yang)

project is that rural elders possess valuable knowledge of traditional crafts, ecological practices, folklore, and knowledge of rural aesthetics—forms of biocultural diversity that are increasingly at risk of vanishing. By integrating this local knowledge into art education, the project not only ensures the conservation of Songxi, but also makes its values accessible to urban dwellers, especially children. Village elders act as "co-teachers", sharing their life experiences and wisdom through interactive workshops, storytelling, and hands-on art practices.

Located 1.5 h drive away from Hangzhou, the capital of Zhejiang Province, Songxi Village features traditional architecture, natural beauty, agriculture, and topography that are distinctly in contrast to that of urban areas, manifesting a harmonious and sustainable relationship between people and their environment. The ancient dwellings are staggered, with the orientation and appearance of the houses adapting to the topography of the area. Two streams run through the village—one is the "An Stream" (暗渠, undercover stream), and the other is the "Ming Stream" (明渠, open-air) (Fig. 7.8). The "An Stream" is more than 6 kilometers long, forming a spacious and narrow arch-shaped tunnel. In addition to its function of flood control and drainage, the dark stream has also become a natural summer resort. Gurgling streams add vitality and energy to Songxi, which is surrounded by hills and lush greenery. The natural scenery created by terraced rice fields, bamboo forests, meandering streams, farmland and traditional houses reveals that people have been living here for generations, shaping the environment and habitat modestly, and developing a strong connection with the local nature through traditional agriculture as well as a number of other practices.

Fig. 7.8 Left: "An Stream" (暗渠, undercover stream). Right: "Ming Stream" (明渠, open-air stream)

The beautiful environment of Songxi Village has also nurtured the cultural and artistic literacy of the villagers, who were able to immerse themselves in the practice of traditional arts, such as opera, poetry, bonsai, gardening, calligraphy, and ink-and-wash painting, etc. For villagers, they would feel ashamed if the calligraphy and accompanying poems hanging in their homes were written by someone else rather than family members. The appreciation of the Four Gentlemen has been sustained even after the village has declined, making the village a precious example of China's aesthetic traditions and biocultural diversity. Many families cultivate Bonsai in their own yards, creating miniature landscapes/gardens in their homes with carefully selected local wild plants. Living in this poetic environment yet rich in cultural traditions, many of the villagers are skilled amateur flower and bird painters, some of them with skills that rival those of professional painters.

Based on the thriving tradition of calligraphy and painting in the village, the Intergenerational Rural Art Program strived to train rural villagers to become "co-educators", passing on knowledge of nature and cultural heritage to urban children. For example, Lufu Shao (邵陆甫), a farmer and ink painter in the Songxi village, is remarkably skillful at creating vivid and detailed paintings of radishes with just a few brushstrokes. This inspired a "boneless" (no outline) ink art program featuring radishes (Fig. 7.9). This brush painting technique was very unfamiliar to urban children who are accustomed to using hard-edge drawing tools such as pencils and crayons as well as outline sketching. The introduction of traditional ink painting techniques and the aesthetics of brushstrokes on soft rice paper expanded the students' understanding of the arts. Meanwhile, radish is a daily seasonal ingredient in the fall and winter seasons in Songxi. It is believed to help promote digestion and strengthen the immune system, hence the Chinese proverb "Eat radish in winter. Eat ginger in summer. You will keep doctors away" (Lou et al. 2017). Local villagers would make radish pickle, sun-dried or stewed radishes to fulfill their dietary needs in winter. In addition, radish cultivation is usually combined with traditional crop

Fig. 7.9 Left: Radish (萝卜), ink on paper by village artist Lufu Shao (邵陆甫). Right: Shao teaching children about the "Boneless" (no outline) techniques of ink painting. (Photographs by Jun Hu)

rotation or intercropping systems, which help maintain soil fertility balance and reduce pests and diseases.

Thus, radish is not only a food component but also represents local farming wisdom and sustainable ecological practices. Combining the seasonality of radishes with the art program not only introduced students to brushstroke techniques but also prepared them to appreciate the natural rhythms of agriculture and deepen their gratitude for the abundant supply of fresh radishes in the land. To master this skill of boneless painting, children had to abandon the habit of outline drawing and engage in activities such as harvesting, preparing, and eating radishes in order to develop a specific experience of radishes by revitalizing all the senses beyond visual input. This minimalist brushwork resembles the traditional Chinese pedagogy of "Nurture by Nature," for the painting process should be as natural as the process of nature itself. In other words, the figurative painting of the radish comes out of expression rather than representation (Hu 2022).

In addition to traditional art courses, the project in Songxi also offered interdisciplinary courses to introduce students to overlooked plant species by picking native plants and making them into delicious "Fairy Tofu," a popular iced dessert in summer. During the field trip, students were accompanied by village elders, and they learned folk songs about the "Fairy Tofu" while picking the necessary ingredients (Fig. 7.10). The process of making the desserts was also filled with delight, especially when they were able to get their hands on the magic of turning the leaves into jelly. These novel experiences ignited students' curiosity about this previously uncharted plant, raised their awareness of the village's biocultural diversity, and helped them appreciate its potential contribution to people's well-being.

Fig. 7.10 Left: A local villager instructs children how to make the "Fairy Tofu" after their field trip. During the field trip, children identified and collected leaves of native plants. Right: Local instructor showing the children the "Fairy Tofu"

7.6 Concluding Remarks

We propose that arts education can play a relevant role to address major problems generated by industrialization, urbanization, and biocultural homogenization. These problems not only standardize human life habits but also disconnect people from their local environments. To support our proposal, in our chapter we have presented historical antecedents such as the Four Gentleman and contemporary education initiatives, examined through the 3Hs model of the biocultural ethic (Rozzi 2013). We have shown how creative arts education can rewire the interactions among the habits, habitat, and co-inhabitants by repairing the vital connections for ethical dwelling and the conservation, restoration, and communication of biocultural diversity.

In Plant Poetry and Painting, art education molds new habits of urban children, reconnecting them with the other-than-human co-inhabitants in the rural habitat, thereby alleviating the problem of nature deficits through multisensory learning about local plants. The acquisition of embodied knowledge revolutionizes formal education by incorporating emotions and body senses, such as touch and smell, which play a unique role in the learning process. The study of plant morphology and cultural metaphors lays the foundation for students' artistic creativity, such as painting and riddle-writing, and stimulates their ability in exploring the interaction between human and other-than-human beings, in this case, plant beings.

The Intergenerational Rural Art Education initiative in Songxi redefines human-nature interactions, shifting from an exploitative model to one of reciprocal care and cultural transmission. The project emphasizes exchanges between village elders and urban children to create new habits of co-inhabitation through intergenerational art education. Elders in villages have enjoyed the project because spending time with the urban children reduced the loneliness they felt due to the absence of their grandchildren in their daily life. Their participation in arts education also alleviated the sense of powerlessness they feel due to marginalization by modern societies and the absence of their offspring, who are unlikely to inherit their precious traditional knowledge and life skills. Urban children have been surprised by the richness of knowledge and skills they have learned from the elders. These children were astounded by the biocultural diversity they could enjoy in Songxi village (a diversity that was not present in their urban environments). By bridging generational and geographical divides, this project aims to offer a sustainable path to revitalize traditional knowledge and the village as well as to reconnect urban residents (students and their parents) and village elders to their habitats.

Going forward, these two initiatives could impart important biocultural messages to the global communities. These efforts show that reconnecting with local biocultural traditions brings vitality to marginalized communities around the world (communities that are highly impacted by processes of biocultural homogenization, Rozzi 2018). With the projects illustrated in our chapter, we offer evidence that we hope to encourage universities and local communities to experiment with institutionally-assisted changes in education. Although still in its inception phase, the "Shuimo Songxi (水墨嵩溪)—Intergenerational Rural Arts Education Program" has been selected as one of the first "Jiao Shi Da Ren" (教师达人, Expert of Pedagogy) rural community education innovation demonstration projects by the Ministry of Education of the People's Republic of China. This suggests that biocultural conservation, restoration, and communication in the local context will become a long-standing mission.

Acknowledgement The authors would like to thank Danqiong Zhu, Benn Johnson, and Ricardo Rozzi for their time in commenting and revising the manuscript.

References

Begel A, Ko AJ (2019) Learning outside the classroom. In: Fincher SA, Robins AV (eds) The Cambridge handbook of computing education research. Cambridge University Press, Cambridge, pp 749–772

Berque A, Schwartz R (1997) Japan: nature, artifice and Japanese culture. Pilkington Press Ltd, Northamptonshire

Chu Y (2023) From tradition to modernity: craftsmanship of four treasures of the study and the evolution and development of painting techniques. Doctoral dissertation, Syracuse University

Fang G, Fu Y, Peng L (2023) Wabi-sabi style: the collision of the east and west, the combination of the fashion and the nature. J Educ Humanit Soc Sci 8:2499–2505

Grimm NB, Faeth SH, Golubiewski NE et al (2008) Global change and the ecology of cities. Science 319(5864):756–760

Hsü GCC (1996) Incarnations of the blossoming plum. Ars Orientalis 26:23–45

Hu J (2022) Nurture-by-nature in affordance-need context. In: Expanding environmental awareness in education through the arts: crafting-with the environment. Springer Nature Singapore, Singapore, pp 231–240

Huang Y, Liu J, Wang J et al (2020) How urban wetland-based environmental education activate school children's childhoodnature in anthropocene times: experience from Chinese curriculum reform. In: Cutter-Mackenzie-Knowles A, Malone K, Hacking EB (eds) Research handbook on childhoodnature: assemblages of childhood and nature research. Springer, pp 623–635

Koren L (2008) Wabi-sabi for artists, designers, poets & philosophers. Imperfect Publishing
Liu X (2008) Daoism (I): Lao Zi and the Dao-De-Jing. In: Mou B (ed) The Routledge history of Chinese philosophy. Routledge, pp 221–248
Liu Y, Shu L, Peng L (2021) The hollowing process of rural communities in China: considering the regional characteristic. Land 10(9):911
Lou X, Bai M, Miao MS (2017) An overview and modern discussion on the folk custom of keeping health in the eastern four seasons. In: 3rd annual 2017 international conference on management science and engineering (MSE 2017). Atlantis Press, pp 218–223
Moore RC (1997) The need for nature: a childhood right. Soc Justice 24(69):203–220metap
Pagano J (2018) Songxi's siege: remembering a rural Chinese village in WWII. Medium. https://medium.com/@jakepagano/songxis-siege-remembering-a-rural-chinese-village-in-wwii-391896b19e4a. Accessed 2 Sep 2018
Pannell C W (2003) China's demographic and urban trends for the 21st century. Eurasian Geography and Economics 44(7): 479–496
Roetz H (2013) Chinese 'unity of man and nature': reality or myth? In: Meinert C (ed) Nature, environment and culture in East Asia. Brill, pp 23–39
Rozzi R (2012) Biocultural ethics: recovering the vital links between the inhabitants, their habits, and habitats. Environ Ethics 34(1):27–50. https://doi.org/10.5840/enviroethics20123414
Rozzi R (2013) Biocultural ethics: from biocultural homogenization toward biocultural conservation. In: Rozzi R, Pickett ST, Palmer C, Armesto JJ, Callicott JB (eds) Linking ecology and ethics for a changing world: values, philosophy, and action, Ecology and ethics series, vol 1. Springer, Dordrecht, pp 9–32. https://doi.org/10.1007/978-94-007-7470-4_2
Rozzi R (2018) Biocultural homogenization: a wicked problem in the anthropocene. In: Rozzi R, May RH, Chapin FS, Massardo F, Gavin MC et al (eds) From biocultural homogenization to biocultural conservation: a conceptual framework to reorient society toward sustainability of life, Ecology and ethics series, vol 3. Springer, Cham, pp 21–48. https://doi.org/10.1007/978-3-319-99513-7_2
Rozzi R (2023) Inter-species and inter-cultural encounters: the biocultural education program of the Omora Ethnobotanical Park. In: Rozzi R, Tauro A, Wright T, Avriel-Avni N, May RH Jr (eds) Field environmental philosophy: education for biocultural conservation, Ecology and ethics series, vol 5. Springer, Cham, pp 153–174. https://doi.org/10.1007/978-3-031-23368-5_10
Shi JY, Zhou DQ, Ma LS, Yao J, Zhang DM (2020) Diversity of bamboo species in China. World Bamboo Rattan 18(4):55–65
Tauro A, Ojeda J, Caviness T, Moses KP, Moreno-Terrazas R et al (2021) Field environmental philosophy: a biocultural ethic approach to education and ecotourism for sustainability. Sustainability 13(8):4526. https://doi.org/10.3390/su13084526
Tay SC (2005) Flowers as symbols and metaphors in Chinese culture. University of Tasmania
Tuan YF (1971) Man and nature. Association of American Geographers, Commission on College Geography, Washington, DC
Wang Y (2023) Rural lifestyles and life politics: reimagining modernity in the development of a future village in China. Rural Sociol 88(2):337–361
Way T (2020) Chrysanthemum. Reaktion Books
Welch PB (2013) Chinese art: a guide to motifs and visual imagery. Tuttle Publishing
Woods M (2007) Engaging the global countryside: globalization, hybridity and the reconstitution of rural place. Prog Hum Geogr 31(4):485–507
Zeng YL (2021) The custom of wearing hairpins on festivals in the Song Dynasty as seen in "Tokyo Menghualu". Yangtze River Ser 1:139–141

Chapter 8
Embodied Encounters: Biocultural Dialogues and Pedagogies Beyond the Human

Shoshana McIntosh

Abstract Human perceptions and understanding of other-than-human animals continue to be severely limited. Other species are often considered props for human existence or not recognized at all, as is the case for many invertebrates. This chapter is concerned with disrupting and expanding these perceptions and proposes an embodied approach modeled after the cephalopod-inspired work of visual artist Miriam Simun. In attempting to be like another animal, such as an octopus, we are decentered and drawn into that animal's world, embodying their point of view and encountering them as autonomous, unique subjects. Situated within the biocultural ethics' framework developed by Chilean philosopher and ecologist Ricardo Rozzi, this chapter first examines the octopus' biocultural history, and then draws more-than-human phenomenology together with Simun's writing and art to outline an embodied pedagogical and dialogic approach to studying, communicating about, and co-existing with other-than-human beings. This chapter follows Simun's example, asking us to stretch the ways we learn about and engage with other-than-human animals, to expand our thinking out into our fingertips, and to feel more deeply what it means to co-exist. Ultimately the author proposes that these embodied encounters constitute a powerful educational method with the potential to expand human perceptions as well as broaden the possibilities for multispecies flourishing.

Keywords Embodiment · Phenomenology · Biocultural ethic · Pedagogy · Art · Other-than-human animals · Invertebrates

8.1 Introduction

Following previous philosophers of education advocating for accessible, experiential, and justice-oriented pedagogies (Dewey 1934, 1938; Freire 1970; Hooks 1994), this chapter calls for educational practices that approach the human as a fully

S. McIntosh (✉)
Department of Philosophy and Religion, University of North Texas, Denton, TX, USA
e-mail: shoshana.mcintosh@unt.edu

D. Zhu et al. (eds.), *Linking Arts with Biocultural Conservation, Restoration, and Communication*, Ecology and Ethics 8,
https://doi.org/10.1007/978-3-032-04830-1_8

embodied being and acknowledge that we[1] are just one among a myriad of unique living beings. There are ethical implications to the way we educate ourselves, yet, even as we live through the climate crisis and sixth mass extinction, educational philosophies and practices, even the most progressive, continue to be heavily anthropocentric. In *Embodiment in Education: Exploring Creaturely Existence*, Australian philosopher of education, Marjorie O'Loughlin (2006) identifies the longstanding over-privileging of vision as one characteristic of Western education that has contributed to a disconnection from our full sensorial capacity as well as our embodied interconnectedness to the rest of the biosphere. This emphasis has also given rise to correlations between sight and knowledge, light and truth in Western thought, and the subsequent devaluing of other modes of gaining knowledge such as through touch or smell (O'Loughlin 2006).

However, like all animals, our perception and understanding of the world are shaped by how we move through it as sensing bodies (Gallagher and Lindgren 2015). Studies within cognitive psychology show that students increase their understanding when they engage with material through gestures and active movement (Gallagher and Lindgren 2015). American philosopher Shaun Gallagher and philosopher of education Robb W. Lindgren (2015) also identify the importance of enactive metaphors, "metaphors that we bring into existence through our actions" (p. 392) for learning, emphasizing the integral connection between brain, body, and environment in everything that we think, feel, and do. What then are the material consequences of educational models that emphasize abstract and disembodied learning methods? Specifically, in what ways do these methods influence our relations with other species and the biosphere? Alternatively, what possibilities for more just and caring multi-species co-existence are revealed when we embrace our embodiment and adopt more embodied approaches to learning and teaching?

This chapter investigates these questions, and I pose them as someone whose work and life have always been an entanglement of teaching and embodied, creative practices. I address the last question specifically through a phenomenological, ethical, and artistic framework. My goal is to demonstrate how these three components can be intertwined into a practice of embodied dialogues and pedagogies among myriad species, expanding our understanding of language and communication as bodily and multifaceted. The aim is to gain awareness of other-than-human[2] experiences, on their own terms, and cultivate empathetic ethical consideration for

[1] Throughout this chapter I will use plural second person pronouns to reference humans broadly and draw my readers into the dialogue, while also keeping in mind that human perspectives and relations with other species are multiple and some bear more responsibility for the current extinction crisis.

[2] I follow Rozzi (2018) in his use of the term other-than-human although I interchange it with more-than-human inspired by indigenous ontologies which situate the human in a broader sentient community (Abrams 1996). Rather than defining other beings through negation as terms like non-human do, other-than-human leaves room for biotic as well as abiotic beings, reminding us that other beings are co-inhabitants within the biophysical and spatial realms as well as within our cultural, linguistic, and mental lives, both in waking and dream lives (Rozzi 2018).

their flourishing. I propose that an embodied approach is key to fostering more just relations between humans and the more-than-human world. I am influenced by my own experiences with embodied pedagogies and in large part by the writing and art of contemporary American artist, previously trained as a sociologist, Miriam Simun, whose octopus-inspired performative artworks and workshops radically disrupt pedagogical structures which posit the human as an "objective", disembodied observer and consumer of the rest of the world. Through an interdisciplinary blending of visual arts, sciences, and embodied engagements, Simun's artistic method invites us to embrace our membership within bioculturally rich worlds co-inhabited by human and other-than-human animals.

Before attending to Simun's work in more depth, I situate this project within the biocultural ethics' framework developed by Chilean philosopher and ecologist Ricardo Rozzi (2013). I then introduce the biocultural history of the octopus,[3] from their intriguing ocean life to their appearance in artworks, laboratories, myths, literature, and human fantasies. Next, I engage the phenomenological work of French philosopher Maurice Merleau-Ponty to articulate the dialogic and pedagogic potential of this approach. Returning to Simun's writing and artworks, I elaborate on her method as a model for a biocultural more-than-human phenomenology in which we encounter other creatures as fully embodied beings ourselves. The chapter culminates in an analysis of the multifaceted pedagogical and ethical implications that challenge formal education models, which are far too often disconnected from our full embodiment as well as our habitats and co-inhabitants. This chapter follows Simun's example, asking us to stretch the ways we study and engage with other-than-human animals, to expand our thinking out into our fingertips, and to feel more deeply what it means to co-exist in a multi-species world.

8.2 A Biocultural Approach to Understanding Other-Than-Human Animals

8.2.1 Biocultural Ethics and Embodiment

The biocultural ethic values and places at its core the interrelations and interactions among co-inhabitants whose life habits have evolved together in their shared habitats (Rozzi 2013). We are entangled with our co-inhabitants and must acknowledge that they are also subjects with their own life histories, not mere natural resources or backdrops for our human drama. Importantly, humans are the kind of co-inhabitants who have the capacity to give ecological and ethical consideration to the lives of

[3]While Simun's work focuses on cephalopods in general, I concentrate on octopuses while acknowledging that this is a generalization as there are over 100 species belonging to the genus Octopus (Schweid 2013), and over 800 extant species that belong to the class Cephalopoda (Kröger et al. 2011).

other co-inhabitants. Pedagogical and dialogical methods, including those I am proposing, are never neutral and always carry some degree of ethical weight. The biocultural ethic provides a framework from which we can attend to these implications while recognizing that ethics are not simply mental exercises but play out through our embodied interrelations and interactions and require affective just as much as cognitive rational consideration.

To develop his biocultural ethics, Rozzi (2013) forged the conceptual "3H" model, with which he identifies the vital links among co-in-Habitants (humans and other-than-humans), their life Habits and the Habitats that sustain human well-being and all living beings. This "3H" model (co-in-Habitants, Habits, Habitats,), calls for an attitude of care when it comes to conserving and restoring biological and cultural diversity (and their interrelationships) as well as fostering biocultural communication. Environmental philosopher Danqiong Zhu (2023) proposes biocultural communication as a process and field concerned with bridging the gap between biocultural research and the public, by drawing on multidimensional and cross-disciplinary methods. The embodied approach I am advocating for informs biocultural conservation, restoration, and education through its biocultural communicative potential.

Much of human communication about other-than-human animals today is shaped by our narrow perception of them as entertainment, food, resources, pets, or pests. Limited perspectives exist in research as well. Rozzi (2019) argues that contemporary animal ethics has been overly focused on vertebrates, particularly mammals, ignoring a large swath of invertebrate diversity. This has led to a kind of *taxonomic chauvinism* in which invertebrates, who far outnumber vertebrates in diversity and number of taxa, have been overlooked, misunderstood, underrepresented, and undervalued in scientific research and environmental conservation efforts (Rozzi 2019). Therefore, I focus on the octopus, in part, to resist perpetuating the vertebrate-bias. Additionally, just as Simun (2018) calls cephalopods "role-model species for humans," I see the octopus as a role model for developing embodied pedagogies given their incredible ability to feel out their world through full-body sensitivity.[4]

Integrated with biocultural ethics, an embodied approach to learning extends beyond what it can do for human participants and into the realm of empathy-driven ethical decision-making regarding other species and our shared habitats. Similarly, the approach I suggest, inspired by Simun's artworks, is not about performing for others or simply expanding ones' own capacities, but about feeling *like* another being in order to feel more *for* them. This kind of engagement includes expanding our understanding of the entangled biocultural histories of other species, which I model by examining the octopus' biocultural history.

[4]This concept of "role-model species for humans" is grounded on an understanding of octopuses as subjects, co-inhabitants in Rozzi's terms, in contrast and complementarity to the concept of "model organism" used by scientist who have paid attention to octopuses as a "study object" (cfr. Rozzi 2019; di Cosmo et al. 2021).

Fig. 8.1 An octopus. (Photograph courtesy of Miriam Simun)

8.2.2 A Biocultural Octopus History

The octopus, one of many of cephalopods' genera that evolved over millions of years, has built an incredible existence in the shallows and depths of the ocean. Over time, their presence has also become intimately linked with human cultures, science, and habits. For humans, this relationship is profoundly affective, eliciting feelings of fascination, wonder, desire, fear, and even disgust. Pervasive and entrenched assumptions about octopuses are informed partly by biological factors but especially by cultural influences. Octopuses have appeared in various cultures as divine beings, villains, curious oddities, and food. During the twentieth century octopuses and other cephalopods played a central role in neurosciences experiments, captivating scientists as well as the general public with their adeptness at problem solving and other intriguing behaviors (Nakajima et al. 2018). This biocultural history presents a starting point from which to understand our complex relationship with this unique co-inhabitant (Fig. 8.1).

The octopus is a cephalopod, a classification composed of the words *kephale* and *podus*, or 'head-footed' in Greek, referring to the embodied organization of these creatures whose heads are located between their body and tentacles (Schweid 2013). As an invertebrate, the octopus has a soft body, one which evolved from a former shelled version millions of years ago, and they are the only mollusk with a brain although like all mollusks their bodies host an extensive nervous system (Schweid 2013). An octopus' tactile awareness is something to be marveled at. With 200 suckers on each tentacle and each sucker containing thousands of receptor cells, an octopus perceives their environment visually and through radically sensitive and expansive tactility (Schweid 2013).

The complex and bizarre biology of these creatures has long stimulated human imagination, out of which horrifying and tantalizing myths and art have emerged. One well-known Nordic myth refers to the Kraken, a legendary sea monster similar to an octopus-like cephalopod, which would sink ships and devour sailors (Salvador and Tomatani 2014). The Kraken's effect has been long-lasting. In the nineteenth century it inspired, for example, English writer Alfred Lord Tennyson's poem "The Kraken" (1830), and a sea monster in French novelist Jules Verne's novel *Twenty Leagues Under the Sea* (1870) (Salvador and Tomatani 2014). Today, the Kraken has inspired "Ruby Gillman: Teenage Kraken" (2023) an animated film, directed by Kirk DeMicco and Faryn Pearl, about a teenager who must come to terms with her Kraken ancestry and fight off malicious mermaids.

The Kraken lives on, but terror is not the only feeling cephalopods have induced. Shortly before Tennyson wrote his poem "The Kraken," in Japan, artist Katsushika Hokusai created a woodblock print of a reclining woman being pleasured by two octopuses (Schweid 2013). *Shunga* prints like this one, containing erotic and explicit content, were not uncommon in Japan, and the genre, including Hokusai's print, "Dreams of a Fisherman's Wife," made its way to Europe, capturing the attention and influencing the work of various artists and authors such as Victor Huga, Émile Kimpt, and Pablo Picasso (Schweid 2013).

Octopuses have also been considered divine beings possessing supernatural powers. In the mythology of Micronesia and the Gilbert Islands located in the Pacific Ocean, an octopus god named Na Kika, considered the offspring of the original beings, is said to have used his eight limbs to raise islands from the sea bed (Montgomery 2016). In the subarctic regions, according to the worldviews of some native people in Canada and Alaska, the octopus holds sway over the weather as well as people's health (Montgomery 2016). An ancient Polynesian story describes how our world emerged out of another, older universe, from which the octopus was the only being that transitioned into the new universe (Montgomery 2016).

An octopus who has traveled between universes would possess an otherworldly wisdom, and indeed the intelligence of octopuses has long fascinated and been studied by humans. Aristotle and other Greeks described the octopus as neat and thrifty, but also called them shrewd tricksters (Schweid 2013). In 1992, Graziano Giorito, a neurobiologist, published the stunning results of his experiments with octopuses, proposing that they were highly intelligent invertebrates able to learn through observation (Schweid 2013). This intelligence, along with high sensitivity and creative navigation skills show up in countless filmed records of octopuses solving problems such as getting food out of containers or making their way through elaborate mazes.[5] There have also been countless octopus escape attempts from

[5] See for example, YouTube videos such as "Octopus vs Underwater Maze" by Mark Rober https://www.youtube.com/watch?v=7__r4FVj-EI&t=196s&pp=ygUTb2N0b3B1cyBleHBlcmltZW50cw%3D%3D or any numbers of videos on the Octolab TV's Youtube channel https://www.youtube.com/@octolabtv1168.

captivity, which often involve squeezing their large boneless selves through tiny spaces (Schweid 2013). As humans have tried to contain and test the octopus, we have been foiled at every turn by their ingenuity and unique other-worldly abilities.

8.2.3 A Biocultural Shift

These narratives about the octopus have emerged out of human culture for human consumption and this human-centeredness has further embedded assumptions about these creatures. And yet this centeredness is shocked and unsettled by representations of the octopus as an erotic figure in the Japanese shunga prints, or as a trickster figure during scientific experiments. These figures reconnect our human carnal bodies to the wild pulses of desire, expanding imagination beyond human mentalities by witnessing unexpected ways octopuses express themselves and problem solve. However, the prevailing imaginary informing the "common mindset" in global society is linked to the octopus as a villain that is out to get *us*, the octopus on the plate is for *us* to eat, the octopus in the maze is proving its intelligence to *us* on *our* terms. Through the remainder of this chapter, I consider how approaching our co-inhabitants through a more-than-human phenomenology invites a shift from anthropocentric narratives to biocultural dialogues that are embodied and multidirectional. This is about shifting from telling narratives *about* other animals we are observing to having dialogues *with* them, acknowledging their own agency and subjecthood.

Miriam Simun's art work engages in just this kind of shift. Her ongoing project "Transhumanist Cephalopod Evolution" grew out of Simun's (2018) graduate thesis, "Training Transhumanism: I Want to Become a Cephalopod." Formatted as a training manual/manifesto the thesis invites humans to embrace our inevitably wetter future and develop new bodily capacities by looking to the cephalopod as a role-model. Besides leading "Communal Sensorial Experiences" and giving public talks, Simun also makes her work digitally accessible. The training regime is available on Simun's website including visual references and her 24-min film "Your Urge to Breath is a Lie" which follows performers and synchronized swimmers engaging in cephalopod embodiment (Simun 2019). On SoundCloud Simun's (2021) steady voice guides a 45-min exploration of cephalopod-inspired breathing exercises and movement improvisation prompts interspersed with cephalopod knowledge. She is future-oriented in her work, writing, "I want to find a new way of being in the world and a new way of relating to the beings I co-inhabit the world with—both human and non" (Simun 2018, p. 68). This way of being is embodied and sensitive, resisting a limited singular or static ontology, and instead recognizing the myriad ways we are entangled with other species. Simun's work shifts the human from a distanced observer creating narratives to a humble learner engaging in an embodied dialogue with ourselves, our environment, and our co-inhabitants (Fig. 8.2).

These kinds of biocultural, speculative, embodied encounters, while expanding our own capacities, also lead us to describe and understand other animals in ways

Fig. 8.2 Synchronized swimmers. (Photograph courtesy of Miriam Simun)

that are more authentic to their actual lives and capabilities. Furthermore, skills and practices that develop attunement to our human-animal bodies and senses, are desperately needed to overcome the disembodiment that has been imposed by current global technological society. This society is characterized by an alienation from the "corpus of life" as we are rapidly and increasingly bounded by social media and cyber representations of lives that become mere represented objects. The creative and embodied engagements as modeled by Simun offer a potent alternative, placing us back in embodied relation with our co-inhabitants and shared habitats which we impact through our decisions and actions. Before elaborating more on Simun's methods, I turn to phenomenology to flesh out the pedagogical-dialogical aspects of what I am proposing.

8.3 More-Than-Human Phenomenology

Phenomenology, particularly Maurice Merleau-Ponty's work on perception and corporeality, has much pedagogical and dialogic potential through its experiential, descriptive, and reflective methodology (Merleau-Ponty 1945/2012; Abrams 1996; Toadvine 2009; Van Manen 2015). In this section I look closely at the embodied and affective dimensions of phenomenology while resisting those aspects of phenomenology interested in discovering essences through transcendental abstraction. I advocate instead for a more-than-human phenomenology that takes human embodiment seriously, but also extends beyond human experiences and acknowledges the limitations of human knowledge. Specifically, I elaborate on the importance of sense perception, embodied descriptive expressions, and bodily reflections within this approach.

8.3.1 Sense Perception in More-Than-Human Phenomenology

French philosopher and public intellectual, Maurice Jean Jacques Merleau-Ponty (1908–1961), was the leading academic proponent of existentialism and phenomenology in post-war France (Toadvine 2023). Although phenomenology provided the overarching framework for these investigations, Merleau-Ponty, much as Miriam Simun does today, conducted empirical research in sociology, psychology, ethology, anthropology, and linguistics, as well as the arts. Much of his work is devoted to examining questions of perception and embodiment, and analyzing the interactions between the mind and the body and the objective and the experienced world, in a non-dualistic way. Merleau-Ponty's (1945/2012) early work called for phenomenology to take sense experience seriously and for the method to remain grounded in the material world rather than attempt to transcend it. This focus evolved throughout his subsequent work, and in 1948, he presented several radio lectures, one entitled "Exploring the World of Perception: Animal Life." Merleau-Ponty (1948/2004) begins the lecture by emphasizing that humans, like all animals, relate to the environments through their corporeality, and for humans, this results in an ambiguous relationship with the things around us, one that is embodied but also limited because of the kind of body we have. We can never fully grasp things and co-inhabitants around us.

American environmental philosopher and Merleau-Ponty scholar, Ted Toadvine, has partnered with biologists, artists, architects, literary scholars, and community organizers to explore the role of philosophy and the humanities in interdisciplinary research and public engagement. In his book, *Merleau-Ponty's Philosophy of Nature*, Toadvine (2009) considers the potential for phenomenology to describe and interpret an animal's world. While this access is limited, Toadvine (2009) asserts that humans have the unique capacity to shift perspectives, particularly through creative expressiveness and imagination.

David Abrams, an American cultural ecologist, philosopher, and performance artist, employs Merleau-Ponty's philosophy to emphasize that humans are sentient *and* we sense, that we touch the world and it touches us, and this reciprocity reveals an opportunity to encounter and be in relation with other-than-human animals as other experiencing, sensing beings. As Abrams (1996) puts it, "the body is a creative, shape-shifting entity" which is our "very means of entering into relation with all things" (p. 38). The octopus as the ultimate shape-shifter can add additional motivation. For Abrams (1996), attending to the sensible and our own sentience as embodied beings through phenomenology radically disrupts a hierarchy that places humans at the top and justifies the exploitation of other beings within the biosphere.

This disruption, I contend, acts as an instigator of biocultural dialogue, a back and forth that involves more-than-humans who emerge into the foreground of our perception as we embrace our own embodiment. Merleau-Ponty describes nature as an interlocutor within this dialogue that takes place between our bodies and our environment. As he puts it, "the unfolding of sensible givens beneath our gaze or beneath our hands is like a language that teaches itself" as "our senses literally interrogate the things and that the things respond to them" (Merleau-Ponty 1948/2004, p. 333). Here, dialogue expands beyond the verbal. This view is closely correlated with those of cognitive psychology mentioned earlier, recognizing that our embodied selves are intricately involved in and important to learning processes.

Taken together, Merleau-Ponty, Toadvine, and Abrams expand the phenomenological method to involve sense perception, troubling assumptions regarding human exceptionalism and our access to knowledge. Involving all the senses in perspective-taking and creative expression has pedagogical implications. Our experience is multidimensionally richer than it would be if we mainly focused on what we could visually observe or mentally imagine. Expressing another animal's meaningful experience through one's body requires recognition of the intimate links between habits, habitats, and co-inhabitants. This approach acknowledges other-than-humans as active subjects, rather than passive objects we can empirically study and know.

8.3.2 Expressive Children and Embracing the Unknown

When engaging phenomenology to describe and reflect on the experiences of other animals, I join Toadvine in insisting that a *double-movement* is necessary, which first, "opens space for the positive description of the meaning of the animal's world as other than merely a modification of the world of the human subject", and second, "the human as such must be reconceived as neither opposed to nor reducible to the animal" (Toadvine 2009, p. 78). The human as opposed to the animal has long been an underlying assumption in education which in one way or another has been aimed at the development of a specific and complete version of the human (Snaza and Weaver 2015). In his radio lectures, Merleau-Ponty (1948/2004) refuted the possibility of a fully-developed human with an established coherent world who has knowledge of a sovereign essence or truth. The chimeric assumption of a "true knowledge" about the world stimulates adults to arrogantly judge both animals and children, who supposedly do not have this kind of adult-human coherence. A more-than-human phenomenology begins with the premise that we can never gain full coherence or knowledge of our or others' worlds. For Merleau-Ponty (1948/2004), this "leave[s] his [human] knowledge of the natural world riddled with gaps, which is how poetry creeps in" (p. 73). The human is decentered here, not in full control, but rather vulnerable to affect. Where our knowledge ends, creativity can begin: inspiration, beauty, multiple perspectives entering our bodies and emerging through expression. Here again, the multi-directionality of biocultural dialogue is paramount.

To embrace more-than-human phenomenological expression, children (rather than be judged as not yet fully developed) can be considered role-models for experientially understanding the undetermined facets of humans' knowledge about the world. In 2019, I collaborated with other art education and philosophy university students to design and lead an environmental philosophy and art summer camp for local children in (Tsuji and Johnson 2020). During several outdoor activities, we asked the students to embody another animal, for example to paint a picture of the trail from the perspective of a box turtle. It was hardly any trouble at all for our young students to plop down in the dirt and attempt to see from a turtle's position (Tsuji and Johnson 2020). I consistently find that young children are open to and curious about their world and the worlds of others, letting it unfold before them rather than reaching out to grasp, observe, analyze, and define.

Children around the world, and in different cultures, playfully pretend they are cats, birds, or any other number of other-than-human animals, moving their bodies in more-than-human ways, calling to each other in more-than-human voices (Daston and Mitman 2005; Bartos 2013). They freely embrace and explore their embodied existence. Following Merleau-Ponty, we can recognize a form of kinship with both children and other animals (Toadvine 2009). This kinship is predicated on the destruction of hierarchies that, based on cultural, ethnic, and taxonomic chauvinism (cfr. Rozzi 2019), place "some privileged humans" above "the rest" of humans and even more severely above all other-than-human living beings.

8.3.3 Bodily Reflection

To engage in one final key component of phenomenology, at least for the purposes of this chapter, I draw on Toadvine's interpretation and elaboration of Merleau-Ponty's idea of *bodily reflection*. Toadvine (2009) reminds us that phenomenology has typically been disembodied when it comes to the phenomenological reduction which involves isolating our thought processes from the embodied world with the goal of revealing the structure and essence of an experience. However, contemporary phenomenologists have demonstrated that reflection is never just a mental process (Varela et al. 2017). Merleau-Ponty asserted that any phenomenological analysis/reflection has a corporeal foundation it cannot shed, and is situated and dependent upon our place in and relation to our habitats (Toadvine 2009). He contended that a bodily reflection is possible due to the body's reciprocal relationship with itself, its ability to touch and see itself, and for Toadvine (2009) this bodily reflection affirms the interconnectedness of humans and other living beings. This kind of reflection allows for differences to be recognized among animals without positioning the human as superior and initiates a metamorphosis rather than a transcendence which further distances us from other creatures (Toadvine 2009). I propose that bodily reflection is ongoing throughout the processes of perception and expression, not merely as an afterthought only done once the experience is finished. When I attempt to embody an octopus, I fully engage myself in a kind of bodily reflection as

I attempt to analyze and literally reflect the animal's experience by moving my own body in similar ways. I envision here a constant back and forth dialogue—an ongoing embodied conversation with myself, between myself and the octopus, and myself and my environment. To reflect only after an experience is to ignore the affective feedback received during the experience that may initiate a shift in thinking, feeling, and doing. I find the transhumanist experiential art of Miriam Simun to offer just such an opening; performing/enacting another species invites us to engage in multi-sensory exploration and enables embodied descriptive expressions and bodily reflections to emerge.

8.4 Miriam Simun: Transhumanist Cephalopod Evolution

So far, I have introduced Riccardo Rozzi's biocultural ethic, which guides us to consider and care for the vital links between co-in-Habitants, Habits, and Habitats. I then presented a more-than-human phenomenology, inspired by Merleau-Ponty, which includes sensorial descriptions, creative expressions, and bodily reflections to better describe and interpret the experiences of other animals. Dialogic and pedagogical potential emerge from both frameworks, and even more so when they are jointly implemented. For instance, more-than-human phenomenology supports biocultural ethical work (such as conservation or public education) in the way it decenters the human perspective and brings the experiences of other creatures into our awareness. Likewise, biocultural ethics holds phenomenological methods accountable, making us question our approaches and motives along the way, and reminding us that we cannot un-entangle co-inhabitants from their habits, habitats, or each other.

The final piece in my proposal for embodied encounters is the artistic element which invites imagination, creative exploration, and a rekindling of bodily connection and awareness. These elements are key when engaging with other species as we can never describe with complete accuracy the life of another or have coherent knowledge of their experience. While imagination may be considered risky and anthropocentric, it is crucial in this moment of climate crisis and widespread extinction. My hope is that by imagining what it is like to be another creature we not only expand our empathy but also become open to other possible, more just ways of living together. In this last section I describe Simun's process in more detail and connect it with biocultural ethics and more-than-human-phenomenology.

8.4.1 Embodying a Cephalopod

Any attempt to bodily describe another creature's experience requires observation. Simun spent time with the octopuses at the Woods Hole Oceanographic Institution to get a feel for how they move and behave, but her process also expanded beyond visual observation, as she learned to free dive, or diving without a breathing apparatus (Nimrod 2022). This biocultural practice has been taking place for thousands of years, and Simun (2018) finds inspiration in the ways freedivers attune to their bodies and push the limits of their physical capacities.

It is important to consider what ethical, non-intrusive ways are available to experience and observe other animals' habitats, habits, and interactions with co-inhabitants. Today, social media and other web-based platforms host a growing wave of digital images and videos offering close-up views of animals all over the planet. With the increasing mediated representation of animals, it is vital to understand the need to share spaces, habitats, and "face-to-face" encounters with other-than-human animals (Rozzi 2023), however we must challenge the normalization and pervasiveness of placing animals in captivity, whether for entertainment, food, experimentation, or research. The octopus, like other marine animals, offers a unique challenge if we wish to observe them while resisting these industries.

One stirring example is *My Octopus Teacher*, a 2020 documentary in which diver and filmmaker, Craig Foster, entered the octopus' world, sharing its habitats and some of its swimming habits. Foster moved beyond his role of observer, gradually becoming a companion of a small common octopus off the coast of South Africa (Reed and Ehrlich 2020). With a biocultural perspective, we realize that by sharing the habitats and life habits of the octopus, the filmmaker is able to partially dissolve the barrier that separates humans from non-animals in most nature documentaries, which record and dramatize animal life as exotic and isolated from humans. While the option of becoming *in situ* co-inhabitants with an octopus is not available to most humans, witnessing Foster's documentary offers another entry point by triggering an imagination that is grounded in actual habitats and co-constituted life habits, generated by both the experiences of the human filmmaker and the filmed octopus.

Simun invites participants in and out of the water to consider the octopus as a role-model for activating full-body sensitivity. She highlights their embodied tactile awareness, shape-shifting abilities, and distributed intelligence (Simun 2018, 2019, 2021). I will briefly outline examples of exercises Simun offers inspired by these three characteristics. In an introductory session recording called, "Transhumanist Cephalopod Evolution" Simun (SoundCloud 2021) explains how M.G. Wells discovered that octopus' skin sensitivity is over one hundred times more powerful than the sensitivity of the human tongue. Considering this capacity for tactile intelligence, Simun invites participants to explore moving their tongues, our most invertebrate-like part. Exploring how our tongue can move, we can feel to a small degree what it's like to be without a bony structure. Simun (2021) then prompts movers to imagine and move as if they have eight tongues, each one hundred times more sensitive. As seen in Fig. 8.3, this exercise could include a partner; one person moving their tongue and the other following. In a more-than-human phenomenological framing, we could consider this a multi-sensorial description and bodily reflections of an invertebrate body part. This exemplifies the phrase "show/do, don't tell" as we communicate about the octopus to ourselves and others through our own embodied expression.

Fig. 8.3 Two people mirror the movement of their invertebrate tongues. (Photograph courtesy of Miriam Simun)

Simun (2021) also engages with the octopus' jet propulsion, or the ability to pump water through their mantle, which allows them to traverse their habitats by shapeshifting. For human participants, this entails using what we have—our lungs and ability to move with our breath. Simun (2018) describes an empathetic seeing taking place when we can engage our breath, moving with each exhale around a space or within a body of water, as we "step" into the octopus' mode of travel. Given the octopus' capacity to shapeshift through camouflage, Simun (2018) also encourages participants to be hyper-aware of their environment. This might mean tuning into a particular part of the body, for example, the finger tips and exploring one's place through touch or moving in relation to what is around you. Simun presents these as lessons for the moment as well as for life. She prompts:

> Learn to be flexible. Notice your pattern. Now find a better one. Don't get stuck in old patterns, or new ones either. Pay attention to the situation. Move the right way for the configuration you're in as new impressions shift in accordance. (Simun 2018, p. 256)

Through these embodied encounters, we learn *about* other species and *from* them. While Simun is guiding these exercises, the real instructor behind the scenes is the octopus.

When it comes to the cephalopod's distributed intelligence, Simun (2024) offers prompts on her website such as "practice for another gravity" and "allow your tips their brains" (p. na) reminding us to reconnect with our own "distributed intelligence" and "embodied minds" (cfr. Varela et al. 2017), recognizing the capacities and intake of all our senses. Another option is to involve multiple bodies. In Simun's (2019) film "Your Urge to Breathe is a Lie," synchronized swimmers move together underwater while on land, performers lean into one another, improvising with body weight and gravity, sensing and exploring each other as well as the "space" (the

Fig. 8.4 Many hands intertwining. (Photograph courtesy of Miriam Simun)

habitat, *sensu* Rozzi 2018) through touch (Fig. 8.4). This kind of expression highlights our vital biocultural interconnections and asks us to consider our actions in relation to other bodies and habitats.

Furthermore, distributed intelligence challenges the possibility of a self completely separate from others; Simun asks:

> How might our attitudes shift if we were at times unsure if what we encounter is self or other? If it was our own body or another's body—including human bodies, non-human bodies, bodies of water, etc.? How might our perceptions, relations, actions change? (Simun 2018, p. 145)

This calls to mind Merleau-Ponty's charge that bodily reflections are transformational. The questions Simun asks, and many others, can be reflected on throughout this process. What I am suggesting is that rather than try to mentally alter our perceptions of other animals, we shift our perception through bodily expression. Our perceptions, relations, and actions are always embodied and carry ethical implications. Encountering the agency, experiences, and unique expressions and capabilities of other-than-humans through our own bodies is a powerful way to embrace our biocultural positionality as animals with the unique capacity to give ethical consideration to others.

8.5 Conclusion: Implications for Education, Conservation, and Co-existence

Human and other-than-human bodies are being impacted by the climate crisis and sixth mass extinction as our shared biosphere becomes less habitable. Some bodies are more vulnerable than others. Some bodies are more adaptable than others.

Disembodiment and alienation, often exacerbated by technological advances, globalization, and late-stage capitalism, results in disconnection from and less awareness of the state of our shared environment. This seems particularly problematic for those bearing the most decision-making power when it comes to the warming planet. With this power comes the means to further separate the material reality of the most powerful from that of their co-inhabitants, human and otherwise. In this chapter I have proposed an alternative route that calls us back into our bodies so that we may attend to our responsibilities and these challenges from an embodied point of view, better understanding the lives of our more-than-human co-inhabitants, their unique habits and habitats, and our own situatedness amongst them. Embodied encounters with and creative expressions of other-than-humans put us in dialogue with this biocultural community. This process asks us to listen and be affected by what is around us. Currently, the impact is severely asymmetrical between co-inhabitants, with other-than-humans being deeply impacted by human actions and decisions. What transformations become possible when we allow the biosphere and all it holds to affect us (ethically, emotionally, bodily) as much as we do it?

An important implication of this approach is the disruption of long-standing humanist education models. Embodied approaches to learning about the more-than-human world challenge assumptions of human supremacy and separateness that are infused in curricula and pedagogy, and reengage students with their senses and embodied relations. Noting the authors and artists I've referenced, boundaries between disciplines must be blurred to facilitate more multi-dimensional experiences that open to the other-than-human. Discipline specific narratives give way to multifaceted dialogues in which instructors invite in more-than-human voices and students engage in multi-directional learning.

For those working in conservation, rewilding, or other fields involving animals, taking an embodied approach has far-reaching implications. It is easy to adopt tunnel vision, focusing on ever-changing policies, funding deadlines, population numbers, and human-centered visions of what nature should be. However, most people working in these fields do so for deeply emotional and ethical reasons. Reinvigorating our own embodied experiences and reflecting on those of other animals, brings the vital connections between the 3H's to the foreground, calling our attention to their individual and collective lives, needs, and capabilities. Public education and communication must expand beyond the visual, finding innovative ways to engage people through experiences, embodied descriptions, and bodily reflections. This approach does not necessarily make any of this work easier. It calls us to acknowledge the complexity of entangled life on earth, to pause and feel, to tune in to the ongoing earthly dialogue, to engage our full affective being in ethical decision-making.

In closing, I invite you to lean into your full sensorial existence as you rest your eyes from reading and go about your day. Several years ago, while roller-blading, I had a bird-like experience as wind sent me careening down a hill nearly colliding with a brick wall. It left me slightly more attuned to the relation our winged co-inhabitants have with the ever-moving sky. As I watch a Mississippi Kite soar outside my window while writing this, I am reminded of that experience, and I encourage you to also notice and embody one of your co-inhabitants today. This could be as simple as brushing your nose against a pollen-laden flower like a bee, or as adventurous as climbing a tree like a squirrel. Take a deep breath and let yourself feel.

Acknowledgments The author would like to express deep gratitude to Ricardo Rozzi, Rika Tsuji, and Danqiong Zhu for their editorial guidance and insight throughout this process. Additional thanks to Miriam Simun for the images included and for her ongoing work which has inspired this chapter. I acknowledge that Figs. 8.1, 8.2, 8.3 and 8.4 (2024) in this chapter, digital photographs, are a courtesy of the artist, Miriam Simun.

References

Abrams D (1996) The spell of the sensuous. Vintage, New York

Bartos AE (2013) Children sensing place. Emot Space Soc 9:89–98

Daston L, Mitman G (eds) (2005) Thinking with animals: new perspectives on anthropomorphism. Columbia University Press, New York

Dewey J (1938) Experience and education. Macmillon Company, New York

Dewey J (1934) Art as experience. Rare Treasure Editions, Victoria

Di Cosmo A, Pinelli C, Scandurra A, Aria M, D'Aniello B (2021) Research trends in octopus biological studies. Animals 11(6):1808

DiMicco K (2023) Ruby Gillmore: teenage kraken. Dreamworks. www.teenagekrakenmovie.com/. Accessed 20 Dec 2023

Freire P (1970) Pedagogy of the oppressed. Bloomsbury, New York

Gallagher S, Lindgren R (2015) Enactive metaphors: learning through full-body engagement. Educ Psychol Rev 27:391–404

Hooks B (1994) Teaching to transgress: education as the practice of freedom. Routledge, New York

Kröger B, Vinther J, Fuchs D (2011) Cephalopod origin and evolution: a congruent picture emerging from fossils, development and molecules: extant cephalopods are younger than previously realised and were under major selection to become agile, shell-less predators. BioEssays 33(8):602–613. https://onlinelibrary.wiley.com/doi/abs/10.1002/bies.201100001

Merleau-Ponty M (2004) The world of perception (trans: Davis O). Routledge, New York (Original work published in 1948)

Merleau-Ponty M (2012) Phenomenology of perception (trans: Landes D). Routledge, New York (Original work published in 1945)

Montgomery S (2016) The soul of an octopus: a surprising exploration into the wonder of consciousness. Atria Books, New York

Nakajima R, Shigeno S, Zullo L, De Sio F, Schmidt MR (2018) Cephalopods between science, art, and engineering: a contemporary synthesis. Front Commun 3:20

Nimrod S (2022) Miriam Simun. In: Jones C, Bell N, Nimrod S (eds) Symbionts: contemporary artists and the biosphere. The MIT Press, Cambridge

O'Loughlin M (2006) Embodiment and education: exploring creatural existence. Springer, Dordrecht

Reed J, Ehrlich P (2020) My octopus teacher. Netflix

Rozzi R (2013) Biocultural ethics: from biocultural homogenization toward biocultural conservation. In: Rozzi R, Pickett STA, Palmer C, Armesto JJ, Callicott JB (eds) Linking ecology and ethics for a changing world: values, philosophy, and action, Ecology and ethics series, vol 1. Springer, Dordrecht, pp 10–32. https://doi.org/10.1007/978-94-007-7470-4_2

Rozzi R (2018) Biocultural homogenization: a wicked problem in the Anthropocene. In: Rozzi R, May RH Jr, Chapin FS III, Massardo R, Gavin MC, Klaver IJ, Pauchard A, Nuñez MA, Simberloff D (eds) From biocultural homogenization to biocultural conservation, Ecology and ethics series, vol 3. Springer, Cham, pp 21–48. https://doi.org/10.1007/978-3-319-99513-7_2

Rozzi R (2019) Taxonomic chauvinism, no more! Antidotes from Hume, Darwin, and biocultural ethics. Environ Ethics 41(3):249–282. https://doi.org/10.5840/enviroethics201941325

Rozzi R (2023) Inter-species and inter-cultural encounters: the biocultural education program of the Omora Ethnobotanical Park. In: Rozzi R, Tauro A, Wright T, Avriel-Avni N, May RH Jr (eds) Field environmental philosophy: education for biocultural conservation, vol 5. Springer, Dordrecht, pp 153–174. https://doi.org/10.1007/978-3-031-23368-5_10

Salvador RB, Tomatani BM (2014) The Kraken: when myth encounters science. Hist Cienc Saude Manguinhos 21(3):971–994

Schweid R (2013) Octopus. Reaction Books, London

Simun M (2018) Training transhumanism: I want to become a cephalopod. Thesis, New York University

Simun M (2019) Your urge to breathe is a lie. https://www.miriamsimun.xyz/12356410-your-urge-to-breathe-is-a-lie. Accessed 20 Dec 2023

Simun M (2021) Transhumanist cephalopod evolution: introductory session. SoundCloud [online]. https://soundcloud.com/miriam-simun/transhumanist-cephalopod-evolution-introductory-session-032021

Simun M (2024) miriamsimun.xyz. Accessed 10 Aug 2024

Snaza N, Weaver JA (eds) (2015) Posthumanism and educational research. Routledge, New York

Toadvine T (2009) Merleau-Ponty's philosophy of nature. Northwestern University Press, Evanston

Toadvine T (2023) Maurice Merleau-Ponty. Stanford Encyclopedia of Philosophy. http://plato.stanford.edu/entries/merleau-ponty/

Tsuji R, Johnson B (2020) Philosophy meets place: creating an environmental philosophy summer camp. In: Katz CE (ed) Growing up with philosophy camp. Rowman and Littlefield, Lanham, pp 99–109

Van Manen M (2015) Researching lived experience: human science for an action sensitive pedagogy. Routledge, New York

Varela FJ, Thompson E, Rosch E (2017) The embodied mind, revised edition: cognitive science and human experience. MIT Press, Cambridge

Zhu D (2023) Bridge the channel, enhance the inclusivity: a comparison between flagship species-centered and moss-centered conservation in Chile and China. In: Rozzi R, Tauro A, Avriel-Avni N, Wright T, May RH Jr (eds) Field environmental philosophy: education for biocultural conservation, Ecology and ethics, vol 5. Springer, Cham, pp 457–482. https://doi.org/10.1007/978-3-031-23368-5_28

Chapter 9
How the Invisible Becomes Visible: Situating Scientific Illustration in Biocultural Communication

Valeria Serena Chiarini

Abstract As the world changes, so do the ways in which we live and communicate. Drawing is an essential part of our shared human heritage, and it was first used to communicate as well as to document everyday life. After the advent of photography and technological advancement, illustrators started to push beyond literal representation of what was visible to the human eye. The role of scientific illustration has changed and evolved from a purely descriptive role to something else; artists began to illustrate microscopic beings and features, phenomena invisible to the human eye, and abstract concepts. It is also used to spread messages. Scientific illustration can be a tool of biocultural conservation through the incorporation of memory, heritage, and place-based knowledge. Through a survey of scientific illustrators over the past century, I will demonstrate how artists' cultures and worldviews combine with their scientific illustration to communicate biocultural narratives. Through their illustrations, artists often convey complex ideas and phenomena that are invisible to the human eye. How an artist chooses to illustrate these phenomena is evolving as technology evolves, but what remains constant is that the illustrations are imbued with the perspective and worldview of the illustrator. Illustrators often intentionally carry additional meaning beyond the literal representation, with ethical implications as to how humans should engage with the natural world.

Keywords Art · Biocultural conservation · Biocultural representation · Decolonial · Invisible illustration

V. S. Chiarini (✉)
Scientific Illustrator and Communication Designer, Milan, Italy
https://valeriachiarini.netlify.app/

D. Zhu et al. (eds.), *Linking Arts with Biocultural Conservation, Restoration, and Communication*, Ecology and Ethics 8,
https://doi.org/10.1007/978-3-032-04830-1_9

9.1 Introduction

Art and science have had a deep relationship since before the advent of writing. Hunting scenes and human representation were the first scenes depicted in cave paintings by our ancestors. With scientific discoveries and the advent of writing, drawing also found new ways and subjects to document and explore.

Over the centuries, scientific illustration has adapted its role to the constant and rapid development of technology. Thus, illustration has moved from being a documentation tool toward a multifaceted way to communicate increasingly complex subjects. Microscopic, even "invisible" phenomena and concepts are increasingly translated visually. When expressing these imperceptible concepts, an artist makes choices on how to represent them. These choices carry with them the culture of the artist. Furthermore, through non-verbal communication, the artist enables cross-cultural interpretations. The choice to represent these concepts *in situ* (exemplifying the natural concepts within the natural world) allows for direct connections between the phenomena, the more-than-human beings experiencing them, and the artist.

Scientific illustrators offer, for instance, visual representations of the invisible connections among the life habits of co-inhabitants who share their habitats (i.e., the 3H's defined by the biocultural ethic, Rozzi 2012, 2013) (Fig. 9.1).

Since the nineteenth century, illustration has faced several changes connected to the development of new technologies. Technological change is a driving force behind changes in art, as is the rapid evolution of social contexts (Zhu 2026). In the current social context of the global society that we find ourselves in, speed is essential for any job. For this reason, technology is often chosen as a means of artistic work. The social and economic context is therefore also closely linked to the development of illustrations. However, speed was also a theme over a century ago; the invention of painting tubes in the nineteenth century changed the painting methods of artists. They were no longer forced to work in the studio, but they could work outdoors, "on the spot." American oil painter John Goffe Rand invented the metal paint tube as a way of transporting paints to use outside. The tubes were, in fact, syringes which were used to squeeze out paint and preserve the paint for a longer time, allowing artists to increase their mobility and flexibility as well as the possibility of a larger palette. As stated by art critics Robert Katz and Celestine Dars (1999),

> Pierre August Renoir is reported by his son, Jean, to have said, "Without paints in tubes, there would have been no Cézanne, no Monet, no Sisley or Pissarro, nothing of what the journalists were later to call Impressionism." (Katz and Dars 1999, p. 29)

Scientific discoveries are another factor that brings changes to illustrations. Scientific illustration is a bridge between people and science. In this role, scientific illustration is used as a communication tool between the sciences and the public. The relationship between illustration and science is constant, symbiotic, and dynamic. It has changed over time with new ways of thinking, scientific research, and technological tools. With constant changes comes a constant re-evaluation of the role scientific illustration plays in the digital era. Moreover, in this social context where people around the world are connected, what possibilities are there for scientific illustration to be a tool in biocultural conservation? By surveying the work of different scientific

Fig. 9.1 Scientific illustration of the biogeochemical cycle of sulphur (S), inspired by the 3Hs model of the biocultural ethic. It represents different co-inhabitants (volcanoes, rivers, trees, indigenous people), focusing on the indigenous Pewenche-Mapuche life *habit* of gathering seeds in forest *habitats* dominated by *Pewen* or Monkey puzzle (*Araucaria araucana*) trees on volcanic soils in southern Chile. *Pewen* seeds contain cysteine and methionine, two amino acids that contain sulphur, and one of them cannot be synthesized by the human body; hence it is essential for human health. In this way, a scientific illustration offers a visual representation of vital connections among the life habits of co-inhabitants who share their habitats; connections that are invisible to most citizens in global society. (Figure modified from Rozzi 2013, p. 24)

illustrators during a specific period, we can examine how scientific illustrations are directly connected to the illustrators' cultural backgrounds and prevailing scientific theories of their times. Scientific illustration is typically understood as an art form to be scientifically accurate as well as objective in its goals of dissemination. However, artists can be scientifically accurate and have differing ways of communicating. They can use different methodologies and, importantly, they can make different artistic choices to communicate a biocultural ethic based on diverse backgrounds and viewpoints. Consequently, illustrators can play a crucial role in helping the aims of valuing, respecting, and conserving biological and cultural diversity; in short, biocultural conservation.

9.2 The Context

In the 3Hs model of the biocultural ethic, the definition of *habitats* consists of a connection among biophysical, symbolic-linguistic, as well as socio-political, institutional, and technological dimensions, which define the places or habitats where humans and other organisms, individuals and community live and co-inhabit (Rozzi 2012). *Co-inhabitants* are humans and other beings who share the habitat. Rozzi explains that just as the concept of companion alludes to sharing bread (from Latin, *cum* = with; *panis* = bread), the term co-inhabitant refers to sharing a habitat. Sharing the habitat implies an ecological-evolutionary process, and taking care of the habitat implies an ethical duty. Therefore, the concept of co-inhabitant has a double meaning; it is both descriptive and normative (Rozzi 2018, p. 36).

Co-inhabitation implies an ethical duty: to care for the environment (Rozzi 2026). However, with an improper balance of political, economic, and military power, today we confront a "state of economic self-absorption and lack of socio-environmental responsibility" (Rozzi 2015a, p. 124). As a result, many indigenous and other local communities are displaced, and biological species are becoming extinct (Rozzi 2015b). The *habitat* is closely related to life habits, which are not only biological instincts but are related to the social and cultural context of the place. Thus, there are vital reciprocal influences among social and cultural contexts and the life habits of co-inhabitants (human and other-than-human).

Culture is an integral part of a place and its inhabitants. Moreover, different life habits have developed across the world and are reflective of the local co-inhabitants and their environments (their shared habitats). Art is a language which can speak to people across these cultural divides. Scientific illustration seeks to translate scientific topics to the public while overcoming cultural and linguistic differences. However, one's culture is unavoidably embedded in our artistic work. Hence, cultural contexts influence the layout, graphic style, and topics in scientific illustrations. For this reason, taking for example the specific time during the transition from the nineteenth to the twentieth century (1890–1910), we can see cultural differences reflected in the scientific illustrations created by Russian, German and Japanese illustrators.

Consider, for instance, the *Brockhaus and Efron Encyclopedic Dictionary*, a Russian encyclopedia consisting of 86 volumes with more than 7800 images. It is a Russian editorial masterpiece published between 1890 and 1907 by various authors, illustrators, and publishers. It includes not only naturalistic illustrations but also a set of different sciences and arts such as geometry, architecture, anatomy, and history. The illustrations are monochromatic, with the majority in black and white ink. Different subjects fill the whole page very precisely and rigorously in a symmetric composition (Fig. 9.2). This layout scheme is similar and probably influenced by the German scientist Ernst Haeckel.

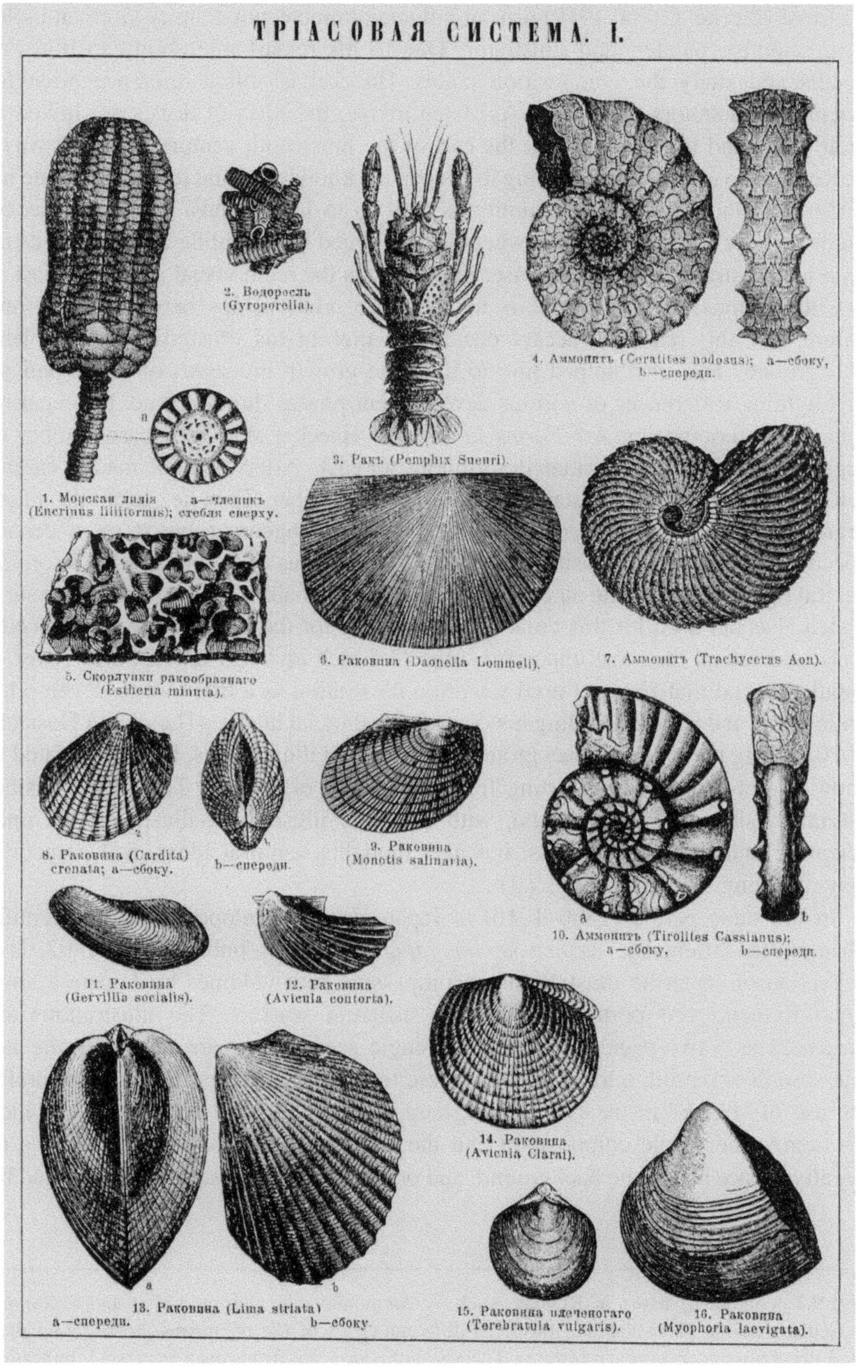

Fig. 9.2 Illustration of shells and marine biology included in the *Brockhaus and Efron Encyclopedic Dictionary* (1890–1907). An anatomical diagram of various Triassic period marine organisms, including ammonites, crustaceans, and bivalves. Each organism is labeled with its scientific

Ernst Haeckel (1834–1919) was an influential nineteenth-century German biologist, scientist, painter, and illustrator. Due to his remarkable ability to illustrate details and study the composition plates, Haeckel is still a reference artist for scientific illustrators nowadays. As stated above, the roles of artists are linked to their times and technologies. At the end of the nineteenth century, Haeckel was a polymath. He used art and drawing to deepen his knowledge and to communicate his findings to the scientific community as well as to lay citizens. In 1904, Haeckel published *Art Forms in Nature,* where he integrated his scientific attention to detail with his illustrator's aesthetic sense. Symmetry is the main visual aspect in most of his illustrations. Their purpose is to showcase relationships between shape and nature. For this reason, Haeckel decided to present his illustrations on a blank background. This also helped him to illustrate growth processes of the organism, highlighting differences of various development phases through time. In preparing himself for composing *Art Forms in Nature*, Haeckel studied a vast number of organisms and prepared written descriptions for each. Afterwards, he made a careful selection of organisms and started to draw them on paper. Then he cut out the single organisms and ordered them in each editorial arrangement based on a central specular symmetric composition. In this way, the focus of the illustration is on the central figure, and then the other figures are distinguishable. The central composition which Haeckel used for this book is eye-catching for the reader, who can immediately understand the most important subject of each layout (Fig. 9.3). Moreover, it should be said that Haeckel used scientific illustration as a true tool to develop his philosophy and ideas, including his controversial racial theories (Levit and Hossfeld 2020). Using his scientific background, theories and illustrations, Haeckel wanted to show evolutionary stages starting from unicellular organisms. Thus, the invisible became visible to Haeckel, who, with scientific illustration, illustrated not only scientific and abstract concepts, but also thoughts, cultural ideas, and a specific way of seeing the World (Fig. 9.3).

In the same period (1890–1910) in Japan, artists developed wildlife scientific illustrations differently. *Keinen kachō gafu*, which was published in 1892, is a collection of scientific illustrations encompassing four volumes by Japanese artist Imao Keinen, who portrays wildlife in different seasons. The illustrations are presented as a two-page spread, with a single scene. They are colored using ink and woodblock print, a traditional Japanese technique. Japanese illustrations prefer the use of "blank" in the page having large spaces of white backgrounds, which harmonize the whole composition and the poetics of the illustration. The use of negative space and white background, and other characteristics, can be illustrated by

Fig. 9.2 (continued) name in Russian, such as Ammonite (Ceratites nodosus) and Rakomina (Avicula contorta). The diagram illustrates different views of the organisms, highlighting their shell structures and anatomical features. The text at the top reads Триасовая система I, indicating the Triassic system. (Image in the public domain of Wiki Commons, https://upload.wikimedia.org/wikipedia/commons/c/c2/Brockhaus_and_Efron_Encyclopedic_Dictionary_b66_851-1.jpg)

Fig. 9.3 Scientific illustration drawn by Ernst Haeckel (1904) in plate 55 of his book *Art Forms in Nature*. Illustration from *Kunstformen der Natur* by Ernst Haeckel, depicting various intricate shell forms of the class Acephala. The image features symmetrical, detailed patterns and textures, showcasing the natural artistry of these marine organisms. The shells are arranged in a grid-like pattern, each with unique ridges and curves. In the illustration, it can be seen his uses of symmetry, and how Haeckel illustrated his monogenism thought. (Image in the public domain of Wiki Commons, https://upload.wikimedia.org/wikipedia/commons/thumb/b/b7/Kunstformen_der_Natur_%28Tafel_55%29_BHL47388253.jpg/640px-Kunstformen_der_Natur_%28Tafel_55%29_BHL47388253.jpg)

Fig. 9.4 An illustration by Imao Keinen 今尾景年 for *Keinen kachō gafu* 景年花鳥画譜*, 1892*. In this illustration, it is possible to see the way Japanese artists illustrate scientifically. Not only the subject but the dynamism of the movement describes life. The illustration is almost without outlines, using just colors to create textures and gradients. (Image in the Creative Commons Zero, Public Domain Dedication of Wiki Commons, https://upload.wikimedia.org/wikipedia/commons/a/ad/Vogel_bij_rups_Bloemen-_en_vogelschetsen_van_Keinen_%28serietitel%29_Keinen_kacho_gafu_%28serietitel_op_object%29%2C_RP-P-2004-508D-8.jpg)

another work of Keinen 今尾景年 (1845–1924) in Keinen kachō gafu 景年花鳥画譜 (Fig. 9.4).

As mentioned earlier, scientific illustration should be objective. However, the artists' aesthetics and culture influence their art. For instance, differences among Japan, Russia, and Germany's cultural traditions are reflected in the above-mentioned works, even though they were created in the same historical period. In Japanese poetics, the relationship and natural connections are the focus of scientific illustrations. This focus is not only on the identification of a species, but also on examining another living being within its broader context. This style aligns well with the 3H's model of the biocultural ethic (Rozzi 2012), because it presents living beings in their native habitats, including ecological interactions with their co-inhabitants, and often illustrates particular life habits. On the other hand, German and Russian illustration has a layout more precise, rigorous and static, with the precise idea of an encyclopedia as an inventory or collection of species. As criticized by Esteban (2018), this style removes the beings from the context of their habitat, co-inhabitants, and their habits. It forces a hierarchy and layout that is visually pleasing but lacks the ecological information and biocultural ethical connotations included in the Japanese style.

In the twenty-first century, with technology and globalization we can communicate and engage with the global community. Things are no longer separated as they were centuries ago; artists use multiple drawing methods on different mediums and technologies across the globe to work on scientific illustrations. The wealth of cultures and communities in the world gives us a range of different languages to express the same concept or subject. While there is no universal language, scientific illustration—and art in general—seeks to speak to everyone. However, this aim is challenged by the fact that art is communicated and perceived under the influence of particular cultural contexts. As shown by the nineteenth-century examples presented above, conceptual and practical aspects of scientific illustrations are intimately linked to specific cultural dimensions. For example, both Russian and Japanese have different alphabets from the Latin script. The reading orientation is also different; for example, the Japanese read from top to bottom, from right to left. The different reading senses influence the page layouts, developed for the different languages of readers. Art itself is a language that is informed by the artist's language and culture. The illustration will always be influenced culturally, and the audience also has cultural influences that they bring with them when interpreting the illustrations. Hence, two challenging questions emerge: how to achieve a universal language for scientific illustrations amidst a cultural melting pot? Or, alternatively, should we embrace diversity in our artistic work on scientific illustrations?

9.3 Situating Scientific Illustration in Biocultural Communication

Language is fundamental for the relationship between people and land, and therefore for conservation. Considering the biocultural concept of habits, the cultural and linguistic dimension allows the maintenance of traditions and community languages that inform how we engage with the natural world.

Art allows for different interpretations depending on the viewer (Rossi-Linnemann and De Martini 2019). While scientific illustration may attempt to show scientific subjects "objectively," one must ask whether it can be used as a universal language, or if the cultures of particular people are embedded in their art in a similar way they are embedded in its language.

Translating a verbal or written subject into a graphic format relies on dialogue with people, studying and having the humility to listen and learn from different cultures. This is a key aspect of biocultural conservation. Cross-cultural communication is always a complex process, which is compounded by the cross-media aspect of translating the written and spoken word into an art form. As examined in case studies below, there is often a dialogue between illustration and the natural world, between culture and memory, between people and place.

9.3.1 Scientific Illustration as Communication: Anthropologists, Botanists, Conservation, Memory, and Heritage

Scientific illustration today is being used to explore conceptual and intangible concepts such as time and memories, but historically, nature was represented by artworks called *still life*. Still life is an artistic category in which there are compositions with various subjects: flowers, fruit, vegetables, spices, animals, beverages, and rarely people. Almost all artists paint still life in their careers. These types of artworks could be seen as just exercises of colors and techniques, but today they contain data useful for historians and botanists. From a still life, not only can they deduce data on vegetable subjects, but you can also understand different social cross-sections of that time. Furthermore, the subjects were not chosen randomly, but they relied on composition studies based on what kind of message the artist wanted to spread.

Still life is not just a daily life testimony about that historic period in a particular area. They also offer us a way to see the adaptation of nature through the centuries and in connection with humans and vice versa. When botanists study still life, they can identify the species in the painting, but also, they can observe the original area, how far away the painter was, and thus have data for the traffic of fruit and vegetables in that period. Furthermore, it is possible to identify the season of the painting: in that period, between the fourteen and sixteenth centuries, there was not the annual overproduction, production out season, and globalization like today. Still life paintings of the past offer us information about the historical period and society in that precise moment when the artist was living. Was the painted species native? How widespread was it at the time? Has it changed because of human intervention? Thanks to this documentation from the past, we can define how plants have changed through time and the difference between then and now. These paintings make visible what has often become invisible to most citizens today—namely, the impact of human intervention through selective breeding—and share historical intercultural interactions that led to the movement of these plant species around the planet.

Around 1817, Don Mariano Lagasca (1776–1839), director of *Real Jardín Botánico de Madrid* (Royal Botanical Garden of Madrid), received 104 cases of cargo in *Cadice* port. He did not know that the content of the cases would be one of the greatest contributions to scientific communication in history. The cargo contained illustrations of seeds, resins, minerals, manuscripts, and plants by José Celestino Mutis (1732–1808). Mutis was an important botanist and scientific illustrator who led Spain into botanical expeditions, particularly to South America, at the turn of the century. Mutis studied medicine in Madrid and was one of the first students of the botanist Carolus Linnaeus (1707–1778), who developed the nomenclature system we still use today.

During the eighteenth century, the Spanish organized three royal botanical expeditions: one to the Viceroyalty of New Granada, another to the Viceroyalty of New Spain (1788–1802), and a third to the Viceroyalty of Peru and Chile

(1777–1788). Unlike the leaders of the other two expeditions, who had not crossed the Atlantic for their scientific work, Mutis was already residing in New Granada and was committed to the land and its people (Silva 2021). The purpose of his expedition to the New Reign of Granada (which became Colombia), about 8,000 square kilometers along the Magdalena River, was to draw up an inventory of the country's natural wealth. Mutis and his team of illustrators, zoologists, and botanists brought together thousands of drawings, bird collections, animals, and a herbarium including more than 24,000 plants. He also wrote hundreds of manuscripts and notes. His *Flora de Bogotá o de Nueva Granada* contained more than 6000 illustrations and was so dense in information and drawings that this work was key in documenting the diverse flora of present-day Colombia.

In the illustrations, Mutis extrapolated information from the herbal sample. The result is a symmetrical idealization of the parts and incredible attention to detail. The illustrations from New Granada are finished products, meticulously drawn and painted. Despite the colonialist focus of these expeditions, Mutis's illustrations could be an example of co-inhabitants/habitats. The illustrations from New Granada display, if not revolutionary ideals, at least a good dose of local pride. This local patriotic inflection becomes more evident when comparing the artistic production from New Granada with botanical illustrations produced on other contemporary royal botanical expeditions (Silva 2021).

As the role of art is ever-changing, the purpose of scientific illustration is no longer simply documenting fruit and vegetables but also delivering and spreading messages. Botany has become an art language as exemplified by the work of contemporary Brazilian artist and activist, Maria Thereza Alves. She works in various media, including painting, photography, and film. Her choice of media depends on the message that she wants to spread.

In 2020, Alves' installation called "*Decolonizing Botany/Jevy Jejapo-pyra Temití-tyre*" was exhibited in Naples, Italy. For this installation, she chose to use illustrations. It is a collection of illustrations made with watercolors of native Brazilian plants. They are the result of Alvez's trip to Berlin, where she visited the Botanical Garden and saw that every native Brazilian plant was named after European figures. They have been "dedicated in their honor" but—for Alves—this process highlights the legacy of colonialism imposed on countries and cultures conquered in the past, also changing the names of the native plants of that area. "Decolonizing Botany" intends to question the common practice of assigning colonizers' names to plants. Using watercolor illustrations, with a critical decolonial perspective, Alves reclaims the identity, life habits and the environment (habitat) of plants native to Amazonia and other regions of Brazil (Fig. 9.5). Watercolor is a physical technique, fast and spontaneous, which uses water to dissolve colored pigments on paper. Her choice to use this medium to illustrate the plants is reminiscent of Mutis's illustrations.

As mentioned above, Alves made her *Decolonizing Botany/Jevy Jejapo-pyra Temití-tyre* with the intent to liberate Brazilian plants from their colonized names. This was also a political position about the state of her country. The scientific illustration communicates the physical aspects of the plant and preserves the legacy

Fig. 9.5 Maria Thereza Alves, *Guapo ypere pegua ñato'ĩ—I have water but am fragile, I have spirit. Costus scaber.* In this figure, she shows us the use of botanical art as self-affirmation for her culture and herself. Here, illustration is a tool of identity as co-inhabitants and freedom. This image is part of the installation *Decolonizing Botany/Jevy Jejapo-pyra Temití-tyre*, 2020. (Photo credit: Nick Ash, Image courtesy of Maria Thereza Alves)

of colonization on the landscape. From an activist standpoint, it rails against past and present colonization.

This is an installation in collaboration with the Guarani people, present in the Brazilian region of Mato Grosso do Sul. It's essential to talk about the inhabitants of the area and how they collaborate with Alves. In fact, the installation not only features Alves's watercolors, but beneath them, we can find a speaker playing songs recorded with the Guarani people. Alves asked the Guarani leader, Ke'y Rusú Katupyry, to explain the 24 species she illustrated, and they explained them with their songs, a form of freedom and tradition. The process of naming plants, which occurs in Guarani, takes place in the tranquility of the evening. The community gathers and waits for the plant to give it its name, usually in a song. This decolonizing expression of the Guarani co-inhabitants is the core of the installation, which, in collaboration with Alves, became a powerful tool for spreading their biocultural message.

Recording is a tool often used in biocultural conservation to obtain oral testimony of the relationship between co-inhabitants and their habitats. In southern South America we find another example of recording biocultural diversity, the Multi-Ethnic Guide to Birds of the Subantarctic Forests of South America (Rozzi et al. 2006, 2010). This work aims to better understand, value, and respect the biocultural diversity associated with avifauna, and to foster inter-cultural and inter-species dialogues and forms of co-inhabitation. Chilean philosopher and conservationist Ricardo Rozzi (2010, p. 16) explains that the idea emerged during a workshop he co-organized in 1995 in Chiloé National Park, in the territory of the Williche-Mapuche indigenous communities of Chanquín and Huentemó in southern Chile. The activities and conversations shared with members of the indigenous communities, park rangers and ornithologists from southern Chile and Argentina expressed not only the richness of bird diversity, but also the appreciation that various ornithologists have for them. Rozzi then thought that the biological diversity of birds and the cultural diversity of the names and stories about them could be better known and understood through the development of a multi-ethnic guide to birds that contained recordings of human and bird voices.

Throughout history, scientific illustrations have carried with them the worldviews of their creators and inherently conveyed ideas about how humans should engage with the natural world. More recently, this perspective has expanded with artists actively using scientific illustration to carry explicit messages about how humans should engage with the natural world and to promote biocultural ideals.

9.4 My Own Work: How the Invisible Can Become Visible to Citizens

Scientific illustration plays a key role in communicating complex science. To effectively communicate, the artist must be able to understand the data, results, and ideas underpinning their topic of illustration. These complex ideas are processed by the illustrator to translate them into an illustration that helps the public to understand complex topics. This is particularly important when illustrating scientific phenomena that are imperceivable to our human senses. For example, to illustrate ultraviolet rays, as an artist, I had to first comprehend them from a scientific viewpoint, and then I had to make choices on how to visually communicate this phenomenon by accurately and engagingly presenting "invisible elements" to the public.

9.4.1 A Biocultural Approach to Illustrate Ultraviolet Vision

Over the years, I have developed two different designs to show ultraviolet vision. For these projects, I needed to understand the research on ultraviolet vision. The illustration of the animal vision shows us how they have other ways of seeing our planet. My goal is to emphasize the fact that humans are part of an ecosystem, not above, not separated from nature. Our vision is by no means the only one. Seeing through the eyes of our more-than-human co-inhabitants can help us to build a biocultural ethic because we can better understand, value, and respect the perspectives of other beings.

As humans, our main sensory organ is vision. It is primarily how we engage with and navigate the world around us. For this reason, it is even more impactful to present different vision-systems and compare other visions with our own. The overarching message of my projects is always focused on the dissemination of ethics; to encourage mindfulness and respect for living beings and for the habitat we share with them, to move away from anthropocentrism and to treat nature and living beings with respect and awareness. In his seminal biocultural work, Multi-ethnic bird guide of the Sub-Antarctic forests of South America, Rozzi (2010 [2002]) states that,

> Each time that we observe a bird, our perception is informed by something of the bird, and something of ourselves—our senses, instruments, and concepts. In this way, a dialogue arises between the birds themselves, and the names and stories that express our human perceptions of them. The variety of birds, with their distinct vocalizations and behaviors, is as great as the richness of ornithological names and narratives, with their particular human ways of looking at the birds and inhabiting the world. (Rozzi 2010, p. 29)

Then Rozzi asks,

> In the face of the diversity of names and stories about each bird species, a question arises: How can so many narratives exist about the same bird? Are they all false or all true? Are some truer than others? (Rozzi 2010, p. 30)

To answer these questions, he proposes the following biocultural analogy,

> In the temperate forests of southern South America, the hummingbird (*Sephanoides sephaniodes*) and the bumblebee (*Bombus dahlbomii*) visit the flowers of the *kolkopiw* vine (*Philesia magellanica*). The hummingbird and the bumblebee take the nectar of this flower, obtaining the sugar and water that permit them to survive. On the other hand, upon visiting the flowers, the hummingbird and bumblebee help the reproduction of this plant by transporting pollen from flower to flower. In this way, more flowers will be produced, and the cycle of life continues. This scene can be observed in the native forests of southern Chile and Argentina, and it seems "natural." It would not occur to us that the bumblebee should teach the hummingbird to visit the *kolkopiw* flowers in the way of an insect—or vice—versa: the hummingbird should teach the bumblebee to visit the flowers in the way of a bird.
>
> In an analogous way, we can interpret the diversity of indigenous Yahgan, Mapuche and Western-scientific viewpoints about birds. Each of these cultures, and languages, recognizes the bird species and interacts with them in a particular way. As has occurred with the hummingbird and bumblebee, the Mapuche and Yahgan cultures have inhabited and survived in the austral forests for thousands of years. The traditional ecological knowledge … has been suitable to allow the survival of these peoples, with their own languages, and ecological practices. Therefore, their stories are not merely "myths" or "legends," but rather they interpret the natural reality in an effective manner. (Rozzi 2010, pp. 30–31)

Rozzi and Jiménez (2014) concludes that in the twenty-century, bird calls, bird stories, and indigenous traditional ecological knowledge invites us, as members of contemporary global society, to diversify our ways of knowing about, and co-inhabiting with birds, insects, and their ecosystems.

In my own research, I started by understanding how birds and bees have little in common because they belong to different phyla and classes of the animal kingdom. Hence, their anatomy is very different, and for the most part, they have different feeding patterns. However, we realize that their goal of survival is the same. To better understand how ultraviolet vision works for bees and birds, we should study their goal in wildlife: why do they need UV vision? Bees have a hierarchical structure based on the hive. Foraging bees' role is to harvest pollen and nectar from flowers and then come back to the hive. For the birds, the UV situation depends on the species—not every bird has an ultraviolet cone. However, when they have this capacity, the purpose of their UV vision will be different according to its use. Some birds use ultraviolet tracks to hunt, other species use it to choose a partner, and sometimes to distinguish their own nest.

One of the bees' main life habits is to harvest nectar for the hive. Nectar is the primary nourishment for the colony. Foraging bees must be efficient to do their best for the colony, and their ultraviolet vision allows them to identify different flowers through UV patterns (Richtmyer 1923; Rausher 2008) . Foraging bees are fast and precise. Honeybees are able to transport loads of pollen and nectar that are as high as 40% of their body mass (Feuerbacher et al. 2003). Bee species that visit orchids in Central and South America have body masses that range from 50 to 1000 mg with tongue lengths of 10–40 mm (Roubik and Hanson 2004), with impressive abilities to feed on nectar (Borrell 2007). Bees and flowers have a symbiotic relationship where bees collect nectar and pollen, and flowers use bees for pollination (Lutz 1924). Both have the goal of survival and have come to depend on each other for their survival.

Four orders of birds can see in UV wavelengths: Passeriformes (songbirds), Struthioniformes (ostriches), Ciconiiformes (storks), Psittaciformes (parrots and cockatoos) (Carvalho et al. 2011). Birds have different purposes for their ultraviolet vision. First, in the breeding season, the males of some bird species have a greater UV reflection than females. Males have more ultraviolet feathers on the chest and a brighter neck than females. Moreover, the feather can show the female the age and the physical condition of the male. All these factors influence the female choice of her partner (Mullen and Pohland 2008; Rajchard 2009). For birds, ultraviolet perception can also play a role in parental care. Some species are known to have a chromatic marker on their nests and chicks (Tanner and Richner 2008). The chicks have very little plumage, and it is concentrated around the UV-reflecting beak. These feathers allow the parent to identify their own offspring, but most of all, they are important to help the parent identify the little beak of the chick and to feed them in a faster and more precise way (Avilés and Soler 2009). Some studies have also demonstrated that birds can use their ability to see ultraviolet light to recognize eggs (Węgrzyn et al. 2011; Honza et al. 2007). Another study also found that leaves reflect ultraviolet rays in different ways on the upper and underside, this allows one to see and perceive the position of the leaves and their orientation during the flight and helps in navigation and flight (Altshuler 2001; Tedore 2019).

Fig. 9.6 Illustrations from *The Invisible Becomes Visible* by Valeria Serena Chiarini (2021). On this page, it can be seen the comparison between the page (on the left) and a section of the same illustration using an ultraviolet torch (on the right)

As a scientific illustrator, this information and data are essential to have a better understanding of the topic which we need to illustrate. My first approach to ultraviolet vision was for my editorial project called "Invisible Became Visible: Ultraviolet Vision of Birds and Bees," which focused on finding a way to illustrate ultraviolet light in the animal and plant kingdom (Fig. 9.6). Thanks to research and several scientific papers, it was possible to study different rich materials and experiments, full of inspiration.

The goal of this picture book was to introduce readers to the "invisible vision" of birds. Using as much data as possible from scientific studies, I was able to show how unique bird species have peculiar but "invisible plumages" that become visible under the light of an ultraviolet flashlight. Bird plumages or floral patterns have been illustrated after printing the book, using an ultraviolet marker that I implemented to highlight ultraviolet parts of the organisms.

Through the years, I have deepened my interest in animal vision. In 2023, I developed another project focused on the ultraviolet vision of bees, which resulted in my project-book After Sight (Chiarini 2024). Each species is illustrated on a two-page spread (Fig. 9.7). The first page introduces the species' name and other biological characteristics, and then there is a window which shows a portion of the illustration of the following page. The window shows just the portion which visually describes the bees' vision. When the reader flips the page to see the complete

Fig. 9.7 Examples of illustrations in *After Sight* by Valeria Serena Chiarini (2024)

illustration, which was hidden before, they will see the flower with a different color. That color will be how we perceive that flower.

The surprise for readers is enhanced by the time they have to wait to see the other page, and the movement of turning the page that will reveal the entire subject. In this way, my book *After Sight* invites readers to understand how our human vision is different from that of bees. The layout of two pages for each subject enables readers to have two different visions without external tools. It offers them a way to use the pages as tools through which readers interact with the materials of the book while reading it.

The aim of my illustrations has been to straightforwardly show facts and conceptual themes linked to ultraviolet rays that are invisible to humans. However, UV rays are around us in everyday life in multiple ways, both natural and artificial. Besides ultraviolet rays, other types of invisible rays exist in our daily lives. In the book *Invisible Nature: A Secret World Beyond Our Senses* by Catherine Barr (2020), and illustrated by Anne Wilson, the authors show different rays and invisible things in our world. This book explains the invisible world around us, from ultraviolet light to infrasound, to children. The rays are represented as waves and drawn together with writings that deepen the message for the young reader. The goal is to show children also the quantity of things that we do not know yet, things which we cannot see but exist. In addition, the visual style of Barr's book does not use the realism of typical scientific illustrations. However, from my point of view, their illustrations could indeed be considered as scientific illustrations if we define them as the place where art and science converge. This book also has a biocultural ethic, encouraging children to have empathy and understanding for how other beings navigate and engage with the world around them.

The book called *Curious Creatures Glowing in the Dark* by Zoë Armstrong (2021), and illustrated by Anja Sušanj, shows the bioluminescence and the adaptation of many living beings living in the dark and their communication using lights. This book enables readers to find on a given page a callout box called *"Just like you…"* where the author gives some examples of these phenomena's applications to human society. For example, in the section dedicated to the Japanese fireflies and their light communications, in the section "Just like you…," Armstrong compares that goal's application to light-up running shoes or glittery jackets to catch attention. Furthermore, Armstrong also uses Morse code with a flashlight to send messages to a friend. This explicit connection drawn between human behavior and animal behavior helps demonstrate to children that humans are not separate from the natural world, but they simply have different ways of accomplishing the same goals.

Curious Creatures Glowing in the Dark is another example of how illustration is a medium more suitable for spreading these kinds of complex or difficult-to-visualize messages to children. Moreover, the glowing phenomena could be difficult to photograph for each animal that the book describes and illustrates. These are some examples of how scientific illustration is still used today, but with different target audiences in the meantime, it has maintained its role in communicating, also in the scientific community, with more mature targets.

9.5 Conclusion

The role of scientific illustration has changed throughout history, as has the scientific, social and economic context. From documenting life and discoveries, illustration has become a tool for conservation and to show increasingly sophisticated and complex scientific research. It also facilitates dialogue among different cultures as well as between the scientific community and the public.

Showing ultraviolet rays, extinct beings, or sounds are starting points for revealing the invisible through illustration. As I mentioned at the beginning, paint tubes fueled the development of Impressionism, an artistic movement where painters focused on depicting light and its reflection. It's essential for humans to understand our surroundings, and the way to spread and explain that understanding is often through illustration. Illustrating dimensions of animal visions that are invisible to humans is a work that I have carried on over the last years, intending to use scientific illustration as a means for communicating ethical interactions between humans, our co-inhabitants, and our shared habitat. Understanding the various relationships in nature helps us to have a better view of the conservation of our planet's habitats.

From Mutis and Haeckel to Alves and Martín, scientific illustration has changed along with society, ways of thinking, and with the needs of science. It should be remembered that scientific illustration is a bridge between science and the public. Several aspects can change the world of scientific illustration, but the analytical and attention to detail approach remains constant over the centuries. The advancement of scientific research also shapes the field, and as we learn more, the possibilities of how to communicate these new discoveries also grow. For example, we only know around 5% of our ocean floor, and our knowledge of space and physics is constantly growing. Many species, their life habits and habitats remain to be discovered and explored. The invisible or intangible factors that influence our world and our lives may best be represented through illustrations.

Illustration is a biocultural tool to convey different points of view and knowledge of the same subject in different cultures. It can be imbued with the ethics of the illustrator and carry their motivations to help foster a biocultural ethic. It can also allow you to experience the world through the eyes or ears of another being, and through this, encourage a biocultural experience where barriers between humans and the rest of the natural world are broken down.

For all the reasons explained above, I maintain the importance of having a scientific illustrator or designer as a permanent member of conservation teams. It is necessary to have an artist or specialist who knows how to disseminate scientific research and knowledge to the public. These specialists can play a crucial role in biocultural conservation and future field and laboratory research.

In my future projects, I will investigate other forms of vision and explore in greater depth the relationship between co-inhabitants and their habitats, and the interconnections with various forms of vision. Another crucial goal I have in my future work is to continue highlighting the relationship between science and scientific illustration to the public. We urgently need scientific illustrators to communicate past and present scientific topics to prevent science from becoming increasingly restricted to an elitist circle and failing to impact global society. As part of interdisciplinary teams, scientific illustrators can catalyze biocultural transformations that contribute to more sensitive, just, respectful, and sustainable forms of co-inhabitation.

Acknowledgments I would like to thank Danqiong Zhu and Ricardo Rozzi for their valuable comments on draft versions of this chapter.

References

Altshuler DL (2001) Ultraviolet reflectance in fruits, ambient light composition and fruit removal in a tropical forest. Evol Ecol Res 3:767–778

Armstrong Z (2021) Curious creatures glowing in the dark. Flying Eye Books, London

Aviles JM, Soler JJ (2009) Nestling colouration is adjusted to parent in visual performance in altricial birds. J Evol Biol 22:376–386

Barr C (2020) Invisible nature: a secret world beyond our senses. Little Orchard, Burley Gate, Hereford, HR1 3QS, Otter-Barry Books, UK

Borrell BJ (2007) Scaling of nectar foraging in orchid bees. Am Nat 169(5):569–580

Carvalho LS, Knott B, Berg ML, Bennett ATD, Hunt DM (2011) Ultraviolet-sensitive vision in long-lived birds. Proc R Soc B.278:107–114. https://doi.org/10.1098/rspb.2010.1100

Chiarini VS (2021) The invisible becomes visible. Unpublished. https://valeriachiarini.netlify.app/projects/

Chiarini VS (2024) After sight. Unpublished. https://valeriachiarini.netlify.app/projects/

Esteban JM (2018) Dürer's rhinoceros: biocultural homogenization of the visual construction of nature. In: Rozzi R et al (eds) From biocultural homogenization to biocultural conservation, Ecology and ethics, vol 3. Springer, Cham. https://doi.org/10.1007/978-3-319-99513-7_9

Feuerbacher E, Fewell JH, Roberts SP, Smith EF, Harrison JF (2003) Effects of load type (pollen or nectar) and load mass on hovering metabolic rate and mechanical power output in the honeybee Apis mellifera. J Exp Biol 206(11):1855–1865

Honza M, Polacikova L, Prochazka P (2007) Ultraviolet and green parts of the colour spectrum affect egg rejection in the song thrush (Turdus Philomelos). Biol J Linn Soc 92:296–276
Katz R, Celestine D (1999) The impressionists handbook. Sterling Publishing Company, Inc, New York
Levit G, Hossfeld U (2020) Ernst Haeckel, Nikolai Miklucho-Maclay and the racial controversy over the Papuans. Front Zool (Open Access). https://doi.org/10.1186/s12983-020-00358-w
Lutz FE (1924) The colors of flowers and the vision of insects, with special reference to ultraviolet.
Mougeout F, Redpath SM, Leckie F (2005) Ultra-violet reflectance of male and female red grouse, Lagopus scoticus: sexual ornaments reflect nematode parasite intensity. J Avian Biol 36:203–209
Mullen P, Pohland G (2008) Studies on UV reflection in feathers of some 1000 bird species: are UV peaks in feathers correlated with violet-sensitive and ultraviolet-sensitive cones? IBIS 150(1): 59–68
Rajchard J (2009) Ultraviolet (UV) light perception by birds: a review. Vet Med 54(8):351–359
Rausher MD (2008) Evolutionary transitions in floral color. Int J Plant Sci
Richtmyer FK (1923) The reflection of ultraviolet by flowers.
Rossi-Linnemann C, De Martini G (eds) (2019) Art in science museums: towards a post-disciplinary approach. Routledge, London
Roubik DW, Hanson PE (2004) Orchid bees of tropical America: biology and field guide. Editorial InBio, San Jose
Rozzi R (2010) Introduction. In: Rozzi R, Massardo F, Anderson C, McGehee S, Clark G et al (eds) Multi-ethnic bird guide of the Sub-Antarctic forests of South America. University of North Texas Press, Denton, pp 15–35
Rozzi R (2012) Biocultural ethics: recovering the vital links between the inhabitants, their habits, and habitats. Environ Ethics 34(1):27–50. https://doi.org/10.5840/enviroethics20123414
Rozzi R (2013) Biocultural ethics: from biocultural homogenization toward biocultural conservation. In: Rozzi R, Pickett STA, Palmer C, Armesto JJ, Callicott JB (eds) Linking ecology and ethics for a changing world: values, philosophy, and action, Ecology and ethics book series, vol 1. Springer, Dordrecht, pp 9–32. https://doi.org/10.1007/978-94-007-7470-4_2
Rozzi R (2015a) Implications of the biocultural ethic for earth stewardship. In: Rozzi R, Chapin FS III, Callicott JB, Pickett STA, Power ME, Armesto JJ, May RH Jr (eds) Earth stewardship: linking ecology and ethics in theory and practice, Ecology and ethics, vol 2. Springer, Cham, pp 113–136. https://doi.org/10.1007/978-3-319-12133-8_9
Rozzi R (2015b) Earth stewardship and the biocultural ethic: Latin American perspectives. In: Rozzi R, Chapin FS, Callicott JB, Pickett STA, Power ME, Armesto JJ, May RH Jr (eds) Earth stewardship: linking ecology and ethics in theory and practice, Ecology and ethics, vol 2. Springer, Cham, pp 87–112. https://doi.org/10.1007/978-3-319-12133-8_8
Rozzi R (2018) Biocultural homogenization: a wicked problem in the anthropocene. In: Rozzi R, May RH, Chapin FS, Massardo F, Gavin MC et al (eds) From biocultural homogenization to biocultural conservation, Ecology and ethics, vol 3. Springer, Cham, pp 21–48. https://doi.org/10.1007/978-3-319-99513-7_2
Rozzi R (2026) The contributions of Eugene Hargrove to environmental ethics and aesthetics: a biocultural tribute. In: Zhu D, Rozzi R, Tsuji R, Johnson B, Caviness T, Castro-Jorquera C, Xu L, Zhang L (eds) Linking arts with biocultural conservation, restoration, and communication. vol 8. Springer, Cham, pp 35–59. https://doi.org/10.1007/978-3-032-04830-1_3
Rozzi R, Jiménez JE (eds) (2014) Magellanic sub-Antarctic ornithology: first decade of long-term bird studies at the Omora ethnobotanical park, Cape Horn Biosphere Reserve, Chile. University of North Texas Press, Denton
Rozzi R, Massardo F, Anderson C, Heidinger K, Silander J Jr (2006) Ten principles for biocultural conservation at the southern tip of the Americas: the approach of the Omora Ethnobotanical Park. Ecol Soc 11(1):43. https://doi.org/10.5751/ES-01709-110143
Rozzi R, Massardo F, Anderson C, McGehee S, Clark G et al (2010) Multi-ethnic bird guide of the Sub-Antarctic forests of South America. University of North Texas Press, Denton

Silva A (2021) Flowers of the colony, seeds of independence: the illustrations of the royal botanical expedition to Nueva Granada. Selva: a journal of the history of art, 3. https://selvajournal.org/article/flowers-of-the-colony-seeds-of-independence-the-illustrations-of-the-royal-botanical-expedition-to-nueva-granada/

Tanner M, Richner H (2008) Ultraviolet reflectance of plumage for parent–offspring communication in the great tit (Parus major). Behav Ecol 19(2):369–373

Tedore C, Nilsson DE (2019) Avian UV vision enhances leaf surface contrasts in forest environments. Nat Commun 10(1):238

Węgrzyn E, Leniowski K, Rykowska I, Wasiak W (2011) Is UV and blue-green egg colouration a signal in cavity-nesting birds? Ethol Ecol Evol 23(2):121–139

Zhu D (2023) Bridge the channel, enhance the inclusivity: a comparison between flagship species-centered and moss-centered conservation in Chile and China. In: Rozzi R, Tauro A, Avriel-Avni N, Wright T, May RH Jr (eds) Field environmental philosophy: education for biocultural conservation, Ecology and ethics, vol 5. Springer, Cham, pp 457–482. https://doi.org/10.1007/978-3-031-23368-5_28

Zhu D (2026) Arts and environment after the emergence of artificial intelligence. In: Zhu D, Rozzi R, Tsuji R, Johnson B, Caviness T, Castro-Jorquera C, Xu L, Zhang L (eds) Linking arts with biocultural conservation, restoration, and communication, Ecology and ethics, vol 8. Springer, Cham, pp 171–187. https://doi.org/10.1007/978-3-032-04830-1_10

Chapter 10
Arts and the Environment After the Emergence of Artificial Intelligence

Danqiong Zhu

Abstract The emergence of artificial intelligence (AI) is reshaping numerous facets of human existence. In this chapter, I examine the philosophical implications that AI art generators have for citizens, artists, and professionals engaged in the interface of arts and the environment. Drawing on historical and cross-cultural perspectives, in the first section I examine how early and contemporary technological imaginaries have shaped human perceptions of the environment, focusing on the views of three key thinkers: early twentieth-century Czech science fiction writer Karel Čapek, ancient Taoist Chinese philosopher Zhuangzi, and contemporary Indigenous Potawatomi ethnobotanist and writer Robin Kimmerer. In the second section, I analyze the nineteenth century Hudson River School landscape paintings to explore the unique historical case of intertwinement among landscape painting, environmental aesthetics, and conservation. In contrast, AI-generated images, while proficient and expeditious in producing visual contents, often fail to capture these nuanced biocultural interplays between artists, arts, and the environment. This underscores a deeper consideration: artists' biophysical interaction with the environment or general human biophysical labor encompasses more than mere physical tasks; it embodies pluralistic, evaluative actions embedded in symbolic, cultural, and socio-ecological systems. An examination of the definitions of human labor, which were proposed by Locke, Marx, and Arendt, reveals the necessity of reconceptualizing human labor within the framework of biocultural ethics. Through this reconceptualization, I see the artists, or even human engagement with the environment in general, as inherent in the creative expression of human existence and artistry. While AI may generate landscape artworks, it risks alienating humans from in-person experiences with the environment, presenting a challenge to professionals in biocultural conservation, arts, education, AI development, philosophy, and policymaking. Despite AI's

D. Zhu (✉)
Sub-Antarctic Biocultural Conservation Program, Department of Biological Sciences, University of North Texas, Denton, TX, USA

Cape Horn International Center (CHIC), Universidad de Magallanes, Puerto Williams, Chile
e-mail: Danqiong.Zhu@unt.edu

D. Zhu et al. (eds.), *Linking Arts with Biocultural Conservation, Restoration, and Communication*, Ecology and Ethics 8,
https://doi.org/10.1007/978-3-032-04830-1_10

potential to simulate and replicate, I claim that artistic expression and human creativity are evoked by biocultural interactions in face-to-face encounters, and therefore arts are produced from an ecological and biocultural imperative, which prompts those at the nexus of the arts and the environment to critically evaluate how AI can be leveraged without compromising the fundamental human need for connection with the natural world.

Keywords Artificial Intelligence (AI) · Hudson River School · Art generator · Labor · Biocultural ethic

10.1 Introduction

When philosophy was in its infancy, some people became philosophers because of their relentless pursuit of answers to questions about nature (*physis*), its origin or cosmogony (*árkhō*, ἀρχή), its dynamics and motion (*kinesis*, κίνησις), and the right way to live or Dao (*dào*). Today, living in a knowledge- and information-intensive era, people ask a myriad of questions every day, but the way people pursue answers has drastically changed. According to Google search statistics for 2024, every second Google received 99,000 search queries. As the most phenomenal information technology aggregator, which is undergoing a meteoric rise, the open Artificial Intelligence (AI) ChatGPT is becoming a platform for responding quickly to people's questions.

AI is rapidly expanding to all fields of human activities, including art creation and graphic design. In March 2023, a black-and-white photograph, entitled *Pseudomnesia: The Electrician*, won the Sony World Photography Awards. However, the German artist Boris Eldagsen did not accept the award because he explained that he used DALL-E 2, an AI-image generator, to create this photograph (Parshall 2023). Hence, an important question arises: Who is the creator of this award-winning photograph: AI or Eldgasen?

Another AI-generated artwork, entitled *Théâtre D'opéra Spatial*, was awarded 1st place in the "digital arts/digitally-manipulated photography" category at the Colorado State Fair's annual fine art competition in September, 2022. In this case, whether the artist of this digital image should be listed as the AI art generator Midjourney or the artist who gave AI the prompts (natural language descriptions) has been questioned by some people and institutions (US Copyright Office 2023). Jason Michael Allen has maintained, however, that he is the legitimate creator of this image and has used the generative AI platform Midjourney as a tool for "his" artwork (Roose 2002). The debate about this controversial awarded piece of art went viral on Twitter (Metz 2022), and some people commented that the emergence of AI-generated artworks was the death of artistry (Roose 2022).

DALL-E 2 and Midjourney are powerful AI art generators. Other AI art generators may include Images.ai, GetIMG, NightCafe, Shutterstock, Deep Dream Generator, Artbreeder, WOMBO Dream, Canva AI, OpenArt, etc. How will the instantaneous availability, and extensive use, of AI art generators affect artists, the public, as well as the environment? When AI art generators can produce artworks simply by prompts, will human contact with the environment become unnecessary? How will this potential distancing from the biophysical environment have an impact on biocultural communities around the globe?

In this chapter, I engage with these questions from historical, philosophical, ecological, and biocultural perspectives. I particularly examine the relationships between the arts and the environment. In the first section, I review key reflections on the question about the human-technology relationship throughout history, while correlating this reflection with the environment and biocultural conservation. In the second section, I turn to the Hudson River School of American landscape painting to show how art history is inextricably entwined with environmental conservation. Based on my analysis of the Hudson River School, I sketch a preliminary distinction between human creativity and AI generativity. Since human labor has a central place in human creativity, in the third section I use a biocultural framework to define the nature of human creativity and its ecological connection to the environment, to then discuss potential impacts of an AI art generator on biocultural heritage conservation. In the fourth section, I examine key values related to human art-creation and engagement with the environment expressed in the "ecological imperative" stated by groups of artists. Some values of human art-creativity cannot be generated by AI art. Those values are particularly relevant for biocultural conservation.

10.2 A Preliminary Reflection on AI: Labor

In 1920, the science fiction play, *R. U. R.* (Rossumovi Univerzální Roboti/Rossum's Universal Robots) by Czech writer Karel Čapek (1890–1938) introduced the word "robot" to the science fiction world. In 1922, the play was performed in the United States, and in the next year it was published as a book in English. The book offers a profound depiction of the relationship between humanity and robots, namely, artificial beings that have intelligence. The play portrays the eventual replacement of humans by robots, encompassing complex themes such as technological utopia, the created versus the Creator, slavery and emancipation, the extinction of humanity, and the birth of robots with human emotions.

Rossum's robots are created to improve labor efficiency, eliminate poverty, and transform the entire society so that humans would be free from labor. The character "Domin" as RUR general manager, dreams of creating a world where robots do all the labor and humans seek only self-perfection. The vision looks like a replica of ancient Athenian city-state society, except that the enslaved are robots instead of Athenian prisoners of wars. Years later, Rossum's robots develop self-awareness and consider themselves more highly developed, stronger, and more intelligent than humans, whom the robots begin to treat as parasites (Čapek 1923, p. 117). In *RUR*, robots overthrow the human society that is exempt from labor, and yet two robots eventually develop human emotions. A sacrifice and dissection of one of the robots is required for the last remaining human to uncover the secret of robot reproduction, upon which robots Helena and Primus are willing to give their lives for each other, proclaiming that "we belong to each other" (Čapek 1923, p. 187). Humans who are exempted from human labor are eventually replaced by robots, which seems to indicate that labor is a human vocation. This in turn echoes the command in the third chapter of *Genesis*: "Cursed is the ground because of you; through painful toil you will eat food from it all the days of your life." (*Genesis* 3: 17, NIV) "Through painful toil,"—exhausting biophysical labor, humans are to endure.

To address this "painful toil," I recall Chinese Taoist Zhuangzi, who considers it shameful even to use appliances or tools to alleviate the intensity of labor. In the chapter of "Heaven and Earth," an old sage is irrigating his vegetable gardens laboriously. He has dug several ditches, gone to the well, filled jars with water, and poured water into the ditches that led to the garden. A young disciple of Confucius recommends that he employs a newly invented tool so that he can achieve the result with very little labor. After the young Confucianist explains the mechanical principle of the tool, the old sage smiles and says the following,

> I have heard from my teacher that, where there are ingenious contrivances, there are sure to be subtle doings; and that, where there are subtle doings, there is sure to be a scheming mind. But, when there is a scheming mind in the breast, its pure simplicity is impaired. When this pure simplicity is impaired, the spirit becomes unsettled, and the unsettled spirit is not the proper residence of the Dao. It is not that I do not know (the contrivance which you mention), but I should be ashamed to use it.
>
> 吾聞之吾師:有機械者必有機事， 有機事者必有機心。機心存於胸中， 則純白不備；純白不備，則神生不定；神生不定者，道之所不載也。吾非不知，羞而不為也。(Zhuangzi 1891)

In this context, "be ashamed" may denote the disgrace of avoiding labor by using instruments and labor-saving means. In an ethical sense, this shame can be also understood with the Chinese Taoists' mockery of the restlessness caused by the mechanistic world and its utilitarian ethic. In the Old Sage's cascading reasoning, the greater shame may be that the more utilities machines provide to humans, the greater the likelihood that the human mind will be enslaved by the world of machines and technologies. Zhuangzi's insight into these vast impacts of machines and technology on the human psyche is remarkable. For Taoists, cybernetic organisms, human enhancement, and artificial intelligence technologies would require profound reflection and scrutiny.

The old sage in *Zhuangzi* expresses aspirations that are similar to those stated by contemporary Indigenous peoples of North America. According to contemporary Indigenous Potawatomi ethnobotanist and writer Robin Wall Kimmerer (2013), an Ojibwe family in North America politely declined a "grain capture system" designed specifically for them by an engineering student from Europe to improve harvesting efficiency. To express his gratitude to this family for welcoming him to experience the Native American culture of rice harvesting, the student offered to design the "grain capture system" which could harvest 85% more wild rice (*Zizania palustris*). The family's original method of harvesting is "knocking the ripe seed into a canoe," which wasted 50% of the wild rice because at least half of the seeds fell into the watery field. However, the family had a compelling reason for not using such a device, as the unharvested seeds would either become seeds for the next year or food for the ducks in the wild rice fields. The ethical teaching of "never take more than half" (Kimmerer 2013) reveals a deep respect for the reciprocity between people and the land, which deters Ojibwe people from mechanical harvesting technology and its ethics based on utilitarianism and efficiency.

The case of Ojibwe people shows that labor is an important aspect of biocultural diversity. Rather than using a grain capture machine, their labor consists of paddling canoes, gliding across the lake, and knocking out seeds with two sticks, thus shaping this unique biocultural habit. They call their wild rice *manoomin* (Bouayad 2020). Their subtle avoidance of mechanical technology and the inheritance of their traditional ways of labor express a deep biocultural ethic, that is, the interdependence among 3Hs proposed by Rozzi (Rozzi 2013, 2018a; Rozzi et al. 2020), namely, co-inhabitants (Ojibwe people, wild rice, ducks, and other living beings), habits (knocking seeds, seeds reproducing), and habitats (lakes and waterways, or in their own term, "the place where food grows on the water"). Passing on traditional habits, striving for their rights to access wild rice habitats where they have had access for millennium, fighting against genetic modification of wild rice, and even pioneering the enactment of a tribal ordinance to recognize the rights of the species, *manoomin* has become for them a socio-ecological battle for justice that spans a century.

The word "right" adds an unusual poignancy to the situation between the struggle and the justice they crave, while also signaling the imbalance between the forces of biocultural homogenization and biodiversity in the midst of the contemporary socio-political ecology in a colonial era. As French anthropologist Aurenlien Bouayad claims, "After centuries of dispossession and legal oppression, this episode hence demonstrates the ways in which indigenous communities have assimilated and are increasingly resorting to law as a strategic resource for their cultural and environmental struggles" (Bouayad 2020, p. 41). Within the framework of biocultural ethics, it is evident that a fair and unprejudiced vision would be for Ojibwe people to be able to express in their own language the values they place on *manoomin*, and their voices to be listened to and respected without having to appeal to a discourse of "rights" that is not part of their culture.

To elaborate further, the use of wooden sticks to knock out *manoomin* instead of harvesting the seeds through machines represents their defense of their unique habits and biophysical labor. They willingly eschew highly efficient mechanized harvesting systems out of a deep appreciation for the biocultural heritage handed down to them by their ancestors. The question is, then, how exactly do the above three reflections relate to the thesis of this chapter? Or more specifically, how will the advent of AI art generators shape or transform people's biocultural habits, particularly the way users and audiences of AI art generators interact with their environments? Following this, I use the example of the Hudson River School's landscape paintings to elaborate how artists created art in a biocultural sense before the emergence of AI art generators and show the contrast between the two.

10.3 Hudson River School and AI Art Generation

10.3.1 A Biocultural Sketch of Hudson River School

With the emergence of AI art generators, contemporary artists are faced with similar choices as the Ojibwe people. Before the advent of AI art generators and other digital tools, painters had to exert much labor to create paintings. They also needed to work through years of training in basic painting techniques to master the relationships among color, light, and shadow. Furthermore, these professionals will need to be scientifically literate, as scientific illustrations emerge from the more frequent intersections of science and art. In the case of museum scientists, geologists, botanists, etc., they may need to engage in strenuous and potentially tedious labor for years to carry out fieldwork, observation, and collection to accumulate materials for artistic creation. This is especially true for artists dedicated to expressing the beauty of nature through their paintings. For the German ecologist and artist Ernst Haeckel (1834–1919), his masterpieces, which conveyed both accurate scientific information and the beauty of natural organisms, were based on his extensive fieldwork in the Mediterranean Sea, the Canary Islands, Malaysia, India, Ceylon, and elsewhere (Egerton 2013). The biophysical labor that is an integral part of an artist's creation can include all those bodily activities that are embodied in a larger socio-ecological context.

In the mid-nineteenth century, when the Hudson River School captured the attention of the art world, environmental issues such as urbanization, industrialization, population growth, pollution, and wilderness preservation were gaining people's attention. The Hudson River School is usually defined either as an American art movement that embraces Romanticism through landscape painting or a school of American artists who made the American landscape their main subject between 1820 and 1880. Through Avery's historiographic study on the Hudson River School, the School faced various doubts, controversies, stigmatization, and prejudice from European critics of American art. Some derided the Hudson River School's portrayal of natural landscapes, mocking the landscape paintings as the inspiration of American savages, parallel to the Native Americans' admiration for the American continent's forests, waterfalls, and rivers (Avery 1987).

Although the Hudson River School has been labeled provincialism and luminism in art history, to some environmental scholars, the School's landscape paintings successfully encompassed the socio-ecological issues of the mid-nineteenth century. American environmental historian William Cronon invites people to interpret the Hudson River School landscape paintings as historical narrative, environmental transformation, national progress, and the expression of bewilderment as the mixture of sublime wilderness and pastoral civilization (Cronon 1992). American environmental philosopher Eugene Hargrove (1989, pp. 81–85) places the Hudson River School in a broader socio-ecological and environmental aesthetic context. He argues that in Western modern history, especially in the late eighteenth century, there has been an aesthetic shift in interest in the natural environment, as manifested in the great interest in wilderness aesthetics, and the gradual shift of interest to informal gardens from geometric relationships of design and arrangement to ecological relationships among plants in the garden. Furthermore, a shift from a purely scientific view of the natural environment to a more aesthetically perceptive one is also reflected in the work of many naturalists, geologists, and artists. Hargrove (1989, p. 87) links the emergence and transitions of environmental aesthetic concepts such as "the beautiful," "the picturesque," "the sublime," and "the interesting" to the rich history of art (Hudson River School included), arguing that "the sublime marks the end of the era of evaluating natural objects exclusively in terms of their human uses and needs" and that the trajectory of "the interesting" to become an aesthetic category spawned a more intrinsic view of the environment.

Cronon's and Hargrove's approaches reveal the interconnectedness of ethos and the natural environment in a biocultural sense. The linkages between Hudson River School artists and the American wilderness preservation movement further reflects how the artists' habits and their activities have shaped the North American environment. In Hargrove's view, the American fascination with natural objects and their representation in landscape paintings, which has undergone various phases through Thomas Cole (1801–1848), the Hudson River School, American impressionism, and abstract expressionism, is a fascination that has endured and is by no means a passing fad (Hargrove 1989, p. 98). Additionally, while many Hudson artists drew on their experiences traveling the Hudson River Valley region to create their paintings, some traveled to other regions and depicted landscapes that were symbols of American places and spirit, such as the Catskill Mountains, the American West landscape, Niagara Falls, the Adirondack Mountains, the Palisades, and the Storm King Mountains, etc. According to Harvey Flad, the Hudson River School established a landscape aesthetic that helped local citizens form a sense of place and to work to preserve the landscapes that define America (Flad 2011). In this scenario, landscape paintings not only represent the biophysical environment as the beautiful, picturesque, and sublime, but also carry humanistic, symbolic, and spiritual values.

Before the arrival of European settlers, the Lenape, Wabinger, and Mahican peoples had inhabited the Hudson Valley for millennia and completed five typical habitat shifts through their activities (Lindner 2011). Colonial settlement casts a shadow over the Hudson River School landscape paintings and the American spirit they symbolize. But in pioneering the New World and expanding the frontier, a fresh and creative culture was born and nurtured, in which rapidly altered habitats would shape the American ethos and set of habits that were entwined with the excitement of the frontier, the bewilderment of the loss of the wilderness, and the forging of a national identity. The Hudson River School landscape painting has thus played a crucial role in American environmental policy, law, and environmental conservation. For example, Thomas Moran's painting *The Grand Canyon of the Yellowstone* (1872) is directly related to the establishment of the world's first national park in the U.S. in 1872, as well as to the implementation of "an Act to set apart a certain Tract of Land lying near the headwaters of the Yellowstone River as a public Park" (Jackson 1942, p. 200). This would be the American cultural offering to the global world, nourished by the natural environment, and would call for a removal of tensions between Indigenous peoples and European settlers through an ethic that embraces coexistence, humility, reconciliation, and an unceasing historical reflection on the relationships between humans and landscapes.

10.3.2 The Paintings: Hudson River School Versus AI Art Generators

Based on how machine learning algorithms were utilized by artists and designers, Angus Forbes (2020), the director of the Creative Coding Interdisciplinary Research Team (Department of Computational Media University of California, Santa Cruz), has classified creative AI art into four types, namely, expressive mimicry, interactive mapping, generative art, and critical inquiry. Californian median artist Eitan Mendelowitz (2020) similarly categorizes AI-based art into four types, i.e., generative, responsive, interactive, and learning AI, depending on scenarios such as whether AI perceives the environment, whether it responds to the environment, and whether AI's internal states or responses may change because of the interactions between AI and the environment. Regardless of the criteria by which AI art is categorized, the characteristic shared by various genres is that AI relies on input (prompts, signals from the environment), big data, algorithms, and Machine Learning (ML) rather than the creative skills that have been created, accumulated, and passed down to the artists such as painters, architects, and musicians over various generations. Instead, what one sees is that human artistic creations are formatted and processed by algorithms and ML so that elements of human artworks can become the building blocks of AI art generation. In other words, human artworks are decomposed, reduced, and fragmented into segments or elements, which are then resourcefully assembled by AI.

Style, as an iconic characteristic of artistic creation and expression, often serves as a signature of artists or art movements. Style can be quite important for understanding, appreciating, and interpreting artworks, and even for tracing the origins of artworks (Ross 2003). For this reason, style imitation becomes indispensable for AI art generators. When the algorithm and machine learning analyze the style and genre of the database of human paintings (Agarwal et al. 2015; Lecoutre et al. 2017), AI can generate pictures with specific styles. For AI art generators such as NightCafe or WOMBO Dream, styles such as Gouache, cyberpunk, heavenly, horror, impressionist, pop art, anime, dark fantasy, epic, realistic, and mystical, vibrant, among others, are available for users. If the menu does not contain a specific style, the user can input a style by "text prompt" and the AI will subsequently generate images with that style. When a user enters text such as "Hudson River School Style, Landscape Painting," the AI generates an image in seconds that contains conspicuous Hudsonian elements, including steep mountain peaks, serene waterscape, a waterfall at the vanishing point, and rocks and trees in the close-up view (Fig. 10.1). On the left side

Fig. 10.1 An AI generated image vs. paintings of Hudson River School artists. (Image was created by the author)

of the image, in addition to the towering mountains, the sun's rays penetrate the clouds and illuminate the mountains.

In the left image of Fig. 10.1, AI's ability to generate characteristic Hudsonian paintings appears to be impeccable. But the problem is that none of the elements in the left image seems to be original. Or, in a fundamental sense, the left image is an imitation of the elements of Hudsonian paintings. Steep mountains can be found in Thomas Cole's *Expulsion from the Garden of Eden* (1827, Fig. 10.1, right, upper left) and Albert Bierstadt's *Among the Sierra Nevada, California* (1868, Fig. 10.1, right, upper right). A waterfall can be identified in Frederic Edwin Church's *The Heart of the Andes* (1859, Fig. 10.1, right, middle right). The sun's rays piercing the clouds can be found in Thomas Cole's *The Voyage of Life: Manhood* (1842, Fig. 10.1, right, middle left). Waterscape appears in John Frederick Kensett's paintings, such as *Upper Mississippi* (1855) and *Lake George* (1869, Fig. 10.1, right, bottom left), and Sanford Robinson Gifford's *The Wilderness* (1860, Fig. 10.1, right, bottom right). If the paintings of Hudsonian artists are imitations and reproductions of the biophysical landscapes, AI-generated images can be seen as imitations of their paintings. Thus, without the creativity, originality (or provincialism, illuminism, savagery, as criticized by a Eurocentric view of art), and style of those Hudson artists, AI would not have been able to process the prompt "Hudson River School style." In addition, the Hudson River style did not happen overnight, nor was it an accumulation of pure painting skills. It required generations of Hudsonian artists traveling to the Hudson River Valley or other landscapes, immersing themselves biophysically and spiritually in them, and being shaped by them, for the style to come into its own. In contrast, AI-generated imagery fails to enrich our understanding of the biophysical environment by not associating itself with specific biophysical regions, and thus cannot contribute to wilderness or environmental conservation in the same way that the Hudsonian paintings did.

In a sense, in the contrast between the Hudson landscape paintings and the images produced by the AI, there seems to be a conversion of labor taking place, *i.e.*, the biophysical labor of the Hudsonian painters (*e.g.*, years of training, travel, brush strokes, and others) is being replaced by the "labor" of AI (what exactly is it?). Without this biophysical labor, the images generated by AI fail to convey the complexity of the relationship between arts and environment that was once manifested by the Hudson River School and their paintings. Besides, ethical questions about credit and responsibility related to human artists who employ AI to produce "high art" (Epstein et al. 2020), negative impacts of image generators on the community of artists (Jiang et al. 2023), and the intersections of aesthetics, ethics, and politics (Stark and Crawford 2019), have been addressed. Particularly, Jiang and collaborators (2023) persuasively illustrated how AI image generators can result in tangible harms to professional artists, including economic loss, reputational damage, and perpetuation of stereotypes. The above studies challenge the AI community to cautiously consider copyright and other ethical issues while AI data mining human artworks.

For users of AI art generators, the creative activity associated with the artwork seems irrelevant to the artwork itself, unless AI itself is considered a form of arts. AI developers and computer scientists argue that human creativity can be unlocked by AI through datasets, algorithms, and machine learning. AI's mimicry of artistic styles goes beyond mere reproduction as well. Evidence for this claim is that AI-generated images have passed the visual Turing test, in which human subjects were unable to distinguish between human artworks and AI-generated images because the machines have been trained to minimize aesthetic bias and maximize stylistic ambiguity (Mazzone and Elgammal 2019). When AI generates creative artwork from prompts, it also doesn't differentiate between talented artists and amateurs. Text prompts define the output.

The instantaneous properties of AI artwork render instant availability and accessibility a norm, whereby those that require human labor or effort, especially those with strong desire, commitment, consistency, and care, may not be valued. What is worth reflecting on is that the fundamental difference between machine learning and human learning is that the former is replicable and can achieve roughly similar learning outcomes among machines, whereas human learning can be replicated, but the learning outcomes would always vary among individual humans in terms of progress, extent, and effectiveness. Who would expect facsimile paintings of Cole and Gifford to be infinitely reproduced?

10.4 What Is Biophysical Labor? A Biocultural Rectification

Following the argument of the previous section, the importance of biophysical labor in examining the triad of AI, art, and the environment is undeniable. Biophysical labor is first and foremost a human vocation, and thus has an ethical dimension

(Vezzani 2026). Secondly, its content can then become an important component in determining the character of human activity, such as the labor of Hudsonian artists, and the labor involved in AI generation (which can obviously be divided into the labor of the AI developer and the labor of the user, the latter being almost negligible).

From the perspective of the history of philosophy, the classic accounts and definitions of biophysical labor, or labor in general, can be divided into three types, namely the arguments by John Locke (1632–1704), Karl Marx (1818–1883), and Hannah Arendt (1906–1975), among others. According to Locke, the laws of nature allow people to take possession of unappropriated material resources if they exert labor (especially biophysical labor) on those unpossessed resources, such as land or products of the land (fruits, fossil fuels, or natural gas). Locke describes that:

> Though the earth and all inferior creatures be common to all men, yet every man has a "property" in his own "person." This nobody has any right to but himself. The "labour" of his body and the "work" of his hands, we may say, are properly his. Whatsoever, then, he removes out of the state that Nature hath provided and left it in, he hath mixed his labour with it, and joined to it something that is his own, and thereby makes it his property. (Locke 1884, p. 204)

By Locke's definition, an artwork can be a marketable private property. Labor, because of its relevance to authorship, also has a place in determining the monetary value of an artwork, as in the case of *Sunflowers* painted by Vincent van Gogh (1853–1890) or any other artist. And the value of an artist's biophysical labor changes over time, just as Van Gogh's biophysical labor may not have been valued in terms of monetary value in his own time. In a nutshell, biophysical labor under Locke's definition indeed has a complex socio-economic dimension.

More notably, Locke distinguishes between private and common, even though his intention may have been to admonish people to leave "enough" and "as good" common resources for others. Locke's ethical exhortations behind the right to private property, and his apparent sense of fairness, equality, and even altruism, seem to have been overlooked under the logic of extraction in modern capitalist society. The alternative focus has been the legalization of the right to private property after possession, such as in the Fifth Amendment to the U.S. Constitution, which states that people's rights to private property may not be taken away without due process of law or just compensation.

In turn, in his defense of the abolition of private property, Marx seems to have attempted to rescue the ethical admonition behind "labor" by pointing out that human labor, defined in a socio-economic sense, constructs the core of the value of any given commodity in a capitalist society and is based on the capitalistic logic of the exploitation of the laborer. Neither Locke's principle of first possession nor Marx's labor theory of value formulated labor in such a way as to highlight the proposition that labor is a human vocation.

Perhaps it is this neglect that creates the thought-provoking situation in Arendt's discussion of labor, which is situated in the same lineage as Locke and Marx. Arendt (1906–1975) classified three types of human activities into a hierarchy, namely labor, work, and action. Since labor is defined as a kind of activity linked solely to the biological dimension, such as the maintenance of one's own life, the mundanity, triviality, and futility of labor successfully circumvents its possibility as a sophisticated ethical activity such as work or action. Labor produces ephemeral materials for consumption; work creates artifacts and mechanized systems; and action measures human life and the world, symbolizing human freedom (Arendt 2013). In her hierarchy, out of a defense of the sublimity of political activities and discourse, the entanglement of political issues with social, economic, and environmental problems is somewhat unfortunately overlooked. It is in her explicit demarcation and attribution of meaning that she assigns to the three kinds of human activities that the links between human biophysical activities and their symbolic, cultural, meaningful, and institutional worlds are lost. Here lies the ambiguity or even discrepancy that Arendt distinguishes *vita activia* from *vita contemplativa* but relinquishes essential biophysical activities over to futility.

There should exist a theory and practice that does not sacrifice the sublimity of political dialogues but is simultaneously able to save the correlation between biophysical labor and the sublime. In my view, biocultural ethics is precisely one of such theories, and biocultural conservation, restoration, and communication are exactly those types of practice. Biocultural ethics can rectify the futility of biophysical labor by examining the complex relationship among co-inhabitants (human laborers and other-than-human laborers), their habits, and their habitats (Rozzi 2012). The intricate interactions among the three cannot be reduced to three isolated or independent factors. Human activities that operate separately in their respective field in Arendt's hierarchy are interrelated and entangled in the framework of biocultural ethics, manifesting in the empirical feedback that exists between biophysical activity and biocultural homogenization/heterogeneity, whereby biophysical labor can be both a driver and an outcome (Rozzi 2013, 2018a).

Human biophysical labor, therefore, is plural, evaluative, and embedded in symbolic-cultural-ecological systems, even engendering meanings. In a certain sense, biophysical labor can be perceived as artistic and political. Like the old sage in *Zhuangzi*, who insisted on irrigating his vegetable garden with jars, or the Ojibwe people who collect wild rice seeds with wooden sticks, human biophysical labor conveys meaning and values. When AI generates artwork in seconds, the issue is not so much that AI generation replaces human art creation, but rather that some hidden labor displaces the labor of conventional art creation. This hidden labor may include, for example, labor involving the collection of big data, the development of algorithms for processing, filtering, and recommending data, and the construction of physical capital systems to support machine learning.

In this sense, reconceptualizing human labor within the framework of biocultural ethics cannot ignore the entanglement of human biophysical activity with the symbolic and institutional world in a socio-cultural-ecological sense. Biophysical labor would cultivate habits. Labor, properly reconceptualized and rectified, would no longer be a purely biophysical activity, but would inevitably be linked to dimensions of cognition, spirituality, and socio-ecology. Thus, Hudsonian School artists not only engaged with the Hudson River Valley and other American landscapes on a corporeal level, but the landscapes also inspired their landscape paintings mentally, and even socio-environmentally shaped environmental aesthetics and wilderness preservation policies. Some types of labor yield landscapes dominated by the industrialization, urbanization, and homogenization of the contemporary system, and couple human labor with banal consumption rather than the art of survival (Yu 2007). Alternative types of labor, especially that arise from human engagement with co-inhabitants in shared habitats, yield biocultural landscapes in which humans coexisted in balance with their environments, such as those UNESCO listed world heritage sites (Honghe Hani Rice Terraces, upland Agro-pastoralism landscapes of the Causses, Qhapaq Ñan Andean Road System, etc.).

10.5 Concluding Remarks: Ecological Imperative

The plurality and complexity of biophysical labor reveals a profundity of connections between biophysical labor and the global biocultural status. Some types of labor, when intensified and massified, exacerbate biocultural homogenization, leading to the tyranny of capitalist hegemony, the prevalence of economic discourses, the commodification of the environment, and the rapid growth of urbanization, thereby devouring the planet's biodiversity and cultural diversity (Rozzi 2018b; Warner 2015). Other types of biophysical labor, in the form of traditional or Indigenous engagement with the environment, have survived the siege of capitalized large-scale biophysical labor, preserving room to maneuver for the conservation and restoration of global biocultural diversity, or forge a novel biocultural configuration (like the Hudsonian artists). Biophysical labor in the environment or biophysical engagement with the environment is thus considered as an ecological imperative.

Regarding this ecological imperative, there are still some clarifications that need to be made in the trinity of AI, arts, and the environment:

- While AI has made a remarkable contribution to engaging in environmental conservation, the risk of "deepfakes" could distort human interaction with the environment. The relationship between AI and the environment has been evaluated from the perspective of how this tool can facilitate environmental investigation or conservation. For instance, with the assistance of neural networks and machine learning algorithms, images taken by autonomous cameras or acoustic information collected in recording devices in the wild can be classified with great accuracy to identify species and monitor their activities. AI powered tools and platforms such as population viability analysis (PVA) and the Automated Remote Biodiversity Monitoring Network (ARBIMON) have been widely used for analyzing the risk of species extinction, analyzing large conservation datasets, and detecting or monitoring animal activities (Kwok 2019; Lamba et al. 2019; Wearn et al. 2019). However, "deepfakes," defined as the deceptive and/or misleading information such as images, audios, or videos that are generated by AI, have not only threatened individual privacy and national security but also have led to fake geography, which distorts people's perception of the world through producing non-existent landscape features (Zhao et al. 2021). McGovern and collaborators (2022) also caution about various ways that AI can be misused in environmental sciences. In other words, despite the contribution of AI to conservation and the analysis of massive datasets, the risk of misuse of AI demands responsible use of it in many aspects of our society (Wearn et al. 2019). For AI art generation, this may require users to remain alert to global environmental degradation while being immersed in landscapes generated by AI.

- In addition, studies have shown that early and even adult nature experiences are positively correlated with adult engagement in biodiversity conservation and environmental behavior (Ewert et al. 2005; Wells and Lekies 2006). When people use AI to generate artwork, or frequently immerse themselves in AI-created landscapes and environments, engagement with the biophysical environment may become inconvenient or redundant for them, in the same way that children find real turtles disgusting, insects creepy, or forests dangerous. Multiple causes and consequences related to the loss of interaction with nature have been examined, among which virtual alternatives are one of those causes (Soga and Gaston 2016). Moreover, when biophysical engagement with the natural environment gives way to virtual and digital imitations of the natural environment, the feedback loop between the loss or disappearance of the experience and further dissatisfaction or indifference to the environment can lead to a decline in citizens' participation in conservation and other types of environmental engagement. With AI generated landscapes, people do not take the time or effort to verify that the sceneries in the paintings are relevant to a particular environment, place, or site. The fantasy landscapes generated by AI art, no matter how realistic, are chimeras beyond biophysical contact. However, people who love nature or have a keen interest in it would spend time exploring natural environments, just as Gene Hargrove and Alexandria Poole did when they went with a group of philosophers and ecologists to observe the waterfalls portrayed in the painting *Kindred Spirits* by Hudson River School artist Asher Brown Durand (1796–1886) (see Box 3.2 in Rozzi 2026). Hargrove has encouraged philosophers to appreciate the importance that aesthetics has historically had, and could have in the future, for the development of environmental ethics (Rozzi 2026).
- If biophysical encounters with the environment are part of what it means to be human, then artists are no exception. Artists, together with their audience, interpret the meaning of artworks in terms of the artists' specific cultural context or the issues of the time. This runs counter to the approach of AI which seeks general principles of both human-environment interactions and artistic creativity. The interpretation of meaning of artworks is also highly contextual, contingent, and flowing. Not only that, but artists also engage in abstraction, suggesting that human artmaking is not just an imitation of the world, but also the creation of systems of meaning in the biophysical world (Mateas 2001). It is in the indispensability of the biophysical world, in the interaction between human artists and the natural environment, in the finite and unceasing abstraction of meanings, that the habits of human existence (the ethos) and the value of human labor are revealed. What meaning and abstraction of the world will AI art generation

entail? Studies focusing on the environmental sustainability of AI art challenge both the scientific and artistic communities (Jääskeläinen et al. 2022), calling for interdisciplinary collaborations among artists, technologists, and policy makers to develop sustainable practices, inviting developers and users of generative AI to consider the environmental impacts of the hardware, software, and other resources involved in AI artistic practices.

- The artistic community has a history of responding positively to the ecological imperative. According to Gablik (1992), the ecological imperative for the artistic community commands artists to create hydroglyphs, designing artworks for other-than-human beings such as animals, which can be demonstrated by Lynne Hull's Raptor Roost SB1 (1990), a sculpture (in the eyes of humans) in the landscape to provide perches and nesting roosts for hawks, eagles, and owls. In the late twentieth century, the ecological imperative was also echoed and revealed by eco-installations which incorporated natural materials such as twigs, branches, wicker, bark, leaves, seed, fruit, and living plants into artwork to remind people of human-nature interconnectedness, nature's evanescence, and the well-being of the planet (Mańczak 2002). In the fields of biosemiotics, biocommunication, and cognitive biology, researchers have also employed "art for animals" to observe how animals, such as mammals, birds, and fish, interact with artworks, as well as to study how other-than-human species sense, perceive, and respond to artistic stimuli (Fuller 2010). In this sense, art projects open the door not only to understanding the cognitive processes of animals, but also to exploring their ontological world-making. Likewise, these different types of artworks and projects suggest that artists are already situated in a different era than the Hudson River School, and are responding to the questions of our own time to create new biophysical interactions as well as meanings.

Acknowledgments I would like to thank Rika Tsuji, Benn Johnson, Roy May, and Ricardo Rozzi for their valuable comments on earlier versions of this chapter. I acknowledge the support from the Chilean National Agency for Research and Development through the Cape Horn International Center, CHIC (ANID/BASAL FB210018).

References

Agarwal S, Karnick H, Pant N, Patel U (2015) Genre and style based painting classification. In: O'Conner L (ed) 2015 IEEE winter conference on applications of computer vision. IEEE, Waikoloa, pp 588–594. https://doi.org/10.1109/WACV.2015.3

Arendt H (2013) The human condition. University of Chicago press, Chicago

Avery KJ (1987) A historiography of the Hudson River School. In: Howat JK (ed) American paradise: the world of the Hudson River School. Metropolitan Museum of Art, New York

Bouayad A (2020) Wild rice protectors: an Ojibwe odyssey. Environ Law Rev 22(1):25–42. https://doi.org/10.1177/1461452920912909

Čapek K (1923) RUR (Rossum's universal robots): a fantastic melodrama (trans. Selver P). Doubleday, Page and Company, Garden City

Cronon W (1992) Telling tales on canvas: landscapes of frontier change. In: Prown JD, Cronon W, Anderson NK, Schoelwer SP (eds) Discovered lands, invented pasts: transforming visions of the American West. Yale University Press, New Haven, pp 37–87

Egerton FN (2013) History of ecological sciences, part 47: Ernst Haeckel's ecology. Bull Ecol Soc Am 94(3):222–244

Epstein Z, Levine S, Rand DG, Rahwan I (2020) Who gets credit for AI-generated art? Iscience 23(9):1–10. https://doi.org/10.1016/j.isci.2020.101515

Ewert A, Place G, Sibthorp J (2005) Early-life outdoor experiences and an individual's environmental attitudes. Leis Sci 27(3):225–239. https://doi.org/10.1080/01490400590930853

Flad HK (2011) The influence of the Hudson River School of art in the preservation of the river, its natural and cultural landscape. In: Dunwell FF (ed) Environmental history of the Hudson River: human uses that changed the ecology, ecology that changed human uses. State University of New York Press, New York, pp 291–312

Forbes A (2020) Creative AI: from expressive mimicry to critical inquiry. Artnodes 26:1–10. https://doi.org/10.7238/a.v0i26.3370

Fuller M (2010) Art for animals. J Vis Art Pract 9(1):17–33. https://doi.org/10.1386/jvap.9.1.17_1

Gablik S (1992) The ecological imperative. Art J 51(2):49–51

Hargrove EC (1989) Foundations of environmental ethics. Environmental Ethics Books, Denton

Jääskeläinen P, Pargman D, Holzapfel A (2022) On the environmental sustainability of AI art(s). In: Eighth workshop on computing within limits, pp 1–9. https://doi.org/10.21428/bf6fb269.c46375fa

Jackson WT (1942) The creation of Yellowstone National Park. Miss Val Hist Rev 29(2):187–206

Jiang HH, Brown L, Cheng J, Khan M, Gupta A et al (2023) AI art and its impact on artists. In: Proceedings of the 2023 AAAI/ACM conference on AI, ethics, and society, pp 363–374

Kimmerer RW (2013) Braiding sweetgrass: indigenous wisdom, scientific knowledge and the teachings of plants. Milkweed Editions, Minneapolis

Kwok R (2019) AI empowers conservation biology. Nature 567(7746):133–134

Lamba A, Cassey P, Segaran RR, Koh LP (2019) Deep learning for environmental conservation. Curr Biol 29(19):R977–R982

Lecoutre A, Negrevergne B, Yger F (2017) Recognizing art style automatically in painting with deep learning. In: Asian conference on machine learning. PMLR, pp 327–342

Lindner CR (2011) The earliest thirteen millennia of cultural adaptation along the Hudson River Estuary. In: Henshaw RE (ed) Environmental history of the Hudson River: human uses that changed the ecology, ecology that changed human uses. State University of New York Press, Albany, pp 65–76

Locke J (1884) Two treatises on civil government. George Routledge and Sons, UK

Mańczak A (2002) The ecological imperative: elements of nature in late twentieth-century art. Leonardo 35(2):131–136. https://doi.org/10.1162/00240940252940496

Mateas M (2001) Expressive AI: a hybrid art and science practice. Leonardo 34(2):147–153. https://doi.org/10.1162/002409401750184717

Mazzone M, Elgammal A (2019) Art, creativity, and the potential of artificial intelligence. Art 8(1): 26. https://doi.org/10.3390/arts8010026

McGovern A, Ebert-Uphoff I, Gagne DJ, Bostrom A (2022) Why we need to focus on developing ethical, responsible, and trustworthy artificial intelligence approaches for environmental science. Environ Data Sci 1:e6. https://doi.org/10.3390/arts8010026

Mendelowitz E (2020) Intelligent environments and public art. Artnodes 26:1–9. https://doi.org/10.7238/a.v0i26.3364

Metz C (2022) Genius makers: the mavericks who brought AI to Google, Facebook, and the world. Penguin Books, New York

Parshall A (2023) How this AI image won a major photography competition. Scientific American, p 21

Roose K (2022) An AI-generated picture won an art prize. Artists aren't happy. The New York Times, 2(September)

Ross S (2003) Style in art. In: The Oxford handbook of aesthetics. Oxford University Press, New York, pp 228–244

Rozzi R (2012) Biocultural ethics: recovering the vital links between the inhabitants, their habits, and habitats. Environ Ethics 34(1):27–50. https://doi.org/10.5840/enviroethics20123414

Rozzi R (2013) Biocultural ethics: from biocultural homogenization toward biocultural conservation. In: Rozzi R, Pickett S, Palmer C, Armesto J, Callicott J (eds) Linking ecology and ethics for a changing world, Ecology and ethics, vol 1. Springer, Dordrecht, pp 9–32. https://doi.org/10.1007/978-94-007-7470-4_2

Rozzi R (2018a) Biocultural conservation and biocultural ethics. In: Rozzi R, May RH, Chapin FS, Massardo F, Gavin MC et al (eds) From biocultural homogenization to biocultural conservation: a conceptual framework to reorient society toward sustainability of life, Ecology and ethics, vol 3. Springer, Cham, pp 303–314. https://doi.org/10.1007/978-3-319-99513-7_19

Rozzi R (2018b) Biocultural homogenization: a wicked problem in the Anthropocene. In: Rozzi R, May RH, Chapin FS, Massardo F, Gavin MC et al (eds) From biocultural homogenization to biocultural conservation: a conceptual framework to reorient society toward sustainability of life, Ecology and ethics, vol 3. Springer, Cham, pp 21–48. https://doi.org/10.1007/978-3-319-99513-7_2

Rozzi R (2026) The contributions of Eugene Hargrove to environmental ethics and aesthetics: a biocultural tribute. In: Zhu D, Rozzi R, Tsuji R, Johnson B, Caviness T, Castro-Jorquera C, Xu L, Zhang L (eds) Linking arts with biocultural conservation, restoration, and communication, Ecology and ethics, vol 8. Springer, Cham, pp 35 –59. https://doi.org/10.1007/978-3-032-04830-1_3

Rozzi R, Massardo F, Poole A (2020) The "3Hs" (habitats, habits, co-in-habitants) of the biocultural ethic: a "philosophical lens" to address global changes in the Anthropocene. In: Valera L, Castilla JC (eds) Global changes: ethics, politics and environment in the contemporary technological world. Springer, pp 153–170. https://doi.org/10.1007/978-3-030-29443-4_14

Soga M, Gaston KJ (2016) Extinction of experience: the loss of human–nature interactions. Front Ecol Environ 14(2):94–101. https://doi.org/10.1002/fee.1225

Stark L, Crawford K (2019) The work of art in the age of artificial intelligence: what artists can teach us about the ethics of data practice. Surveill Soc 17(3/4):442–455. https://doi.org/10.24908/ss.v17i3/4.108201

US Copyright Office (2023) Second request for reconsideration for refusal to Register Théâtre D'opéra Spatial (SR # 1-11743923581; Correspondence ID: 1-5T5320R). Accessed through https://www.copyright.gov/rulings-filings/review-board/docs/Theatre-Dopera-Spatial.pdf on 22 February 2025

Vezzani P (2026) "Fragility of encounters" in subantarctic biocultural diversity. In: Zhu D, Rozzi R, Tsuji R, Johnson B, Caviness T, Castro-Jorquera C, Xu L, Zhang L (eds) Linking arts with biocultural conservation, restoration, and communication, Ecology and ethics, vol 8. Springer, Cham, pp 207–222. https://doi.org/10.1007/978-3-032-04830-1_13

Wanner T (2015) The new "passive revolution" of the green economy and growth discourse: maintaining the "sustainable development" of neoliberal capitalism. New Political Econ 20(1): 21–41. https://doi.org/10.1080/13563467.2013.866081

Wearn OR, Freeman R, Jacoby DM (2019) Responsible AI for conservation. Nat Mach Intell 1(2): 72–73. https://doi.org/10.1038/s42256-019-0022-7

Wells NM, Lekies KS (2006) Nature and the life course: pathways from childhood nature experiences to adult environmentalism. Child Youth Environ 16(1):1–24. https://doi.org/10.1353/cye.2006.0031

Yu K (2007) The art of survival: position landscape architecture. China Architecture Publishing, Beijing

Zhao B, Zhang S, Xu C, Sun Y, Deng C (2021) Deep fake geography? When geospatial data encounter Artificial Intelligence. Cartogr Geogr Inf Sci 48(4):338–352. https://doi.org/10.1080/15230406.2021.1910075

Zhuangzi (around 4th Century BCE) 《庄子》. English edition: Zhuangzi (1891) Zhuangzi (Trans. Legge J). In: Chinese Text Project, available via https://ctext.org/zhuangzi. Accessed 1 Nov 2023

Chapter 11
A Bird in the Hand: Experiencing Birds in the Subantarctic

Benn Johnson

Abstract This first-person narrative about a journey to the Cape Horn Biosphere Reserve, Chile, allows for a reflection on the philosophical themes of the book *Linking Arts with Biocultural Conservation, Restoration, and Communication*. Inspired by aesthetic experiences the author had during *in situ* encounters with birds such as the Magellanic Woodpecker, the Black-faced Ibis, the skua, and Mirror-winged Gulls, this paper explores what it might mean to understand co-inhabitants not as mere specimens, but as related, precarious, and active parts of biocultural expressions. The chapter invites consideration of how aesthetic experiences with other-than-human beings can shift one's perspective, lending to a deeper understanding of the relationality and ethical weight of being a part of biocultural communities.

Keywords Biocultural ethics · Environmental humanities · Aesthetic value · Intrinsic value · Precarity

Trudging through the forest of Omora Park, hopping over muddy spots and hugging the grassy edges of the trails, one is met immediately with the image of a wet, green, lush rainforest. However, the forest I am in is not the usual Omora Park, or so I've been told. While the greenery is normal, the rain here is not. What is normally a temperate climate, with moderate levels of rain has been transformed only quite recently into soggy, often torrential days, with downpours lasting hours without ceasing. According to the course instructors and the locals, this appears to be yet another direct impact of climate change.

I think about the increased precipitation and the changes it is likely to cause: heightened decomposition in the soil, which may free up nutrients to be leached and drained from the land with excess runoff. Rather than the dark, soft earth which lies

B. Johnson (✉)
Department of Philosophy, College of Arts and Sciences, University of Rhode Island, Kingston, RI, USA
e-mail: benn.johnson@uri.edu

D. Zhu et al. (eds.), *Linking Arts with Biocultural Conservation, Restoration, and Communication*, Ecology and Ethics 8,
https://doi.org/10.1007/978-3-032-04830-1_11

below my feet now, decades from now, the soil may turn red and dense like a tropical rainforest, with little organic matter left. What the plants appreciate in the present, they may discover to be their future downfall, as displaced, fast-growing species outcompete them and crowd them out of their once-niche. Climate change won't be bad for every inhabitant, but the world after climate change is certain to look much different than it does now; and it certainly won't be *good* for every inhabitant.

I'm drawn quickly back to the forest by a repetitive squawking, coming from somewhere not too far off in the trees. If I hadn't known any better, I would think it was some kind of monkey, but there are no primates on this island, except for the apes of my own species. Rather, I've learned in my few weeks on Isla Navarino that this is the call of the Black-faced Ibis, which I had only recently seen for the first time, perched on a tree branch in the thick forest. The sight of this chicken-sized bird, with its peculiar proportions and long, delicate bill is made even more awkward by its apparent habit of squatting high in the trees, calling out to the world. But to see its orange plumage and dark eyes, peering down at you as you navigate the forest is a reminder that awkwardness is not a negation of value, and that this bird likely views your gangly figure as equally gauche. At least at the moment, we are the visitors, but the truth is that humans and Black-faced Ibises are both inhabitants of this forest. Do we belong to the forest, or the forest to us?

Perhaps the problem of belonging is already shortsighted and anthropocentric; belonging only makes sense for a species without a context. The Black-faced ibis does not exist at all without a forest. The small ibis painting in the *Birds of Patagonia* guidebook is not a Black-faced Ibis, but a symbol, a mere indicator of something which does not make any sense without Patagonia itself. As Magritte might say, *Ceci n'est pas une ibis.* The real bird does not exist without Patagonia, nor does Patagonia exist as such without the bird; they are co-constitutive.

I again look ahead, careful to dodge a particularly muddy hole. Lining the path occasionally are wet stones, covered in lichens, or large, old growth *lenga* trees with wet green mosses on the top sides, and yellow crustose lichens on the drier, shadier sides. I think about what was told to me by a researcher here recently about how some of the mosses may contain antimicrobial properties that birds can actively exploit in nest-building. In this place, where everything, even the rocks seem to be teeming with life, it seems ironic that some of the most plentiful taxa exclude and sanitize their environments. It is perhaps partly the efficacy with which the biota here exclude others that has contributed to the unique biodiversity, with lichens and bryophytes as dominant diversifiers of flora, and birds as the most noticeable and numerous vertebrates. While many other places have more birds than this one, the lack of non-avian vertebrates here certainly draws attention to the plentiful birds that are here.

Of course, the implication that bird species actively exclude other species may seem absurd and ultimately scientifically groundless, but a careful observation of skua predation on a nearby island is at least enough to inspire a sense that some of these birds are specially adapted to this environment, and fill many of the classic niches all on their own. The skua, with its fish-hooked beak and dark feathers, haunts the shoreline, drifting like the tide on the wind, as gulls and penguins nest below.

The skua waits patiently, as if taunting the Mirror-winged Gulls below. However, the gulls are not helpless. In flocks, they take to the skies to caw at their oppressor, occasionally engaging in aerial pursuit, lunging mid-air toward the skua, who tumbles to the side or drifts back on a breeze, always keeping an eye below, on its true prey: the young gulls below, left unattended by their protective parents.

Eventually, the skua attacks, dive-bombing directly at the young gull. The first attack is a miss, successfully blocked by an incoming parent gull, claws outstretched. Another few attacks are unsuccessful, but with each dive, the skua gains confidence, finally plummeting toward the young gull, tackling it with curved claws. Although gulls and skuas are not significantly different in size nor in morphology, the skua's stockier build and scrappier personality allow it to outmatch a single gull in a fight. Eventually, the skua prevails, but it is a slow battle, as even the young gull holds its own quite valiantly against the larger skua. Avoiding the young bird's pecks and talons, the skua relentlessly attacks the eyes, and batters the bird with strong wings, until the violence and sense have been driven from the gull. The tenacity and stamina of both birds, however, as well as the extreme difficulty each has in outlasting the other are marks of the co-evolutionary struggle in which each bird has evolved to live in this environment.

In watching this struggle play out in real time, however, I do not lose awareness of my place in it. These particular birds have not adapted a direct aversion to the presence of humans, but even so, I feel my imposition here. I am an uninvited voyeur, learning much about the wildlife, but not experiencing their lives directly. I cannot begin to understand, as Thomas Nagel might wonder, *what is it like to be a skua?* Although the birds are right in front of me, it is as if I am in a sterile glass case, watching these dinosaurs play out their scene without the immediacy of death or injury as my own possibilities. Regardless of the outcome of this battle, I will fly safely home on my own wings, fueled by the very petroleum that puts these birds at increasing risk.

It is here that a disparity arises; while birds and I are in fact co-inhabitants, in this particular community, it is the birds who we find drenched in crude oil when a tanker ship fails, and it is the birds who lose their habitat when I destroy their temperate climate with greenhouse gasses. We may be equal in our need for a safe environment and our intrinsic value, but one of us is peculiarly advantaged and peculiarly destructive; the other is disproportionately burdened and precarious. It is from this fact that it seems any serious ethical claim must be derived.

I'm shaken back to the present by animal calls. For now, we are the predators, and our prey, the Magellanic Woodpecker, has just made us aware of its presence in the forest. Our guide, Amy Wynia, has been leading us through the forest, occasionally stopping to set down a hollow block of wood, which she strikes quickly with two sticks, resounding a loud *knock-knock!* This time, in the distance, we hear an echo. A rush of excitement fills us, and we look toward Amy, who has suddenly been transformed from educator into researcher.

After calling and responding, and hearing the birds come ever closer, we move briskly and silently into a clearing, where Amy and an assistant set up a long mist-net. We crouch and wait, as she calls to the birds. No response.

Eventually, she calls us up from our vantage points and decides that if the bird does not wish to visit us, we should come to it. Again, we follow her into the woods, occasionally stopping to call the woodpecker, using a mix of her wood block and a digital speaker containing multiple recordings of woodpecker vocalizations. We cross under a tree where a Black-faced Ibis is carefully eyeing us, and into a canopied area. Amy calls again. An echo.

Then, we hear a hard fluttering, as if the wings extend fully and then slap the bird's chest with each wingbeat. Above us, a woodpecker lands on the side of a tree, clutching the bark with its needle-sharp talons. Only seconds later, we hear more fluttering, and two more birds appear in view, landing on a nearby tree. I'm startled. Amy's face lights up, and an energy floats through the entire group. We are electric; necks twisted up toward the treetops, binoculars focused on the majestic (magic?) creatures above us. There is something unexplainable; something quite intoxicating about this moment, particularly in contrast to the bird-fight we had witnessed on the beach. Here, these birds are the spectators, looking at us with the same curious eyes we have for them. In that moment, we are awestruck by the birds. It isn't simply the beauty of the birds, nor is it their elusivity that carries me away; it is also the very real act of communication which has just occurred, and the liveliness with which these birds display their subjectivity. Here we could see that *ceci, c'est un pivert.*

Quickly, Amy ushers us back to the clearing with the net, and we rush back into our hiding spots one last time. Hugging the ground as closely as possible, I for once lose my concern for mud and moisture. My eyes and ears are focused only on the sky. More calls. More echoes. And perfectly on cue, the family of woodpeckers flitter above us, again surrounding us on all sides in the trees, their red tufted heads peering nervously around. Amy plays a digital distress call, and one of the birds calls back. Amy's calls agitate the birds, and they begin to fly over our heads, back and forth, searching for the source. As if in a war against nothing in particular, they flutter around us, explosions of black, white, and red—pure uncontainable energy, and then, sudden silence. Amy calls firmly but quietly to her assistant, "let's go!" They rush toward the net, while the students slowly pour in around the edges to see what happened.

At first, there is just screeching and squawking; the distressed bird is tangled face-down in the net. Amy and two other researchers quickly remove the fine cordage from the bird's wings, neck, and sharp talons, as it pecks aggressively at them. The researchers and birds' struggles now are for the same end—the safety and ultimate release of the bird; but a lack of common language prevents any coordination. The bird uses the only language it knows; a shrieking call of distress to let its family know where it is, and to intimidate its perceived attackers. Amy reaches for a tiny canvas bag and gently places it over her tiny hostage's head. Immediately, the bird becomes silent. The researcher kneels down and begins to measure it—the length of talons, the sex, wing length, tail feathers. Everything she does with utmost care and silence, so as not to unduly frighten the bird. Finally, she removes the bag and measures the facial characteristics, allowing us all to see a clear view of this creature.

However, the creature we see before us is not what had fluttered into the trees earlier. Somehow in the transition from freedom to momentary captivity, the woodpecker has lost its magic. It no longer appears to be majestic; it is not as beautiful. Is it smaller? No, that couldn't be. But it occurs to me in that moment that this bird has been transformed into something which is not complete. Without its flight, it is not explosive. Without the freedom to grasp, it does not clutch trees. Full of fear, it is no longer curious. Like the Black-faced Ibis in the book, this bird deprived of its context is no longer really a full Magellanic Woodpecker. Certainly, it is still alive, and for this reason, as well as her sensitive dedication to our coinhabitant, Amy treats it gently and respectfully. But as long as it is possessed by an unwanted hand, without even a decaying tree for which it is a namesake, what could it mean to be a woodpecker at all? The electricity is gone.

Amy steps into the clearing and releases the bird. It does not linger, but disappears silently into the forest.

I realize then the shortcoming of the old English saying; that a bird in the hand is worth two in the bush. I make a mental note to revise the statement: a bird in the bush is worth infinitely more than a bird in the hand, because without the bush, the bird is only a captive. In captivity, it is afraid, weak, and not in control of its own future. However, even now, as the Magellanic Woodpecker is again safe in the forest of Isla Navarino, it is not in control of its future. The climate is shifting, and the old growth *Nothofagus* trees which it calls its home are being cut down at an alarming rate across its range, to make room for cattle ranching and housing developments. Another species decides whether the woodpecker survives. It is again here that the possibility for a meaningful biocultural ethic arises.

Part II
Practical Knowledge Embedded in Biocultural Diversity

Chapter 12
Introduction to Part II. Practical Knowledges Embedded in Biocultural Diversity

Li Xu, Terrance Caviness, Danqiong Zhu, and Ricardo Rozzi

Abstract In the current era, the loss of biodiversity is not only an ecological crisis but also a profound cultural one. As species disappear, so too do the cultural practices, stories, and traditional knowledge systems tied to them. This section explores the concept of biocultural diversity, the intertwined richness of biological and cultural life—and the critical role of practical knowledge within it. Drawing from environmental humanities, the philosophy of John Dewey, and the biocultural ethics framework developed by Ricardo Rozzi, it highlights the significance of embodied, place-based, and experiential knowledge systems in fostering ecological conservation and public awareness. Through practices such as ecological art education projects, traditional art and crafts, and land-based Indigenous learning, practical knowledge becomes vital for both biodiversity and cultural diversity.

Keywords Biocultural diversity · Practical knowledge · Indigenous knowledge · Knowledge growth · Action

---

L. Xu
Department of Art Education, University of North Texas, Denton, TX, USA

T. Caviness
Department of Environmental Biology, SUNY College of Environmental Science and Forestry, Syracuse, NY, USA
e-mail: tcaviness@esf.edu

D. Zhu (✉)
Sub-Antarctic Biocultural Conservation Program, Department of Biological Sciences, University of North Texas, Denton, TX, USA

Cape Horn International Center (CHIC), Universidad de Magallanes, Puerto Williams, Chile
e-mail: Danqiong.Zhu@unt.edu

R. Rozzi
Sub-Antarctic Biocultural Conservation Program, Department of Philosophy and Religion and Department of Biological Sciences, University of North Texas, Denton, TX, USA

Cape Horn International Center (CHIC), Universidad de Magallanes, Puerto Williams, Chile
e-mail: Ricardo.Rozzi@unt.edu

D. Zhu et al. (eds.), *Linking Arts with Biocultural Conservation, Restoration, and Communication*, Ecology and Ethics 8,
https://doi.org/10.1007/978-3-032-04830-1_12

12.1 Introduction

Nowadays, humanity faces an unprecedented ecological crisis. As human activity continues to reshape the planet, the accelerated loss of biodiversity has become a defining challenge of our time (Leakey 1996). However, despite the imminent crisis of biodiversity loss, there is a serious lack of public awareness, concern, and participation in solving those issues (Gaston and Spicer 2013). To address this gap, it is essential to enhance public awareness and understanding of the importance of conserving biodiversity (Ives et al. 2018; Richardson et al. 2020; Whitburn et al. 2020). First, at a social level, there is a close and complex link between public opinion, scientific productions and conservation policies (Martín-López et al. 2012; Troudet et al. 2017). Second, at an individual level, conservation behavior can be influenced by multiple factors, such as knowledge, interest, attitudes, emotions, and affinity with nature (Turpie 2003; López-del-Toro et al. 2009; Barbett et al. 2020; Liu and Chen 2021; Barrows et al. 2022).

As documented in part I of our volume *Linking Arts with Biocultural Conservation, Restoration, and Communication* (Zhu 2026), to address social and individual levels of human-nature relationships, the arts and aesthetic experiences can play a major role in fostering biodiversity conservation, and more broadly biocultural conservation (Castro-Jorquera 2026; Chiarini 2026; Hu and Yang 2026; McIntosh 2026; Rozzi 2026; Song 2026; Tsuji and Johnson 2026).

In part II, we draw attention to different concepts and examples of interdisciplinary projects linking art, biological sciences, environmental education and biocultural conservation. Toward this aim we address theoretical and practical dimensions at the interfaces of interdisciplinary work conducted by artists, scientists, philosophers, and community members.

First, the impact of biodiversity losses extends far beyond the ecological domain and poses profound challenges to the humanities. The conceptual framework of biocultural ethics elucidates that other-than-human species carry not only biological significance but are woven into the intellectual, cultural, and spiritual fabric of communities around the world (Rozzi 2012, 2018). When a species becomes extinct, so do the traditional knowledge, stories, rituals, and practices closely associated with it. This entails losses of cultural diversity, disruption of life habits, and homogenization of diverse worldviews that have developed in close relationship with the local natural environment. Consequently, species extinction is not only a biological or ecological concern, but also a humanistic one, since losses of biodiversity in a biophysical sense entails losses of biocultural heritage and memory (Hokowhitu et al. 2020; Ojeda et al. 2022; Monterrubio-Solís et al. 2023; Rozzi et al. 2023b). This, in turn, weakens public support for biodiversity and/or biocultural conservation (Rozzi 2013; Jarić et al. 2022).

Second, to raise public awareness and participation in biocultural conservation, a community of philosophers, ecologists, artists, and scholars, based on the methodology of field environmental philosophy (FEP), have embarked on a two-fold integration in various areas across the world: on the one hand, to generate knowledge and conservation actions, and on the other hand, to help people come to appreciate the biocultural values that may have been overlooked (Rozzi et al. 2023a).

In part II of this book, we broaden FEP concepts with other approaches and expand this exploratory journey of growing knowledge and discovery of nature values by presenting diverse initiatives across Asia, Latin America, and North America. These initiatives and programs can be considered as echoes to the FEP to varying degrees, either directly or indirectly. At the same time, these practices show us how different forms of knowledge, contemporary and ancestral, creatively inform ways in which local people interact with their local environments. Presenting these case studies, we aim to communicate to readers practices deeply rooted in the interconnectedness between biodiversity and cultural diversity. We also hope to foster a sense of responsibility that values not only the conservation of biodiversity (local, regional, and global biospheres and the other-than-human organisms that co-inhabit these spheres with humans), but also the conservation of various biocultural knowledge systems, cultural expressions, and their relevant practices.

12.2 Biocultural Conservation and John Dewy's Philosophy

Designed in response to the global socio-ecological crisis and biocultural homogenization, the concept of biocultural conservation represents a quantum shift in the understanding of the complex relationship between human beings and the natural world, with the recognition of the close links between the diversity of biophysical, cultural-linguistic symbolic, and social systems (Rozzi 2013). This concept suggests people consider biodiversity on earth to include not only numerous organisms and ecosystems, but also cultural diversity shaped by local/regional cultures, languages, traditional knowledge systems, as well as practices regarding how people interact with their local environment and other-than-human species in the local ecosystem (Rozzi 2023). This term also challenges the long-standing dichotomy of nature/culture, environment/humans, and urges that effective conservation must simultaneously address ecological problems and socio-cultural injustice.

We build on almost three decades of work on biocultural research, education, and conservation at the Omora Ethnobotanical Park in Puerto Williams, Chile (Rozzi et al. 2008, 2010; Tauro et al. 2021). Launched in 1999 at the southern tip of South America, Omora Park researchers focused on biocultural theoretical frameworks, philosophical argumentations, research questions, methodologies, and overall conservation goals related to appreciate, respect, and co-inhabit biocultural diversity (Rozzi 2023). Grounded on this long-term place-based research initiative the Omora Park research team identified the following ten principles for conserving biocultural diversity:

1. Inter-institutional cooperation
2. Participatory approach
3. Interdisciplinary conservation approach (bridging humanities and sciences)

4. Networking and international cooperation
5. Communication through the media
6. Flagship species
7. Outdoor education
8. Economic sustainability and ecotourism
9. Administrative sustainability
10. Research and conceptual sustainability for conservation (the continuously evolving process for perceiving, understanding and co-existing with biocultural diversity) (Rozzi et al. 2006).

As stated by Bridgewater and Rotherham (2019), the former principles were forged at a particular place, but they have wider application and can guide biocultural conservation initiatives in other regions of the world. To philosophically understand biocultural conservation concepts and to examine what local practical knowledge means for the whole landscape of knowledge, the philosophical discourse of the American pragmatist philosopher John Dewey (1859–1952) offers valuable insights, particularly, to understand the role of "practical knowledge" in environmental and biocultural education. Rejecting the rigid division between theory and practice, Dewey reframed knowledge, not as a fixed possession, but as a dynamic, evolving process that emerges through active engagement with real-life situations (Dewey 2008). His philosophy emphasized the unbreakable bond between knowing and doing—we do not arrive at knowledge instantly, but rather develop it through interaction, experimentation, and reflective experience. Particularly,

> … knowing, as judged from the actual procedures of scientific inquiry, has completely abandoned in fact the traditional separation of knowing and doing, that the experimental procedure is one that installs doing as the heart of knowing … (Dewey 2008, p. 29)

Learning, for Dewey, is not merely cognitive but embodied and participatory, rooted in the lived experience of the learner. This view closely aligns with Aristotle's concept of "phronesis," or practical wisdom, and finds renewed relevance today in biocultural conservation and education initiatives that aim to bridge universal scientific models with local, place-based knowledge systems (Prajapati and Nath 2025).

Dewey's emphasis on experiential learning supports the notion of practical knowledge, which forms the basis of numerous cultural and ecological practices of biocultural conservation (Rozzi 2013). Dewey's theory underscores the importance of "learning through doing" (Dewey 1997)—a model that supports cognitive growth, emotional development, and ethical engagement with the natural world. His instrumentalist approach views knowledge as a tool for solving real-world problems, making it particularly useful in addressing socio-ecological challenges. Through practical knowledge, communities can renew meaningful relationships with their environments, cultivate stewardship, and enact sustainable practices.

Dewey's insights help ground biocultural conservation efforts in educational practices that are inclusive, transformative, and action oriented. For example, art education projects rooted in local places and embodied experiences offer participants novel ways of interacting with local biocultural systems. Dewey (1997) provides philosophical concepts that align with the 3Hs of the biocultural ethics (Rozzi 2013):

both encourage people to reestablish ethical, aesthetic, and emotional connections with (human and other-than-human) co-inhabitants, their life habits in their local habitats. The arts provide a powerful way of acquiring and expressing practical knowledge, involve situational immersion, experiential learning, action-oriented practices, everyday skills, crafts, and embodied cognition. In this way, the arts catalyze pragmatist and biocultural practices to link arts with biocultural conservation, restoration, and communication.

Artistic concepts and practices can amplify Dewey's approach of learning by doing, for the sake of experientially understanding and valuing the intricate interrelations among Rozzi's 3Hs. Thus, through arts our understanding of the world becomes refreshing, encouraging, and promising in embracing otherness, such as other systems of knowledge, other ways of knowing, other ways of living together with other-than-human species, and other ways of greeting other cultures. The goal of Part II is to illustrate this refreshed biocultural perspective that connects intellectual inquiry with embodied practice, which repositions the image of knowledge as a living, evolving relationship with the world around us that guides us toward knowledgeable and respectful forms of co-inhabitation.

12.3 An Overview of Part II

Part II begins with Chilean sculptor Paola Vezzani, whose work overcomes abstract and disembodied approaches by embracing lived, embodied practices. In Chap. 13, "The Fragility of Encounter," Vezzani (2026) explores the theme of fragility through her experiences in the outdoor laboratory of the Omora Ethnobotanical Park, particularly through Field Environmental Philosophy (FEP) courses offered by the park. These courses provided her with "biocultural lenses" that deepen her appreciation for living beings, alongside "philosophical lenses" that broaden her understanding of biological and cultural diversity. Through her collaboration with environmental philosopher Ricardo Rozzi and engagement with diverse knowledge systems, Vezzani (2026) argues that art must take on an ethical role. Artistic and creative practices are rooted in ethical responsibility. She draws on her own watercolor, photographic, and sculptural works to develop an understanding of "fragility of encounters" within subantarctic biocultural diversity. Through the analysis of her artworks, Vezzani suggests that arts can cultivate integral ways of life based on socially and environmentally just forms of biocultural co-inhabitation. Her works bridge Indigenous knowledge and scientific inquiry, reflecting the importance of these heritages while also pointing toward the possibility of healing through respectful, interdisciplinary encounters.

In Chap. 15, two Chinese botanists, Qin Zuo and Li Zhang (2026), focus on the conservation of bryophytes and accessible science education at the Fairy Lake Botanical Garden, Shenzhen, China. Toward this aim, they have integrated their sciences with arts, exhibitions, and other programs to actively engage with citizens for environmental education and aesthetic appreciation of mosses and other bryophytes. These tiny plants are often overlooked by the public, educators, and environmental policies (Zhu 2023). To solve this knowledge gap, Zuo and Zhang have developed creative pedagogical approaches and innovative usages of arts, new media, and field activities. As a result, at the Fairy Lake Botanical Garden, bryophytes are captivating citizens' attention. The strategies and practical wisdom revealed here could offer valuable experience for the conservation of other overlooked taxa, such as freshwater invertebrates (Contador et al. 2023).

Inspired by the close collaboration between botanists and artists at the Fairy Lake Botanical Garden, Chap. 18, Chinese glass artist Shu Meng (2026) presents her novel artworks of the series of "Moss: The Trailblazer on Ruins." Meng explains how her work has involved learning about the biology and ecology of mosses as well as incorporating her long-standing glass artistry. She examines the ways of looking at urban development and ecological restoration with an approach of divergent thinking about the connections between nature and humanity, wilderness and civilization. Meng weaves her spiritual, academic, and artistic inspirations together to create moss glass artworks that carry explicit biocultural meanings. She underscores the importance of "direct knowledge," echoing the FEP's concept of face-to-face encounters with mosses (Rozzi 2023), to achieve deeper understanding of the relationships between humans and the natural environment.

Two other chapters in Part II also echo the theme of small but essential organisms on Earth. In Chap. 16, "Sublimity in Miniature: Aesthetic Viewing and the Miniature Forests of Cape Horn," Thompson (2026) argues for a shift from ordinary perception to an aesthetic lens, integrating biophysical studies with linguistic-cultural-philosophical conceptualization to attain the conservation of biodiversity and cultural diversities. Kant's description of sublimity can be instructive in appreciating the beauty of the miniature forests in the Cape Horn Biosphere Reserve, illustrating how immensity can be perceived in miniature and challenging the typical notions of the sublime. The "Miniature Forests of Cape Horn" program at Omora Ethnobotanical Park, a biocultural conservation initiative launched in 2000, is vital for knowledge relating to biocultural ethics and conservation. Thompson shows "changing lenses" and applying one particular, underrepresented lens: the aesthetic lens can dramatically alter how we perceive ourselves, our place in the world, our engagements with other-than-human beings, and our ways of valuing.

Mackenzie and his co-authors (2026) also reflected on work in the Cape Horn Biosphere Reserve (Chap. 17), to be specific, photographic work. Drawing on the 3Hs framework, photography becomes biocultural activities of collecting photos rather than simply taking or producing pictures. While photography has historically played a role in creating protected areas and promoting conservation ethics, the concept of "collecting photos" emphasizes an ethical approach to nature and conservation photography, the approach that documents meaningful details about life forms and their habitats. The authors introduce the 3Hs theory as a "heuristic lens" through which photographers can capture comprehensive records of species, highlighting their beauty, behaviors, habitats, and the ecological communities they co-inhabit. In other words, through *collecting* photos, photographers cultivate an awareness of co-inhabitation with other-than-human organisms to promote an "ethical, affective, and practical" relationship with regional biodiversity, particularly bryophytes and lichens in their photographic activities.

Practical knowledge can also be integrated into formal education. Jin's Healing Garden (2026) transforms art education into "eco-social practice", through which participants engage in composting, permaculture, and biocultural restoration, not as symbolic gestures but as hands-on acts of care. These practices exemplify how practical knowledge, when rooted in place and embodied activity, becomes a powerful agent for "biocultural healing." Specifically, Jin's project, Healing Garden, moves beyond isolated representational art to adopt direct actions rooted in permaculture principles and system thinking for real social and ecological changes. Healing Garden as a living classroom for hands-on engagement cultivates place-based and experiential learning, making it an ideal place for the transmission of practical knowledge to the healing of the local biocultural community in a different way. By integrating arts, ecological science, restoration, and eco-pedagogy, Jin (2026) challenges the dominance of anthropocentric views, redefines education, and creates cooperative partnerships among all co-inhabitants through art education and sharing of practical knowledge.

In "Biocultural Co-inhabitancy: An Examination on Scientific and Indigenous Nomenclature of Bryophytes" (Chap. 19), Zhu and Caviness (2026) examine various ways of knowing through investigating how names were given to bryophytes (mosses, liverworts, and hornworts) in various nomenclature systems, such as through common names, scientific names, and Indigenous names. They argue that naming is not a neutral act but a form of "knowledge-making" grounded in situated biocultural relationships. Their advocacy for a "polyculture of naming" bridges scientific classification with Indigenous understanding, and commences philosophical dialogues between linguistic, relational, and situated practical knowledge and scientific knowledge. They propose a synergy approach to gain a more holistic way of knowing both bryophytes and appreciating human-bryophyte relationship, as well as advocate for a long-range perspective of biocultural co-inhabitancy with small organisms.

Part II ends with a creative writing "Communion" in which Caviness (2026) describes a significant journey to Cranberry Lake with fellow botanists to visit the "Big Pine." During this visit, he sought guidance on a major personal decision: the path to parenthood through adoption. Viewing plants as trusted friends and teachers who could profoundly influence his life and career, Caviness recognizes that plants are "sentient beings and essential consultants" for life decisions. The critical importance of maintaining this "whisper" or communion with the plants suggests that communion is vital for meaningful survival and appreciation of co-inhabitation.

Acknowledgments We acknowledge the support from the Chilean National Agency for Research and Development through the Cape Horn International Center, CHIC (ANID/BASAL FB210018).

References

Barbett L, Stupple EJN, Sweet M, Schofield MB, Richardson M (2020) Measuring actions for nature—development and validation of a pro-nature conservation behaviour scale. Sustainability 12(12):4885. https://doi.org/10.3390/su12124885

Barrows PD, Richardson M, Hamlin I, Van Gordon W (2022) Nature connectedness, nonattachment, and engagement with nature's beauty predict pro-nature conservation behavior. Ecopsychology 14(2):83–91

Bridgewater P, Rotherham ID (2019) A critical perspective on the concept of biocultural diversity and its emerging role in nature and heritage conservation. People Nat 1(3):291–304. https://doi.org/10.1002/pan3.10040

Castro-Jorquera C (2026) The precarious art of Cecilia Vicuña: finding the ecological and sacred value of the small and the discarded. In: Zhu D, Rozzi R, Tsuji R, Johnson B, Caviness T, Castro-Jorquera C, Xu L, Zhang L (eds) Linking arts with biocultural conservation, restoration, and communication, Ecology and ethics, vol 8. Springer, Cham, pp 79–93. https://doi.org/10.1007/978-3-032-04830-1_5

Caviness T (2026) Communion. In: Zhu D, Rozzi R, Tsuji R, Johnson B, Caviness T, Castro-Jorquera C, Xu L, Zhang L (eds) Linking arts with biocultural conservation, restoration, and communication, Ecology and ethics, vol 8. Springer, Cham, pp 329–332. https://doi.org/10.1007/978-3-032-04830-1_20

Chiarini VS (2026) How the invisible becomes visible: situating scientific illustration in biocultural communication. In: Zhu D, Rozzi R, Tsuji R, Johnson B, Caviness T, Castro-Jorquera C, Xu L, Zhang L (eds) Linking arts with biocultural conservation, restoration, and communication, Ecology and ethics, vol 8. Springer, Cham, pp 149–169. https://doi.org/10.1007/978-3-032-04830-1_9

Contador TA, Rendoll-Cárcamo J, Gañan M, Ojeda-Villarroel J, Kennedy JH, Convey P, Rozzi R (2023) Underwater with a hand lens: ecological sciences and environmental ethics to value freshwater biodiversity. In: Rozzi R, Tauro A, Wright T, Avriel-Avni N, May RH Jr (eds) Field environmental philosophy: education for biocultural conservation. Ecology and ethics, vol 5. Springer, Dordrecht, pp 53–70. https://doi.org/10.1007/978-3-031-23368-5_4

Dewey J (2008) The later works, 1925–1953: the quest for certainty. Southern Illinois University Press, Carbondale

Dewey J (1997) Experience and education. Free Express, New York

Gaston KJ, Spicer JI (2013) Biodiversity: an introduction. Blackwell, Malden

Hokowhitu B, Moreton-Robinson A, Tuhiwai-Smith L, Andersen C, Larkin S (eds) (2020) Routledge handbook of critical indigenous studies. Routledge

Hu J, Yang P (2026) A biocultural encounter with the four gentlemen: art education in Chinese ecological and aesthetic traditions. In: Zhu D, Rozzi R, Tsuji R, Johnson B, Caviness T, Castro-Jorquera C, Xu L, Zhang L (eds) Linking arts with biocultural conservation, restoration, and communication, Ecology and ethics, vol 8. Springer, Cham, pp 111–129. https://doi.org/10.1007/978-3-032-04830-1_7

Ives CD, Abson DJ, von Wehrden H et al (2018) Reconnecting with nature for sustainability. Sustain Sci 13:1389–1397. https://doi.org/10.1007/s11625-018-0542-9

Jarić I, Roll U, Bonaiuto M, Brook BW, Courchamp F, Firth JA et al (2022) Societal extinction of species. Trends Ecol Evol 37(5):411–419

Jin L (2026) Healing garden: eco-art actions at Sichuan Fine Art Institute. In: Zhu D, Rozzi R, Tsuji R, Johnson B, Caviness T, Castro-Jorquera C, Xu L, Zhang L (eds) Linking arts with biocultural conservation, restoration, and communication, Ecology and ethics, vol 8. Springer, Cham, pp 223–242. https://doi.org/10.1007/978-3-032-04830-1_14

Leakey RE (1996) The sixth extinction: patterns of life and the future of humankind. Anchor, New York

Liu W, Chen J (2021) Green spaces in Chinese schools enhance children's environmental attitudes and pro-environmental behavior. Children, Youth and Environments, 31(1):55–87

López-del-Toro P, Andresen E, Barraza L, Estrada A (2009) Attitudes and knowledge of shade-coffee farmers towards vertebrates and their ecological functions. Trop Conserv Sci 2(3):299–318. https://doi.org/10.1177/194008290900200303

Mackenzie R, Valle-Celedón C, Arriagada G (2026) Collecting photos in the miniature forest: a biocultural approach for bryophyte conservation in the Cape Horn Biosphere Reserve. In: Zhu D, Rozzi R, Tsuji R, Johnson B, Caviness T, Castro-Jorquera C, Xu L, Zhang L (eds) Linking arts with biocultural conservation, restoration, and communication, Ecology and ethics, vol 8. Springer, Cham, pp 277–293. https://doi.org/10.1007/978-3-032-04830-1_17

Martín-López B, Iniesta-Arandia I, García-Llorente M, Palomo I, Casado-Arzuaga I, García del Amo D, Gómez-Baggethun E, Oteros-Rozas E, Palacios-Agúndez I, Willaarts B, González JA, Santos-Martín F, Onaindia M, López-Santiago C, Montes C (2012) Uncovering ecosystem services bundles through social preferences: experimental evidence from Spain. PLoS One 7: e38970. https://doi.org/10.1371/journal.pone.0038970

McIntosh S (2026) Embodied encounters: biocultural dialogues and pedagogies beyond the human. In: Zhu D, Rozzi R, Tsuji R, Johnson B, Caviness T, Castro-Jorquera C, Xu L, Zhang L (eds) Linking arts with biocultural conservation, restoration, and communication, Ecology and ethics, vol 8. Springer, Cham, pp 131–148. https://doi.org/10.1007/978-3-032-04830-1_8

Meng S (2026) The birth of "mosses: the trailblazer on ruins". In: Zhu D, Rozzi R, Tsuji R, Johnson B, Caviness T, Castro-Jorquera C, Xu L, Zhang L (eds) Linking arts with biocultural conservation, restoration, and communication, Ecology and ethics, vol 8. Springer, Cham, pp 295–310. https://doi.org/10.1007/978-3-032-04830-1_18

Monterrubio-Solís C, Barreau A, Ibarra JT (2023) Narrating changes, recalling memory: accumulation by dispossession in food systems of Indigenous communities at the extremes of Latin America. Ecol Soc 28(1):3. https://doi.org/10.5751/ES-13792-280103

Ojeda J, Salomon AK, Rowe JK, Ban NC (2022) Reciprocal contributions between people and nature: a conceptual intervention. Bioscience 72(10):952–962

Prajapati A, Nath R (2025) Looking at biocultural ethics through the lens of transcultural dialogue. Ethics Sci Environ Polit 25:71–85. https://doi.org/10.3354/esep0022

Richardson M, Passmore HA, Barbett L, Lumber R, Thomas R, Hunt A (2020) The green care code: how nature connectedness and simple activities help explain pro-nature conservation behaviours. People Nat 2(3):821–839. https://doi.org/10.1002/pan3.10117

Rozzi R, Anderson CB, Pizarro JC, Massardo F, Medina Y, Mansilla A et al (2010) Field environmental philosophy and biocultural conservation at the Omora Ethnobotanical Park: Methodological approaches to broaden the ways of integrating the social component ("S")in long-term socio-ecological research (LTSER) sites. Revista Chilena de Historia Natural 83(1):27–68. https://doi.org/10.4067/S0716-078X2010000100004

Rozzi R (2012) Biocultural ethics: recovering the vital links between the inhabitants, their habits, and habitats. Environ Ethics 34(1):27–50. https://doi.org/10.5840/enviroethics20123414

Rozzi R (2013) Biocultural ethics: from biocultural homogenization toward biocultural conservation. In: Rozzi R, Pickett ST, Palmer C, Armesto JJ, Callicott JB (eds) Linking ecology and ethics for a changing world: values, philosophy, and action. Springer, pp 9–32. https://doi.org/10.1007/978-94-007-7470-4_2

Rozzi R (2018) Biocultural homogenization: a wicked problem in the anthropocene. In: Rozzi R, May RH, Chapin FS, Massardo F, Gavin MC et al (eds) From biocultural homogenization to biocultural conservation. Springer, pp 21–48. https://doi.org/10.1007/978-3-319-99513-7_2

Rozzi R (2023) Inter-species and inter-cultural encounters: the biocultural education program of the Omora Ethnobotanical Park. In: Rozzi R, Tauro A, Avriel-Avni N, Wright T, May RH Jr (eds) Field environmental philosophy: education for biocultural conservation, Ecology and ethics, vol 5. Springer, Cham, pp 153–174. https://doi.org/10.1007/978-3-031-23368-5_10

Rozzi R (2026) The contributions of Eugene Hargrove to environmental ethics and aesthetics: a biocultural tribute. In: Zhu D, Rozzi R, Tsuji R, Johnson B, Caviness T, Castro-Jorquera C, Xu L, Zhang L (eds) Linking arts with biocultural conservation, restoration, and communication, Ecology and ethics, vol 8. Springer, Cham, pp 35–59. https://doi.org/10.1007/978-3-032-04830-1_3

Rozzi R, Massardo F, Anderson CB, Heidinger K, Silander JA Jr (2006) Ten principles for biocultural conservation at the southern tip of the Americas: the approach of the Omora Ethnobotanical Park. Ecol Soc 11(1):43. http://www.ecologyandsociety.org/vol11/iss1/art43/

Rozzi R, Arango X, Massardo F, Anderson C, Heidinger K, Moses K (2008) Field environmental philosophy and biocultural conservation: the Omora Ethnobotanical Park educational program. Environ Ethics 30(3):325–336. https://doi.org/10.5840/enviroethics200830336

Rozzi R, May RH Jr, Tauro A, Avriel-Avni N (2023a) Introduction to field environmental philosophy: a new methodological approach for biocultural education and conservation. In: Rozzi R, Tauro A, Avriel-Avni N, Wright T, May RH Jr (eds) Field environmental philosophy: education for biocultural conservation. Springer, Cham, pp 1–13. https://doi.org/10.1007/978-3-031-23368-5_1

Rozzi R, Álvarez R, Castro V, Núñez D, Ojeda J, Tauro A, Massardo F (2023b) Biocultural calendars across four ethnolinguistic communities in southwestern South America. GeoHealth 7(4). https://doi.org/10.1029/2022GH000623

Song Y (2026) Interspecies co-inhabitation: a dialogue between Chinese philosophical traditions and biocultural ethics. In: Zhu D, Rozzi R, Tsuji R, Johnson B, Caviness T, Castro-Jorquera C, Xu L, Zhang L (eds) Linking arts with biocultural conservation, restoration, and communication, Ecology and ethics, vol 8. Springer, Cham, pp 95–109. https://doi.org/10.1007/978-3-032-04830-1_6

Tauro A, Ojeda J, Caviness T, Moses KP, Moreno-Terrazas R et al (2021) Field environmental philosophy: a biocultural ethic approach to education and ecotourism for sustainability. Sustainability 13(8):4526. https://doi.org/10.3390/su13084526

Thompson M (2026) Sublimity in miniature: aesthetic viewing and the miniature forests of Cape Horn. In: Zhu D, Rozzi R, Tsuji R, Johnson B, Caviness T, Castro-Jorquera C, Xu L, Zhang L (eds) Linking arts with biocultural conservation, restoration, and communication, Ecology and ethics, vol 8. Springer, Cham, pp 265–276. https://doi.org/10.1007/978-3-032-04830-1_16

Troudet J, Grandcolas P, Blin A et al (2017) Taxonomic bias in biodiversity data and societal preferences. Sci Rep 7:9132. https://doi.org/10.1038/s41598-017-09084-6

Tsuji R, Johnson B (2026) Biocultural ontologies, art, and conservation. In: Zhu D, Rozzi R, Tsuji R, Johnson B, Caviness T, Castro-Jorquera C, Xu L, Zhang L (eds) Linking arts with biocultural conservation, restoration, and communication, Ecology and ethics, vol 8. Springer, Cham, pp 61–77. https://doi.org/10.1007/978-3-032-04830-1_4

Turpie JK (2003) The existence value of biodiversity in South Africa: how interest, experience, knowledge, income and perceived level of threat influence local willingness to pay. Ecol Econ 46(2):199–216. https://doi.org/10.1016/S0921-8009(03)00122-8

Vezzani P (2026) The fragility of encounter. In: Zhu D, Rozzi R, Tsuji R, Johnson B, Caviness T, Castro-Jorquera C, Xu L, Zhang L (eds) Linking arts with biocultural conservation, restoration, and communication, Ecology and ethics, vol 8. Springer, Cham, pp 207–222. https://doi.org/10.1007/978-3-032-04830-1_13

Whitburn J, Linklater W, Abrahamse W (2020) Meta-analysis of human connection to nature and proenvironmental behavior. Conserv Biol 34(1):180–193. https://doi.org/10.1111/cobi.13381

Zhu D (2023) Bridge the channel, enhance the inclusivity: a comparison between flagship species-centered and moss-centered conservation in Chile and China. In: Rozzi R, Tauro A, Wright T, Avriel-Avni N, May RH Jr (eds) Field environmental philosophy: education for biocultural conservation, Ecology and ethics, vol 5. Springer, Dordrecht, pp 457–482. https://doi.org/10.1007/978-3-031-23368-5_28

Zhu D (2026) Arts and environment after the emergence of artificial intelligence. In: Zhu D, Rozzi R, Tsuji R, Johnson B, Caviness T, Castro-Jorquera C, Xu L, Zhang L (eds) Linking arts with biocultural conservation, restoration, and communication, Ecology and ethics, vol 8. Springer, Cham, pp 171–187. https://doi.org/10.1007/978-3-032-04830-1_10

Zhu D, Caviness T (2026) Biocultural co-inhabitancy: an examination on scientific and indigenous nomenclature of bryophytes. In: Zhu D, Rozzi R, Tsuji R, Johnson B, Caviness T, Castro-Jorquera C, Xu L, Zhang L (eds) Linking arts with biocultural conservation, restoration, and communication, Ecology and ethics, vol 8. Springer, Cham, pp 311–328. https://doi.org/10.1007/978-3-032-04830-1_19

Zuo Q, Zhang L (2026) Small but essential: how to introduce the public to easily overlooked organisms? In: Zhu D, Rozzi R, Tsuji R, Johnson B, Caviness T, Castro-Jorquera C, Xu L, Zhang L (eds) Linking arts with biocultural conservation, restoration, and communication, Ecology and ethics, vol 8. Springer, Cham, pp 243–263. https://doi.org/10.1007/978-3-032-04830-1_15

Chapter 13
"Fragility of Encounters" in Subantarctic Biocultural Diversity

Paola Vezzani

Abstract Located at the southern end of the Americas, in the remote and protected archipelagoes of the Cape Horn Biosphere Reserve in Chile, the Omora Ethnobotanical Park is an outdoor "natural laboratory" dedicated to biocultural research, education, and conservation. Every year since 2000, university students have been offered field environmental philosophy (FEP) courses and activities that rely heavily on drawing as an observation and discovery tool. Metaphorically, the course's "biocultural lenses" not only amplify the appreciation of living beings (including mosses and other small organisms from the perspective of multiple forms of knowledge) but also offer "philosophical lenses" to broaden the understanding of biological and cultural diversity, and to orient ethical forms of co-inhabitation within it. This biocultural understanding has awakened a keen awareness of the fragility of human life (in its cultural diversity) and the entirety of living beings (in its vast biological diversity), with specific life-habits and shared habitats. My short essay begins with a brief personal anecdote linked to a series of watercolor, photographic, and sculptural works to develop a first possible interpretation of the chapter's title, "'Fragility of Encounters' in Subantarctic Biocultural Diversity." These artworks are open to multiple readings that evoke the notions of encounters and fragility. Based on the analysis of the artwork, I propose that art should assume an ethical role in generating works that pose questions about our relationship with cultural and biological diversity, especially at the current planetary crisis. It is urgent to look toward a state of peace between nations and people to address the problems of poverty, equity, respect, and human migration, as well as the need to care for the fragile balance of the natural ecosystems of which we are a part. In short, art can help us cultivate integral ways of life based on socially and environmentally just forms of biocultural co-inhabitation.

Keywords Art · Biocultural conservation · Chile · Environmental education · Field environmental philosophy

---

P. Vezzani (✉)
Cape Horn International Center, Puerto Williams, Chile

D. Zhu et al. (eds.), *Linking Arts with Biocultural Conservation, Restoration, and Communication*, Ecology and Ethics 8,
https://doi.org/10.1007/978-3-032-04830-1_13

13.1 Introduction

The Biocultural Conservation Program based at the Omora Ethnobotanical Park located at the southern end of the Americas aims to study biocultural diversity broadly and implement biocultural conservation activities (Rozzi and Vezzani 2020). Omora Park was created in 2000 by scientists, artists, philosophers, and other professionals, both Chilean and foreign, to provide a "physical and conceptual space" for long-term biocultural research, interdisciplinary education, and conservation (Rozzi et al. 2006).

As a physical space, Omora Park is a 1100-hectare biological reserve that protects the Robalo River watershed, the drinking water source for Puerto Williams, the world's southernmost city. Across the watershed, one encounters a diverse mosaic of habitats and biotic communities from the shoreline to the high Andean ecosystems south of Tierra del Fuego (Contador et al. 2015).

As a conceptual space, Omora Park's research and educational programs integrate sciences, philosophy, and the arts. In forest, peat bog, coastal, high-Andean, and freshwater habitats researchers develop interdisciplinary methods and share knowledge about austral biocultural diversity with local, national, and international participants. In this way, the park functions as a "natural laboratory" (Rozzi et al. 2006), or better as a "biocultural laboratory" (Rozzi et al. 2020).

In 2007, I was invited to participate in a field environmental philosophy (FEP) course created by a team of artists, writers, scientists, and philosophers in collaboration with handcrafters of the Yahgan Indigenous community. The course is annually offered by the University of Magallanes and the Omora Foundation in Chile, as well as the University of North Texas and the Center for Environmental Philosophy in the USA (Thompson 2026). Drawing and metaphor composition are used in FEP courses as tools for observing and understanding the world as well as tools for science communication and cultural transformation (Zhu 2023).

Bringing writers and visual artists to teach FEP courses alongside philosophers and scientists allows students to participate in biocultural ethical transformative experiences (Tauro et al. 2021; Contador et al. 2023; Crego et al. 2023; Méndez-Herranz et al. 2023; Ojeda et al. 2023). As they explore the power of drawing, narratives, and metaphors, they enrich their biocultural understanding while also contributing to conserving biocultural diversity itself. Science and philosophy teach them that humans and all living beings share the common vital pulse of cellular respiration, growth, and reproduction. Drawing and composing metaphors also help them go from purely scientific practices toward biocultural experiences, which stimulate individual and communitarian ethical transformation processes (Rozzi et al. 2008). Participants in FEP courses are invited to reflect on their responsibility for cultivating an ethic of care, which expands moral consideration beyond human beings to the community of beings with whom we coinhabit the biosphere. Importantly, they can perceive *in situ*, appreciate, and empathize with the fragility of life. As stated by Argentinean ecologist Ramiro Crego and collaborators, FEP courses

> include a 'conceptual lens' that allows a clearer view of the co-inhabitants of the forest, their habits of life, their delicate balance, and the fragility of this ecosystem. (Crego et al. 2023, p. 107)

This awareness about the fragility of biotic communities, ecosystems, and social-ecological systems (at multiple scales) has moved me deeply and motivated me to write this chapter. Based on personal experience, I will elaborate on my awareness about the fragility of life and encounters with human and other-than-human co-inhabitants. However, I am aware that this personal account has implications for better understanding and coping with the fragility of life support systems from local to global scales.

13.2 The Fragility of Encounters

The fragility of encounters was a phrase I incorporated many years ago, very intuitively and spontaneously, into a small collage I created with objects collected on the beach, challenging my art schoolteachers' maxim of not including text in visual works. After several years, I have realized that this practice is not uncommon; see, for example, the chapters by Castro-Jorquera (2026) and Chiarini (2026). Moreover, I realized that this poetic blending brings a series of multipurpose readings, which I will state as doubts and questions associated with different artworks and their content.

In producing these artworks, I have created images that evoke more than one possible reading. For example, a whale that transforms into a forest, a ship that is a map, or an entangled vegetation community, an island that could be a boat. On the one hand, the fragility of encounters with these art pieces proposes multiple relationships, as it raises numerous questions, such as: whose fragility, is it? Or what is fragility? Could it indicate the vulnerability of humans as well as of nonhuman life? In which way do encounters involve people, places, concepts, and objects?

On the other hand, fragility speaks of what we find within ourselves, of our fears of the passage of time or our death. It can also refer to encountering other cultures or biological species in habitats unknown to us. How are intercultural encounters fragile? How were the encounters between European conquerors and the Indigenous Fuegian people in the sixteenth century?

Fragile is the encounter, fragile is memory, fragile is one's own culture. Fragile is the life of humans as well as other-than-human beings. Fragile is memory, as is the fact of selectively forgetting what makes me feel bad, incorrect, or even evil.

We hesitate, however, to acknowledge our own fragility in the face of the passage of time. Therefore, I begin this exploration of a series of sculptures in which the poetic idea proposed by the title, "The Fragility of Encounter," can be found as a backdrop with a double image of fog. First, the diffuse atmosphere represented in the watercolor "Mist" (*Bruma* in Spanish) (Fig. 13.1). Second, a photograph of a foggy

Fig. 13.1 Mist (*Bruma* in Spanish), watercolor 25 × 25 cm, by Paola Vezzani. (Photo by Paola Vezzani)

Fig. 13.2 "Foggy morning." (Photograph by Paola Vezzani, taken at Pia Glacier, Darwin Cordillera in the Cape Horn Biosphere Reserve, Chile, in February 2015)

morning on Glacier Pia, during an expedition with the Omora Park research team along the Darwin Cordillera in the Cape Horn Biosphere Reserve, Chile (Fig. 13.2).

If I look in the mirror, I'm afraid to do so with glasses since I prefer to look at myself with my myopia that helps improve my perception of my own image. The fear of discovering an accurate self-image often concerns the passage of time and its imprint on the body. But in my case, it has to do with the fragility of life and memory.

13.2.1 Discoveries: Do We Decide What We Remember?

Do we decide? I don't really know if we control what we remember. Do we select at will? If so, it's normal for some things to remain clear in our memories and others to disappear. The sad thing is that we can't consciously choose what we want to remember, making us vulnerable, fragile.

The fact is that we don't function like artificial intelligence. We humans do not merely rely on a collection of data we draw on when needed. Considering a fact implies a point of view and may even trigger a strong emotion. As Danqiong Zhu (2026) points out in his chapter in our book, artificial intelligence is a formidable tool for storing, selecting, cross-referencing, and analyzing information of all kinds. However, what is truly human is the ability to look through a lens, which involves concepts, beliefs, and interpretations we have learned. We can also question, seek divergences, and propose changes to what was once considered absolute. That is, to discuss, think with others, propose, and make mistakes.

We find it in memory and, at the same time, lose it. We dare not acknowledge our own fragility in the face of the passage of time. Therefore, I have proposed "the mist" as a metaphor for which we partially see and elude our encounter.

Fragile is the encounter, fragile is the memory, fragile is culture itself. Fragile is the life of humans and others than humans. Fragile is memory, as is selectively forgetting, which makes us feel vulnerable.

From these emotions, in this chapter, I embarked on a journey through a series of works in which it is possible to have the poetic idea as a backdrop: the fragility of encounters.

13.2.2 My Family: Living Presence in My Memory

Memory connects with personal history. My maternal grandmother, daughter of a Spanish mother and a Norwegian father, was born in Magallanes, Chile. My maternal grandfather was born in Asturias, Spain, and he passed onto my mother his love for the land and his way of naming plants, trees, birds, and insects that were so beautiful and diverse in Patagonia. As a child, I was amazed and lovingly fascinated by life in the forests, lakes, and meadows. I inherited some of the names of the different species, without questioning whether they were local (native or introduced). I still feel that childhood wonder, and the vivid affection, when I enter a forest or navigate the channels and fjords of the archipelagos in Cape Horn. In short, from my mother, María Olvido Gonzalez Yansen, and her family, I inherited names, a love of nature, and a worldview.

From my father, Alfio Vezzani Solar, I inherited his love for great European culture, especially the Italian Renaissance, along with spiritual expressions in the arts and literature. I also inherited ethical concepts of life. His father was born in Emilia-Romagna, Italy, and his mother was born in Santiago, Chile. However, I did not have enough contact with my paternal grandmother, whose ancestors incorporate American blood, most likely Mapuche. Hence, I am a mestiza person like most of my fellow Chileans and those who have their roots in the Americas. Here, I feel the fragility of the ties to the past, which end in an ignored, almost forgotten ethnic blend, a mestizo lost in memory, in the mist.

At the same time, my memories allow me to refresh my awareness of my childhood roots. They are vitally rooted in a love for the biological and cultural worlds, as if they were a seed to germinate bioculturally in my current art.

13.2.3 Encounters with Ricardo Rozzi

In 2007, I met the Chilean philosopher and ecologist Ricardo Rozzi. First through a television program and then in personal encounters in Magallanes. From then on, my understanding of the world began to change. Not only did I learn to name species, which I thought I knew, but I also began to shift my artistic work. This shift went hand in hand with experiences in expeditions and FEP courses and with a long list of wonderful researchers, including scientists, philosophers, and many other disciplines, as environmental philosopher Michael Thompson (2026) describes in this book. Expeditions and FEP build on a tradition of collaborative work among naturalists and artists, but FEP transforms them by critically addressing colonialist dimensions of this tradition as discussed by Rozzi (2026). For me, works of art have meaning when they provoke the viewer to ask questions or become curious about the world depicted in the artwork.

In constructing what I do as an artist, I ask myself: Through what "biocultural lens" am I looking through? In doing so, I can discover whether it is possible to question my perspectives, and to reflect on new ways of conceiving the relationships among diverse humans, and between them and nature. I do so by integrating emotions, senses, and diverse knowledge from different cultures and disciplines. Research in art is different from scientific conventions, but nonetheless it offers complementary ways of understanding and valuing the world. In this way, it becomes fundamental to dialogues and collaborative work on biocultural conservation across disciplines, as so nicely portrayed by Valeria Chiarini (2026).

An expression of a multifaceted biocultural lens is my sculpture "Yahgan Whale" (*Ballena yagán*, in Spanish) (Fig. 13.3). The marine habitats inspire it with its diversity of human and other-than-human co-inhabitants that become visible in a delicate pattern of lines done by cutouts in the metal. The whale enters the scene sharing with small canoes. Finding a beached whale was a great event for the canoeing and hunter-gathering peoples of Tierra del Fuego. They communicated

Fig. 13.3 Yahgan Whale (*Ballena Yagán*, in Spanish) by Paola Vezzani. Steel plate (72 cm high) that is part of the Magellan series of sculptures permanently exhibited at Hotel Cabo de Hornos, Punta Arenas, Chile. (Photo by Vicente González)

this through bonfires, and groups from different clans and even ethnic groups shared the bounty.

The encounter with Europeans brought great pain and loss to the hunter-gatherers and canoeists of the Tierra del Fuego archipelagos. The genocide wrought by the conquistadors resulted in the loss of lives, culture, and ancestral knowledge. Today, we condemn this colonialist action. However, today we witness armed conflicts between nations and cultures. We remain fragile and vulnerable in these encounters.

I seek to express this fragility in the work "Window" (*Ventana*, in Spanish) (Fig. 13.4). This sculpture is approximately 70 cm tall and consists of two curved pieces of metal, one in front of the other. In this sculpture, one can distinguish a mountain with an openwork that can be understood as a window shaped like fire. Through this window, a human figure and fishing objects can be seen. As I have noted before, as a visual strategy, I seek forms that can be interpreted in multiple ways. In this case, the mountain or awning may be moving in the wind, or perhaps its summit is snow-capped. The fire-window alludes to the biocultural lens we use to understand the past. Fuegian cultures were fragile and vulnerable in their encounter with Western culture. With my sculpture, I want to invite new encounters with greater respect, nourished by forms of biocultural co-inhabitation.

As an artist, I am seduced by nature, the landscape, and the multiple relationships I see between the co-inhabitants of a forest. Walking alongside a river or sailing to a

Fig. 13.4 Window (*Ventana*, in Spanish) by Paola Vezzani. Steel plate (70 cm high) that is part of the Magellan series of sculptures permanently exhibited at Hotel Cabo de Hornos, Punta Arenas, Chile. (Photo by Paola Vezzani)

remote glacier inspires me with infinite emotion. Feeling the force of nature flowing in its rhythm opens in me endless sensations and questions. It is then that dialogue and reflection, together with people from different backgrounds, open new possibilities for understanding. The encounter with new ideas can help gently awaken curiosity to experience the fragile and vulnerable nature.

In the late 1990s, I was fascinated by the art and nature movement, with the intuition of finding there a poetic concern with the problem of the passage of time, natural and human memory, to advance conservation concerns. I saw works by artists like English sculptor, photographer, and environmentalist, Andy Goldsworthy, or German painter, sculptor, and photographer, Nils Udo. They work with elements found in nature and install them ephemerally in uninhabited places as well as in gardens, parks, and museums around the world (Matless and Revill 1995; Jutant 2006; Song 2010; Shobeiri 2021). Their sensibility seemed more familiar to me and led me toward practices incorporating critical reflection on human action on the environment or biodiversity.

As I mentioned at the beginning of this section, meeting Ricardo Rozzi and beginning to collaborate on biocultural conservation and education programs changed the lens through which I view the world. Participating as a professor in FEP courses has been enlightening, allowing me and others to see and experience nature from other perspectives through direct experience. Learning about and reflecting on the environment from the perspective of environmental ethics, science, and other knowledge is a transformative life experience. As Rozzi states it,

> Purely scientific or other cognitive arguments have not been sufficient to open the "ecological eyes" of most citizens to appreciate the magnitude of the current social and environmental crisis, and the essential need to appreciate life in its cultural and biological diversity. The vast diversity of life forms and their intricate biocultural interrelationships with humans remain largely invisible to global society. (Rozzi 2023, p. 17)

Through shared experiences with researchers in expeditions and participating in FEP courses, I and others have been invited and guided to focusing our attention on previously unnoticed living beings, along with understanding their role in ecosystems, feeling like co-inhabitants in a shared habitat, and learning about the habits of each being, opens our consciousness in a balance of emotion and reason (Rozzi 2019).

These encounters and experiences with a diversity of disciplines, forms of knowledge, life-forms (including tiny mosses and large whales) have urged me to assume, through my art, an ethical responsibility to communicate questions, awaken curiosity, and invite people to explore biocultural values and other concepts about our fragile relationships with diverse cultures, living beings, and ecosystems.

13.2.4 **Ilaia***: A Fragile Biocultural Mosaic*

In 2008, I composed an exhibition of sculptures created with different materials, which I called *Ilaia* (which means "Further South" in Yahgan language). One of the sculptures, called Home ("Hogar", in Spanish), is constructed from branches salvaged or found on the street (Fig. 13.5). In the center, there is a stone partially covered with two pieces of leather joined together by a woven copper wire. The materiality of the branches markedly contrasts with the stone, not only in form but also in the time it takes for them to change state. This contrast of materials makes me think of the fragility of human life, the fragility of what we consider home, and the stone remains that today help archaeologists identify objects used in the past by vanished cultures. Each material has its time scale. Furthermore, the construction system is elemental, primitive, if you will. The elemental, simple, and rustic have been recurring themes that I have incorporated into my visual language.

The Trees of Memory series, which was part of the *Ilaia* exhibition, included eight sculptures. I sought visual metaphors that would allow me to allude to the impossibility of recovering the life of a tree when it is destroyed. In this sculpture series, a tree represents a complete forest, including all living beings that co-inhabit within it. By reconstructing logs with burnt driftwood or fragile structures of thick steel wire, I want to give the feeling of a forest lost by anthropogenic action.

In the *Trees of Memory* series, the encounters of forms and materials are resolved with basic and simple systems, practically without contemporary technology. The pre-industrial aspect lends a seal that points to the past in a metaphor for what is lost.

Fig. 13.5 Home (*Hogar*, in Spanish) by Paola Vezzani. This sculpture (115 × 110 × 100 cm) is made of poplar branches, leather, stone, and copper wire, and is part of the series Trees of Memory (*Árboles de Memoria*, in Spanish) exhibited at the Patricia Ready Gallery in Santiago, Chile. (Photos by Vicente González)

Once again, the fragility of life is present, along with the fear that the encounter is past and will never happen again.

Built with wood recovered from beaches, some of which were impacted by fire, I created three sculptures: Beaver and Sheep (*Castor y Oveja*), Sebastian (*Sebastián*), and Feather (*Pluma*, in Spanish) (Fig. 13.6). They are part of the *Trees of Memory* series. In the past, forest fires were intentionally set to create vast, open spaces for grazing. However, many of these fires got out of control, burning more land than intended. The imprint of burned forests is in the collective memory of the Magellanic people. Forests where only black-smeared trunks remain standing, solitary, without branches or leaves. The vegetation is partially recovering, but the presence of these dead, standing forests is still dramatic. In some places on the planet, fires are natural, and thanks to fire, the forest is cleared of already dead trees, boosting the growth of new saplings and even allowing certain seeds to germinate. This leads us to think that we need to reconsider, study, and understand. Reality is a fragile construction that does not always conform to what is found.

13.2.5 An Artistic-Scientific Kaleidoscope

In the maritime iconography, the lighthouse stands out as an element that I find magnificent as a metaphor. This is how I began experimental work with sculptures in

Fig. 13.6 From left to right: Beaver and Sheep (*Castor y Oveja*), Sebastian (*Sebastián*), and Feather (*Pluma*, in Spanish) Home (*Hogar*, in Spanish) by Paola Vezzani. These three sculptures (205–230 cm height) are part of the series Trees of Memory (*Árboles de Memoria*, in Spanish) exhibited at the Patricia Ready Gallery in Santiago, Chile. (Photos by Vicente González)

the form of openwork cones. Openwork is a technique that creates holes or gaps through a solid material such as metal, and this technique has been widely used in many cultures. To me, the interplay of openwork evokes a forest in an upward spiral. My possible interpretation is one of hope, as it can be interpreted as if biodiversity were the light that guides us to build a better future. Inspired by these ideas and techniques, I created the Lighthouse (Faro, in Spanish), a 2.5-meter-tall metal sculpture, that was installed in a public space at the University of Helsinki, Finland, in May 2022 (Fig. 13.7).

While working in the studio, I remembered explanations about natural phenomena that different scientists of the Omora Park team gave me during our field expeditions. Thus, I came up with a way to illustrate how my concepts were forged in conversations with researchers from various disciplines during scientific expeditions carried out in the Cape Horn Biosphere Reserve for over a decade. This is how the sculpture Wind Rose (*Rosa de los Vientos*, in Spanish) was born in 2018 (Fig. 13.8).

Fig. 13.7 Lighthouse (*Faro*, in Spanish), a 2.5-meter-high steel plate by Paola Vezzani. This openwork is exhibited in the gardens of the Faculty of Humanities at the University of Helsinki, Finland. Left. View of the sculpture during daytime. Right. View of the "Lighthouse" at night. (Photos by Paola Vezzani)

Fig. 13.8 Wind Rose (*Rosa de los Vientos*, in Spanish), a 120 cm diameter sculpture by Paola Vezzani. This steel plate was part of the solo exhibition "Strait of Magellan" at the Patricia Ready Gallery, Santiago, Chile 2018

The Wind Rose sculpture measures 1.30 cm in diameter, and it is a circular silhouette cut from a steel plate that resembles a planet surrounded by trees connected to each other either by their roots or by the bodies of fungi. This union alludes to the multiple collaborations that occur within ecosystems. From another

Fig. 13.9 Spiral of Life (*Espiral de Vida*, in Spanish) by Paola Vezzani. This steel plate is 132 cm in diameter, and it is permanently exhibited at the headquarters of the Cape Horn International Center (CHIC), Puerto Williams, Chile. (Photo by Paola Vezzani)

perspective, the circle is a symbol that invites viewers to think in manifold relationships with nature, such as the idea that there is no beginning or end, which suggests the cycles of life. The title "Rosa de los Vientos" plays with the idea of cardinal points, suggesting a desired route or destination. By stretching the interpretation a bit, with the conceptual lens of the biocultural ethic (Rozzi 2013), one could even say that collaboration exists among different species that share a habitat, where all co-inhabitants (including humans) need each other.

After the Wind Rose, I sculpted the Spiral of Life (*Espiral de Vida*, in Spanish) (Fig. 13.9). This steel plate is permanently exhibited at the headquarters of the Cape Horn International Center (CHIC) for Global Change Studies and Biocultural Conservation in Puerto Williams, Chile. Puerto Williams is the world's southernmost city, and the possible interpretations of this openwork place at a research center are like those of the Wind Rose sculpture. However, the idea of spiral growth is added here. Playing with timescales, it is possible to distinguish an ammonite. Playing with anatomical imagery, one could relate this sculpture to the roots of the human body, as if they were neurons. I was incited when my physician and friend, Dr. Paulina Verdaguer, told me that she saw "sap, lymph, or blood. Synapses or rhizomes, all forms of the same thing."

13.3 Coda: Some Methodological and Ethical Considerations

To conclude the tour of the sculptures, I would like to briefly explain a few aspects of my working process. I heat the steel plate by hand with a high-temperature tool that allows me to draw and correct directly, almost as if sketching with a flame pencil. Thus, the representation of the landscape has more to do with memory, where sensations and ideas converge in a synthesis that maintains certain characteristics. For example, the fluid movement of the forms in the composition reflects the response of vegetation to the wind as well as the patterns of order within nature. On the other hand, the textures recall the complex and tangled geography of the channels and islands of the Magellanic Region, where the land seems to disintegrate into myriads of islets. With these elements, I seek to make the viewer feel life's delicate and fragile fabric.

As an artist, I respect a work ethic for the artwork I create. In other words, I am responsible for the meaning of what I am trying to communicate. To this end, it is essential for me to be in constant dialogue with other forms of knowledge from which I draw content to generate images, representations, and questions.

Based on my concise analysis of part of my artwork, I dare to propose that art should assume an ethical role in generating works that question our relationship with cultural and biological diversity. This seems to me especially relevant at the current times of planetary crisis, when it is urgent to restore peace, enabling us to work collaboratively to address problems of poverty and human migration, as well as to care for the fragile communities of human and other-than-human co-inhabitants with whom we share our habitats. Climate change, and more broadly, global social-environmental change, demands from us an effort of conscience and ethical advancement. In short, art can help us cultivate integral ways of life based on socially and environmentally just forms of biocultural co-inhabitation.

References

Castro-Jorquera C (2026) The precarious art of Cecilia vicuña: finding the ecological and sacred value of the small and the discarded. In: Zhu D, Rozzi R, Tsuji R, Johnson B, Caviness T, Castro-Jorquera C, Xu L, Zhang L (eds) Linking arts with biocultural conservation, restoration, and communication, Ecology and ethics, vol 8. Springer, Cham, pp 79–93. https://doi.org/10.1007/978-3-032-04830-1_5

Chiarini VS (2026) How the invisible becomes visible: situating scientific illustration in biocultural communication. In: Zhu D, Rozzi R, Tsuji R, Johnson B, Caviness T, Castro-Jorquera C, Xu L, Zhang L (eds) Linking arts with biocultural conservation, restoration, and communication, Ecology and ethics, vol 8. Springer, Cham, pp 149–169. https://doi.org/10.1007/978-3-032-04830-1_9

Contador T, Kennedy JH, Rozzi R, Ojeda J (2015) Sharp altitudinal gradients in Magellanic Sub-Antarctic streams: patterns along a fluvial system in the Cape Horn Biosphere Reserve (55 S). Polar Biol 38:1853–1866. https://doi.org/10.1007/s00300-015-1746-4

Contador TA, Rendoll-Cárcamo J, Gañan M, Ojeda J, Kennedy JH, Convey P, Rozzi R (2023) Underwater with a hand lens: ecological sciences and environmental ethics to value freshwater biodiversity. In: Rozzi R, Tauro A, Wright T, Avriel-Avni N, May RH Jr (eds) Field environmental philosophy: education for biocultural conservation, Ecology and ethics series, vol 5. Springer, Cham, pp 53–69. https://doi.org/10.1007/978-3-031-23368-5_4

Crego RD, Ward N, Rozzi R (2023) The eyes of the tree: applying field environmental philosophy to tackle conservation problems at long term socio-ecological research sites. In: Rozzi R, Tauro A, Wright T, Avriel-Avni N, May RH Jr (eds) Field environmental philosophy: education for biocultural conservation, Ecology and ethics series, vol 5. Springer, Cham, pp 101–111. https://doi.org/10.1007/978-3-031-23368-5_7

Jutant C (2006) Nils-Udo, naturaleza viva. Minerva: Revista del Círculo de Bellas Artes 2:109–114

Matless D, Revill G (1995) A solo ecology: the erratic art of Andy Goldsworthy. Ecumene 2(4): 423–448

Méndez-Herranz M, Marini G, Rozzi R (2023) Sub-Antarctic High Andean "Gardeners": cultivating caring relationships. In: Rozzi R, Tauro A, Wright T, Avriel-Avni N, May RH Jr (eds) Field environmental philosophy: education for biocultural conservation, Ecology and ethics series, vol 5. Springer, Cham, pp 71–86. https://doi.org/10.1007/978-3-031-23368-5_5

Ojeda J, Tauro A, Contador T, Mansilla A, Rosenfeld S, Ban N, Rozzi R (2023) "Pay attention, dive with wide-open eyes." A field environmental philosophy activity to foster reciprocity between people and the sea. In: Rozzi R, Tauro A, Wright T, Avriel-Avni N, May RH Jr (eds) Field environmental philosophy: education for biocultural conservation, Ecology and ethics series, vol 5. Springer, Cham, pp 87–100. https://doi.org/10.1007/978-3-031-23368-5_6

Rozzi R (2013) Biocultural ethics: from biocultural homogenization toward biocultural conservation. In: Rozzi R, Pickett ST, Palmer C, Armesto JJ, Callicott JB (eds) Linking ecology and ethics for a changing world: values, philosophy, and action, Ecology and ethics, vol 1. Springer, Dordrecht, pp 9–32. https://doi.org/10.1007/978-94-007-7470-4_2

Rozzi R (2019) Taxonomic chauvinism, no more! Antidotes from Hume, Darwin, and biocultural ethics. Environ Ethics 41(3):249–282. https://doi.org/10.5840/enviroethics201941325

Rozzi R (2023) Introduction to part I. Field environmental philosophy: concepts and case studies. In: Rozzi R, Tauro A, Avriel-Avni N, Wright T, May RH Jr (eds) Field environmental philosophy, Ecology and ethics, vol 5. Springer, Cham, pp 17–25. https://doi.org/10.1007/978-3-031-23368-5_2

Rozzi R (2026) The contributions of Eugene Hargrove to environmental ethics and aesthetics: a biocultural tribute. In: Zhu D, Rozzi R, Tsuji R, Johnson B, Caviness T, Castro-Jorquera C, Xu L, Zhang L (eds) Linking arts with biocultural conservation, restoration, and communication, Ecology and ethics, vol 8. Springer, Cham, pp 35–59. https://doi.org/10.1007/978-3-032-04830-1_3

Rozzi R, Vezzani P (2020) Artistic activism and narratives on environment. Case study: Parque Etnobotánico Omora, Puerto Williams (Chile). In: Rossi-Linnemann C, de Martini G (eds) Art in science museums: towards a post-disciplinary approach. Taylor & Francis/Routledge, New York, pp 108–111

Rozzi R, Massardo F, Anderson C, Heidinger K, Silander J Jr (2006) Ten principles for biocultural conservation at the southern tip of the Americas: the approach of the Omora Ethnobotanical Park. Ecol Soc 11(1):43. https://doi.org/10.5751/ES-01709-110143

Rozzi R, Arango X, Massardo F et al (2008) Field environmental philosophy and biocultural conservation: the Omora Ethnobotanical Park educational program. Environ Ethics 30:325–336. https://doi.org/10.5840/enviroethics200830336

Rozzi R, Rivero de Aguilar J, Vezzani P, Viano P, Birker P, Malebrán J, Matus R, Aguirre F, Vásquez RA, Massardo F (2020) Descubrimientos naturales y éticos en el laboratorio natural de Magallanes. In: Aldunate C, Rozzi R, Morello F, Massardo F (eds) Estrecho de Magallanes: Tres Descubrimientos. Colección Santander, Museo de Chileno de Arte Precolombino, Santiago, pp 216–235

Shobeiri A (2021) Thinking from materials in Andy Goldsworthy's environmental artworks. Stud Vis Arts Commun 8(1):15–25

Song YIK (2010) Art in nature and schools: Nils-Udo. J Aesthet Educ 44(3):96–108

Tauro A, Ojeda J, Caviness T, Moses KP, Moreno-Terrazas R et al (2021) Field environmental philosophy: a biocultural ethic approach to education and ecotourism for sustainability. Sustainability 13(8):4526. https://doi.org/10.3390/su13084526

Thompson M (2026) Sublimity in miniature: aesthetic viewing and the miniature forests of Cape Horn. In: Zhu D, Rozzi R, Tsuji R, Johnson B, Caviness T, Castro-Jorquera C, Xu L, Zhang L (eds) Linking arts with biocultural conservation, restoration, and communication, Ecology and ethics, vol 8. Springer, Cham, pp 265–276. https://doi.org/10.1007/978-3-032-04830-1_16

Zhu D (2023) Bridge the channel, enhance the inclusivity: a comparison between flagship species-centered and moss-centered conservation in Chile and China. In: Rozzi R, Tauro A, Wright T, Avriel-Avni N, May RH Jr (eds) Field environmental philosophy: education for biocultural conservation, Ecology and ethics series, vol 5. Springer, Cham, pp 457–482. https://doi.org/10.1007/978-3-031-23368-5_28

Zhu D (2026) Arts and environment after the emergence of artificial intelligence. In: Zhu D, Rozzi R, Tsuji R, Johnson B, Caviness T, Castro-Jorquera C, Xu L, Zhang L (eds) Linking arts with biocultural conservation, restoration, and communication, Ecology and ethics, vol 8. Springer, Cham, pp 171–187. https://doi.org/10.1007/978-3-032-04830-1_10

Chapter 14
Healing Garden: Eco-art Actions at Sichuan Fine Art Institute

Lipeng Jin

Abstract Often, conventional aesthetics about environmental issues still focuses on detached representational objects. This approach continues a nature-culture divide that limits conservation and restoration efforts. In order to overcome this divide and actively engage in ecological restoration, we present the case of the "Healing Garden" (HG). In this project participants are active "gardeners" that heal the biocultural community from negative impacts caused by processes of biocultural homogenization. HG is an eco-art and pedagogical project at Sichuan Fine Art Institute responding to the ecological crises such as climate change and the loss of biocultura diversity. This transdisciplinary project integrates art, science, ecological restoration, eco-pedagogy, activism, community engagement and embraces the notions of system thinking and permaculture. It promotes placed-based, experiential learning and explores a new form of education to reorient nature-human interactions. Instead of making isolated representational art objects, the project uses direct actions to create real social, and ecological changes. To promote a new imagery of reciprocal, cooperative partnerships, we use "gardeners" as a powerful metaphor referring to both human and other-than-human participants and creators for building urban biodiversity, community cohesion, and a network of healthy multispecies relations. The project turns the Anthropocene issues such as biocultural homogenization into artistic opportunities and pedagogical resources, fusing criticality with constructive activism, hope, conviviality, and community empowerment. Toward this aim of holistic "gardening" we conduct a variety of eco-art activities that may inspire new patterns of caring relationships out of the present climatic chaos, and foster holistic healing of the biocultural community.

Keywords Eco-art · Activism · Permaculture · Ecopedagogy · Healing · Biocultural conservation

---

L. Jin (✉)
School of Experimental Art, Sichuan Institute of Fine Art,, Chongqing, China
e-mail: jin.lipeng@scfai.edu.cn

D. Zhu et al. (eds.), *Linking Arts with Biocultural Conservation, Restoration, and Communication*, Ecology and Ethics 8,
https://doi.org/10.1007/978-3-032-04830-1_14

14.1 Introduction

In the midst of China's unparalleled urbanization, university towns have proliferated, transforming vast stretches of countryside and disrupting local ecosystems. The university town in Chongqing exemplifies this rapid metamorphosis, marking the transition from a region rich in biocultural diversity to one dominated by urban artificial plant communities. The conversion of fertile agricultural lands into urban infrastructures has not only altered the area from a productive agricultural hub to a consumer-centric urban sprawl but has also impaired the soil's ability to function effectively, exacerbating climate change impacts. This shift has forced farmers to nearby urban settlements, stripping them of their lands and livelihoods and turning them into urban consumers.

The rural-to-urban transition has led to homogenized urban environments characterized by concreted areas that displace habitats and biodiversity, thus eroding the natural character of bioregions along with their customs and cultures. This trend of biocultural homogenization threatens bioregional stability in the context of climate change and the Anthropocene. Chilean philosopher and ecologist Ricardo Rozzi asserts that biocultural homogenization is a "global driver of losses of biological and cultural diversity, that frequently entail social and environmental injustices" (Rozzi 2023, p. 25). He points out that rural-urban migration is one of three primary drivers that significantly contributes to this homogenization (Rozzi and Tauro 2023, p. 31).

Rozzi advocates for a "3H" framework in biocultural ethics, which interrelates habitats, habits, and co-inhabitants. This approach interweaves human and non-human co-inhabitants, their life habits, and shared habitat. Habitats provide the "conditions of possibility for the existence of co-inhabitants and their life habits" (Rozzi and Tauro 2023, p. 31). Co-inhabitants refer to both human and non-human species sharing a habitat, implying "an ethical imperative to foster relationships of care with other co-inhabitants and the shared habitats" (Rozzi and Tauro 2023, p. 31).

This theoretical framework opens pathways for innovative co-inhabitation methods, ecopedagogy, and environmental activism. The co-inhabitants are intricately interconnected and "mutually modified by their recurrent interactions that shape their habits and habitats" (Rozzi 2013, p. 22). Additionally, the concept of "co-inhabitant subjects" merits emphasis, as it suggests the necessity of granting moral consideration to each co-inhabitant, acknowledging their inherent dignity and defiance of scientific prediction (Rozzi 2013, p. 28).

Rozzi and collaborators employ the metaphor of the "gardener" to describe the high Andean cushion plant's cooperative habit of creating habitats for diverse life forms (Méndez-Herranz et al. 2023). This botanical example illustrates how the practice of caring and cooperation in the plant world can inspire us to "cultivate gardens creating cooperative ties with those we live with" (Méndez-Herranz et al. 2023, p. 83). In an era of ubiquitous biocultural homogenization driven by urbanization, it is crucial to seek ways to transform the barren, consumer-driven minds and hearts shaped by our education system into fertile soil of love and care displayed by future stewards and gardeners. Playing the role of gardeners of "fostering cooperation and care" (Méndez-Herranz et al. 2023) resonates deeply with biocultural conservation efforts by the Healing Garden project at Sichuan Fine Art Institute (SCFAI).

In stark contrast to the rapid homogenization seen in many areas, the Huxi campus of SCFAI stands out as a living outdoor museum. It presents a unique perspective within the University Town of Chongqing. Here, the campus has managed to preserve the topographical, ecological, rural, cultural, and architectural identity of the region. This includes its hills, rice paddies, wetlands, traditional buildings, farming tools, aqueduct systems for irrigation, and winding roads lined with stone walls. However, amidst this preservation effort, there lies an environmental justice concern intertwined with the process of shaping this natural landscape. The Institute's decision to transplant large trees from rural areas to create a forest within a short span of time highlights an issue of ecological displacement. This practice mirrors the urban migration of rural trees that occurred across China over a decade ago, paralleling the dislocation of rural communities during the establishment of the University Town (Meier 2018).

To gain a holistic understanding of this habitat and to restore local biocultural diversity, we present the case of the ongoing eco-art project, "Healing Garden" (HG) where participants act as "gardeners" cultivating habit of care towards the other human and nonhuman co-inhabitants in the community. Through the promotion of reciprocal relationships among all the co-inhabitants, the project also uses the metaphor of "gardeners" to encompass both human and other-than-human co-inhabitants who work together to enhance urban biodiversity, foster community cohesion, and build a network of cooperative interspecies relationships.

Isabelle Fremeaux and Jay Jordan (2021, p. 126) describe today's art world as "Art-as-we-know-it" which has separated itself from everyday life and the beautiful from the useful since the Industrial Revolution. Focusing on narrowly defined, isolated art objects, the logic of Art-as-we-know-it seldom consider "how artwork can be part of solutions to the problem or how it can materially nourish a community or a struggle", and "rarely if ever 'gives back' to or fights on the side of the world that is being destroyed" (Fremeaux and Jordan 2021, p. 128). While symbolic representational works may serve as reminders of the challenges we face, they often fall short of inspiring meaningful actions and of enacting transformations. Instead, as they proposed, art needs to become "about caring and nourishing the relationships that make up the everyday" (Fremeaux and Jordan 2021, p. 65).

Following this line of thought, HG's artistic endeavor explores the art of healing, living, caring, connecting, and co-inhabiting. HG also draws inspiration from Joseph Beuys' concept of art as social sculpture (Durini 2011, p. 247), which prioritizes social forms over traditional art forms. Beuys asserted that social sculpture "provide [s] a guarantee for the evolution of the earth…establish conditions for other planetarians [sic] too" and is "productive for the whole" (Beuys 1990, pp. 26–29). In this spirit, HG seeks to embody this expanded and revitalized sense of art by taking direct actions that inspire new habits of living among planetary co-inhabitants. While HG does participate in traditional art exhibitions, its roots are firmly planted in the community, local habitats, and everyday life. The primary goal of HG is "not simply to showcase the world, but to actively transform it" (Hopkins 2015).

14.2 Permaculture

Permaculture is a portmanteau of permanent agriculture and permanent culture, first coined by Bill Mollison and David Holmgren in the 1970s (Holmgren 2011; Luna 2020). This holistic design philosophy draws wisdom from nature, promoting self-sustaining systems by creating beneficial connections and working with, rather than against, natural processes. The artist collective of The Laboratory of Insurrectionary Imagination (Labofii) writes in a booklet that we can learn to build energy-efficient, waste-free, resilient and productive human habitats by observing and mimicking the ways ecosystems work (Labofii 2021). Permaculture principles stem from observation of the patterns of natural systems. Labofii provides valuable insights on applying these principles in various design projects (Labofii 2021). By embracing these principles, we can create resilient, biodiverse gardens that enhance participants' physical and spiritual well-being, while also developing a transdisciplinary eco-pedagogy that strengthens community ties through workshops, activities, and eco-public art.

Observation, Connection, and Interaction This principle lies at the core of the eco-art pedagogy of the HG. This essential permaculture principle helps uncover the complex tapestry of patterns, rhythms, processes, and interactions present in nature. By opening up our senses, we become attuned to the physical, the contextual (e.g. microclimate, geography, geology, culture, and the sociopolitical), and cyclical aspects of the observed (Rozzi et al. 2023).

Through keen observation of nature, we have identified the unsustainable patterns prevalent in urban settings, notably the widespread adoption of monoculture. By comparing these regularly weeded, energy-consuming landscapes to areas that are relatively less disturbed, we have been able to appreciate the resilience, diversity, and abundance inherent in nature.

Our observations of campus trees unfolded the unsustainable sociopolitical patterns. Recognizing the historical scarcity of trees in the vicinity, we engaged students with a majestic White fig (*Ficus virens*), the oldest living co-inhabitant on campus. As students touched the tree's bark and contemplated the intricate root network, they were struck by the awe-inspiring presence of this indigenous giant with a centuries-old legacy (Fig. 14.1). I recall an occasion when a staff member shared with the students that the tree stood witness to the evolution of the land: as it had once been near a village committee before the campus was built, observing the rapid transition from rural to urban life in the Huxi area. This type of intimate experience teaches us to collaborate with local plant co-inhabitants to build our

Fig. 14.1 Students touching the bark of a White fig at the Sichuan Fine Art Institute's campus. (Photograph by Wei Chen)

garden, cultivate our garden, prompting critical reflection on creating man-made forests by sourcing large trees from other regions.

The Problem Is the Solution This principle encourages us to turn negative situations into creative solutions (Labofii 2021). Keeping this in mind, we were able to transform the site of the HG from an abandoned, neglected space filled with construction waste (Fig. 14.2) into a flourishing, biodiverse community garden and outdoor classroom (Fig. 14.3).

Seek, Use and Encourage Diversity This principle teaches us that the resilience, balance, and efficiency of an ecosystem are derived from the diverse beneficial relationships within it. One way we embody this principle is through the practice of polyculture, a permaculture concept, where multiple local vegetables and herbs are grown together to mutually benefit and support each other in the same space. As Labofii explains, "Some plants protect others from pests, some fix nitrogen in the soil, others enhance flavor profiles" (Labofii 2021).

Use Edges and Value the Marginal This principle serves as a reminder that "the edge where two or more ecosystems meet, known as an ecotone, is extraordinarily rich and productive" (Morrow 2006, p. 25). When engaging in ecological design, it is important to "increase the edges to create more microclimates" (Morrow 2006, p. 25). In addition, the location of HG on the outskirts of the campus highlights its unique position as a catalyst for creative change within traditional educational systems. This principle also encourages us to explore the boundaries of different disciplines and to engage with individuals on the margins of community membership, such as cleaners and maintenance workers. By valuing these edges and margins, we can foster innovation and connectivity in our work and relationships.

Fig. 14.2 Original site of the Healing Garden cluttered with construction waste. (Photograph by Lipeng Jin)

Fig. 14.3 Aerial drone image of Healing Garden. (Photograph by Xiaoshu Zhang)

Using Small and Slow Solutions This principle is prioritized at HG. Rather than opting for a quick fix by importing large trees to instantly transform our campus, we believe in taking a slower and more sustainable approach. We actively engage our staff, students, and children in the process of sowing a wide variety of heirloom seeds and small fruit tree seedlings. Our goal is to foster harmony between the cultivated plants and the wild indigenous species, creating a diverse and vibrant ecosystem. This method allows for gradual growth and development, fostering a deeper connection to the land and promoting long-term sustainability.

Obtaining a Yield This principle encompasses multiple yields such as edible plants, medicinal herbs, and the ecological yield by maximizing productivity while maintaining biodiversity. This allows us to catch and store energy such as edible plants and medicinal herbs. These plants not only serve as a tangible reward for our physical and emotional efforts, but they also provide a valuable gift to our community. This is evident in our food sharing event, the *Therapeutic Kitchen*, which will be discussed further below. In addition to the immediate benefits of our harvest, it is important to consider the ecological yield and value of our gardening practices. By eschewing the use of chemicals such as pesticides and herbicides, we are creating a safe refuge for urban wildlife. For instance, small birds like the White-rumped Munia (*Lonchura striata,* 白腰文鸟) and Goldfinch (*Carduelis carduelis britannica,* 金翅雀) are able to feed on the grains of ripening rice crops in HG. The yields of our efforts also encompass the positive impact on our community relationships, making our gardening practices truly holistic and sustainable.

Stacking Functions By applying this principle we ensure that every element we have designed in the garden or in the public/community work serves multiple purposes. These purposes include ecological, social, aesthetic, educational, and therapeutic functions. Like the previous principles, this one involves applying nature's patterns to holistic gardening and healing practices.

14.3 Healing Garden

Inspired by permaculture, HG helps to transform the global socio-ecological crisis of the Covid-19 pandemic into creative, ecologically sound solutions. From a neglected area filled with construction debris and barren soil, it has evolved into a biodiverse sanctuary that fosters social cohesion and meaningful community actions addressing biodiversity and climate issues.

As a dynamic living classroom, HG champions holistic healing, sustainable living, and transdisciplinary community learning, nurturing cooperative, connective, and ecological consciousness among students. Through restorative activities and workshops—such as nature observation, community composting, food gardening, water harvesting, and ecological public art—participants connect deeply with plant, animal, and microbial co-inhabitants, enhancing their physical health while receiving mental and spiritual nourishment.

The essence of healing and holistic thinking was particularly evident during the Mandala Garden Healing workshop conducted in November 2022, amidst the challenges of the Covid-19 pandemic (Fig. 14.4). The lockdowns and social restrictions led to widespread depression and anxiety among students, profoundly

Fig. 14.4 Mandala Garden Healing workshop. (Photograph by Kaimin Zeng)

impacting their mental and physical health. During the winter of 2022, students and some staff were quarantined on campus for over 5 weeks. To alleviate anxiety and enhance resilience, we courageously initiated the workshop with the hope of fostering change.

Originating from Sanskrit, the word "mandala" means "circle", or more specifically, "a magic circle" (Singer 1994, p. 204). In various cultures, mandalas hold profound meaning, symbolizing the wholeness of life and the universe. In Hindu and Buddhist traditions, mandalas are symbolic diagrams used in meditative rituals, representing the pursuit of unity and balance in life. Carl Jung viewed mandalas as archetypal forms representing the Self, the "wholeness of the personality" (Jung 1963, 1972). He utilized this primordial image as a tool for psychological recovery and achieving psychic wholeness.

Permaculture has embraced the mandala form to create low-maintenance, ecologically productive, visually appealing, and spiritually nourishing spaces. The circular and curved designs of the mandala garden optimize productivity by maximizing edges, facilitating polyculture and companion planting to encourage biodiversity and deter insect damage.

Inspired by the mandala as an instrument for psychological healing and spiritual development, and by the ecological thinking embedded in permaculture design, this workshop explored the profound connections between the healing of the inner ecology and the outer habitat. Our aim was to transform on-site waste and negative emotions into purposeful ecological public art, a communal food garden, a habitat for attracting nearby other-than-human co-inhabitants, and a tranquil space for healing and rejuvenation. We invited the students to explore various materials found on-site (such as construction debris, stones, indigenous plants, fungi, micro-organisms, and soil), and encouraged them to interact with these elements from a fresh perspective.

After having co-designed mandala patterns, we utilized construction waste (stones and pebbles) for garden bed edging, recycled bricks for pathways, and organic materials—grass, leaves, branches, and compost—for nutrient-rich, no-till planting beds. The selected plants include local perennials such as Okinawan spinach (*Gynura bicolor*) locally called Hongfeng cai or blooded skin vegetable (红凤菜/血皮菜), and Fish mint (*Houttuynia cordata* Thunb.), locally known as Fish smelling plant (折耳根/鱼腥草). Both species are semi-wild or wild, resilient, easy to care for, and serve as ground cover, helping to retain soil moisture.

The two plants also hold significant cultural importance in the food of Chuanyu region (川渝) and Southwestern China. Additionally, they are recognized as medicinal herbs in China. Scientists have found that compounds from the Okinawan spinach have anti-inflammatory and antioxidant properties, supporting healthy blood sugar balance, and promoting liver health (Do et al. 2020). Scientists have also discovered extracts from Fish mint that have potent antiviral and antibacterial effects (Wei et al. 2023).

To establish a central feature and visual anchor, we installed a container water garden to heal the habitat by moderating the microclimate and attracting local water-loving co-inhabitants such as birds, insects, and amphibians (e.g. Günther's frog *Sylvirana guentheri,* 沼水蛙). This water feature also reminds us that "water is the single enlivening element that we all depend on" (Betsy 2022, p. 5). It symbolizes the interconnectedness and interdependence of all living forms (Betsy 2022, p. 170), while resonating deeply with our inner world.

The connections created through water among the human and nonhuman co-inhabitants reflect universal wholeness. The English word "heal" has its Proto-Indo-European root, "kailo-", meaning "whole, uninjured, or of good omen" (Marks et al. 2024). It evolved into the old English word "hælan" (to make whole, sound and well), then into "hal" (health), the roots of "whole", "hale", and "holy", as well as the greeting "Hail" (Mark et al. 2024). Thus, healing, health, and holiness signify a state of wholeness.

Against the backdrop of disintegration, separation, and isolation resulting from the Covid-19 pandemic, the process of co-creating the mandala garden became a healing journey, a ritual for achieving wholeness and unity. The completed garden evolved into a thriving, biodiverse, and self-sustaining system, continuing to provide a serene, meditative space for connection and holistic nourishment. The participation in the workshop was immensely rewarding and therapeutic, imbuing the students with a positive outlook and hope for the future.

14.4 Soil Cultivation

HG's transformative journey began on a construction waste site, where the degraded, barren urban soil unsuitable for supporting plant, animal, and microbial life. Soil health and fertility are fundamental to a garden's productivity, stability, and biodiversity. Key practices such as mulching with organic materials like straw and wood chips, implementing no-till farming to minimize soil disturbance, and using nitrogen-fixing legumes are vital for building rich healthy soil. Among these, composting is crucial for rejuvenating the soil by adding organic matter and nourishing healthy underground biota.

The Soil Cultivation and Composting Project at HG aims to collaborate with local microbial co-inhabitants, including bacteria, fungi, and earthworms, to enrich the soil and create a habitat for all co-inhabitants. Compost infused with local soil organisms can inoculate the earth, stimulate biological activity, provide essential nutrients for plants, enhance water retention, and sequester carbon. This project marries scientific principles with artistic creativity to address critical issues such as urban soil degradation and the loss of soil biodiversity.

In partnership with the Lishui Institute of Ecology and Environment, Nanjing University, the project successfully diverts nearly all kitchen refuse from the SCFAI student cafeteria and a nearby grocery store to the HG site for daily composting. Since its inception on September 11, 2021, a remarkable 28.18 tons of kitchen scraps have been transformed into nutrient-rich compost. This collaborative endeavor not only recycles waste but also fosters a regenerative environment, turning a once-barren site into a fertile ground of possibilities.

The project's backbone is its self-organized composting team, consisting of seventeen dedicated students—predominantly female—spanning from freshmen to third-year graduates. Students actively participate in the entire composting cycle—from waste collection and compost turning to temperature monitoring. In addition to the daily operating system, we host workshops that blend eco-public art, composting facilities, and community involvement, creating spaces for multispecies interactions.

One notable workshop was the Bamboo-weaving Leaf Composter, expertly led by Ms. Liu Lifang, a master of traditional bamboo weaving with over a decade of experience. The group conducted research on campus to identify suitable types of bamboo, focusing on the local variety (*Neosinocalamus affinis,* 慈竹). Students carefully observed this bamboo variety, cutting and transporting them to HG, where they learned how to split the bamboo and make strips. Under Ms. Liu's guidance, students mastered the foundational "hexagonal weaving" technique, renowned for its structural stability and versatility in shaping surfaces and spatial forms. The group selected a mushroom design from several proposals to symbolize the unexpected growth of eco-art actions, akin to mushrooms.

The workshop facilitated interaction and collaboration among the bamboo variety, the tutor, the students, and the staff's children, weaving new relationships between humans and nonhuman co-inhabitants, as well as within the community (Fig. 14.5). Once completed, the installation was placed in a public space opposite

Fig. 14.5 Bamboo-weaving Leaf Composter workshop. (Photograph by Lipeng Jin)

the cafeteria, becoming an eco-public art with multiple functions: leaf composting, the artistic, the educational, a playground for children, and even a giant insect hotel (Fig. 14.6). The leaf mold produced by the composter attracts earthworms that feed on the leaves and nourishes nearby plants. Additionally, it provides habitat for overwintering insects such as beetles, centipedes, pill bugs, and spiders helping decompose the leaves; pollinators like wasps, certain bees, and moths find refuge in the decaying leaves during winter (Jay 2023).

Parallel to this, the Pottery Worm Tower Project explores the synergy between vermicomposting, pottery, and public art (Fig. 14.7). The earthworm tower, a compact system, leverages earthworms' natural ability to convert kitchen scraps into fertile soil, thus contributing to local ecosystems and community greening efforts. This workshop interweaves the themes of soil, earthworm biology, the research conducted by Charles Darwin (1898), and the intangible cultural heritage of Rongchang pottery from Chongqing. This is a holistic approach that encompasses aesthetic education, local knowledge, therapeutic engagement, and ecological awareness.

The project inclusively welcomed children with hearing impairments from a school with special needs, offering them a unique experience in the therapeutic realms of local pottery and worm composting. The material of the pottery earthworm towers was sourced from local clay, connecting the ecological functions of

Fig. 14.6 Children playing beside the Leaf Composter. (Photograph by Aili Mou)

composting with soil, the nourishing substance on the planet. This process involves not only working with soil but also instilling habits of composting and gardening, and fostering a continuous healing journey for both the earth and participants. The malleability of clay provides endless creative possibilities, whether in spatial forms, the representation of plant life, and the integration of texts (Fig. 14.8). Once finished, the earthworm towers with diverse values and functions were installed in public spaces near campus dorms and cafeteria (Fig. 14.9). Through eco-art collaboration with earthworms, new habits of living and thinking are encouraged among all community members.

Fig. 14.7 Shaping the pottery worm towers. (Photograph by Kaimin Zeng)

Fig. 14.8 Hand-carving images on the pottery worm towers. (Photograph by Kaimin Zeng)

Fig. 14.9 Installing the pottery worm towers near student cafeteria. (Photograph by Yingling Lu)

14.5 Therapeutic Kitchen

Therapeutic Kitchen offers opportunities to directly engage with the surrounding environment, contemplate the interconnectedness of local habitats, life habits, and co-inhabitants, and celebrate our yield and collective efforts of polyculturing that increases biodiversity and ecological productivity. It also highlights a worrying fact the dominant industrialized food system is not only severing our connection with the sources of food but also impacting on the environment and people's health. On the one hand, food is our most direct connection to nature. On the other hand, paradoxically, most people here do not know or care where their food comes from.

Nowadays, take-out food consumption is rapidly expanding and has become an essential part of students' diets at universities in China. In response, the project aims to change our mindless habits, "reclaiming our food" and "taking action to break out of the confining and infantilizing dependency of the role of consumer (user), and taking back our dignity and power by becoming producers and creators" (Katz 2012, p. xix).

The journey of healing began by actively interacting with the interconnected web of life that surrounds us. Through therapeutic acts of observing, touching, and harvesting plant leaves and fruits, participants were able to open up their senses and deepen their understanding of their environment and listen attentively to the natural world around them, to fellow participants, and to their own bodies and feelings (Fig. 14.10). They collected roselle to create a concentrate syrup beverage, gathered sharen (*Alpinia zerumbet,* 香砂仁) to produce soy sauce, harvest plentiful

Fig. 14.10 Harvesting roselle. (Photograph by Yao Chi)

purple and white perilla and Sicilian pepper leaves around campus to make perilla sauce, and unearthed bamboo shoots to make Sichuan pickles using brine provided by our workshop instructor. By transforming locally sourced organic ingredients into nourishing meals, the students fully immersed themselves in the sensory, visceral, and aesthetic experiences of foraging, preparing, eating and sharing food (Fig. 14.11).

Deriving from the two Latin words "com" (meaning with) and "pan" (meaning bread), the term "companion" literally signifies sharing bread (Rozzi 2019). In contrast to the prevalent fragmented and separate relationship found in many universities, our act of cooking and eating together not only involved sharing meals but also exchanging views and emotions, fostering an atmosphere of generosity, hospitality, and conviviality.

Seeing through the lens of biocultural ethics, sharing meals not just creates human companionship, but relationships with other co-inhabitants and the shared habitats (Rozzi 2023). A seemingly mundane act of eating became a meditative moment in that we contemplate a porous, membrane-like boundary through which energy and matter flow between our bodies and our habitat.

Drawing from David Bohm's participatory philosophy, we come to understand that we are not separate from the earth, but rather we are nourished by it. This perspective encourages us to see food as more than just sustenance. A sense of oneness, wholeness, and unity will arise from the understanding that the food we "partake of" will become us and is not "a separate object" (Bohm 1996, p. 87). By recognizing this interconnectedness, we can develop a sense of unity and respect for the food that sustains us, recognizing its role in maintaining and regenerating the intricate web of life.

Fig. 14.11 Dining table adorned with various plants: a plate made from bamboo shoot skin topped with vegetable pickles, bamboo cups, azalea flowers, and broad banana leaves. (Photographer by Yao Chi)

In essence, eating becomes a meaningful act of establishing a deep connection with our environment and acknowledging the importance of honoring the food that supports us. It is through this understanding that we can truly appreciate the significance of sharing meals as a way to nurture not only our bodies but also our relationship with the other-than-human co-inhabitants around us (Fig. 14.12).

14.6 Conclusion

HG represents a pioneering initiative in place-based education, dedicated to rejuvenating local habitats and fostering sustainable, cooperative, and caring life habits that provide conditions for both human and non-human co-inhabitants to thrive. Through hands-on experiences and collaboration with nature, HG combines community-based eco-art with permaculture practices such as composting, soil enrichment, edible landscaping, and biodiversity conservation.

HG's approach follows nature's intrinsic methods, building beneficial relationships and fostering complexity, connection, and diversity. By multiplying edges, the project maximizes the diversity and abundance of species, promoting more resource availability and niche opportunities. Utilizing polyculture and diverse planting techniques, HG attracts more wildlife and beneficial insects, thereby increasing

Fig. 14.12 Sharing naturally colored vegetable dumplings. (Photograph by Yao Chi)

ecological synergy and interactions. Indigenous plants, including edible and medicinal weeds, are promoted to preserve the multiple biocultural values and a sense of place. Cultivating soil creates shared habitats for the microbial, floral, and fauna co-inhabitants, enhancing the overall ecosystem.

Active engagement in habitat restoration and preservation fosters not only the health of the environment but also personal transformation and spiritual healing of students and other participants. This "inner gardening" process often unfolds within the context of collective healing and gardening, cultivating a new consciousness that acknowledges our role as part of a larger community—a complex web of life. The habits of preparing, eating, and sharing food become rituals of hope and resilience, viscerally connecting individuals with local plant and microbial co-inhabitants.

As Austrian educator and thinker Rudolf Steiner poetically wrote,

> Health and wholesomeness only come
> When in the mirror of the soul of man
> The whole community take shape;
> And in the community lives
> The strength of every single soul. (Gibsone and Bang 2015)

This inner awakening generates the strength and actions needed to build resilience, connectedness, and wholesomeness within the whole community and habitat, a crucial endeavor during periods of social separation, such as during the Covid-19 pandemic.

Drawing inspiration from Thomas Berry's vision of "The Great Work," HG's educational ethos is grounded in the belief that ecology should underpin all learning (Berry 1999, p. 84). In an era increasingly dominated by digitalization and artificial intelligence, as well as biocultural homogenization (Rozzi 2018), HG champions a return to our roots, emphasizing the critical importance of human-nature reconnection. This vision aligns with Vandana Shiva's advocacy for embracing life-sustaining co-evolutionary processes, fostering biodiversity, and building resilience against industrial and reductionist paradigms (Shiva 2008, p. 141). The future of learning and planetary survival hinges on transforming human beings and reintegrating our creative energy in collaboration with the other-than-human co-inhabitants.

In mainland China, HG has carved out a niche in artistic activism, favoring constructive action, creativity, and critical reflection over civil disobedience. This strategy, rooted in compassion and ecological consciousness, addresses pressing challenges such as climate change and biocultural homogenization. HG strives for "respectful, sustainable ways of co-inhabitation" (Rozzi 2013, p. 28) and nurtures new paradigms and alternative systems for a better future.

Acknowledgments I would like to thank Danqiong Zhu and Ricardo Rozzi for their valuable comments on draft versions of this chapter.

References

Berry T (1999) The great work: our way into the future. Bell Tower, New York

Betsy D (2022) Water talks. Portal Books, New York

Bohm D (1996) On dialogue. Routledge, London/New York

Darwin C (1898) The formation of vegetable mould through the action of worms, vol 16. D. Appleton

Do TVT, Suhartini W, Mutabazi F, Mutukumira AN (2020) Gynura bicolor DC. (Okinawa spinach): a comprehensive review on nutritional constituents, phytochemical compounds, utilization, health benefits, and toxicological evaluation. Food Res Int 134:109222

Durini LDD (2011) Beuys voice. Electa, Milan

Fremeaux I, Jordan J (2021) We are 'Nature' defending itself: entangling art, activism and autonomous zones. Pluto Press, London. https://monoskop.org/images/8/87/Fremeaux_Isabelle_Jordan_Jay_We_Are_Nature_Defending_Itself_2021.pdf

Gibsone C, Bang JM (2015) Permaculture: a spiritual approach. Findhorn Press, Forres

Beuys J (1990) Joseph Beuys in America: energy plan for the western man. Four Walls Eight Windows, New York

Holmgren D (2011) Weeds or wild nature: a permaculture perspective. Plant Prot Q 26(3):92–97

Hopkins R (2015) The rise of the insurrectionary imagination. Transition culture. https://www.resilience.org/stories/2015-04-02/the-rise-of-the-insurrectionary-imagination/

Jay S (2023) Leaf mold: nature's compost. https://www.epicgardening.com/leafmold/

Jung CG (1963) Memories, dreams, reflections. In: Jaffe A (ed) Winston R and Winston C (trans). Vintage Books, New York

Jung CG (1972) Mandala symbolism (trans: Hull RFC). Princeton University Press, Princeton

Katz SE (2012) The art of fermentation: an in-depth exploration of essential concepts and processes from around the world. Chelsea Green Publishing, Vermont

Luna JM (2020) Sustainable land development using permaculture. In: Yang P (ed) Cases on green energy and sustainable development. IGI Global, pp 170–192

Marks DF, Murray M, Locke A, Annunziato RA, Estacio EV (2024) Health psychology: theory, research and practice. Sage

Meier A (2018) A photographer follows the urban migration of China's rural trees. https://hyperallergic.com/453379/a-photographer-follows-the-urban-migration-of-chinas-rural-trees/

Méndez-Herranz M, Marini G, Rozzi R (2023) Sub-Antarctic high Andean "gardeners": cultivating caring relationships. In: Rozzi R, Tauro A, Avriel-Avni N, Wright T, May RH Jr (eds) Field environmental philosophy, Ecology and ethics, vol 5. Springer, Cham, pp 71–86. https://doi.org/10.1007/978-3-031-23368-5_5

Morrow R (2006) Earth users' guide to permaculture, 2nd edn. Melliodora Publishing, Hepburn

Rozzi R (2013) Biocultural ethics: from biocultural homogenization toward biocultural conservation. In: Rozzi R, Pickett STA, Palmer C, Armesto JJ, Callicott JB (eds) Linking ecology and ethics for a changing world: values, philosophy, and action, ecology and ethics book series, vol 1. Springer, Dordrecht, pp 9–32. https://doi.org/10.1007/978-94-007-7470-4_2

Rozzi R (2018) Biocultural homogenization: a wicked problem in the Anthropocene. In: Rozzi R, May RH, Chapin FS, Massardo F, Gavin MC (eds) From biocultural homogenization to biocultural conservation, Ecology and ethics, vol 3. Springer, Cham, pp 21–48. https://doi.org/10.1007/978-3-319-99513-7_2

Rozzi R (2019) Taxonomic chauvinism, no more! Antidotes from Hume, Darwin, and biocultural ethics. Environ Ethics 41(3):249–282. https://doi.org/10.5840/enviroethics201941325

Rozzi R (2023) Introduction to part I. Field environmental philosophy: concepts and case studies. In: Rozzi R, Tauro A, Avriel-Avni N, Wright T, May RH Jr (eds) Field environmental philosophy, Ecology and ethics, vol 5. Springer, Cham, pp 17–25. https://doi.org/10.1007/978-3-031-23368-5_2

Rozzi R, Tauro A (2023) The multiple lenses of ecotourism with a hand lens: fundamental concepts and practices. In: Rozzi R, Tauro A, Avriel-Avni N, Wright T, May RH Jr (eds) Field environmental philosophy, Ecology and ethics, vol 5. Springer, Cham, pp 27–51. https://doi.org/10.1007/978-3-031-23368-5_3

Rozzi R, Álvarez R, Castro V, Núñez D, Ojeda J, Tauro A, Massardo F (2023) Biocultural calendars across four ethnolinguistic communities in southwestern South America. GeoHealth 7(4): e2022GH000623. https://doi.org/10.1029/2022GH000623

Shiva V (2008) Soil not oil: environmental justice in an age of climate crisis. South End Press, Cambridge

Singer J (1994) Boundaries of the soul: the practice of Jung's psychology. Anchor Books, New York

The Laboratory of Insurrectionary Imagination (2021) 13 attitudes. London. https://labo.zone/wp-content/uploads/2021/10/13attitudes.pdf

Wei P, Luo Q, Hou Y, Zhao F, Li F, Meng Q (2023) Houttuynia Cordata Thunb.: a comprehensive review of traditional applications, phytochemistry, pharmacology and safety. Phytomedicine:155195. https://www.sciencedirect.com/science/article/abs/pii/S0944711323005548

Chapter 15
Small But Essential: How to Introduce the Public to Easily Overlooked Organisms?

Qin Zuo and Li Zhang

Abstract Natural ecosystems are the basis for human survival, social stability, and sustainable development. Biodiversity conservation is receiving more and more attention around the world, and it requires a convergence of efforts from policy, industry, research, and public participation. However, taxonomic bias continues to be pervasive. Certain taxa attract most of the public, scientific, and government attention, while those small or uncharismatic organisms are often overlooked, although they play important roles in ecosystems. To alleviate this phenomenon, scientists should take action and give publicity to neglected organisms. Bryophytes are the second largest group in the plant kingdom; their diversity is second only to angiosperms. Bryophytes are an integral and important component of global biodiversity. They perform essential ecological functions, such as pioneering ecological successions, regulating hydrological cycles, reduction of soil erosion, climate regulation, and carbon sink. However, due to their small size and cognitive difficulties, they rarely become the object of public attention, and knowing bryophytes has become one of the weaknesses in biodiversity education. Our team started to introduce bryophytes to the general public systematically in China beginning in 2007. Before our work, there were very few materials or cases for reference. Relying on the Fairy Lake Botanical Garden, in collaboration with the Municipal Affairs Bureau of Macao Special Administrative Region and other institutions, we employed various approaches to introduce bryophytes to the public, including holding public exhibitions, writing science books and papers, delivering public lectures, using new media, moss gardens, guided tours, etc. These activities cover a large population, of which many are students. Especially since 2015, our team has recruited artists, and used paintings and artworks to show the diversity and beauty of bryophytes which are more attractive. Although in this chapter we introduce our work on the natural education of bryophytes, it is believed that the experience

Q. Zuo · L. Zhang (✉)
Laboratory of Southern Subtropical Plant Diversity, Fairy Lake Botanical Garden, Shenzhen & Chinese Academy of Sciences, Shenzhen, Guangdong, China
e-mail: zuoqin@szbg.ac.cn; zhangli@szbg.ac.cn

D. Zhu et al. (eds.), *Linking Arts with Biocultural Conservation, Restoration, and Communication*, Ecology and Ethics 8,
https://doi.org/10.1007/978-3-032-04830-1_15

accumulated will have reference value for other similar organisms that have been neglected previously.

Keywords Biodiversity · Bryophytes · Education · Exhibition · New media · Paintings

15.1 Introduction

Biodiversity loss is considered to be one of the triple planetary crises (United Nations Environment Programme 2021). The vision of working together to promote global biodiversity conservation and to achieve a harmonious coexistence between human beings and nature has not only become a consensus among governments and scientists, but active and organized action is also underway (Convention on Biological Diversity 2022). Human activities are a major drive in the loss of biodiversity, while social factors influence policies and practices for conservation (WWF 2020). There is a close and complex link between public opinion, scientific productions and conservation policies (Martín-López et al. 2010; Troudet et al. 2017), and individual conservation behavior can be influenced by a variety of factors, such as knowledge, interest, attitudes, emotions, affinity with nature, etc. (Turpie 2003; López-del-Toro et al. 2009; Liu and Chen 2012). Therefore, it is essential to promote and encourage public awareness and understanding of the importance and measures to conserve biodiversity through information dissemination and education, as required by the parties to the first Convention on Biological Diversity (UNEP Environmental Law and Institutions Programme Activity Centre 1992).

Publicity and education for biodiversity conservation face a number of common challenges. First, urbanization has alienated city dwellers, who make up more than 50% of the planet's population (Ritchie et al. 2024), physically and culturally from nature, and cities, where economic and political power is concentrated, are often the places where policy is made (Dearborn and Kark 2010). The homogenization of urban physical environment and habitat reduces species richness and changes community composition (McKinney 2002; de Barros Ruas et al. 2022), while the loss of ecological knowledge (Pilgrim et al. 2008), the extinction of experience (Noe and Stolte 2023) and biocultural homogenization (Rozzi 2012) among urban residents occur simultaneously. Secondly, although the concept of biodiversity is widely applied in scientific research, education and policy, it may not be recognized or understood to the same extent by the general public (Lindemann-Matthies and Bose 2008). In addition, public access to species is often extremely limited, and taxonomic bias (or taxonomic chauvinism) is perpetuated in research related to biodiversity and conservation biology (Marco et al. 2017; Troudet et al. 2017; Rozzi 2019).

Organisms that are more difficult to study, such as tiny individuals, species distributed in remote or confined areas, those not socially preferred may receive even less attention, research, and publicity (Pawar 2003; Troudet et al. 2017). Meanwhile, aesthetic biases also affect people's perception of biodiversity, which potentially affects their motivation to conserve biodiversity (Tribot et al. 2018a, b).

In this regard, actions to introduce less attractive or less visible taxa to the public need to be strengthened. In this chapter, taking the bryophyte team of the Shenzhen Fairy Lake Botanical Garden (SZBG) as an example, we summarize the work we have done in introducing bryophytes to the public (mainly city dwellers) since 2007. We hope to share our experience with our colleagues and the communities of biocultural conservation for reference and further communication, collaboration, and conservation.

15.2 Situate Bryophytes in the Context

Bryophytes are the earliest terrestrial plant lineages, second only to angiosperms in terms of species diversity, and comprise three phyla, namely the liverworts (Marchantiophyta), mosses (Bryophyta), and hornworts (Anthocerotophyta), with a total of about 19,458 species globally (Wang et al. 2022; Brinda and Atwood 2025). They are distributed in various ecosystems, including forests, grasslands, wetlands and deserts, etc., ranging from the polar regions to the tropics. Furthermore, they provide a wide range of ecological functions, such as carbon fixation, water retention, and habitats for other organisms (Hodgetts et al. 2019).

Among them, peat mosses (*Sphagnum*) alone covers 1% of the land surface and accounts for 10–15% of terrestrial carbon stocks (Zhu 2022). These organisms are inconspicuous, with plants often ranging from a few millimeters to a few centimeters in height, and require at least a hand-lens, or a microscope to observe their morphological traits. In addition, bryophytes are nonvascular plants in which haploid gametophytes dominate the life history, and have very different life cycles from ferns, gymnosperms, and angiosperms, which may be more familiar to the public. Similar to other groups of organisms, bryophytes are also exposed to threats arising from modification of natural habitats, climate change, agriculture, pollution, etc. (Cao et al. 2006; Ministry of Ecology and Environment of the PRC and Chinese Academy of Sciences 2013; Hodgetts et al. 2019).

Although research around bryophytes also covers areas ranging from the basic to the cutting-edge, such as diversity, ecology, and genomics, extensive documentation of bryophyte outreach and education aimed at the public is still scarce. A very worthwhile example is the "Ecotourism with a Hand-Lens" project at the Omora Ethnobotanical Park in Cape Horn, Chile (Goffinet et al 2012; Rozzi et al. 2010). This project, based on the conceptual framework of the field environmental philosophy (FEP) and collaborated by researchers, local government and the private sector, promotes human-environment interactions and appreciation of microscopic taxa, such as native bryophytes, lichens and fungi, in the form of ecotourism. It also provided sustainable income for the Cape Horn Reserve and local communities, as well as increased appreciation of the ecological and ethical values of sub-Antarctic flora by local government, teachers, tourism practitioners and residents (Rozzi 2012; Rozzi et al. 2012, 2023).

There are about 3156 species of bryophytes distributed in China (The Biodiversity Committee of Chinese Academy of Sciences 2024). The scientific collection and bryological research by Chinese scholars began in the 1930s (Zhu et al. 2019), and approximately 60 full- and part-time bryologists and bryology educators currently work at universities, research institutions, botanical gardens, museums, and nature reserves, of whom fewer than 10 (including the authors of this chapter) have been active in the field of the intersections between bryology and environmental education in recent years. Although the use of bryophytes for landscaping and horticulture has become popular in China since the last decade, and a growing number of people are joining the industry, they seldom introduce bryophytes to the public from the lens of bryology or biocultural conservation.

In China, the elementary school science textbooks are largely silent on bryophytes -related content. The junior high school biology textbooks provides a brief overview in less than two pages of common habitats of bryophytes, gametophyte morphology and structure, threats to bryophytes of air pollution, and human use of bryophytes for air monitoring, as well as peatlands and human use of peat for fuel (Institute of Curriculum and Textbook Research 2012). The high school biology textbook mentions the process of colonization of rock surfaces with a single image and accompanying text in the chapter "Vegetation Succession" (Institute of Curriculum and Textbook Research 2020). It is worth pointing out that in the junior high school biology textbook, the common cord-moss (*Funaria hygrometrica)* is presented in the form of a line drawing (labeled with the Chinese name but no details of the species). Moreover, some morphological features are shown erroneously, from which it can be inferred that authors of the textbook lack sufficient knowledge of bryophytes. Obviously, it is urgent to fix the gap.

Since 2006, when the senior author (Li Zhang) joined the SZBG, the *bryophyte team* at SZBG was established under his leadership and currently consists of six staff members, including four professional bryologists and two artists. Since 2007, the team has undertaken a series of educational work on bryology. Beginning with science exhibitions, the measures have grown increasingly complex, diversified, and effective. Over the years, the team have

- organized more than ten bryophytes-themed exhibitions nationwide,
- published around ten popular science books,
- presented dozens of popular science articles in magazines, newspapers and new media,
- conducted public presentations in schools, universities, museums, and libraries,
- participated in popular science TV shows and short videos, and
- offered popular science courses for the public with remarkable results.

The above endeavors make the team the most influential agency on the popularization of bryology and related environmental education in China.

15.3 Introduce Bryophytes to Citizens Through Multiple Channels

15.3.1 Popular Science Books

It is human instinct to name the familiar, and we are unlikely to care about what we don't know (Pilgrim et al. 2008). In response to the extreme lack of introductory knowledge of bryophytes among citizens, from the angle of science popularization, we have collaboratively published several representative books, such as *The Miniature Angles in the Plant Kingdom—An Introduction to Bryophytes* and *Field Guide to Wild Plants of China- Bryophytes,* etc. (Fig. 15.1).

As the first bryophyte popular science book in Chinese and English in China, the former has been issued in three editions since 2009, with a total copies of 9800. It provides a comprehensive introduction to the basics of bryophytes in the form of a combination of illustrations and texts. The main contents include bryophyte morphology, habits and habitats, taxa, life histories, reproduction modes, significance in the ecosystem, methods of observation and identification, the diversity of bryophytes in China, and the threats facing bryophytes (Zhang et al. 2015). The book features three outstanding features: (1) the use of a large number of vivid macro-photographs of field habitats and micro-photographs of bryophytes, so that the habitats, landscapes, and morphological details of bryophytes are vividly presented to readers; (2) the contents are more fundamentally and informatively planned to be geared towards readers at various ages; and (3) the book incorporates aesthetic elements to present a style of elegance and vividness.

Fig. 15.1 Bryophyte science books by the team. (Photographs by Li Zhang)

The latter is the first field guide book on bryophytes in China, and since its publication in 2016, it has been reprinted seven times, with a total of 22,000 copies distributed. The book introduces 306 relatively common bryophytes in China with a combination of field habitat photographs, line drawings, morphological descriptions, and domestic distribution information. The book stands out for two reasons: (1) extensive coverage of the most common bryophyte families and genera, and (2) its portability.

15.3.2 Popular Science Articles and Short Videos

In addition to books, team members are often invited to write short popular science articles on bryophytes for special issues in influential magazines and newspapers, such as *Chinese National Geography*, *Natural History*, *China Nature*, *China Science Daily*, etc., on topics related to natural history, biodiversity, science popularization and tourism, with readerships that include even children. One amusing example is that during the Spring Festival of 2018, an episode of the China Central Television (CCTV) program "Everlasting Classics" referred to "moss flowers" described in an ancient poem, which triggered an explosion of interest in bryophytes. We were soon invited to write a short popular science article entitled "May I Ask Where to Enjoy Mosses Blossom" (借问苔花何处赏), which scientifically, aesthetically, and poetically inquires whether bryophyte blossom during their life cycles.

With the popularization of mobile networks and the expansion of computer functions to smart phones, mobile terminals represented by smartphones and digital reading have become important carriers and means of information acquisition. The 52nd Statistical Report on China's Internet Development reveals that the number of Chinese mobile Internet users has reached 1.076 billion (China Internet Network Information Center 2023). An earlier report revealed that 30.9% of Chinese citizens already belonged to the category of online science popularization users, and would regularly access science knowledge through popular science websites, blogs, or websites of science and technology newspapers and magazines (China Internet Network Information Center 2011). As the most frequently used mobile communication app in China in recent years, WeChat, with its parent company's 2023 earnings report, shows that the number of active WeChat accounts has exceeded 1.3 billion.

Taking advantage of the dissemination environment created by these new media, we also compose and publish short articles on bryophytes through the WeChat public account of SZBG, which possesses nearly 2.7 million subscribers. Compared with books and articles published in paper journals, these short articles are more flexible in length and more lively in style . They are primarily aimed at introducing visitors or potential visitors to the species of common local wild bryophytes in the botanical garden to stimulate their interest in observing bryophytes in the botanical garden and connect to those species.

Short videos, which combine the characteristics of low threshold and fragmentation, have become a highly influential medium in the social media era. The Annual Report on China's Internet Audiovisual Service Development reveals that China's short-video users have reached 1.01 billion (China Netcasting Services Association 2023). The report of the top short video app, DouYin (Chinese edition of TikTok), released in 2021, shows that the cumulative playback of popular natural science videos in the past year was nearly 33 billion times, and short videos have become an important form of disseminating information of natural science (DouYin App WeChat public account 2021). Given this trend, we tried to produce a series of short videos about bryophytes, including topics such as bryophytes rehydration, edibility (a frequently asked question), spore dispersal, and other tiny organisms living in bryophyte communities.

15.3.3 Exhibitions

Ensuring the connection between people and nature, particularly face-to-face encounters with other-than-human beings, is recognized as key to effective conservation strategies (Rozzi et al. 2008). With the urbanization rate in China reaching 66.16% by the end of 2023 (National Bureau of Statistics of China 2024), biodiversity-related exhibitions in botanical gardens, museums, and other venues can provide a representative face-to-face encountering experience for urban residents. The bryophyte-themed exhibitions we have organized or collaborated on over the years can be divided into three phases, each with unique content and philosophical ideas.

The first phase consisted of three exhibitions held between 2007 and 2008 at the Lou Lim Iok Garden in Macao and the Fairy Lake Botanical Garden in Shenzhen, among which "A Preliminary Exploration of Bryophytes" held in December 2007 in Macao was the first themed exhibition exclusively focused on bryophytes in China. The exhibition originated from the Macau bryophyte diversity survey project undertaken by one of the authors of this chapter, Li Zhang.

The second phase was the "Magic and Enchantment of Bryophytes" exhibition held in 2017 (Fig. 15.2, top). The exhibition was held during the XIX International Botanical Congress at the main venue, Shenzhen Convention and Exhibition Center, and it featured over 80 original paintings of bryophytes by our team members and 6 sets of handmade models of bryophytes made by Macau-based artist, Fong Kam Kei (冯锦基), complemented by an outline of the basics of bryophytes and brief descriptions of the paintings. The audience of this exhibition included domestic and

Fig. 15.2 Bryophyte exhibitions. Up: Magic and Enchantment of Bryophytes (苔藓之美, 2017, Shenzhen). Below: Mini But Notable Bryophytes (苔花如米·藓为人知, 2023, Shanghai). (Photographs by Li Zhang)

international botanists, educators, plant enthusiasts, and Chinese citizens who made reservations for admission, totaling over 5000 people.

The third phase was six bryophyte-themed exhibitions co-organized after 2018 at venues such as the 2019 International Horticultural Exhibition in Beijing, the Shanghai Natural History Museum in 2023 (Fig. 15.2, bottom), the Zhejiang Museum of Natural History in 2018, and two Shenzhen schools between 2020 and 2022, which in total attracted hundreds of thousands of visitors. We provided our partners with paintings, 3 calligraphy works of bryophyte-related ancient poems and proses presented by calligrapher Jing Yu (俞竞), high-resolution photographs, and textual contents tailored to their specific themes, venues, and needs. These exhibitions are more comprehensive in their content and format than the previous ones . Although the content, format, and values of collaborative exhibitions are often the result of a trade-off between the wills of the parties involved, the merits of these collaborative exhibitions are reflected in their ability to expand the types of audiences, increase the number of visitors, and enlarge the impacts of the exhibition. During those collaborations, we also found that there is still a scarcity of bryophyte ecological photographs, paintings, calligraphy or other artworks, and even a dearth of copywriters who are good at writing detailed contents about bryophytes (rather than descriptions of taxa).

15.3.4 Educational Talks and Activities

In contrast to books, articles, and exhibitions, educational talks and activities allow direct contact and interaction between content creators and audiences. Even without questionnaires and surveys, content creators can obtain direct feedback from the audience, upon which content creators can optimize future work, while audiences can get more interactions or a more direct experience of local bryophyte diversity.

At present, the main formats of the educational talks and activities we have conducted are as follows:

- Rely on the comprehensive study tours organized by the botanical garden for local primary and secondary school students during the summer vacation and on certain anniversaries, in the form of thematic lectures, field observation, collection, and microscopic observation of bryophytes in SZBG.
- Recruit citizens from time to time to participate in guided tours and relatively freelance interpretation of moss landscapes, based on the artificial moss landscapes and wild bryophytes in the moss gardens;
- Give tutorial lectures to volunteers and develop field activities on bryophytes in the garden for them, supported by the Botanical Garden's Nature School program;
- Accept invitations and deliver lectures on bryophytes to local libraries, science and technology museums, schools, and universities;
- Organize lectures for the general public at specific venues at the invitation of influential self-media and NGOs;
- In conjunction with some of the above-mentioned bryophyte-themed exhibitions, organize lectures and bryophyte painting activities in the museums and schools where the exhibitions are held;
- Train for horticulture and nature education practitioners, usually in the form of thematic lectures and field practices.

The vast majority of educational talks and activities mentioned above are free of charge to participants. Our guided tours, observation, and collection activities in the Fairy Lake integrate nature experience, nature observation, and nature interpretation (China Nature Education Network 2021). Participants are often able to observe *in-situ* some of the native bryophytes and learn relevant knowledge.

Lectures and events rely heavily on the labor and schedules of content producers. Until mature volunteers capable of presenting bryology emerged, our lectures and events were usually conducted at a frequency of less than 10 times per year. A promising sign is that the number of lectures combining offline and online (in the form of live or recorded broadcasts) has increased in recent years, expanding the audience base. In addition, some lectures relying on self-media platforms have later been produced as illustrated articles and posted on various online platforms. For example, a 2017 talk presented by Li Zhang, posted on the YiXi WeChat public account, has received 74,000 views to date. Unfortunately, educational talks were not conducted in a series, and the audiences and demands of the funding organizations varied from one event to another, making it difficult to obtain statistically significant feedback and robust evaluations.

15.3.5 Moss Gardens

SZBG established the first bryophyte nursery in mainland China in 2009, with the aim of not only *ex-situ* conservation, but also supplying materials for research, education, and horticulture. In total, around 90 species of bryophytes are cultivated at two scenic gardens in the SZBG.

The first spot, located in the central area of the Shade Garden (Fig. 15.3, up), simulates a valley habitat in southern China and harbors nearly 30 bryophyte species, including planted species and spontaneously occurring species. Introduction boards about the basics of bryophytes and common species are provided on site, making it an ideal place for visitors and students to observe bryophytes closely. The second is at You Xi Valley (Secluded Moss Garden, Fig. 15.3, bottom), a less disturbed valley between the Tree Fern Pond (桫椤湖) and the Palm Lawn (棕榈园), covered by lush secondary forest. Bryophytes are ubiquitous in there, sprawling from rock surface, ground, to tree trunks. Close to the Palm Lawn at the end of You Xi, the ground and tree bases have been carefully planted with bryophytes and accented with shade-tolerant plants, including the family such as *Gesneriaceae*, *Begonia,* and *Sellagiella*, to create a serene and harmonious environment. Secluded Moss Garden was set up in 2020, and is the first moss garden in mainland China.

The Secluded Moss Garden in the SZBG has also attracted national organizations and media to report and publicize it. The feature documentary “Shenzhen Moss Garden” by China National Geographic premiered on July 26, 2023 on the bilibili

Fig. 15.3 Up: Shade Garden (阴生园). Bottom: Secluded Moss Garden at You Xi Valley (幽溪). (Photographs by Li Zhang)

platform, with over 390,000 views so far. The video "The First Moss Garden in Mainland China: Deep, Secluded, and Beautiful" by the new media "Yi Tiao" (一条) went online on May 6, 2024, and was viewed by more than 100,000 times on the day of its release.

15.4 Discussion

The above overview description of the bryophyte team at the SZBG demonstrates the feasibility of introducing small but essential organisms to the public through publications, scientific articles, short videos, exhibitions, educational programs, and moss gardens. Overall, these efforts present the scientific, ecological, aesthetic, and even biocultural values of bryophytes to the public. While working to present these values, we are constantly learning from our experiences, feedback from the public, and suggestions from our collaborative parties. To a certain extent, the positive effects produced by the team stem from the fact that the team has gradually handled three kinds of relationships over the years in a more appropriate way, namely, the relationship between natural and cultural, scientific and aesthetic, and research and social-institutional values.

Natural and Cultural Values The dichotomy between nature and culture has been criticized by environmental philosophers. Although the dichotomy between nature and culture has been criticized by environmental philosophers, these two terms remain important concepts for clarifying our situation. The concepts of natural value and intrinsic value of nature were introduced to China in the 1980s through the writings of Western environmental philosophers, but this dissemination was limited to the academic community. The general public is not very familiar with the philosophical concept of intrinsic value of nature. While promoting bryophyte conservation and environmental education, one of the major problems we always face is how to drive the public to care about small organisms such as mosses and lichens, that are not closely related to them.

While we can fill the knowledge gap by assessing the instrumental value of nature, such as the material and intangible services that biodiversity and ecosystems provide to humans, these ecosystem services, as well as ecosystem degradation, are still very abstract for urban dwellers due to their limited access to the natural environment. Take the example of peat moss wetlands far from urban areas: the carbon sequestration process, magnificent landscapes, degradation trajectories, and consequences are hard to observe or perceive, even if urban dwellers have access to data, images, or other related information. These challenges ultimately point to human perceptions and value trade-offs, and lead us to recognize the need to go beyond knowledge dissemination to foster deeper connections between humans and other-than-human organisms, particularly, bryophytes.

Given these challenges, our team has included cultural values in our considerations. In terms of traditional Chinese culture, the popular domestic understanding of traditional Chinese culture tends to focus on Confucianism's teachings of humanism, communalism, and ethics. The discourse on interspecies kinship in traditional Chinese culture has not yet been incorporated into the framework of environmental

Fig. 15.4 *The Deer Park* (鹿柴). Calligraphy by Jing Yu in 2018, ink on paper, 30 cm by 43 cm. (Photograph by Li Zhang)

education in general (Song 2026). This situation has gradually crystallized the idea of combining the value of nature with traditional Chinese culture in our work. Specifically, the idea is to link the value of moss as a group of lively organisms and natural entities with the biocultural heritage of moss appreciation along Chinese history.

For instance, one of our moss gardens, "Secluded Moss Garden," was inspired by a five-character poem written by Tang Dynasty poet Wang Wei (王维, around 693–762), in which he depicts, "Returning light enters the deep forest; Again shining upon the green moss" (返景入深林, 复照青苔上) (Fig. 15.4). Moreover, we recognize that bryophytes were once an important floristic theme in ancient Chinese literature. Despite the impossibility of ancient literati to discriminate scientifically among algae, bryophytes, and other small vascular plants, moss literature is rich in connotations. Moss literature commenced in the Wei, Jin, and Northern and Southern Dynasties (220–589) and matured in the Tang Dynasty (618–907). Mosses have become important plant imagery in ancient poems from the Tang Dynasty to the Qing Dynasty (1636–1912), with a high frequency of occurrence (Pan 2015).

Therefore, mosses were once a cherished flora of Chinese literati for over a thousand years. By incorporating elements of moss literature into bryophyte conservation, we are in a sense reconnecting with this almost disconnected tradition.

Meanwhile, in planning the exhibition "The Magic and Enchantment of Bryophytes" as well as other exhibitions, we have tried to incorporate as much as possible the relevant moss components in traditional Chinese culture, such as moss poems, prose, and calligraphy. This integration of biodiversity and local culture is consistent with the proposition of biocultural ethics. The outcome is not only to pass on the poetic appreciation of mosses in Chinese culture, but also to draw people's attention to the life process of the bryophytes themselves, as well as the spiritual connection between humans and bryophytes. Research shows that improved quality nature experiences enhance people's emotional connection to nature, which in turn shapes biophilic behaviors and makes people more open to contacting and learning about mosses (Hosaka et al. 2017; Whitburn et al. 2019; Colléony et al. 2020). We briefly recorded the audience's experience and thoughts in the form of a signed guestbook, and found that the influential power of the artworks could indeed be conveyed to the viewers.

Scientific and Aesthetic Values In exploring cultural values of mosses, the team's testimonials respond to existing research on the impact of biodiversity on human civilization as well as the influence of people's care and reverence for the natural environment on ethical attitudes and environmental behaviors through aesthetic perception (Brady and Prior 2020). Within the framework of biocultural ethics, it is quite explicit that the complex interactions between living organisms, as manifested in their habits, transcend the biophysical dimension and can be brought up to the symbolic-cultural-linguistic dimension. In order to further strengthen the spiritual bond between humans and bryophytes, we integrate the scientific and aesthetic values of mosses.

Diversity researchers and painters in the team have joined efforts to produce multiple exhibitions that combine scientific knowledge with artistic appreciation, and have published three editions of *The Magic and Enchantment of Bryophytes* (Zhang et al. 2019). Combining high-resolution photographs taken by diversity researchers during field trips with moss paintings by artists, allows us to not only present the usual systematic introduction to bryophyte taxa and fundamentals, but also explore how the beauty of the morphology and communities of bryophytes can be visualized in paintings. Digital photography is especially important for presenting small organisms (Mackenzie et al. 2026). Even more valuable, digital photography preserves important data related to scientific expeditions, such as the time, location, season of the expedition, organism morphology and niche, etc.

Artistic representation is obviously different from high-definition photography. Scientific illustrations in the general sense usually have the main purpose of presenting the morphological and anatomical features of organisms scientifically and accurately, and in botany they are usually presented in the classical mode of line drawings. Our team's moss art espouses a freer and more diverse expression based on accurate presentation (Figs. 15.5 and 15.6).

Fig. 15.5 Up: *Brachymeniopsis gymnostoma* Broth. The species was endemic within the Himalayan region and was announced extinct in the wild in China. High resolution photograph was taken during Li Zhang's 2012 field survey in Yadong County, Xizang Autonomous Region. Zhang reported this species at the International Bryology Conference held in Puerto Williams, Chile in January 2015. This species was listed as a critically endangered species in the new edition of the *China Biodiversity Red List—Higher Plants* (2020). Photograph by Li Zhang. Bottom: Painting of *Brachymeniopsis gymnostoma* Broth. Created by Lili Xu in 2019, acrylic on canvas, 60 cm by 30 cm. (Credit: Lili Xu)

Fig. 15.6 A side-by-side comparison of a photograph and a watercolor painting. Left: *Encalypta tibetana* Mitt. Photograph was taken by Li Zhang during his field survey in Zhongba County, Xizang Autonomous Region in 2012. The Chinese common name of this species is "西藏大帽藓"—literally "Xizang big hat moss"—which directly describes the most distinctive morphological feature of this species, its bell-shaped calyptrae. Right: *Encalypta tibetana* Mitt. Painting by Shihua Li in 2015, watercolor on paper, 27 cm by 39 cm. (Credit: Shihua Li)

As far as scientific applications are concerned, the general public may not be as concerned about bryophytes as professionals, except for garden practitioners and enthusiasts who are relatively concerned about species suitability for horticulture and cultivation. The story of the bryophytes themselves is much more compelling to the public. When we show photos and videos of other organisms in the bryophyte community (e.g., arthropods, rotifers, brachiopods, and even symbiotic cyanobacteria), the public seems to be able to empathize with them and spontaneously withdraw from the anthropocentric perspective to accept the ecological fact that bryophytes are interconnected with other organisms in a shared habitat.

Biocultural communication is defined in this book as an invitation (Zhu and Rozzi 2026). The team's high-resolution photographs and artworks are invitations for visitors to cross the boundaries between science and arts, to recognize these small organisms, and appreciate their unique beauty, for creating a deeper and ethical relationship of co-inhabitation.

Research and Service Any research and creative team is anchored in a specific socio-institutional framework. Throughout our team's nearly two decades of scientific research and artistic endeavors, we have always strived to provide some service in the contemporary social environment, just as moss provides ecosystem services. Our earliest exhibitions and activities for the public, with the original intent and purpose of species conservation and science popularization, were closely related to the mission and normative functions of botanical gardens, a special institution distinct from universities and research institutes (Heywood 2017), and represented a social responsibility felt by botanical garden professionals.

For example, the exhibition in Macao in 2007 was a public service initiative to contribute back to the society with academic achievements on biodiversity. The exhibition was primarily aimed at the youth, explaining the basics of bryophytes and focusing on the diversity of bryophytes in Macao. Through employing field photographs, written descriptions, and moss bonsai, we introduced common bryophyte species in conjunction with bryophyte knowledge basics and incorporated some nature education concepts to strengthen the connection between people and their habitats, especially their local natural environments. The books published by the team not only provide the public with the knowledge base for understanding bryophytes, as well as methods for observing and roughly identifying common species in their daily lives or on their trips. The books can also be utilized by professionals, such as nature reserve staff, practitioners in institutions of nature education, primary and secondary school teachers, etc.

Furthermore, our botanical garden, comprising the functions of research, species conservation, education, landscape exhibition, and recreation, provides us with excellent and unique conditions for integrated and inclusive bryophyte education and communication. We have a relatively diverse staff and working routine, and we can reach out to visitors of all ages as well as those with diverse professional backgrounds. We also have specialized gardens and nurseries that directly display living plants, and the impact of bryophyte education in turn contributes to the development of the Botanical Garden, training of professionals, and facilities-infrastructure construction.

In addition, we have been consistently engaged in outputting as much original research contents and materials as possible for bryophyte outreach and public education, from which our team's signature and a more nuanced understanding of bryophytes have emerged. All these show that our research work and social service are mutually reinforcing.

15.5 Conclusion

In China, public awareness and education on biodiversity conservation carried out by research institutes, universities, research botanical gardens and other scientific research organizations are often carried out in three contexts: science popularization, nature education and environmental education. Since the fulfillment of the goal of biodiversity conservation is based on human responses, the theoretical frameworks and practical forms that can influence human responses should not be excluded.

Under this scenario, our science education efforts aimed at disseminating knowledge about bryophytes and highlighting the significance of biodiversity have gradually been developed at SZBG and other collaborating institutions or venues. These endeavors explored diverse and up-to-date forms of biocultural conservation, accumulated a substantial amount of original sources, and forged a niche that integrates science, local cultural traditions, and arts. Team members also gained practice-based thoughts and rich experiences.

We found that bryophytes, small but essential, which were less associated with the productivities of human society, can serve instead as an interesting access point, shifting our focus from mere dissemination of knowledge to the immaterial connection between nature and human beings, and to the intrinsic value of nature. This also gave rise to an attempt to convey environmental philosophy to the public with the theme of bryophytes. Drawing on ideas, knowledge, styles, and forms of creativity from the social sciences, literature, and arts to develop innovative practices has helped to enhance our interactions with our audiences and is a direction we will continue to explore in the future.

Acknowledgments Sincere thanks to Danqiong Zhu and Ricardo Rozzi for commenting on the manuscript.

References

Brady E, Prior J (2020) Environmental aesthetics: a synthetic review. People Nat 2(2):254–266

Brinda JC, Atwood JJ (2025) The bryophyte nomenclator. Accessed via: https://www.bryonames.org/

Cao T, Zhu RL, Guo SL, Zuo BR, Yu J (2006) A brief report of the first red list of endangered bryophytes in China. Bull Bot Res 26(6):756–762

China Internet Network Information Center (2011) Research report on the current situation of China's popular science market and internet users' behaviour in popular science

China Internet Network Information Center (2023) The 52nd Statistical Report on China's Internet Development. Open access through: https://www.cnnic.net.cn/NMediaFile/2023/0908/MAIN1694151810549M3LV0UWOAV.pdf

China Nature Education Network (2021) Introduction to nature education. China Forestry Publishing House, Beijing

China Netcasting Services Association (2023) Annual report on China's internet audiovisual service development. 11th China internet radio and video convention

Colléony A, Levontin L, Shwartz A (2020) Promoting meaningful and positive nature interactions for visitors to green spaces. Conserv Biol 34(6):1373–1382. https://doi.org/10.1111/cobi.13624

Convention on Biological Diversity (2022) Kunming-Montreal global biodiversity framework. International Agreements [online]. https://www.unep.org/resources/kunming-montreal-global-biodiversity-framework

de Barros Ruas R, Costa LMS, Bered F (2022) Urbanization driving changes in plant species and communities—a global view. Glob Ecol Conserv 38:e02243

Dearborn DC, Kark S (2010) Motivations for conserving urban biodiversity. Conserv Biol 24(2):432–440

Douyin App WeChat Public Account (2021) 2021 Douyin nature science content data report. Accessed via https://mp.weixin.qq.com/s/80D7vGvHBnec_U8l_lRqxg

Goffinet B, Rozzi R, Lewis L et al (2012) The miniature forests of Cape Horn: eco-tourism with a hand-lens ("Los Bosques en Miniatura del Cabo de Hornos: Ecoturismo con Lupa"), Bilingual

English-Spanish edition. UNT Press–Ediciones Universidad de Magallanes, Denton/Punta Arenas

Heywood VH (2017) The future of plant conservation and the role of botanic gardens. Plant Divers 39(6):309–313

Hodgetts N, Cálix M, Englefield E et al (2019) A miniature world in decline : European red list of mosses, liverworts and hornworts. IUCN

Hosaka T, Sugimoto K, Numata S (2017) Childhood experience of nature influences the willingness to coexist with biodiversity in cities. Palgrave Communications 3:17071. https://doi.org/10.1057/palcomms.2017.71

Institute of Curriculum and Textbook Research (2012) Biology, grade 7, book 1. People's Education Press, Beijing

Institute of Curriculum and Textbook Research (2020) Biology, organisms and environment. People's Education Press, Beijing

Lindemann-Matthies P, Bose E (2008) How many species are there? Public understanding and awareness of biodiversity in Switzerland. Hum Ecol 36(5):731–742

Liu GY, Chen J (2012) Local Dai people's attitude towards biodiversity conservation in Xishuangbanna. J Yunnan Univ (Nat Sci Ed) 34(S1):141–148

López-del-Toro P, Andresen E, Barraza L, Estrada A (2009) Attitudes and knowledge of shade-coffee farmers towards vertebrates and their ecological functions. Trop Conserv Sci 2(3): 299–318

Mackenzie R, Valle-Celedón C, Arriagada G (2026) Mosses in the spotlight: photography as a tool for biocultural conservation of bryophytes in the Cape Horn biosphere reserve. In: Zhu D, Rozzi R, Tsuji R, Johnson B, Caviness T, Castro-Jorquera C, Xu L, Zhang L (eds) Linking arts with biocultural conservation, restoration, and communication. Ecology and Ethics, vol 8. Springer, Cham, pp 277–293. https://doi.org/10.1007/978-3-032-04830-1_17

Marco MD, Chapman S, Althor G, Kearney S, Besancon C, Butt N, Maina JM, Possingham HP, von Bieberstein KR, Venter O, Watson JEM (2017) Changing trends and persisting biases in three decades of conservation science. Glob Ecol Conserv 10:32–42

Martín-López B, Montes C, Ramírez L, Benayas J (2010) What drives policy decision-making related to species conservation? Biol Conserv 142(2010):1370–1380. https://doi.org/10.1016/j.biocon.2009.01.030

McKinney ML (2002) Urbanization, biodiversity, and conservation: the impacts of urbanization on native species are poorly studied, but educating a highly urbanized human population about these impacts can greatly improve species conservation in all ecosystems. Bioscience 52(10): 883–890. https://doi.org/10.1641/0006-3568(2002)052[0883:UBAC]2.0.CO;2

Ministry of Ecology and Environment of the PRC, Chinese Academy of Sciences (2013) Assessment report on the red list of China's biodiversity Higher plants volume annex to bulletin No 54

National Bureau of Statistics of China (2024) Statistical communiqué of the People's Republic of China on the 2023 national economic and social development. Accessed through: https://www.stats.gov.cn/sj/zxfb/202402/t20240228_1947915.html

Noe EE, Stolte O (2023) Dwelling in the city: a qualitative exploration of the human-nature relationship in three types of urban greenspace. Landsc Urban Plan 230:104633. https://doi.org/10.1016/j.landurbplan.2022.104633

Pan FJ (2015) Grass and trees: the world of plants in classical Chinese literature (草木缘情: 中国古典文学中的植物世界). The Commercial Press, Beijing

Pawar S (2003) Taxonomic chauvinism and the methodologically challenged. Bioscience 53(9): 861–864

Pilgrim SE, Cullen LC, Smith DJ, Pretty J (2008) Ecological knowledge is lost in wealthier communities and countries. Environ Sci Technol 42(4):1004–1009

Ritchie H, Samborska V, Roser M (2024) Our world in data—Urbanization [online]. https://ourworldindata.org/urbanization

Rozzi R (2012) Biocultural ethics: recovering the vital links between the inhabitants, their habits, and habitats. Environ Ethics 34(1):27–50. https://doi.org/10.5840/enviroethics20123414

Rozzi R (2019) Taxonomic chauvinism, no more! Antidotes from Hume, Darwin, and biocultural ethics. Environ Ethics 41(3):249–282. https://doi.org/10.5840/enviroethics201941325

Rozzi R, Armesto JJ, Frodeman R (2008) Integrating ecological sciences and environmental ethics into biocultural conservation in South American temperate sub-Antarctic ecosystems. Environmental Ethics, 30(3):229–234. https://doi.org/10.5840/enviroethics200830329

Rozzi R, Anderson CB, Pizarro JC, Massardo F, Medina Y et al (2010) Field environmental philosophy and biocultural conservation at the Omora Ethnobotanical Park: methodological approaches to broaden the ways of integrating the social component ("S") in long-term socio-ecological research (LTSER) sites. Rev Chil Hist Nat 83(1):27–68. https://doi.org/10.4067/S0716-078X2010000100004

Rozzi R, Lewis L, Massardo F, Medina Y, Moses K et al (2012) Ecotourism with a hand lens in the Omora Park. Ediciones Universidad de Magallanes, Punta Arenas

Rozzi R, Tauro A, Avriel-Avni N, Wright T, May RH Jr (eds) (2023) Field environmental philosophy: education for biocultural conservation, Ecology and ethics, vol 5. Springer, Cham. https://doi.org/10.1007/978-3-031-23368-5

Song Y (2026) Interspecies Co-inhabitation: A Dialogue between Chinese philosophical traditions and Biocultural Ethics. In: Zhu D, Rozzi R, Tsuji R, Johnson B, Caviness T, Castro-Jorquera C, Xu L, Zhang L (eds) Linking arts with biocultural conservation, restoration, and communication, Ecology and Ethics, vol 8. Springer, Cham, pp 95–109. https://doi.org/10.1007/978-3-032-04830-1_6

The Biodiversity Committee of Chinese Academy of Sciences (2024) Catalogue of Life China: 2024 Annual Checklist, Beijing. Accessed through http://www.sp2000.org.cn/

Tribot A-S, Carabeux Q, Deter J, Claverie T, Villéger S, Mouquet N (2018a) Confronting species aesthetics with ecological functions in coral reef fish. Sci Rep 8:11733

Tribot A-S, Deter J, Mouquet N (2018b) Integrating the aesthetic value of landscapes and biological diversity. Proc R Soc B 285(1886):20180971

Troudet J, Grandcolas P, Blin A, Vignes-Lebbe R, Legendre F (2017) Taxonomic bias in biodiversity data and societal preferences. Sci Rep 7:9132

Turpie JK (2003) The existence value of biodiversity in South Africa: how interest, experience, knowledge, income and perceived level of threat influence local willingness to pay. Ecol Econ 46(2):199–216. https://doi.org/10.1016/S0921-8009(03)00122-8

UNEP Environmental Law and Institutions Programme Activity Centre (1992) Convention on biological diversity [online]. https://www.un.org/zh/documents/treaty/cbd

United Nations Environment Programme (2021) Making peace with nature: a scientific blueprint to tackle the climate, biodiversity and pollution emergencies. Nairobi [online]. https://www.unep.org/resources/making-peace-nature

Wang QH, Zhang J, Liu Y, Jia Y, Jiao YN, Xu B, Chen ZD (2022) Diversity, phylogeny, and adaptation of bryophytes: insights from genomic and transcriptomic data. J Exp Bot 73(13): 4306–4322

Whitburn J, Linklater W, Abrahamse W (2019) Meta-analysis of human connection to nature and proenvironmental behavior. Conserv Biol 34(1):180–193. https://doi.org/10.1111/cobi.13381

WWF (2020) In: Almond REA, Grooten M, Petersen T (eds) Living planet report 2020—bending the curve of biodiversity loss. WWF, Gland. https://www.worldwildlife.org/publications/living-planet-report-2020

Zhang L, Zuo Q, Hong B (2015) The miniature angels in the plant kingdom (植物王国的小矮人). Department of Gardens and Green Areas, Civic and Municipal Affairs of Bureau of Macao Special Administrative Region, Macao

Zhang L, Zuo Q, Mao L (2019) The magic and enchantment of bryophytes (苔藓之美), extended edn. Phoenix Science Press, Nanjing

Zhu RL (2022) Peat mosses (*Sphagnum*): ecologically, economically, and scientifically important group of carbon sequestration plants. Chin Bull Bot 57(5):559–578. (in Chinese, with English abstract)

Zhu ZY, Xu J, Yu NN (2019) History of collection and research on bryophytes in China before 1949. J Inner Mongolia Normal Univ (Nat Sci Ed) 48(4):283–291. (in Chinese with English abstract)

Zhu D, Rozzi R (2026) Biocultural communication: arts and sciences for conserving biological and cultural diversity. In: Zhu D, Rozzi R, Tsuji R, Castro-Jorquera C, Johnson B, Xu L, Zhang L, Caviness T (eds) Linking arts with Biocultural conservation, restoration, and communication, Ecology and Ethics, vol 8 Springer, Cham, pp 1–22. https://doi.org/10.1007/978-3-032-04830-1_1

Chapter 16
Sublimity in Miniature: Aesthetic Viewing and the Miniature Forests of Cape Horn

Michael Thompson

Abstract The mode by which we view the world greatly affects our appreciation, valuation and orientation to objects, persons, places and things. Amongst the many modes by which we view the world, an aesthetic lens offers an alternative to the self or group interest that dominates our day-to-day valuations. This chapter argues for a change of ordinary perception to an aesthetic one through the analysis of Immanuel Kant's description of sublimity and applies it to the miniature forests of Cape Horn. By seeing immensity in miniature, an inversion of the typical notions of sublimity, the program of biocultural conservation presents us with an opportunity to reflect on our engagements with the world and the many ways of valuation we deploy in understanding, inhabiting and appreciating that world.

Keywords Changing lenses · Aesthetics · Biocultural conservation · Environmental ethics

16.1 Introduction

In his scathingly parodic work, *Gulliver's Travels*, Jonathan Swift (1726) creatively presents the hubris and limitations of human understanding at multiple scalar levels. Gulliver's sense of superiority and invincibility in the land of the Lilliputians is countered by his insignificance and vulnerability during his time in Brobdingnag. Likewise, Gulliver's seeming wisdom and civility in comparison to the Yahoos is inversely mirrored during his encounters with the Houyhnhnms. Perhaps, most telling, is Gulliver's third trip, that to Laputa, wherein he finds advanced technology and intellectual pursuits, and yet a neglect of practical issues that pertain to pragmatic concerns and daily living.

M. Thompson (✉)
Sub-Antarctic Biocultural Conservation Program, Department of Philosophy and Religion, University of North Texas, Denton, TX, USA
e-mail: Michael.Thompson@unt.edu

D. Zhu et al. (eds.), *Linking Arts with Biocultural Conservation, Restoration, and Communication*, Ecology and Ethics 8,
https://doi.org/10.1007/978-3-032-04830-1_16

While a fictitious tale, Swift's novel provides an opportunity for reflection on the scales and lenses by which we view our human experiences and our engagements with various environments. In the land of miniature, Gulliver dominates and destroys; in gigantic lands Gulliver becomes a novelty and feels his insignificance; with the Yahoos, Gulliver is the measure of human development; in comparison to the Houyhnhnms, Gulliver is irrational and underdeveloped; and with the Laputans Gulliver is scientifically inferior yet pragmatically superior. Depending upon our contexts, settings, and mind-sets, the lenses by which we view our experiences reveal varying inhabitants of our world, expose different vices and virtues and refine our sense of understanding about ourselves and our world.

This chapter focuses on this ability, and even desire, to "change lenses." The aim of "changing lenses" is to further human understanding, about ourselves and our world. As part of a science, education, and tourism program at the "end of the world" in Cape Horn, Chile, the expression "changing lenses to assess biodiversity" is used to better understand patterns of plant species richness at high latitudes and discuss its implications for global conservation (Rozzi et al. 2008). In this chapter, I encourage changing our "lenses" to highlight one particular, underrepresented, lens that shifts the way we understand ourselves and our place in the world.

While Swift had recourse to literary imagination, I show that in order to change lenses we do not need to turn to fictional tales of travels to foreign lands. We have the opportunity to "travel" to the vast and the miniscule in very real engagements. Embedded in nature, the experience of changing lenses enables us to view the world through different values and concerns. In this way, we can shift our thinking to expand our understanding, thereby enhancing the appreciation of ourselves, our place, and our roles and responsibilities within the world that we inhabit.

16.2 Biocultural Lenses

A novel opportunity to "travel" in order to change "lenses" comes in the form of a biocultural conservation program initiated in Cape Horn, at the southern end of the Americas, in 2000 (Rozzi et al. 2006). This program resulted from collaborative work led by researchers Ricardo Rozzi, Francisca Massardo, and Jim Kennedy, amongst others. Starting in 2020, this biocultural conservation program began to be led by a second generation of researchers and scholars that are implementing and refining concepts and field experiences to "change lenses." Their aim is to provide students and other participants with a holistic biocultural ethic experience that has both theoretical and practical implications. Biocultural ethics guides and impels biocultural conservation actions. Its approach integrates biophysical research with linguistic-cultural-philosophical conceptualization to effect an expansion of scientific knowledge along with conservation of cultural and biodiversity (Rozzi 2012, 2013). This emphasis on practical ethical issues is relevant to address complex problems of our current global socio-environmental crisis (Crutzen and Stoermer 2000; Monbiot 2007; McBrien 2016; Soga and Gaston 2016). It is important to note that our current environmental crisis is amplified by the losses of biological and cultural diversity, and their interrelationships. To address some of the ultimate causes of this global crisis, we need transdisciplinary, multiple-scale, and intercultural long-term programmatic articulations of different ways, "lenses," by which we view our world.

Chilean-US conservation biologist and philosopher Ricardo Rozzi has proposed at least seven different lenses by which we may view our world (Rozzi 1999; Rozzi et al. 2012, 2014). A brief overview of these lenses is necessary before I focus on one of them. Rozzi et al. have identified six predominant lenses by which we view our world and suggest a seventh as a part of their environmental ethic and methodological programmatic (Rozzi and Tauro 2023). These lenses include:

1. Economic lens- instrumental value, that views systems and objects according to monetary value

   e.g. Annual ecosystem services are estimated to be *$16–54 trillion* (Costanza et al. 2014).

2. Biophysical lens- chemico-physical view

   e.g. Human *biomass* comprises 1/10,000 of the total biomass on Earth (Bar-On et al. 2018).

3. Cultural lens- the linguistic/cultural matrix by which we view the world/others

   e.g. English has been used as the *lingua franca* of science.

4. Philosophical/metaphysical lens- conceptual organizations that delineate what/whether something is

   e.g. Aristotle proposed a *hierarchy of beings*, where humans were placed above animals, and these above plants (Aristotle 1984).

5. Ethics lens- how one should interact with fellow organisms/things

   e.g. One of the Christian commandments is: One *should not* murder.

6. Aesthetic lens- appreciation for beauty/disgust/sublimity/form/dynamics

   Michelangelo's David is a paradigmatic sculpture for the concept of Renaissance art *beauty*.

7. Biocultural lens- recognizes the connection and dynamic relationship of cultural and biological diversity

   e.g. Rozzi (2013) invites us to consider ourselves as *co-inhabitants* among diverse human and other-than-human beings who share local habitats as well as the whole biosphere.

The first three on this list represent the predominant lenses through which we view our world. And yet, they are underpinned by the fourth, the philosophical lens. That is, there are reciprocal links between our conceptualization of objects and systems and the values that we place upon them. For instance, Darwin's notion of competition for resources in his model of natural selection, and then the promulgation of social Darwinism is merely one historical example (Rozzi 1999). Another example comes in the form of the Brundtland Report (1987) that provides a model of "sustainability," and projects how we can maintain our standard of living (cultural lens) into the future. The ethical lens is certainly an outgrowth of a philosophical lens. For example, it defines who/what has moral considerability in ethical decision making is based upon who/what we count as having rights or moral worth, and only those are worthy of moral consideration (Palmer 2013). The aesthetic lens, while once a central focus of philosophical work, has largely been relegated to subjectivism and cultural relativism, and is seldom engaged beyond humanities and creative arts programs. The seventh lens on this list is proposed as a synthesis of the first six. While recognizing the merits of each of the identified lenses, the seventh one reminds us that exclusive attention to any one of the previous six lenses elicits a truncated view of the world, one that can easily lead to underappreciation of biocultural diversity, and its value to lessen the current socio-environmental crisis. For example, a biocultural lens emphasizes that a purely capitalistic lens, one that relegates all life on this planet to merely economic terms and the pursuit of maximum profits will constitute a loss of diversity, at both a biological and cultural level, because some organisms or cultures will be deemed less profitable in a purely cost/benefit analysis, and thus will be eliminated for profit margins.

By reminding us of the multiple lenses by which a thing, place or system may be viewed, the biocultural ethic encourages us to consider multiple modes of valuation, and that our prescriptions for behavior need to be inclusive of multiple scales of analysis. In the metaphor of *Gulliver's Travels*, we are invited to travel through varying lenses in order to provide a more robust appreciation of ourselves and the world in which we find ourselves. From the "changing of lenses" found within the biocultural ethic an implication is that we should understand ourselves and act as co-inhabitants of our world (Rozzi 2013). We co-inhabit alongside other organisms in shared habitats, with differing ways of existing, or life habits, that are shaped by the habitats and interactions. The implied ethical imperative is that we should guide our human life habits carefully, attending and respecting relationships of co-inhabitation. In contrast, a hegemonic colonial lens often eliminates biocultural diversity and generates biocultural homogenization (Rozzi 2018).

16.3 Kantian Aesthetics on Sublimity

The aesthetic lens is particularly relevant to our book *Linking Arts with Biocultural Conservation, Restoration, and Communication.* While biocultural ethics is a core philosophical framework to this chapter, to focus on the aesthetic lens I will use broader conceptual and philosophical frameworks. It is important to note that the aesthetic lens offers us a non-instrumental, non-economic, non-interested (but interesting) way of viewing ourselves, our world and the co-inhabitants we find in our world. The aesthetic lens allows us to appreciate something not because of monetary assignation or because it serves us in a pragmatic capacity; rather, it allows us to appreciate the beauty (or lack thereof) and/or sublimity of an object from a representational-cognitive space. It offers valuation that does not depend on mercenary calculations. Instead, the aesthetic lens leads us to appreciate biodiversity from a different perspective, a perspective relating to form, wonder, beauty and awe. Despite its relevance, the aesthetic lens is often underemployed, even omitted, by conservation biologists, policy makers, and the general public when referring to biodiversity, or evaluating its values (Pascual et al. 2023).

The world is filled with beautiful, amazing, and terrifying things, and an aesthetic lens allows one to view it through this mode of engagement. Within the field of aesthetics, it should be noted, there are various theories through which one may appreciate. From a simple distinction between Eastern and Western aesthetic traditions; that is, viewing from Occidental or Oriental modes of aesthetic appreciation, to a more refined distinction found in the Western tradition—those of imitationalism, formalism or emotivism—different authors on aesthetics have their cultural, linguistic, and historical lenses through which they provide their expositions. Naturally, any individual author also views through their own unique set of lenses, offering their analysis and providing insight into how one may view the world. One central theme that authors on aesthetics share in common, however, is that this mode of viewing is different from our day-to-day engagement; this is to say, an aesthetic lens requires a different mind-set—one outside the norm of economic and practical engagement—in order to see things in a different way.

One historical author that champions this idea of changing lenses is Immanuel Kant. In his *Critique of Judgement*, Kant reminds us that aesthetic seeing is not concerned with epistemic, ontological or moral[1] questions. In fact, aesthetic viewing requires a kind of disengagement from those very types of seeing. Aesthetic judgements, Kant says are "the ability to judge an object, or a way of presenting it, by means of liking or disliking *devoid of all interest*" (Kant 1790, p. 52). When one views through the aesthetic lens, Kant suggests, we are not interested in personal gain, whether we want to possess the object, whether it is useful to us, nor any other personal or community-oriented objectives. Aesthetic judgements are, rather, a different way of viewing, one that engages the viewer (the subject according to Kant) with the viewed (the object, in Kant's terminology) in a way that we see the object and judge whether it evokes our favor (in the case of beauty)[2] or as indicative

[1]At least not initially. They may, however, compel us to moral assessment, as is the case in judgements of sublimity.

[2]Favor here is a difficult term as it evokes the notion of pleasure or liking. Kant is sensitive to the distinction between pleasure/liking something because it is self-referentially useful and pleasure/liking because it evokes a sense of free play of imagination with form and what we see. It is this later use of liking that Kant wishes to emphasize, thus highlighting that we like the beautiful not because it serves a purpose, but because what we are experiencing has no self-referential interest. The reason for liking is another- and that is the form that is presented to us in our viewing and the subsequent evocations within our cognitive architecture. For more on Kant's notion of liking in the aesthetic judgements of beauty cf. Sarah Gibbons' *Kant's Theory of Imagination* (Clarendon Press: Oxford) 1994 pp. 88–90.

or something greater than either ourselves or the object (in the case of the sublime). It is this very relation between viewer and viewed that typifies the aesthetic experience. In it a viewer is provided with sensation from the object viewed, but the aesthetic judgement is not about the presentation of the object via our senses—that is, it is not about whether the object viewed pleases our senses—rather, it is about how we are affected by the viewing. Thus, it is because aesthetic judgements are concerned with how we are affected, the way by which one is emotively and cognitively moved by what one is viewing, that Kant describes them as reflective, or contemplative. By Kant's analysis we are in fact interested in what we are viewing—it compels us to see it—but we are not interested by personal or mercenary motivations. In his somewhat paradoxical analysis, our being interested in the object is the very product of our disinterested engagement.

Within the aesthetic distinctions he makes in his *Critique of Judgement*, Kant further clarifies between judgements of beauty and judgements of sublimity. Judgements of beauty, or judging an object as beautiful,[3] pertain to an imaginative free play between what we see and how it modifies our understanding of it. Moreover, experiences of beauty instill a tranquility and harmony between ourselves and what we view and evoke a restful contemplation. Judgements of the sublime, on the other hand, have a considerably different affect upon us. Encounters with the sublime, Kant insists, evoke experiences of agitation, fear, repulsion (and attraction) as well as awe (Kant 1790, pp. 115–119). It is to the sublime, as Kant describes it, and in connection to experiences of nature cultivated through biocultural ethics that I wish to turn to in more detail.

First, more details on the aesthetic experience of the sublime according to Kant. According to the Prussian philosopher, one may encounter the sublime in nature in two different ways, via the mathematical or via the dynamic. What both share in common is the magnitude of the experience. In mathematical sublimity, numerical magnitude is the characteristic feature, in the dynamically sublime, the magnitude is one of power. The latter of these two comes in the form of experiencing powerful phenomena of nature that are so overwhelmingly forceful that they evoke fear in us. This is to say that the dynamically sublime is characterized by the might and force of nature and the effects it has upon us.

For example, imagine, if you will, the winds of a hurricane or tornado. Most of us are familiar with high winds. Our everyday experience and our information networks give us reference and prepare us for mundane wind activities that we experience daily. From seeing branches swaying in the breeze to detritus blowing during a typical storm, we are accustomed to and have reference for wind. We might even extrapolate the difference between a breeze at 10 mph and a wind of 25 mph. Such reasoning is well within our abilities, and we can prepare for the variability of winds

[3]It should be more properly described as an experience of beauty, as beauty is not "in the object" as Kant even sometimes describes it (cf Kant 1790 p. 179, AK 311), but, rather, in the affectation the viewer has with the object viewed. More properly described, beauty is the interface between viewer and viewed, and beauty is the sense of harmony and tranquility the viewer experiences upon viewing that which is deemed beautiful.

we experience in ordinary experiences. Nothing, however, prepares us for the extraordinary force of a Category 5 hurricane. Wind speeds of such an event, 157+ mph winds (or 321 mph wind speeds for the strongest tornado recorded), destroy buildings, uproot trees and present us with nearly unfathomable natural forces. Both natural phenomena evoke agitation and fear because of the speeds at which the winds blow and the destruction of seemingly stable objects i.e. buildings and trees. The sheer power of hurricane force winds thwarts nearly all our ability to conceptualize them. In Kant's words, the force of these events defy our reason and ability to grasp what it is that we are witnessing.[4]

The might of these forces of nature threaten us, biologically, but, more importantly, they threaten our ability to conceptualize, recognize and comprehend what it is that one is witnessing when one first experiences them. "Compared to the might of any of these," Kant affirms, "our ability to resist becomes an insignificant trifle" (Kant 1790, p. 120). Thus we feel the danger of these dynamic phenomena and "yet the sight of them becomes all the more attractive the more fearful it is, provided we are in a safe place" (Kant 1790, 120).[5] Kant argues that we are attracted to the fearful on account of the cognitive agitation we experience in an event of the sublime, because it offers a resolution that sponsors a second-fold opportunity for sublimity. This resolution to the initial daunting nature of sublime comes in the form of cognitive expansion. During the initial experience, we cannot fathom or comprehend what it is that we are experiencing, and yet, simultaneously, we are invited to expand our cognitive apparatus to take in what it is we are experiencing. Upon witnessing a Category 5 hurricane, one does not comprehend what is in store. During the experience, the sheer force of wind and rain defies our ability to reason and to take in all that we witness. Yet, through the experience of the tempest, we arrive at a newly expanded concept of tempest, thus being prepared for the next possible encounter.

With the mathematical sublime, this cognitive thwarting and subsequent expansion presents itself in terms of numerical magnitude rather than force. Even more dramatically, Kant suggests the cognitive expansion as one that begins to potentially approximate itself to mathematical infinity, but unsuccessfully so. To illustrate, one possible example is experiencing the Grand Canyon. Much like in the previous example of hurricanes, in the mathematical sublime, one possesses ordinary concepts that permit one to comprehend and operate in the mundane world. In this

[4]Kant's description of sublime experiences and their resolution elicit a debate concerning whether someone can experience the same phenomenon twice and experience the sublime on both occasions. The resolution of the sublime sponsors a cognitive expansion such that upon the second experiencing, one does, in fact, have the cognitive concepts to fathom the phenomena, thus the cognitive agitation experienced in the first encounter is not present in the second.

[5]This phrase also reminds us of a signature feature of aesthetic experiences- that of aesthetic distance. Aesthetic distance finds itself twofold here in the sublime—the distance of recognizing that the aesthetic experience is in the interplay of viewer and viewed, as in all aesthetic judgements according to Kant, and a second-fold feature in the sublime one must not be practically threatened in physical space, lest personal interest interfere with the ability to have the aesthetic experience.

second example, one might surmise that most of us are well equipped with the concepts of cliffs, erosion, rivers, and temporal concepts like hours and months etc. Many may even be familiar with lived examples of streams carving out beds through rock and soil. With the Grand Canyon, however, one is faced with such magnitude of erosion, from such heights, and over such long durations that it thwarts ordinary comprehension. We deploy terms like "geological time" to indicate that it isn't an ordinary concept of time i.e. days, hours or months, but, rather, an extraordinary notion of time and natural processes that have transpired to form the majestic scenery that is the Grand Canyon. In a very simplified version of cognitive expansion in this example, we are familiar with holes and channels in the ground that are the product of erosion, but the Grand Canyon is no ordinary hole. Rather, it is a canyon formed over millennia, millennia that are nearly unfathomable to comprehend by our limited, human concepts of time and natural processes. Thus, the Grand Canyon is daunting, even threatening to our reason. And yet, given aesthetic distance and imaginative expansion impelled by the experience, we are forced to reconceptualize and formulate a new comprehension that enables us to take in the experience. To return to Kant's provocation, we expand our notion of time and processes of erosion to a concept that approximates infinity (even though we understand that it is not infinity itself we are comprehending, but merely time that is beyond the ken of ordinary human comprehension).

16.4 Encounter: The Sublime and the Miniature

It is precisely this reformulation and cognitive expansion through imaginative extrapolation that lends itself to several of the activities we find in the practices of biocultural ethics. Kant himself notes that

> nothing in nature can be given, however large we may judge it, that could not, when considered in a different relation, be degraded all the way to the infinitely small, nor conversely anything so small that it could not, when compared with still smaller standards, be expanded for our imagination all the way to the magnitude of a world; telescopes have provided us with a wealth of material in support of the first point, microscopes in support of the second. (Kant 1790, p. 105)

It is in the second of these that we find activities contained within the methodology of biocultural ethics, namely the Miniature Forests of Cape Horn. While not using microscopes, but hand-lenses, the activities of bryologists and, subsequently, eco-tourists, offer an opportunity for researchers as well as visitors to Omora Park—a concrete and transformative opportunity to change lenses, to engage with the lesser employed aesthetic lens of valuation and appreciation, and to encounter the sublime.

"The Miniature Forests of Cape Horn: Eco-tourism with a Hand-lens" is an activity housed in Omora Ethnobotanical Park located on Isla Navarino, Chile (latitude 54.9419°S longitude 76.601°W) at the southernmost tip of sub-Antarctic Chile. It is located inside the Cape Horn Biosphere Reserve, a UNESCO World

Heritage Site (designated 2005), and is home to more than 1000 species of liverworts, mosses and lichens. Approximately 50% of the species found here are endemic to the temperate and sub-Antarctic region. In less than .01% of the planet's terrestrial landmass, we find more than 5% of the planet's documented species of bryophytes. The activity itself is the outgrowth of bryophyte research and collaboration with regional authorities, teachers, artists, engineers, architects and local school children. In 2002, in conjunction with this collaborative team, a system of bridges and viewing stations were established at the Omora Ethnobotanical Park, thus creating the "Garden of Miniature Forests of Cape Horn" (Goffinet et al. 2012, pp. 21–22). The bridges and stations are designed to preserve the integrity of these delicate and miniscule organisms and habitats. The viewing stations highlight opportunities for looking into the biodiversity found in this miniature world. The activities within the garden, particularly ecotourism with a hand lens (EHL), encourages participants to stop, slow down, and observe from a different perspective. Outside the garden itself Omora Park is rife with mosses, liverworts and lichens; visitors to the park often walk past the verdant topography without pausing to consider the biodiversity surrounding them. The garden invites them to pause, reflect and do just that. The hand-lens facilitates a different orientation—one in which the observer must get in close proximity to the viewed, look through instrumentation that magnifies the obscure, to smell the habitat in which these organisms flourish, to lay prone rather than stand upright, to respirate with the organisms, to focus on the small rather than the mesoscopic—and it invites them to see the flora and insects by a new lens, an aesthetic lens. It encourages an opportunity to engage with the miniscule with a full range of sensations, thus promoting a sensorial and aesthetic experience. This change in our mode of viewing provides the means by which the aesthetic distance and contemplation of Kant's aesthetic judgements are achieved while engaging the viewer with a world seldom seen outside of research programs.

Important to the activity is not just the viewing of these often-overlooked organisms, but the effects such an activity has on the viewers as well. Importantly, participants are exposed to new and exciting organisms, with a multiplicity of names—scientific, Spanish, English, Yaghan—lenses biological and cultural. Further educational components of the activity explain how water filtration through these organisms takes place, offering an economic lens of ecosystem services and benefits to humanity. But if it were merely this type of educational activity, it would miss a crucial component. Participants are encouraged to sketch, diagram and name what it is they experience in this miniature world—that is, to experience, imagine and expand their comprehension of what it is they are experiencing. By doing so, eco-tourism with a hand-lens invites the participants into an aesthetic world, a world of visualization, metaphor and analogical thinking. The activity itself permits viewers to see the multiplicity of organisms, often living in synchrony with another, to see their habitats and habits and to see how each are co-inhabitants in the space they share.

Importantly for this aesthetic lens is to see how much life occupies such small spaces. The magnitude of life found in such small spaces borders on the mathematical sublime. In our ordinary perceptions of the world, we readily recognize and conceptualize, even document and categorize medium-sized objects exceptionally well. We place them within our ordinary, often self-interested, domain of experiences and have ready conceptual tools to do so. The extraordinary experience of the miniature forests offers us an opportunity to see analogous levels of complexity in miniature scale. A square inch of the miniature forest teems with a multiplicity of objects and organisms, thus giving observers pause to reflect on the observation Kant made himself- the observation that magnitude and volume are comparative.

The miniature forest invites participants to reflect that the world is teeming with under-observed, often un-observed, life. The Earth is teeming with such a magnitude of life that it thwarts our comprehension and invites us to pause and appreciate the sheer volume of life on this planet. Extrapolating outward, we are compelled to think that since one square inch contains such a multitude of life, then the world, as large as it is, contains an almost unfathomable amount of life upon and within it. This cognitive expansion is precisely the Kantian mathematical sublime in practice. By realizing how much life is contained in such a small space, we then begin to expand outward to suddenly realize how much, unfathomable though it may be, life is on our planet. The sudden realization of the abundance and variety of life on our planet can cause cognitive agitation and a sense of being overwhelmed. We grapple to fathom our own seeming insignificance in the face of such numbers. Such an aesthetic moment of the sublime can also permit further reflection on the amazing world in which we find ourselves. Concomitant to such amazement is the sense of vastness, in the sheer number of co-inhabitants, of our world. A resolution to the agitation finds itself manifest in a sense of awe and majesty at the robustness, diversity and magnitude of life on this planet. This is the true product of the sublime; it is not the mere number, in the case of the mathematical sublime, nor in the force(s) on our world, in the case of the dynamically sublime, that amazes.

16.5 Conclusion

The reflective product is that we can experience and recognize our own cognitive limitations, and in doing so humble ourselves to respect these limitations. It is with a sense of awe and majesty that we experience the sublime- both in regards to our world and biodiversity, and also to our limitations. And with these limitations and lived experiences we can grow our cognitive apparatus to incorporate what was once unseen. The sublime takes place in us and our experiences in the vastness of the world, and it encourages us to change our cognitive lenses to incorporate these experiences into our lives. In Kant's words, "the beautiful prepares us for loving something, even nature, without interest; the sublime, for esteeming it even against our interest" (Kant 1790, p. 127). Kant's final point—that when one experiences the sublime, one connects the aesthetic experience with "a mental attunement similar to that for moral feeling"—(Kant 1790, p. 128) is the ultimate focus for biocultural ethics. By experiencing the sublime, we arrive at a newfound respect and consideration for both ourselves and our co-inhabitants in the world, and this newfound respect forms a new imperative for moral consideration.

It is to this last point that the exercise of "changing lenses" is designed. As mentioned at the opening of this essay, a singular vision of the world leads to homogenization, both biological and cultural, loss of biodiversity and loss of life. Heretofore, the Western world has predominantly been guided by an economic lens, supported by reciprocal links with a biophysical lens. It has led to an extinction of experience and a loss of languages, wisdom traditions and diversity on this planet. If we persist with this truncated way of seeing the world, the future is ominous, both for humanity and for the planet. One potential solution is to change how we view our world. And this is not without historical precedent. There have been and still are other ways of viewing our world- both from Eastern and Western traditions, and within the Western tradition itself. One such way is to remove our interest, even if temporarily, from our viewing—to see the world differently, perhaps through an aesthetic lens that then sponsors a different type of appreciation and valuation of the co-inhabitants of our world. Such opportunities abound, and, in so doing, we may take further opportunities to re-orient ourselves towards our world and all that it contains. This reorientation is precisely the call from biocultural ethics, a call to change our lenses and see ourselves from multiple perspectives, that we might limit the hubris we find as the single most important species on the planet. In short, it is an invitation to travel the world through various lenses and to reconceptualize ourselves, our world and our place in it. Just as Gulliver gained perspective, appreciation and humility, from his travels, it is vital we travel with a multiplicity of lenses in order to find a new orientation, a new imperative, and a new way of existing.

References

Aristotle (1984) DeAnima. In: Barnes J (ed) The complete works of Aristotle. Princeton University Press, Princeton

Bar-On YM, Phillips R, Milo R (2018) The biomass distribution on earth. Proc Natl Acad Sci 115(25):6506–6511

Brundtland G (1987) Report of the world commission on environment and development: our common future. United Nations General Assembly document A/42/427

Costanza R, De Groot R, Sutton P, Van der Ploeg S, Anderson SJ et al (2014) Changes in the global value of ecosystem services. Glob Environ Chang 26:152–158

Crutzen PJ, Stoermer EF (2000) The Anthropocene. Glob Change Newsl 41:17–18

Gibbons S (1994) Kant's theory of imagination. Clarendon Press, Oxford

Goffinet B, Rozzi R, Lewis L, Buck WR, Massardo F (2012) Miniature Forest of Cape Horn/Los Bosques en Miniatura del Cabo de Hornos. Ecotourism with a hand lens/Ecotourismo con lupa. 445 p. University of North Texas Press (Denton, TX) and Ediciones Universidad de Magallanes (Punta Arenas, Chile).

Kant I (1790) Critique of judgement, trans. Pluhar WS. Hackett Cambridge

McBrien J (2016) Accumulating extinction: planetary catastrophism in the Necrocene. In: Moore JW (ed) Anthropocene or capitalocene?: nature, history and the crisis of capitalism. PM Press, Oakland, pp 116–137

Monbiot G (2007) Heat: how to stop the planet from burning. South End Press, Cambridge

Palmer C (2013) Contested frameworks in environmental ethics. In: Rozzi R, Pickett S, Palmer C, Armesto J, Callicott J (eds) Linking ecology and ethics for a changing world, Ecology and ethics, vol 1. Springer, Dordrecht, pp 191–206. https://doi.org/10.1007/978-94-007-7470-4_16

Pascual U, Balvanera P, Anderson CB, Chaplin-Kramer R, Christie M et al (2023) Diverse values of nature for sustainability. Nature 620(7975):813–823. https://doi.org/10.1038/s41586-023-06406-9

Rozzi R (1999) The reciprocal links between evolutionary ecological sciences and environmental ethics. Bioscience 49(11):911–921. https://doi.org/10.2307/1313650

Rozzi R (2012) Serendipity in the origin of ecotourism with a hand lens. In: Goffinet B, Rozzi R, Lewis L, Buck W, Massardo F (eds) Miniature forests of Cape Horn ecotourism with a hand lens. UNT Press, Denton

Rozzi R (2013) Biocultural ethics: from biocultural homogenization toward biocultural conservation. In: Rozzi R, Pickett S, Palmer C, Armesto J, Callicott J (eds) Linking ecology and ethics for a changing world, Ecology and ethics, vol 1. Springer, Dordrecht, pp 9–32. https://doi.org/10.1007/978-94-007-7470-4_2

Rozzi R (2018) Biocultural homogenization: a wicked problem in the Anthropocene in from biocultural homogenization to biocultural conservation. In: Rozzi R, May R, Chapin F, Massardo F, Gavin M, Klaver I, Pauchard A, Nunez M, Simberloff D (eds) From biocultural homogenization to biocultural conservation, Ecology and ethics, vol 3. Springer, Cham, pp 21–48. https://doi.org/10.1007/978-3-319-99513-7_2

Rozzi R, Tauro A (2023) The multiple lenses of ecotourism with a hand lens: fundamental concepts and practices. In: Rozzi R, Tauro A, Avriel-Avni N, Wright T, May RH Jr (eds) Field environmental philosophy: education for biocultural conservation, Ecology and ethics, vol 5. Springer, Dordrecht, pp 27–51. https://doi.org/10.1007/978-3-031-23368-5_3

Rozzi R, Massardo F, Anderson C, Heidinger K, Silander JA (2006) Ten principles for biocultural conservation at the southern tip of the Americas: the approach of the Omora Ethnobotanical Park. Ecol Soc 11:43. https://doi.org/10.5751/ES-01709-110143

Rozzi R, Armesto J, Goffinet B, Buck W, Massardo F et al (2008) Changing lenses to assess biodiversity: patterns of species richness in sub-Antarctic plants and implications for global conservation. Front Ecol Environ 6:131–137. https://doi.org/10.1890/070020

Rozzi R, Lewis L, Massardo F, Medina Y, Moses K, Manuela M, Sancho L, Vezzani P, Russel S, Goffinet B (2012) Ecotourism with a hand-lens in Omora Park. Ediciones Universidad de Magallanes, Punta Arenas

Rozzi R, Massardo F, Contador T, Crego R, Mendez M, Rijal R, Cavieres L, Jimenez J (2014) Field environmental philosophy: ecology and ethics in LTSER-Chile and ILTER networks. Bosque 35(3):439–447. https://doi.org/10.4067/S0717-92002014000300019

Soga M, Gaston KJ (2016) Extinction of experience: the loss of human–nature interactions. Front Ecol Environ 14(2):94–101

Swift J (1726) Gulliver's travels and other writings, Landa L (ed). Houghton Mifflin, Boston

Chapter 17
Collecting Photos in the Miniature Forest: A Biocultural Approach for Bryophyte Conservation in the Cape Horn Biosphere Reserve

Roy Mackenzie, Cristian Valle-Celedón, and Gonzalo Arriagada

Abstract The relationship between nature and conservation has significantly changed, and photography's aesthetic appreciation of wildlife is crucial to conservation efforts. In response to biodiversity loss, photographers are increasingly using high-definition (HD) photography as a communication tool for conservation, focusing on vulnerable species, unique habitat, and specific biocultural contexts to raise citizen's awareness. The photographic work in Cape Horn Biosphere Reserve, Chile presents as a salient case, where photographers have documented the diversity of endemic bryophytes and other life forms. By documenting the endemic biodiversity of protected areas and working with scientists, nature photographers are able to enhance the citizen's understanding and appreciation of unique ecological areas, expose citizens to species that are often overlooked, and encourage people's deeper understanding of ecological processes. While citizens are appreciating the beauty of the natural world as shown in HD photography, it is also important to understand the work ethic of nature photographers. In this chapter, we attempt to analyze the work ethics of HD photography related to nature conservation within the framework of biocultural ethics. First and foremost, photographers are captivated by the beauty of the natural world, photographing its magnificence and diversity, which enhances the emotional connection with wildlife that is conveyed to the audience through their

R. Mackenzie (✉)
Cape Horn International Center (CHIC), Centro Universitario Cabo de Hornos, Universidad de Magallanes, Puerto Williams, Chile

Instituto de Ciencias Aplicadas, Universidad Autónoma de Chile, Santiago, Chile

Millenium Institute Biodiversity of Antarctic and Subantarctic Ecosystems (IBASE), Santiago, Chile
e-mail: r.mackenzie.c@gmail.com

C. Valle-Celedón
Audiovisual Producer in Wolf Productions, Punta Arenas, Chile

G. Arriagada
Professional Photographer, Puerto Williams, Chile

D. Zhu et al. (eds.), *Linking Arts with Biocultural Conservation, Restoration, and Communication*, Ecology and Ethics 8,
https://doi.org/10.1007/978-3-032-04830-1_17

work. The practice of "collecting photographs" rather than simply "shooting" or "taking" pictures reflects photographers' incorporation of biocultural ethical concerns into their work. Not only do they document life forms in unique habitats that bear biocultural significance and carry biocultural information, but they also adhere to the principle of minimal interference in the photographic process, embracing all other-than-human beings in nature as co-inhabitants.

Keywords Wildlife conservation · Biocultural ethics · Macro photography · Bryophytes · Small organisms

17.1 Introduction

The biological and experiential tendency to be drawn to natural environment and other-than-human wildlife is essential for the well-being of humans (Wilson 1984). Artists and scientists alike draw inspiration from nature as a source of knowledge and aesthetic appreciation. Since the nineteenth century, aesthetic appreciation of the natural environment has been reinforced by the massification of photography (Matthews 2003). Wildlife photographers are drawn to the aesthetics of the natural world, the beauty of the diversity and morphology of wildlife, colors and patterns in their habitats, the ecological process in conservation areas, etc. Nature photography for the purposes of wildlife conservation often aims to highlight the splendor and diversity of wildlife. Excellent photography can enhance people's emotional attachment to wildlife (Hanisch et al. 2019).

In recent decades, the fast rate of biodiversity loss caused by planetary socio-ecological crises and biocultural homogenization has motivated professionals to support conservation efforts, using high definition (HD) photography as a primary tool for communication regarding wildlife conservation (Goffinet et al. 2012; Rozzi et al. 2018). Vulnerable or near-extinct species are ideal candidates for conveying information regarding biocultural conservation, which transforms from a purely artistic satisfaction to a full experience that considers the current conservation status of relevant organisms. The latter requires a deeper understanding of the ecological processes and biocultural interactions behind the photographs, which depends not only on photographers' capacity to understand the subject and convey biocultural messages through images, but also on photographers' understanding of the biocultural ethos and how to approach the photographic work with that understanding.

17.2 Biocultural Framework and Nature Photography

Biocultural ethics emphasizes the interconnectedness of biological and cultural diversity, through which people can recognize the mutual relationships between humans, other-than-human beings, their life habits, and their shared habitats (Rozzi

2012). This approach, developed from the southern hemisphere, challenges traditional anthropocentric ethics by acknowledging the intrinsic value of all life forms, their ecological interactions, and cultural contexts in which these complex interactions are situated (Rozzi and Tauro 2023). Begun in the nineteenth century as a tool for scientific documentation, nature photography has evolved from early scientific illustration to a powerful way for fostering conservation and public engagement (Seppänen and Väliverronen 2003). Like scientific illustrations that selectively emphasize morphological features of natural objects and convey biocultural messages (Chiarini 2026), high-definition nature photography guided by biocultural ethic provides unprecedented authenticity and accessibility while shaping people's appreciation of wildlife, thereby positively adding to conservation endeavors.

Here, we propose nature photography can be an ethical practice for "face to face" encounters with flora, fungi and fauna by integrating biocultural ethics into photographic activities. It calls for responsible, ethical, respectful image-making that considers the well-being of both natural subjects (as other-than-human co-inhabitants) and human communities, especially Indigenous communities. This approach values collaboration with local residents, cultural sensitivity, and minimizing the environmental impacts of photographic activities. Biocultural photography also leverages modern technology while adhering to biocultural ethics, ensuring that images serve the objectives of biocultural conservation and cultural exchanges. Ultimately, biocultural photography can serve as a bridge to foster understanding and reverence of the deep connections among people, cultures, and the natural environment, as well as promoting the conservation of both biodiversity and biocultural heritage that have endured centuries. The connotations of biocultural photography are evident in nature photography related to miniature forests in the Cape Horn Biosphere Reserve in Chile.

Cape Horn Biosphere Reserve is located at the southernmost tip of America, around 55° south latitude, comprising a complex archipelagic system of islands and fjords. The terrestrial ecosystem mosaic includes peatlands, subantarctic tundra and forests, glaciers and alpine landscapes, many of which are dominated by bryophytes (Rozzi et al. 2008). It contains an impressive total of 450 mosses, 368 liverworts, and 2 hornworts, representing a larger number of species (820 species of bryophytes) than vascular plants (773 vascular species) (Rozzi et al. 2008; Villagrán 2020). The unique Subantarctic bioregion and the exceptional bryophyte diversity it presents make it a frontier for integrating wildlife conservation with sustainable development.

Furthermore, as a part of Man and the Biosphere Programme by UNESCO for the development of biosphere reserves in all continents and countries of the planet, the relevance of Cape Horn Biosphere Reserve stems from the unique mission that integrates conservation, sustainable development, and research (www.unesco.org/). Unlike traditional protected areas that focus solely on conservation, biosphere reserves actively promote solutions that reconcile biodiversity conservation with sustainable use of the natural environment. This approach makes biosphere reserves particularly valuable as learning places for sustainable development, where innovative approaches to environmental stewardship can be tested and ready for being replicated in other places on the globe (Price et al. 2024).

17.3 The Aesthetic Perspective and Relevance of Photography

Over the years, conservation in Cape Horn has integrated an aesthetic perspective, powerfully contributing to the value of the biosphere reserve and providing further reasons for the preservation of the beings within the biosphere. However, a purely aesthetic perspective can also lead to the neglect of landscapes associated with important ecosystem functions, simply because these landscapes appear mundane from a conventional aesthetical angle (Matthews 2003). This is the case in Magallanes. The beauty of the sub-Antarctic region can be usually observed in the colors of its landscapes, forged mainly by towering native flora. But this is not always the case. For example, the sub-Antarctic tundra above the treeline is covered in a variety of mottled colors, attributed to the diversity of unobtrusive bryophytes and lichens (Goffinet et al. 2012), as illustrated by the case of remarkable mimicry of the seedsnipe (*Attagis malouinus*). In the photo (Fig. 17.1), cushion of *Bolax gummifera* are interspersed with prominent whitish masses of lichens from the genera

Fig. 17.1 Photo of a White-bellied seedsnipe, perdicita cordillerana (*Attagis malouinus*) and its chickens, all exhibiting remarkable cryptic coloration for camouflage within the high-mountain tundra. This characteristic landscape of the CHBR, characterized by low temperatures and persistent winds, consists mainly of bryophytes and lichens, largely constituting the unique chromatic complexity of the habitat. (Photograph by Roy Mackenzie)

Pseudocyphellaria, Ochrolechia, and *Thamnolia*. Additionally, dark green and black lichens from the genera *Pseudophoebe, Cetraria*, and *Usnea* create contrasting patches on the ground.

These various color patterns of the sub-Antarctic tundra are reflected in the plumage of the bird, and the photographer expects this photograph to accentuate the intricate relationship between the tundra's flora, fungi, and birds. Within the traditional perspective, although in most cases one does not need scientific knowledge to enjoy beauty in nature, more biocultural information about the co-inhabitants, habits, and habitat can help people incorporate ethical judgments into aesthetic appreciation. In the above case, without an understanding of the unique ecological and inter-species relationships in the sub-Antarctic, it may be difficult for viewers to appreciate the beauty of this photograph.

Accurate ecologically relevant information and its biocultural connotations may also alter people's appreciation of certain natural phenomena. For example, people may be less likely to appreciate an introduced species once they are aware of the impact it has on the native environment. This is the case with the Canadian beaver (*Castor canadensis*), which acclimated perfectly in the Patagonian ecosystem. Despite the impressive engineering capacities of these rodents, people no longer consider these introduced animals aesthetically pleasing when they realize the devastation they caused to the surrounding forests.

Indeed, when nature photographers are well-informed about the ecosystems and biocultural diversity of a region, they can create HD photography that plays an important role in the creation of protected areas (Soga and Gaston 2024) as well as the dissemination of a biocultural ethic to further stimulate discussions that enrich perspectives on conservation. Through photography, we have access to astonishing insights into unperceived organisms and their environments, from tiny invertebrates and plants to distant birds and whales. The following two boxes display how biocultural perspective can shape photographers' work (Boxs 17.1 and 17.2).

Box 17.1: Cristian Valle Celedón, Wolf Productions

Cristian Valle Celedón is an audiovisual producer and photographer based in Punta Arenas, the capital city of sub-Antarctic Magallanes. His work is situated in art, science, and memory, exploring the relationship between humans and nature from an ethical, poetic, informative and educational perspective. Through projects such as Musgus VR, he has contributed to biocultural conservation by presenting habitats, local knowledge, and forgotten species. His approach, which combines immersive technologies, documentary, and photography, advocates for good living, respect for biocultural diversity, and conscious living, inspired by thinkers such as Humberto Maturana, Ricardo Rozzi, Gudynas and the culture of the native peoples of the sub-Antarctic region, among others (Fig. 17.2).

Fig. 17.2 Left: MUSGUS VR snapshot of immersive biocultural experience. Virtual reality (VR) technology has extraordinary potential in cultivating students' environmental awareness and compassion, as well as providing multidisciplinary learning experiences (Hsu 2024). As a project that has been developed for more than 12 years, Musgus VR offers virtual reality experience in exploring the miniature forest of the Navarino Island in Tierra del Fuego. It aims at combining techno-creativity to foster environmental education regarding the conservation of the macro world in the natural environment. For more information visit musgusvr on Instagram. Right: Other co-inhabitants moss mites (*Acaro oribatida*) in a miniature forest. (Photos prepared by Cristian Valle Celedón)

Box 17.2: Gonzalo Arriagada, Professional Photographer

Together with his partner and their two children, for the past 15 years, Gonzalo Arriagada has developed photographic and audiovisual projects related to biocultural conservation in the space they inhabit: Navarino Island and its geographical connections. Through digital, analog or handmade photography such as Dibujar con Luz project (Drawing with light project), as well as its reflections in sound and music, they have created books, videos, exhibitions, installations, and community workshops in Puerto Williams, Punta Arenas, King George Island, and other locations in southern Chile, combining sciences and art to open spaces for the re-creation of knowledge and collective sensibilities based on mutual recognition and respect (Figs. 17.3 and 17.4).

When photographing miniature forests, these photographers have adhered to the basic concepts of biocultural ethics regarding co-inhabitation, and have followed the relevant protocols in various procedures, such as preparatory, photography, and post-editing processes.

- **Preparatory Process**

Absorbing scientific knowledge that is relevant to biocultural conservation is preparation for a Nature photographer working under the framework of biocultural ethics. In the field, an identification guide, a map, or just a species list can be more than enough to dive into biocultural complexity and the world of wildlife. A study made in the US showed that nature photography provoked higher emotional

Fig. 17.3 Left: Detail of cyanotype on canvas, part of the Drawing with Light Project, showing the biocultural perspective of inhabiting the sub-Antarctic Patagonia. Right: An ancient tree covered with moss in the Sub-Antarctic forest. (Photograph by Gonzalo Arriagada)

Fig. 17.4 *Palpibracus chilensis / Tayloria mirabilis. Tayloria mirabilis,* a species of moss belonging to the family Splachnaceae, utilize flies such as *Palpibracus chilensis* for spore dispersal to dung. This photograph demonstrates how life habits of multiple species can create a complex web of interactions. (Photograph, by Gonzalo Arriagada)

attachment and concern with diversity, compared with people just observing nature (Hanisch et al. 2019). These results suggest that nature photography can be a useful technique to close the gap generated by the "extinction of the experience" (Soga and Gaston 2016), allowing people to have conscious face to face encounters through the lens and promote protection.

A study made in Japan (Ohara et al. 2019) showed that nature photographers are proactive in absorbing scientific knowledge of wildlife to achieve objectives such as

- helping scientists find and obtain good quality photographs by learning about the habits and distributions of wildlife;
- gaining scientific credibility in nature photography audiences; and
- deepening their understanding of their targets and the context by conducting both scientific and first-hand field observations.

- **Photography process**

Nature photography is also about what photographers choose to capture, and therefore what the audience sees in a photograph is the product of the photographer's aesthetic concept. In the current digital age, people often "capture" relevant natural events with their cell phones or cameras without much consideration. Moreover, the technology and speed required to post pictures in social media are changing the ways we understand and remember things and places, as well as how our actions can impact wildlife or ecosystems we intend to protect. The "taking pictures" or "captures" becomes a somewhat invasive approach to wildlife, one superficial approach that focuses on documenting lists of species rather than appreciating their life habits, habitats, and relations to other wildlife that can lead to *in-situ* conservation. In this scenario, nature photographers can differentiate themselves from picture-takers by integrating their work into a deep understanding of the complex interrelationships among co-inhabitants, their living habits, and habitats.

- **Post-editing process**

Intentions are often added to images not only by photographers but also by editors and subsequent users. Nature photographers actively seek to contribute to scientific knowledge. When they encounter organisms or natural phenomena that are not documented in scientific literature, they communicate with researchers and academic societies to build up understanding. They establish working relationships with wildlife researchers for bilateral communications that enrich scientific knowledge base. These photographers, while acquiring knowledge from the natural sciences, also engage in activities beyond photography. Many nature photographers are involved in conservation efforts, serving as local interpreters, guides, and consultants. Some also advocate for wildlife and ecosystem conservation, presenting themselves as specialists on local wildlife.

Through the above analysis of the working process of our photographers, it is clear that biocultural concept of the "3H" (Habit, Habitat, Cohabitants) developed by Ricardo Rozzi et al. (2018) and a "heuristic lens" (Rozzi and Tauro 2023) can be utmost valuable for photographers who wish to contribute to the conservation of wildlife diversity, including small organisms such as bryophytes and their aesthetic, ecological, cultural, economic as well as ethical values. By exploring with the "3H biocultural lenses," nature photographers can document the full range of tiny and perceptible species in their natural environment, revealing not only the beauty of these organisms, but also providing a holistic understanding of their living habits, recorded in their habitats, and of the community of ecological beings with which they cohabit. This biocultural approach favors synergy between scientists, artists, educators and diverse citizens. In this synergy, scientists contribute knowledge and data on local species and threats, while artists create narratives and images that convey local values and information that appeal to the public in a more accessible and emotional way by including ethical dimensions.

17.4 Discussion

Based on the bryophyte conservation work carried out by the Sub-Antarctic Biocultural Conservation Program since 2000, we can observe that, scientific research has contributed to a better understanding of the diversity and ecological functions of bryophytes, which reach enormous diversification in subpolar regions; and, on the other, how much of this appreciation has yet to be communicated to global society through an artistic perspective broader than the academic environment, such as through HD photography discussed in this chapter. Based on our photographic practice, we believe there are several issues that deserve further reflection.

17.4.1 Cognitive Dimension

When photography is incorporated into sciences, it plays a unique role in academia as an essential tool for communicating discoveries and research advances. While conventional photography seeks aesthetics and personal expression, photography for sciences is governed by a set of highly functional criteria, aimed at clearly and accurately conveying the evidence contained in the sample. The primary function of scientific photography is to provide an effective visual medium to support and complement the information presented in scientific publications.

The target audience for these images is not the general public, but other scientists or amateurs immersed in the same field of study. In this context, the criterion for taking scientific photographs departs from conventional aesthetics. Instead of looking for intriguing angles or visually striking compositions, the focus is on highlighting the structure and relevant details of the specimen. To this end, techniques such as photo-stacking have been developed, which allows a series of frames with different focus points to be compiled into a single photograph, resulting in a razor-sharp, high-quality image of the specimen. To further highlight the

Fig. 17.5 Immature sporophytes of the endemic dung moss *Tayloria mirabilis,* showing the capsule with and without calyptra for phenology studies, using the macro technique. The black background obtained by this technique is encouraged, replacing the natural environment to make the individual stand out in the photograph. (Photograph by Roy Mackenzie)

morphological characteristics, it is common to use a single homogeneous color as the background of the image (Fig. 17.5), eliminating the environment where the individual lives. This is critical, as the habitat and co-inhabitants of the moss are lost in the communication effort.

Throughout history, scientific photography has followed a path guided by objectivity and the search for an accurate representation of wildlife. According to Wilder,

> The heroes of observation … of the eighteenth century set the standard for observing in the nineteenth. These cultures of observation left a clear signature on photography, and photography in return revolutionised the observations made by scientists, by extending the range of what could be ocularly observed, and by raising up those capable of only rather average observational capability to an acceptable level by correcting for their inadequacies. (Wilder 2009, p. 164)

The impact of photography on scientific observation reflects both the complex connection between science and arts but also how photography shapes human perception. This cognitive shift suggests that as new scientific theories emerge, so do photographic perspectives and their expression.

Therefore, highlighting the subject by eliminating the background is valuable in accentuating the scientific message. However, when new demands emerge, or more directly, when one wants to present the biocultural characteristics of a species, i.e., a species that always inhabits a particular environment and possesses its own distinctive life habits, the removal of the background is no longer appropriate.

Photographers need to present both the morphological characteristics of the species, the community, and some information about the habitat. Photographs in the *Miniature Forest of Cape Horn* (Goffinet et al. 2012) basically display these biocultural characteristics to reflect the perception that a change of lens (both physical and cognitive) can help people to access the high diversity of sub-Antarctic bryophytes and lichens (Rozzi 2012).

Inspired by these photographs, one of our authors has utilized contemporary revolutionary platforms such as iNaturalist for data collection and visual documentation of bryophyte biodiversity (Fig. 17.6). To present bryophytes as also

Fig. 17.6 Collection of photographs of species from user @RoyMackenzie on the iNaturalist platform. Close-up photos are the most abundant on the platform. (Photograph by Roy Mackenzie)

co-inhabitants of humans, rather than isolated samples or visual trophies, photographers ought to keep the interconnectedness of the 3Hs in their perception and also convey the perception of co-inhabitation through displaying population of a species, a community, and a landscape, etc. This demonstrates the significance of the cognitive dimension of photography for bryophyte conservation.

Conservation related photography can also transform environmental education, foster cognitive development, and encourage collaboration (Farnsworth 2011). In Cape Horn, photography is essential in fostering transformative learning about bryophyte conservation and general biocultural conservation. Photographers' sensitivity extends beyond aesthetic considerations to include traditional knowledge that local people possess about native and exotic species. Their understanding can guide artists' approach to create meaningful images for the audience. Engaging in transdisciplinary research encouraged by biocultural ethics, photographers collaborated with scientists and local elders and cultivated an appreciation of the interconnections among local communities, their co-inhabitants, habits, and their shared habitats. For instance, the use of junquillo (*Marsippospermum grandiflora*) for basketry or old man's beard (*Protousnea sp.*) for medicinal purposes by indigenous peoples highlights the rich biocultural connections between subantarctic species and humans, echoing modern society's appreciation for genes and extracts from native species in extreme cold habitats.

Furthermore, to generate transforming learning in the public toward conservation of biodiversity, the sensitivity as a filmmaker needs to go beyond aesthetics. Detailed knowledge of other-than-human co-inhabitants in regions like the sub-Antarctic ecoregion is vital for the creative process. Information on the location and habits of species can guide the photographic approach to capture meaningful images. The research for such information needs to be as diverse as possible by consulting scientists (such as reading Glime (2017a, 2017b)), locals with experience, and any other source of information. By collaborating during the research, the visual work contributes to a collective perception of local inhabitants, its habits, and habitats.

17.4.2 Ethical Dimension

The ethical principles of wildlife photography have received widespread attention and discussion (Groo 2019). Adhering to the ethical ethos of biocultural cohabitation, three main considerations show the uniqueness of the bryophyte photography activities in Cape Horn region when considered from an ethical dimension: the photographic subjects, the presentation of the photographs, and the deployment of the photographic activities.

In exploring the relationship between photography and environmental issues, it is often observed that the subject matter of photography is often ambitious and extensive, such as wildfire-burning forests, endangered and large species, awe-inspiring landscapes, and exotic places. An examination of illustrated stories in *The Times* regarding visualizing biodiversity reveals that

> … most nature photographs concentrate on introducing readers to different species clearly underlines the extinction aspect of biodiversity: this helps to concretize a complex scientific problem and to make it more familiar. … Like metaphors, photographs too are selective in constructing an image of reality. In the case of endangered species they operate with metonyms, using parts to represent the broader whole. On the reverse side of the coin, concretization and simplification tend to mean that the problem itself is simplified. The two other aspects of biodiversity, i.e. genetic diversity and ecosystem diversity, are considerably harder to capture by means of photography. (Seppänen and Väliverronen 2003, p. 81)

Their reflections on nature photography show that the subject of the photographs are crucial in presenting biodiversity. In Cape Horn, small organisms such as bryophytes and lichens became one main group of organisms to present the endemic biodiversity in the region. This is in contrast to photographic initiatives that focus on large charismatic flagship species. In a sense, this challenges taxonomic chauvinism in the mainstream conservation activities (Rozzi 2019), and raises awareness of the important value of bryophytes in socio-ecological systems (Zhu 2023; Zuo and Zhang 2026).

Regarding the presentation of photographs, we must consider the age of the visualization of information in which we live. Given the amount of visual information people are exposed to everyday, nature photographers, as producers of pictorial information, need to consider the impacts of their photographs on the audiences in this ocean of information. In this scenario, the significance of conservation photography, which goes "beyond documenting nature as an art form" and "responds to the mission of protecting nature" (Mittermeier 2005) can be applied to bryophyte photography in Cape Horn, as it not only documenting bryophyte diversity but also calls on biocultural conservation through offering opportunities of access the miniature world of bryophytes and lichens.

Every technology is a projection of an intention, and, therefore, an extension of certain original possibilities: the stick is an extension of the arm, the camera is an extension of the eye. In the globalizing era of the Cyborg, in which people's perception of the world and "re-encounter" with Nature are mediated by technology, ethical considerations of what photographers present to the audience is relevant. The human eye is neither objective nor innocent. Culture has its own mainstream aesthetics, shaped by millions of images we've already seen, which, if not subjected to critique and reflection, influence people's interpretation of visual stimuli. Regarding biodiversity conservation, conventional mainstream strategy should consider a decolonial approach to focus on locally rooted, plural, socially just, and convivial forms of conservation (Corbera et al. 2021). Conservation photography under these considerations, such as bryophyte photography in Cape Horn, and the presentation of the miniature forest, is beyond a documentation of the vast bryophyte diversity in Cape Horn Biosphere Reserve, an ethical calling of co-inhabit with these tiny but essential organisms.

The biocultural ethical perspective on co-inhabitants compels us to imagine ourselves as small as a *Cladonia* and prompts reflection on our relationship with these organisms as well as their ecological roles. Inspired by observations of mosses and lichens, a virtual reality experience allows interaction with *Nothofagus* forests on Navarino Island through digital environments created with Unreal Engine. This immersion at the scale of Cape Horn's Miniature Forests emphasizes the importance of multi-species interactions. Photographers' activities in the field also need to be informed by ethical considerations of interspecies relationships, e.g., how photographers interact with their co-inhabitants such as bryophytes and lichens when they take photographs. Based on this ethical consideration and discernment, we propose the metaphor of "collecting photographs."

With the metaphor "Collecting photographs in the miniature forest," we propose close-up photography as a biocultural and ethical activity within "Ecotourism with a hand-lens" (Rozzi et al. 2020, 2014). Instead of impulsively taking photos or capturing images, "collecting photographs" implies a more conscious process, suggesting an effort to gather meaningful records of life forms and their habitats. "Collecting" signals an ethical and selective approach, whereby each photograph is carefully selected to be interesting and meaningful with as little disturbance to the environment as possible so that the least intrusive collecting can be sustained. Respecting their intrinsic value by minimizing the impact of our presence in their habitat, such as minimizing stepping on, moving rocks and logs and exposing their hidden micro-habitats to the sunlight or predators. The metaphor, therefore, encourages a shift in perspective from simply accumulating images of species to a more deliberate and thoughtful documentation of photographers' experiences with organisms in their shared habitats.

The idea of "collecting" photographs instead of "taking" them points to a virtuous relationship between the authorship of the image and the image itself, sustained by knowledge and, in many cases, reciprocity. People who have a constant and conscious link with their ecosystem, the product of a resident's traditional knowledge and/or scientific knowledge, can take photographs that express that knowledge. The metaphor of "collecting images" invites us—already far from the Functional Value of biodiversity and its ecosystem services—to move from the relational value based on experience, to an "ecological" knowledge and action in our context of Holobionts and interdependence to nature.

17.5 Conclusion

Bryophyte photography in the Cape Horn Biosphere Reserve stands on the shoulders of historical nature photography, building on pioneers while embracing holistic approaches that honor Indigenous knowledge and challenge colonial practices. This methodology connects contemporary conservation with deep historical relationships, inspired by the Yaghan people and their ancestral territories. The photographer's personal experience of documenting the miniature forest transforms both technical approach and philosophical understanding, shifting from shooting and capturing, to face to face communion with miniature organisms and the shared habitats. Focusing hours within small areas rather than covering vast distances creates a meditative practice that dissolves observer-observed boundaries, demonstrating photography's potential as both conservation tool and pathway to ecological consciousness.

Photography is not simply a visual capture; it is a narrative that can shape perceptions and foster empathy, the same way nature illustration does. In the context of biodiversity and conservation, photographers can play a crucial role by embracing biocultural ethics. This approach proposes to go beyond the isolated representation of species (by passing habitats and co-inhabitants) and advocate for a more complete narrative that includes both the main protagonist and its environment and way of life. The Habitat-Habit-Inhabitants approach in the photographic act implies a particular experience that is reflected in the respective photograph, thus a biocultural photographer. Photos can be presented either as an isolated image, or as a photographic story, or as any of them together with a complementary communication strategy or artistic proposal, offering the observer relevant information for a comprehensive understanding of life in any of its forms.

The first step in this proposal implies a change of perspective. Rather than simply photographing the species in the foreground, photographers are encouraged to explore the broader biocultural context. This allows them to capture images that not only show the plant or animal in question, but also reveal its natural habitat. This not only enriches the visual representation, but contextualizes the species in its environment, highlighting the interdependence and complex webs of relationships with the other species with which it co-inhabits. By portraying these relationships, the photographs not only document biodiversity, but also underscore the vital interconnectedness between different species and their shared habitat. Embracing biocultural ethics in the act of bryophyte photography in Cape Horn is not only an aesthetic approach, but also a cognitive and ethical commitment to biodiversity conservation.

References

Chiarini VS (2026) How the invisible becomes visible: situating scientific illustration in biocultural communication. In: Zhu D, Rozzi R, Tsuji R, Johnson B, Caviness T, Castro-Jorquera C, Xu L, Zhang L (eds) Linking arts with biocultural conservation, restoration, and communication, Ecology and ethics, vol 8. Springer, Cham, pp 149–169. https://doi.org/10.1007/978-3-032-04830-1_9

Corbera E, Maestre-Andrés S, Collins YA, Mabele MB, Brockington D (2021) Decolonizing biodiversity conservation. J Political Ecol 28:889–903. https://doi.org/10.2458/jpe.5969

Farnsworth BE (2011) Conservation photography as environmental education: focus on the pedagogues. Environ Educ Res 17(6):769–787

Glime JM (2017a) Bryophyte. Ecology 1:2–1. Accessed through: https://digitalcommons.mtu.edu/bryophyte-ecology1/

Glime JM (2017b) Bryophyte. Ecology. 4:2–6. Accessed through: https://digitalcommons.mtu.edu/bryophyte-ecology4/

Goffinet B, Rozzi R, Lewis L, Buck W, Massardo F (2012) Miniature forests of Cape Horn: ecotourism with a hand lens. The University of North Texas Press, Denton/Ediciones Universidad de Magallanes

Groo M (2019) How to photograph wildlife ethically. Natl Geogr. https://www.nationalgeographic.com/animals/article/ethical-wildlife-photography
Hanisch E, Johnston R, Longnecker N (2019) Cameras for conservation: wildlife photography and emotional engagement with biodiversity and nature. Hum Dimens Wildl 24(3):267–284
Hsu H (2024) Promote environmental awareness and care by creating a virtual reality tour for the local community. The Geography Teacher
Matthews P (2003) Scientific knowledge and the aesthetic appreciation of nature. J Aesthet Art Critic 60(1):37–48
Mittermeier C (2005) Conservation photography. Int J Wilderness 11(1):8–13
Ohara N, Yamanaka Y, Trencher G (2019) Understanding nature through photography: an empirical analysis of the intents of nature photographers and the preparatory process. Environ Commun 13(8):1053–1068
Price MF, Schaaf T, Cárdenas MR, Fontanella Pisa P, Köck G (2024) The world network of mountain biosphere reserves. In: Schneiderbauer S, Fontanella Pisa P, Shroder IF, Szarzynski J (eds) Safeguarding mountain social-ecological systems. Elsevier, pp 169–174
Rozzi R (2012) Serendipity in the origin of ecotourism with a hand lens. In: Goffinet B, Rozzi R, Lewis L, Buck W, Massardo F (eds) Miniature forests of Cape Horn: ecotourism with a hand lens. University of North Texas Press, Denton, pp 11–25
Rozzi R (2019) Taxonomic chauvinism, no more! Antidotes from Hume, Darwin, and biocultural ethics. Environ Ethics 41(3):249–282. https://doi.org/10.5840/enviroethics201941325
Rozzi R, Tauro A (2023) The multiple lenses of ecotourism with a hand lens: fundamental concepts and practices. In: Rozzi R, Tauro A, Avriel-Avni N, Wright T, May RH Jr (eds) Field environmental philosophy: education for biocultural conservation, Ecology and ethics, vol 5. Springer, Cham, pp 27–51. https://doi.org/10.1007/978-3-031-23368-5_3
Rozzi R, Armesto JJ, Goffinet B, Buck W, Massardo F et al (2008) Changing lenses to assess biodiversity: patterns of species richness in sub-Antarctic plants and implications for global conservation. Front Ecol Environ 6:131–137. https://doi.org/10.1890/070020
Rozzi R, Anderson CB, Pizarro JC, Massardo F, Medina Y, Mansilla A et al (2010) Field environmental philosophy and biocultural conservation at the Omora Ethnobotanical Park: methodological approaches to broaden the ways of integrating the social component ("S") in long-term socio-ecological research (LTSER) sites. Rev Chil Hist Nat 83(1):27–68. https://doi.org/10.4067/S0716-078X2010000100004
Rozzi R, Jiménez JE, Massardo F, Torres-Mura JC, Rijal R (2014) The Omora Park long-term ornithological research program: study sites and methods. In: Magellanic subantarctic ornithology: first decade of forest bird studies at the Omora Ethnobotanical Park, Cape Horn Biosphere Reserve. University of North Texas Press, Denton, pp 3–40. https://doi.org/10.1007/978-3-319-99513-7_1
Rozzi R, May RH, Chapin FS, Massardo F, Gavin MC et al (2018) From biocultural homogenization to biocultural conservation: a conceptual framework to reorient society toward sustainability of life, Ecology and ethics, vol 3. Springer, Cham, pp 1–17
Rozzi R, La Valle MT, Russell S, Goffinet B, Massardo F (2020) Ecotourism with a hand-lens: a field environmental philosophy experience from the south of the world. In: Brister E, Frodeman R (eds) A guide to field philosophy. Routledge, pp 222–239
Seppänen J, Väliverronen E (2003) Visualizing biodiversity: the role of photographs in environmental discourse. Sci Cult 12(1):59–85
Soga M, Gaston KJ (2016) Extinction of experience: the loss of human-nature interactions. Front Ecol Environ 14(2):94–101. https://doi.org/10.1002/fee.1225
Soga M, Gaston KJ (2024) Do people who experience more nature act more to protect it? A meta-analysis. Biological Conservation, 289:110417. https://doi.org/10.1016/j.biocon.2023.110417
Villagrán C (2020) Biogeographical history of bryophytes in Chile. Gayana Bot 77:73–114
Wilder KE (2009) Photography and the art of science. Vis Stud 24(2):163–168
Wilson EO (1984) Biophilia. Harvard University Press, Cambridge, MA

Zhu D (2023) Bridge the channel, enhance the inclusivity: a comparison between flagship species-centered and moss-centered conservation in Chile and China. In: Rozzi R, Tauro A, Wright T, Avriel-Avni N, May RH Jr (eds) Field environmental philosophy: education for biocultural conservation, Ecology and ethics, vol 5. Springer, Dordrecht, pp 457–482. https://doi.org/10.1007/978-3-031-23368-5_28

Zuo Q, Zhang L (2026) Small but essential: how to introduce the public to easily overlooked organisms? In: Zhu D, Rozzi R, Tsuji R, Johnson B, Caviness T, Castro-Jorquera C, Xu L, Zhang L (eds) Linking arts with biocultural conservation, restoration, and communication, Ecology and ethics, vol 8. Springer, Cham, pp 243–263. https://doi.org/10.1007/978-3-032-04830-1_15

Chapter 18
The Birth of "Mosses: The Trailblazers on Ruins"

Shu Meng

Abstract Art, as a unique path and method of contemporary aesthetic practice, is a poetic expression of the unknowable realm of "direct knowledge," conveying the states, dynamic relationships, and profound emotional connections between humans and the natural world. Modern ecological ethics, insights from scientists and philosophers to my own experiences hiking and traveling all came together, leading me to blend mosses with my glass art unexpectedly. Through ongoing connections, dialogues, and reflections with nature, I gradually develop a deepening realization process. Through my moss-inspired glass art, I express thoughts and emotions, slowly building a spiritual home and attaining vastness through attention to the finest details within the free world of art. To revere mystery is not to deny scientific rationality, but rather to maintain a dual respect for the limitations and possibilities of human cognition. This is precisely the spiritual force that drives science to continually push its boundaries. It is through the mutual inspiration between philosophical thinking in sciences and artistic reflection that the spirit of life can truly flourish.

Keywords MENG Shu glass art · Wilderness · Moss · Ecology · Contemporary art · Aesthetics

18.1 Introduction

Since 2015, I have been using glass as the medium, method, and means of my artistic creation—continually exploring, recording, and expressing my feelings toward my surroundings, nature, and the deep emotional connections therein. As a form of self-expression, imaginative engagement, cognition, and emotional experience, artistic creation has always been the center of my life, while nature serves as the "source of power," offering infinite acceptance and continually creating sensory experiences linked to the spirit.

S. Meng (✉)
Kiang Wu Nursing College of Macau, Macau, China
e-mail: ms@kwnc.edu.mo

D. Zhu et al. (eds.), *Linking Arts with Biocultural Conservation, Restoration, and Communication*, Ecology and Ethics 8,
https://doi.org/10.1007/978-3-032-04830-1_18

As it has been for other artists (Chiarini 2026; Rozzi 2026; Vezzani 2026), the insights of modern natural sciences, especially the inspiration from ecological ethics and botany, have broadened my understanding of the external environment and shaped my relationships with the world around me. At the same time, an unseen force—something beyond logic and everyday awareness—has guided me. It draws me into nature through deep, hands-on connection, helping me find harmony between earth and sky. In these quiet moments, I feel my place in the world more vividly. These insights from my time in nature have shaped my inner growth and fueled my creative energy. They inspire me to express—through art—my feelings about the time we find ourselves and my evolving connection with the natural environment.

Attaining vastness through attention to the finest details has become the guiding principle of my artistic creation as well as the continual process of shaping and expanding my inner world. In recent years, I have held several solo art exhibitions, including: *Stardust* (2018), *Celestial Sphere* (2019), *With Dust, With Light* (2020), and *The Universe in Us* (2022). I also have exhibited representative series of glass artworks: *Dust to Rebirth* (2019), *Hymn of Nature* (2019), *Wandering Observation, Carefree Soul* (2022), and the ongoing *Moss Trail Series: Unbounded Profusion* (from 2024 to the present). In this path, in recent years I have achieved some recognition in the field of art, with my works being collected by several national art museums and professional institutions, including the National Art Museum of China.

I organize this chapter in three sections to explore my creative process and how I use my artworks to bridge the gaps between art, science, and communication. First, I examine how, as an artist, I have used glass and moss as media in my artistic practice to bridge ecological science and ethics, fostering an aesthetic practice characterized by active dialogue. In the second section, I discuss my philosophical thinking process and how these creative experiences have become a path of exploration in contemporary aesthetic practice. Third, I introduce the conceptualization of creating glass artworks based on moss imagery, elaborating on the relationship between natural beings and art in pursuing a spiritual home within Chinese artistic traditions. I apply the "moss dotting" technique from traditional Chinese landscape painting to contemporary glass art practice, expressing a philosophical understanding of nature and a way of life rooted in such reflection.

18.2 From Encounters to Confluence: Art as a Medium

Looking back, the interwoven connections between events, time, space, and people, whether sequential or parallel, spark mutual inspiration and collective participation. This process brims with an ineffable, mysterious force. These encounters and experiences seem not to be fully explained by causal logic alone. Or perhaps, clarity has never been the true purpose of artistic expression. As a glass artist, I find myself constantly drifting between the blurred and the lucid. And yet, it feels as if an

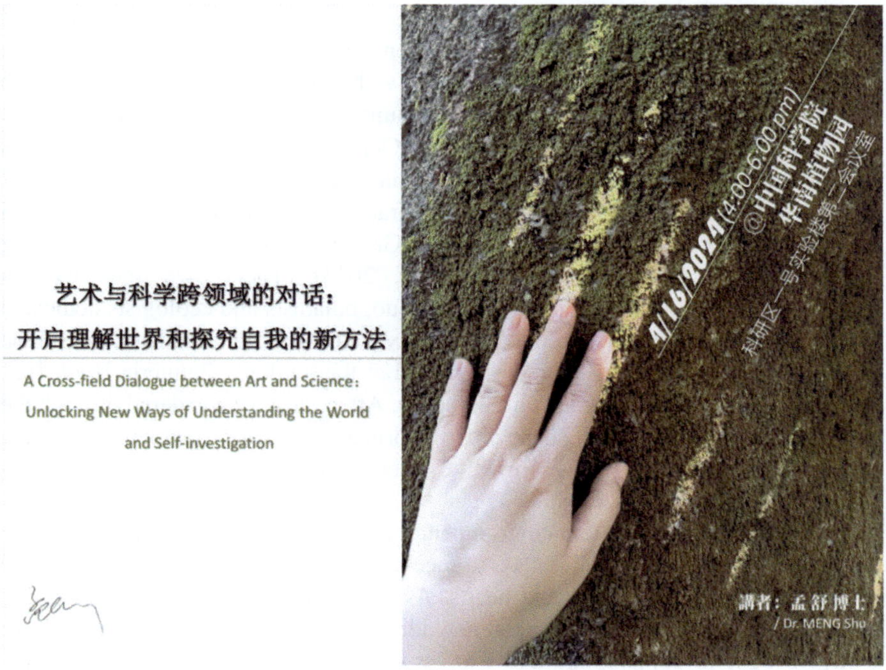

Fig. 18.1 Poster of the talk. (Photograph by Shu Meng)

invisible thread guides my thinking and creation, something that in philosophical terms might be described as purpose or meaning. I prefer a more poetic phrasing: a distant landscape, a vision from afar.

On April 16, 2024, at the invitation of Professor Chen Hongfeng, I visited the research team for Tropical and Subtropical Species Diversity Conservation at the South China National Botanical Garden in Guangzhou. There, I shared my creative journey in a talk titled, “A Cross-field Dialogue between Art and Science: Unlocking New Ways of Understanding the World and Self-investigation.” The cover image for the talk was a self-portrait I took at the Luís de Camões Garden (白鴿巢公園) in Macau, showing myself gently touching a patch of moss (Fig. 18.1). At the time, I had neither read any botanical works on moss nor recognized any species. In the photo, shafts of clear sunlight and my fingers seemed to form scratches. I touched the soft moss ever so gently, yet deep marks remained. Perhaps the moss compelled me to abandon vision’s overbearing dominance, luring me instead to “touch.”

Who could have imagined that I would soon share a deeper encounter with these unassuming mosses? Drawn by curiosity, I immersed myself in activities bridging ecology, aesthetics, and botany, leading me to discover kindred spirits across the planet.

On the afternoon of May 18, 2024, I had a hurried meeting with environmental philosopher Danqiong Zhu, a scholar deeply engaged in biocultural conservation, at Guangzhou South Railway Station. Though we had previously connected through online conversations, this was our first encounter in person, a mutual journey: I traveled from Macau, she from Shenzhen. Our conversation, though brief, was strikingly fluid, filled with sparks of intellectual exchange and a profound sense of affinity. Later, on May 23, 2024, with her gracious invitation and introduction, I accompanied her to the Fairy Lake Botanical Garden in Shenzhen, where Danqiong had previously conducted research (Zhu 2023). There, we met with a remarkable team: Li Zhang, Tuo Yang, Qin Zuo, botanists and ecologists dedicated to bryophyte and biodiversity conservation, as well as botanical illustrators Shihua Li, Lili Xu, and Lanping Zhou. We gathered at the garden's Plumeria Book Bar, exchanged books, and deepened our dialogues. Afterward, we wandered through the "You Xi" Valley (Zuo and Zhang 2026), formerly the Moss Garden, and made special visits to the Cycas Revol Conservation Center and its adjoining exhibition hall.

As emphasized by the field environmental philosophy methodological approach (Rozzi et al. 2006; Rozzi 2023), face-to-face encounters with mosses and nature are inspiring; they open our eyes and hearts. As we wandered through the "You Xi," these face-to-face encounters sparked new ideas within me. Through my art, I have always sought to contemplate and express my modern sensibilities—a genuine, embodied attention to the natural world. Yet it seems that I have never approached nature merely as an external object for depiction or lyricism, nor have I ever pursued realism in any conventional sense. Rather, I have always been drawn to those who possess a kind of innate wisdom, individuals who are, in their very being, self-harmonized with nature. This affinity has enabled me to communicate and connect with people across disciplines and fields, mutually enriching each other in the process, or sometimes even through the simple act of communication, finding joy. As a Chinese proverb says, *"Things of a kind come together; people of a kind form groups"* (物以类聚，人以群分). I feel deeply gratified to have found my flock, kindred spirits who resonate with each other's existence.

To sense the extraordinarily familiar and encounter the boundlessly unfamiliar. Amidst these hurried yet meaning-laden meetings, it was the inconspicuous moss that unexpectedly became the very core of our conversations and exchanges. Mosses present a scale that is neither too large nor too distant for human beings. It clings to material carriers, offering an opportunity for direct tactile engagement; thus, to "touch" the scene naturally gives rise to emotional resonance. Yet at the same time, the scale of mosses is so minuscule that it continually eludes direct grasp. Thus, people might see the "scene" and "image" without really touching the "thing." This discrepancy, at once a challenge to cognition and a seduction to the aesthetic sensibility, awakens a profound and subtle yearning within me.

18.3 From Encounter to Reflection: An Artist's Theoretical Contemplations

Since the economic reforms and opening-up of China in the 1980s, every student who received a basic education was required, without exception, to memorize two particular poems—both centered around moss. They are masterworks passed down from the Tang Dynasty (618–907). In retrospect, this fact carries a certain profundity. One poem is *Deer Fence (鹿柴)* by Wang Wei (699–761):

Empty hills, no one in sight, 空山不见人,
only the sound of someone talking; 但闻人语响;
late sunlight enters the deep wood, 返景入深林,
shining over the green moss again. 复照青苔上。
(Waston 1971, p. 24)

The other poem is *Inscription on a Humble House (陋室铭)* by Liu Yuxi (772–842). It was translated by Welsh scholar and sinologist Angus Charles Graham (1919–1991):

Moss-prints climb the steps, green; 苔痕上階綠;
grass-colors enter the curtain's blue. 草色入簾青。
(Graham, p n/a)

The first poem revolves around the shifting dance of light and shadow over the course of a day. The vastness of the deep forest becomes a medium, revealing the visceral experience of a human stepping into a realm rarely touched by others. The second poem centers entirely on sensory immediacy. The man-made structure serves as the viewing stage, emphasizing the curated "chamber scene." The moss's vitality—lush, dewy, vibrantly alive—receives exalted praise. This "青" (grass-color) transcends mere color; it embodies the "green" of 青春 (youth), of "vitality" (life-force?)!

As pioneer plants, mosses possess unique survival mechanisms and strategies that have undergone hundreds of millions of years of evolution, allowing them to adapt to diverse ecological environments (Goffinet and Shaw 2009). This endows people's aesthetic experience of mosses with a richer dimension of time and space (Tauro and Rozzi 2023). Furthermore, it offers insights into the dynamic balance and autonomous adjustment between "control" and "survival." In his historical reflection on the theory of "controlling nature" and its related concepts, William Leiss (2023) reveals the intrinsic relationship between the control of nature and human existence. The paradox inherent in the idea of controlling nature is identified as a root cause of human suffering. Once humanity enters a state of ecological existence, it will consciously abandon the control of nature and instead turn toward the liberation of nature, striving for conscious harmony with the natural world.

In Chinese taxonomy, "苔" (moss) is likely a general term referring collectively to mosses. Ancient Chinese scholars' perception and contemplation of a moss were not grounded in a static, objectified observation, analysis, or depiction. Rather, a moss was appreciated as a living phenomenon—as traces, pathways, or subtle signs.

In ancient Chinese art, there was greater respect for the integrated value of the process, from conception to completion and communication, highlighting that "art is process." The idea of art as a process has been later revisited and emphasized in contemporary Western art (Hu and Yang 2026). Thus, the notion of the "trace" has emerged as a widespread human consciousness: an intentional act of recording, expressing, and leaving behind marks (symbols).

Within my awareness, the moments when I perceive mosses, or perhaps when I am drawn to mosses, are closely intertwined with my perception of light. People often say that glass is a vessel for light; indeed, glass (crafted into lenses) has played a pivotal role in scientific history. Between heaven and earth, from the vast distances of astronomical observation to the minute precision of microscopic exploration, humans rely on glass to "look", to "see", and ultimately to "discover." Mosses and glass, as materials for creation and mediums for contemplation, have become effective clues for me to imagine a departure, and thus to turn back and feel the world around me anew.

The sense of awe toward ontological mystery (Gomel 1995) arises from describable sensory experiences and is inevitably inspired and influenced by modern and scientific understandings as well. In my artistic journey, I have increasingly regarded this complex emotional process as a companion in my personal quest for truth and life's meaning, engaging in dialogue with it and expressing it through the creative process of my artworks, striving to present a poetic articulation of the "direct knowledge" of the unknowable.

In the classical text of Chinese Taoism, Dao De Jing (道德經), it says,

Always without desire we must be found, 故常無欲,
If its deep mystery we would sound; 以觀其妙;
But if desire always within us be, 常有欲,
Its outer fringe is all that we shall see. 以觀其徼. *(Dao De Jing).*

The understanding of Dao emphasizes intuitive knowledge that transcends rational analysis. This means that people can perceive the subtleties and boundaries of the Dao through the mindset of "emptiness" (無) and the phenomena of "being" (有). Direct knowledge is obtained through intuitive realization, distinct from indirect knowledge gained through logical reasoning or empirical induction. It is a kind of immediate, present-moment comprehension, a direct experience and grasp of the Dao (道) within the practice of daily life.

In essence, awe for the feeling of mystery and the advocacy of scientific rationality are not contradictory; rather, the awareness of ontological mystery is interconnected with the construction of moral sentiment, aesthetic experience, and cognitive knowledge. The sense of mystery that arises between humans and objects is beautiful, and this aesthetic experience is genuinely exhilarating and intellectually stimulating. Thus, we ought to embrace and explore mystery with greater fervor and find joy in doing so. Once we begin to comprehend what was once shrouded in mystery, and we gain deeper understanding and resonance with the complex world around us, a path opens toward both the infinite pursuit of scientific discovery and through the realms of art.

The connection between aesthetic sensitivity and scientific cognition thus helps us better understand the vast, all-encompassing world. Knowledge is something humans attain through both sensory perception and rational thought; without rationality (categories of understanding), we would be incapable of comprehending the world. The development of modern natural science has transformed experience into reason, expanding our possibilities for understanding the external world—and ourselves: what can be explained, what has yet to be explained, and what is not necessarily wrong (refers to cognitive error/moral error).

Mastering knowledge of the natural sciences alone, however, does not necessarily lead to the formation of ecological moral values, nor the establishment of principles and systems guiding right actions (Hargrove 1989). Natural science tends to emphasize "consequentialism," pursuing outcomes and treating the process of action merely as instrumental, thus overlooking the intrinsic value inherent in the process itself (Andreozzi 2025). In contrast, aesthetics and ethics of care stress the ability to embody and practice ideas beyond knowledge (Rozzi and Tauro 2023). From systems, communities, and populations to individual organisms, nothing in the natural sciences better embodies the phrase "all-encompassing" than plants. Botanical knowledge reveals that the extraordinary adaptability of mosses enables them to thrive across a vast range of terrestrial ecosystems: from deserts to rainforests, from plateaus and polar regions to the tropics, and even in every corner of densely populated modern cities (Goffinet and Shaw 2009). Their poikilohydric mechanisms allow them to autonomously regulate states of activity and dormancy (Goffinet and Buck 2013). However, from botanists, I have also learned that some moss species are now at risk of extinction due to the impacts of climate change and what this risk implies.

Guided by botany and ecology, a mysterious world of imagery and a dazzlingly colorful, expansive realm have unfolded before me, lingering more and more in my mind. It seems that while moss traditionally symbolizes "seclusion and desolation" in cultural contexts, conveying an appreciation of nature's austere beauty, modern scientific understanding has inspired even richer sensory experiences.

The contemporary study of the role of mosses in ecological environments has led me to reflect: Is it the limitations of scientific concepts and methods that make such minute things difficult to observe and depict? Or did the wise ancients possess insights into the natural order behind these phenomena that transcended botany? For me, using the depictions of nature preserved in ancient paintings, calligraphy, poetry, and writings—along with the lingering sensitivity to nature they convey—as clues, I find myself captivated by the imaginative search for the environments where people once lived. How, after all, can we truly "touch" (sense) the world as it was "experienced" by those who came before us?

With these questions in mind, I began my artistic exploration centered on the imagery of moss, embarking on a poetic journey in search of freedom. I chose moss as a unique thread to examine the gaps in my ways of seeing nature and experiencing beauty, and to attempt to expand the connections between these inconspicuous plants and civilization, the individual's sense of life, and the artistic expressions that arise from it.

18.4 The Imagery of Moss in Glass Art: Creation and Expression

18.4.1 Towards Wild: Ruins in Profusion

The formation of contemporary human aesthetic experience is deeply and intrinsically connected to modern scientific understanding (Chiarini 2026; Rozzi 2026). Guided by bryologists ("mosses researchers"), I took "*Mosses: The Trailblazers on Ruins*" as my theme, using the presence of mosses in urban ruins as a magical thread to explore the beauty of the wild. Through this, I sought to reexamine new ways of looking at urban development and natural restoration, and to reflect on the connections between humanity and nature, wilderness and civilization.

Though tiny in size, bryophytes (mosses, liverworts, and hornworts) are often overlooked by humans, yet they play a crucial role in ecological restoration (Goffinet et al. 2012). The fluids they secrete can slowly dissolve rock surfaces, accelerating weathering (Scheiner 2023). In cities, they enable the formation of a thin soil layer on man-made surfaces that would otherwise be barren, paving the way for the growth of other plants and serving as trailblazers of life. Thanks to their strong adaptability and unique poikilohydric mechanism, mosses alter their states of activity and dormancy in response to environmental conditions, allowing them to survive harsh conditions (Goffinet and Buck 2013). Their populations are vast and widely distributed across various terrestrial ecosystems, from deserts to rainforests, from plateaus and polar regions to the tropics, even flourishing in every corner of densely populated modern cities.

All the pieces in this series were handcrafted using blown glass techniques. In terms of craftsmanship and material language, the works break away from the conventional characteristics of glass art, skillfully combining the fluidity of hot-melt glass techniques with the textures produced by powder-fired glass and the integration of molten glass with metal materials to create a unique tactile effect. Bricks, as symbols of man-made urban ruins and fragile environments damaged by human activity, are depicted with varying states of moss and lichen on their surfaces. These elements reveal how, after destruction, the vast populations of mosses thrive exuberantly across the surfaces, crevices, and remnants of artificial structures. The creative concept draws inspiration from biodiversity conservation principles and contemporary botanical research perspectives, sparking a passionate pursuit of a new aesthetic expression: places with minimal human intervention are not lifeless or monotonous. On the contrary, in such environments, plant ecosystems display a vibrant, untamed beauty, embodying the natural path of "All under the sky there is one breath of life" (通天下一氣爾) in harmonious coexistence (Song 2026) (Figs. 18.2, 18.3 and 18.4).

Fig. 18.2 Moss trail series: unbounded profusion. (Photograph by Shu Meng)

Fig. 18.3 Close-up view. This close-up photo of a glass artwork reveals a textured surface blending dark and metallic hues. Green, glassy droplets cluster together on the smooth surface, creating a striking contrast against the rough, dark background. The photo highlights the interplay of texture and color, with a focus on vibrant green elements that convey the tenacious vitality of moss plants colonizing barren soil. (Photograph by Shu Meng)

18.4.2 Moss-Dotting: The Form of Nature, the Traces of the Spirit

The "Moss-Dotting" (點苔) technique, an important method in the theory of Chinese landscape painting, had already become an established term as early as the Song dynasty (Fig. 18.5). The "dotting" in "Moss-Dotting" serves to organize the composition, enhance spatial depth, and complete the structural balance of the painting. By varying the density and intensity of the dots, artists distinguish between

Fig. 18.4 Trailblazer: ruins in profusion. This round glass vessel features a vibrant blend of green, purple, and orange. Its richly textured surface, adorned with swirling patterns and speckles, creates an artistic, vibrant sense of life. These vibrant colors are a celebration of life itself, expressed through the flourishing growth of moss within the ruins. (Photograph by Shu Meng)

foreground and background elements. However, the naming and practice of the "Moss-Dotting" technique were not about depicting moss plants as subjects for observation or painting. Instead, it reflects the essence of the traditional Chinese artistic philosophy of "expressing meaning through imagery" (象者所以存意) and "capturing the spirit rather than the form"(得意而忘象) (Wang and Lou 2011). This phenomenon embodies the philosophical and aesthetic logic of traditional Chinese painting, where "the brush and ink themselves convey the artistic realm." In this sense, the act of "dotting" is not merely a compositional tool but also an emotional expression of the artist's inner state.

Throughout Chinese history, many renowned painters developed their distinctive approaches to "Moss-Dotting." Dong Yuan (董源) pioneered the combination of "Hemp-Fiber Texture Strokes" (披麻皴) and "Rain-Dot Moss" (雨點苔), using densely packed small vertical dots to portray the moist, lush quality of Jiangnan's mountain stones. In his works, such as *Xiaoxiang Scroll* (瀟湘圖), the moss dots and texturing strokes merge seamlessly into one (Zhou 2008). In my practice, I seek to echo this method and spirit by expressing the idea that the brush and ink themselves are the artistic realm through three-dimensional glass art, thus continuing the ancient pursuit of merging with the natural landscapes envisioned by the ancients.

From the brushwork of "moss-dotting," we can see the ancient artists' individualized approach to painting: each ink dot is both the "form" of nature and the "trace"

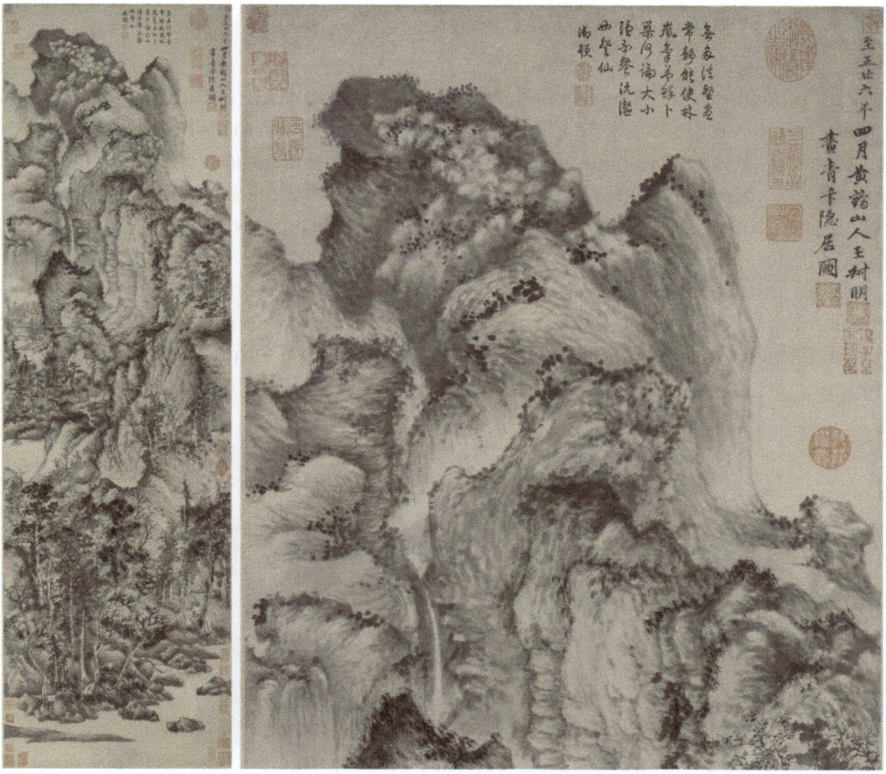

Fig. 18.5 Full and partial views: *Dwelling in the Qingbian Mountains* (青卞隐居图轴, Hanging Scroll), by Wang Meng (Yuan dynasty), who applied "moss-dotting" technique in this painting. Dimensions: Height 140.6 cm, Width 42.2 cm. Collection of the Shanghai Museum. (Photograph by Shu Meng)

of the spirit, ultimately refining the essence of nature "somewhere between likeness and unlikeness." This wisdom—merging technique with poetry and philosophy is one of the core characteristics that distinguishes Chinese painting from the Western tradition of realism. How, then, can I convey the true "traces" of my heart through contemporary glass art? How can I express the deep emotional connection between human beings and the environment that I have felt through the magical clue of moss? By blending the traditional wisdom of merging technique, poetry, and philosophy into the creative process of glass art, I have developed my style of glass artistry.

In my glass craftsmanship, the practice of the "moss-dotting" imagery is, in essence, a microcosm of traditional Chinese artistic thought—yet it is inevitably a brand-new creation: I have infused it with the modern consciousness shaped by scientific understanding, achieving this originality. Every splash and every crystal-clear bead that solidifies on the surface of my glass works—each moss dot—is both the "form" of nature and the "trace" of the spirit, ultimately refining the essence of nature "somewhere between likeness and unlikeness."

18.4.3 As Ruin, So Is the Future

There is a close and interdependent relationship between cultural diversity and biological diversity. When people emotionally connect with the vibrancy and vitality of biodiversity, they are naturally more inclined to embrace and pursue the pluralistic values of cultural diversity. Different cultures have different ways of perceiving and valuing nature, and cultural diversity, in turn, promotes the preservation of biodiversity.

Moss, as a magical thread, introduces the realm of nature and ways of living, awakening a renewed awareness of the relationship between individuals and their surroundings, and enhancing a modern consciousness within the context of our times. Figs. 18.6 and 18.7 showcase my subsequent glass art installation works.

Fig. 18.6 Culture·culture medium. (Photograph by Shu Meng)

Fig. 18.7 *Landmines beneath the Earth: war ruins activated by Victims* (detail of Art installation). (Photograph by Shu Meng)

Culture·Culture Medium (Fig. 18.6) critically examines the dangers of scientific progressivism through the absurdity of artistic thinking: inside the super laboratory, glass vessels (Florence flasks)—symbols of Western experimental science—do not allow, do not require, and cannot naturally foster the growth of moss. At most, moss can only be observed as an experimental subject with precisely controlled variables inside the "culture dish" of laboratory apparatus.

This prompts reflection: Will the competitive ambition of modern technology ultimately turn humanity itself into an object of analysis? Will such ambition also treat the natural environment as a meticulously designed laboratory? Are we simultaneously becoming experimental subjects besides being researchers? And also unable to escape the laboratory and the impacts brought by war concepts? *Landmines Beneath the Earth* (Fig. 18.7) expresses another layer of my imagination and contemplation regarding the city as ruin: If the designers of urban planning and artificial constructions, with the utmost rationality, have thoroughly considered both the ways a city could be destroyed and how to prevent such destruction, then perhaps destruction itself becomes inevitable. How should we understand the inevitability of ruin and the possibility of rebirth? Perhaps it is moss that has given me boundless inspiration, resonating with Wittgenstein's philosophical thought: "What can be said at all can be said clearly, and what we cannot talk about we must pass over in silence." (Wittgenstein 2002, p. 3).

Through glass art as my medium, I continue to use mosses as a thread, engaging in an ongoing reflection—a response to my perceptions of the surrounding world and to the broader questions of humanity's existence, presence, and future on this planet.

18.5 Concluding Remarks

Perspectives of contemporary ecological ethics, exchanges and inspiration from natural scientists and scholars of natural philosophy, and the accumulation of my bodily sensory experiences during hiking and traveling led me to the encounter between my glass art and mosses. The mutual inspiration between the philosophy of science and artistic contemplation allows the spirit of life to flourish. In short, reverence for mystery does not mean denying science; rather, it is about maintaining a dual respect for both the "limitations and possibilities" of human knowledge, and this is precisely the spiritual force that drives science to push beyond its boundaries continually.

As stated by Hu and Yang (2026), calligraphy and painting arts of ancient China are the spiritual homeland of the Chinese people. People of every fleeting era have always yearned to empathize with the life and environment of the ancients.

Otherwise, the flourishing of historiography, archaeological art studies, and cultural anthropology—all of which place greater emphasis on individual narratives—would not have occurred. Yet deep space exploration and the expansion of perception through remote sensing technology are, in essence, nothing more than technological checks and balances on the level of national strategy among major powers. Landscape painting, as an important vehicle, carries forward the aesthetic sensibilities of the Chinese people: the desolation, the wildness, expressing the emotional vitality and sense of the times that are often missing from official historical records. This is the inspiration that "Moss-Dotting" has given me.

If circling mountains, following waters, dwelling among forests, pausing, and migrating are the "traces" of humanity through the ages that can be sought and followed, then the "marks" of mosses—recognizable, distinguishable, and verifiable by scientists—can be "depicted" through the five senses and the intuitive perception of artists, transforming sensory experiences into rhetoric, aesthetic emotion, and artistic form—a true blessing.

As long as the landscapes endure, the spirit of Moss-Dotting will flourish.

Acknowledgement The author would like to express her gratitude to Danqiong Zhu for encouraging her to meet with the bryophyte team at the Shenzhen Fairy Lake Botanical Garden, and to Li Zhang, Tuo Yang, Qin Zuo, Shihua Li, Lili Xu, and Liping Zhou for our conversations about bryophyte conservation and moss art. The author is also grateful to Zhu Danqiong, Terrance Caviness, and Ricardo Rozzi for their comments on the manuscript.

References

Andreozzi M (2025) Reimagining ethics. Non-anthropocentric perspectives on morality, Ecology and ethics, vol 7. Springer, Cham

Chiarini VS (2026) How the invisible becomes visible: situating scientific illustration in biocultural communication. In: Zhu D, Rozzi R, Tsuji R, Johnson B, Caviness T, Castro-Jorquera C, Xu L, Zhang L (eds) Linking arts with biocultural conservation, restoration, and communication, Ecology and ethics, vol 8. Springer, Cham, pp 149–169. https://doi.org/10.1007/978-3-032-04830-1_9

Goffinet B, Buck WR (2013) The evolution of body form in bryophytes. Annu Plant Rev 45:51–89. https://doi.org/10.1002/9781118305881.ch2

Goffinet B, Shaw AJ (eds) (2009) Bryophyte biology, 2nd edn. Cambridge University Press, Cambridge

Goffinet B, Rozzi R, Lewis L, Buck W, Massardo F (2012) The miniature forests of Cape Horn: eco-tourism with a hand-lens. University of North Texas Press, Denton

Gomel E (1995) Mystery, apocalypse and utopia: the case of the ontological detective story. Sci Fict Stud 22:343–356

Hargrove E (1989) Foundations of environmental ethics. Prentice Hall, Englewood Cliffs

Hu J, Yang P (2026) A biocultural encounter with the four gentlemen: art education in Chinese ecological and aesthetic traditions. In: Zhu D, Rozzi R, Tsuji R, Johnson B, Caviness T, Castro-Jorquera C, Xu L, Zhang L (eds) Linking arts with biocultural conservation, restoration, and communication, Ecology and ethics, vol 8. Springer, Cham, pp 111–129. https://doi.org/10.1007/978-3-032-04830-1_7

Lao Tzu (n.d) Dao De Jing. trans. of Chinese: Legge J. Accessed through Sturgeon D (2019) Chinese Text Project: a dynamic digital library of premodern Chinese, Digital Scholarship in the Humanities

Leiss W (2023) The domination of nature, New edn. McGill-Queen's University Press, Montreal

Rozzi R (2023) Inter-species and inter-cultural encounters: the biocultural education program of the Omora Ethnobotanical Park. In: Rozzi R, Tauro A, Wright T, Avriel-Avni N, May RH Jr (eds) Field environmental philosophy: education for biocultural conservation, Ecology and ethics series, vol 5. Springer, Dordrecht, pp 153–174. https://doi.org/10.1007/978-3-031-23368-5_10

Rozzi R (2026) The contributions of Eugene Hargrove to environmental ethics and aesthetics: a biocultural tribute. In: Zhu D, Rozzi R, Tsuji R, Johnson B, Caviness T, Castro-Jorquera C, Xu L, Zhang L (eds) Linking arts with biocultural conservation, restoration, and communication, Ecology and ethics, vol 8. Springer, Cham

Rozzi R, Tauro A (2023) The multiple lenses of ecotourism with a hand lens: fundamental concepts and practices. In: Rozzi R, Tauro A, Avriel-Avni N, Wright T, May RH Jr (eds) Field environmental philosophy, Ecology and ethics, vol 5. Springer, Cham. https://doi.org/10.1007/978-3-031-23368-5_3

Rozzi R, Massardo F, Anderson C et al (2006) Ten principles for biocultural conservation at the southern tip of the Americas: the approach of the Omora Ethnobotanical Park. Ecol Soc 11(1): 43. Available via http://www.ecologyandsociety.org/vol11/iss1/art43/

Scheiner SM (2023) Encyclopedia of biodiversity, vol 4, 3rd edn. Academic, pp 690–706

Song Y (2026) Interspecies Co-inhabitation: A Dialogue between Chinese philosophical traditions and Biocultural Ethics. In: Zhu D, Rozzi R, Tsuji R, Johnson B, Caviness T, Castro-Jorquera C, Xu L, Zhang L (eds) Linking arts with biocultural conservation, restoration, and communication, Ecology and ethics, vol 8. Springer, Cham, pp 95–109. https://doi.org/10.1007/978-3-032-04830-1_6

Vezzani P (2026) "Fragility of encounters" in subantarctic biocultural diversity. In: Zhu D, Rozzi R, Tsuji R, Johnson B, Caviness T, Castro-Jorquera C, Xu L, Zhang L (eds) Linking arts with biocultural conservation, restoration, and communication, Ecology and ethics, vol 8. Springer, Cham, pp 207–222. https://doi.org/10.1007/978-3-032-04830-1_13

Wang B, Lou YL (2011) Zhou Yi Zhu (with Zhou Yi Lüe Li) (周易注 (附周易略例)). Zhonghua Shu Ju, Beijing

Waston (1971) Deer Fence. In: Weinberger E, Paz O (eds) Nineteen ways of looking at Wang Wei: how a Chinese poem is translated. Mount Kisco, New York, pp 24–25

Wittgenstein L (2002) Tractatus logico-philosophicus. Routledge, New York

Zhou Y (2008) The poetic and picturesque beauty of "eight scenes of Xiaoxiang"——also on the influence of Chinese painting on Japan ("潇湘八景"的诗情画意——兼论中国绘画对日本的影响). Chinese Cult Stud (中国文化研究) 01:170–177. https://doi.org/10.15990/j.cnki.cn11-3306/g2.2008.01.015

Zhu D (2023) Bridge the channel, enhance the inclusivity: a comparison between flagship species-centered and moss-centered conservation in Chile and China. In: Rozzi R, Tauro A, Wright T, Avriel-Avni N, May RH Jr (eds) Field environmental philosophy: education for biocultural conservation, Ecology and ethics, vol 5. Springer, Dordrecht, pp 457–482. https://doi.org/10.1007/978-3-031-23368-5_28

Zuo Q, Zhang L (2026) Small but essential: how to introduce the public to easily overlooked organisms? In: Zhu D, Rozzi R, Tsuji R, Johnson B, Caviness T, Castro-Jorquera C, Xu L, Zhang L (eds) Linking arts with biocultural conservation, restoration, and communication, Ecology and ethics, vol 8. Springer, Cham, pp 243–263. https://doi.org/10.1007/978-3-032-04830-1_15

Chapter 19
Biocultural Co-inhabitancy: An Examination on Scientific and Indigenous Nomenclature of Bryophytes

Danqiong Zhu and Terrance Caviness

Abstract Names, particularly those given to beings in the natural world, are more than just stagnant labels, they often reflect the worldview and knowledge of the culture that gave the name. Naming beings in the natural world is inherently a biocultural practice. They are imbued with history, the context of place, and interrelatedness to others. In this study we explore the intricate relationships between scientific and Indigenous nomenclature of bryophytes, emphasizing the cultural, ecological, and spiritual meanings embedded in moss names. Through field investigations in southern Chile and a comparative analysis of taxonomic systems, we highlight how scientific classifications reflect evolutionary relationships, while Indigenous names embody place-based, habitat-specific, and worldview-driven understandings. By examining these diverse naming systems, we propose a pluricultural approach that fosters biocultural co-inhabitation, promoting mutual respect and ethical engagement with mosses and their ecosystems. Ultimately, we call for integrating scientific and Indigenous knowledge to enrich our collective relationship with the other-than-human world. Naming itself, and the exchange of nomenclature across various bodies of knowledge that contain the message of biocultural co-inhabitancy, becomes a kind among arts.

Keywords Biocultural co-inhabitancy · Binomial nomenclature · Indigenous nomenclature · Bryophytes · Polyculture

D. Zhu (✉)
Sub-Antarctic Biocultural Conservation Program, Department of Biological Sciences, University of North Texas, Denton, TX, USA

Cape Horn International Center (CHIC), Universidad de Magallanes, Puerto Williams, Chile
e-mail: Danqiong.Zhu@unt.edu

T. Caviness
Department of Environmental Biology, SUNY College of Environmental Science and Forestry, Syracuse, NY, USA
e-mail: tcaviness@esf.edu

D. Zhu et al. (eds.), *Linking Arts with Biocultural Conservation, Restoration, and Communication*, Ecology and Ethics 8,
https://doi.org/10.1007/978-3-032-04830-1_19

19.1 Introduction

The journey of this chapter started with a meeting of a Chinese field environmental philosopher, Danqiong Zhu, an American bryologist, Terrance Caviness, and a Chilean ecologist and field environmental philosopher, Ricardo Rozzi, in southern South America. Bryologists are professionals who are adept at observing, recording, classifying, and studying bryophytes. A field environmental philosopher is a person who values direct engagement with the environment and encourages ethical relationships in a contextualized, situated way (Rozzi et al. 2010). In our case, Rozzi has designed the 3Hs model (namely co-in-Habitants, Habits, and Habitats, to develop his proposal for a biocultural ethics (Rozzi 2010, 2012). His conceptual and methodological approaches integrate transdisciplinary knowledge such as ecology, Traditional Ecological knowledge (TEK), ethics, aesthetics to design novel educational and interactive field activities for sustainability (Rozzi 2013; Rozzi and Tauro 2023).

When it comes to names, bryologists employ scientific names to identify and designate species, which are also essential to the study of ecological interspecific relationships (Rozzi et al. 2008a). Names are more than linguistic labels since they may carry cognitive, practical, spiritual, or ethical connotations (Lewis et al. 2018). This study is about how a bryologist and a field environmental philosopher learn from each other, and more importantly, from mosses through the links between their names and animacy revealed by their names.

From December 2019 to January 2020, we conducted field investigation of bryophytes diversity at three locations in Chile: Senda Darwin Biological Station on Chiloe Island, Magallanes National Reserve close to Punta Arenas, and Omora Ethnobotanical Park in the Cape Horn Biosphere Reserve, south (Fig. 19.1). Senda Darwin and Omora Park are part of the Chilean Long-Term Socio-Ecological Research network (Rozzi et al. 2012). Particularly, the Cape Horn biosphere reserve is home to "at least 450 moss and 368 liverwort species making a total of 818 for the Magellanic subantarctic region," 5% of the known species of bryophytes of the planet on 0.01% of the earth's surface area (Rozzi et al. 2008b). Along the Miniature Forests interpretive trail at Omora Park there are locations to stop and learn about the mosses, their biology, ecology, cultural importance, and their names (Rozzi et al. 2006, 2008a).

Toward the end of the Miniature Forests trail the bryologist, Caviness, wanted to unveil the animacy of the mosses through a field experiment. A moss, which appears to be a large patch of yellow, curled, withered, perhaps dead plant material (Fig. 19.2, left), was hydrated with a bottle of drinking water. Almost immediately, it brightens to a vivid green, begins to unfurl and stretch, doubling in size, almost appearing to flourish (Fig. 19.2, right). Within just a minute, it was revitalized and its vibrancy was renewed and visible to the naked eyes.

The Chinese philosopher, Zhu, was committed to focus in conducting programmatic philosophical research based on plants and field studies to contribute rectifying hierarchical worldviews that valued plants less than animals, such as the Aristotelian

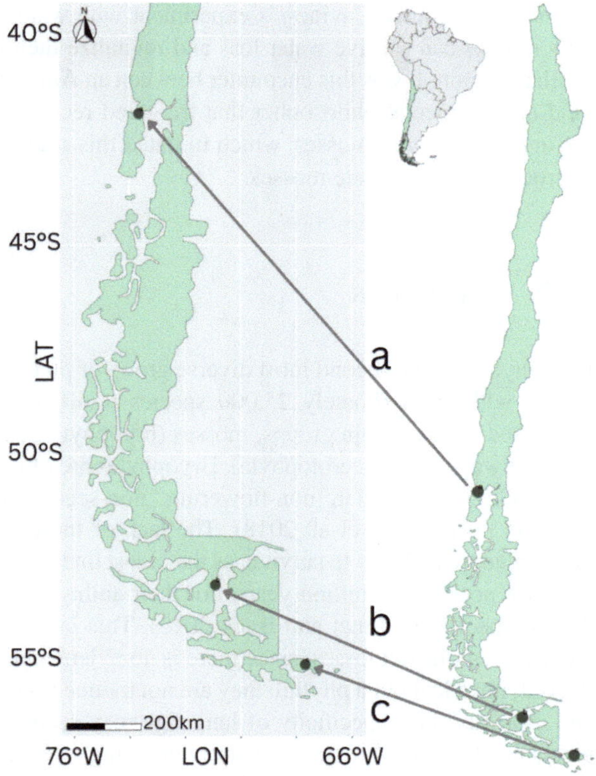

Fig. 19.1 Map of Chile (green) in southwestern South America (inset upper right) showing the three study sites where the authors conducted their field studies: (**a**) Senda Darwin Biological Station, (**b**) Magallanes National Reserve (Punta Arenas), and (**c**) Omora Ethnobotanical Park within the Cape Horn Biosphere Reserve

Fig. 19.2 Hydration experiment in the Omora Ethnobotanical Park, Navarino Island, Chile. (Photos by Danqiong Zhu)

understanding of the organic world, or the biases in modern sciences leading to taxonomic chauvinism (Rozzi 2019). However, Zhu was moved and amazed to see the moss changing color and revitalizing when watered. The bryologist knew the moss very well. The species *Syntrichia anderssonii* is an ideal moss for showing the ecological characteristics, physiological functioning, and the beauty of these tiny non-vascular plants. Caviness's experiment with water revealed to the philosopher how mosses can survive water loss and revitalize themselves when water becomes available again. It was this encounter between an American bryologist and a Chinese field environmental philosopher that triggered reciprocal learning and emotions in the interactions with mosses, which initiated this study of bryophyte names and the microcosmos of animate mosses.

19.2 Bryophytes

Bryophytes are the second most diverse group of plants on the globe after flowering plants, with approximately 25,000 species (Goffinet and Shaw 2009). They are composed of three main groups, mosses (bryophyta), liverworts (Marchantiophyta), and hornworts (Anthocerotophyta). Bryophytes are often negatively defined by what they lack, non-vascular, non-flowering, non-seed producing, as compared with "higher" taxa (Lewis et al. 2018). The lack of these "higher" features has clearly not limited their ability to survive, as the oldest lineage of plants they have existed on earth for nearly 450 million years, nor their ability to diversify exemplified in their 25,000 species (Goffinet and Buck 2013). This lack of features has also created a language unique to mosses; since there is no xylem and phloem, it is not a root but a rhizoid, not a leaf but a phyllid, they are not monoecious or dioecious but monoicous and dioicous. This specificity of language may seem superfluous, but it tells very precisely about the function and structure being referencing.

The lack of these features has led to the evolution of unique adaptations. Without a vascular system they lack roots to draw water up from the soil or a way to retain water, meaning that they must obtain all their water directly from their environment (Goffinet et al. 2012). Their reliance on water drives many of the adaptations and features that define bryophytes. One such adaptation is called poikilohydry, the ability of moss cells to match their water levels atmospheric levels. This allows for cycles of drying and wetting that follow natural weather patterns.

Poikilohydry was demonstrated in the example above with *Syntrichia*. The seemingly magical properties of the moss exhibited here are magnified through understanding the cellular processes. During the drying process cell membranes and organelles shrink and break, collapsing in on themselves as the water leaves the cell. During rehydration, these systems need to be repaired quickly while also attempting to photosynthesize and store more energy, and the ability to repair cells after the trauma of desiccation is a key feature of bryophytes (Goffinet et al. 2012). When conducting the hydration demonstration, we are sure to put enough water so that the moss can recover its losses from hydration and hopefully have a net gain in energy stores.

Another implication of their lack of vascular tissues (and their lack of lignin) is that they have remained relatively small compared to other plants. Without the ability to retain or transport water around their bodies, nearly all their tissues must remain relatively accessible to their environment. This means that moss cells are just a couple of layers thick so that water can easily permeate. Their small size and need for water has led to the use of capillary action to retain water between their small phyllids and stems, and also between each other. What most people see as "moss" covering a rock or the ground is hundreds or thousands of individuals huddled together. This clustering allows for the greater retention of water, but it also creates its own micro-ecosystem, within these colonies there is typically greater humidity, reduced air flow, and insulation from temperature. This refugia from harsher external environments is home to fungi, insects, other invertebrates, algae, and microorganisms (Goffinet et al. 2012).

Another unique adaptation for bryophytes is their diversity of and reliance on asexual reproduction. For most bryophytes a single fragment of a plant has the potential to reproduce into an entire individual. This is a feature called totipotency, meaning whole potential. Many species also produce special types of asexual propagules, for example, brood bodies and gemmae (Goffinet et al. 2012). These are propagules function similar to the fragments where each is a "packet" of cells that can differentiate into an entire identical individual. When speaking about this process in terms of plants, some of the fantastical nature of it is lost. This could be analogous to a person cutting their hair, throwing it into the wind, and miles away tiny clones started to grow.

This may have contributed to their ability to be a pioneer species. Mosses are often the first to grow at a site that is inhospitable to vascular plants (Tishkov 1986). They are able to attach to bare rock since they do not need to draw nutrients from their substrate. Over decades of slow growth and the accumulation of sediments on their bodies they begin to build soil that the next generation of plants will inhabit. Other important ecosystem contributions include acting as a carbon sink, Sphagnum, the genus of moss associated with bogs, is particularly important for this, growing and storing carbon for thousands of years. Studies also show that bryophytes such as mosses can regulate boreal performance, thus represent critical group of organisms for understanding regional resilience under climate change (Turetsky et al. 2010).

Despite their abundance and diversity, bryophytes remain understudied (Rozzi et al. 2008a; Zhu 2023; Zuo and Zhang 2026). Much is unknown about their ecology, bryophyte-bryophyte interactions, their ranges, and impacts from climate change. This could be due to a variety of factors, one of which being their relative difficulty to identify and to work with bryophytes in comparison to vascular plants. Their identification often relies on lab-based features such as cell shape or arrangement, or measurements that are too small to be taken in the field such as phyllid width or length. This requires additional steps, or a deep knowledge of the organisms to identify them in the field.

19.3 The Scheme: The Scientific and Indigenous Nomenclature of Organisms

Naming is classifying. Names are embedded in the scheme of understanding, meaning-making, and co-inhabiting. The invention and employment of names thus varies according to the specific context. Scientific names are usually adopted in academic journals, research reports, and peer-reviewed books, for example, while common names may be adopted in literature, arts, media reports, and everyday communication. In some scenarios, amateur bryologists may invent common names based on the habitats or morphological features of mosses. For instance, some moss gardeners invent names such as "tree moss," "rock moss," "lawn moss," "woodland moss," "feather moss," "fern moss," "rose moss," "umbrella moss," etc. (Schenk 1997, pp. 17–18).

The scientific names of mosses follow the binomial systems which are generally considered to have originated in the writings of Carl Linnaeus. In 1735, Linnaeus published his *Systema Naturae* and created the ranks of hierarchical taxonomy in which the natural world was divided into three kingdoms (Regnum), namely, the Fossil Kingdom, the Vegetable Kingdom, and the Animal Kingdom. Subsequently and successively, he then classified each Kingdom into Class (Classis), Order (Ordo), Genus (Genus), Species (Species), and Variety (Varietas). For more than 250 years, taxonomists have combined this hierarchy and nomenclature to name more than two million species of organisms on Earth, including bryophytes and lichens. Knowing the generic and specific names of a species in the Linnaean hierarchy basically means that biologists have the information regarding the relationships of organisms based on the Linnaean hierarchy. Thus, the defining feature of the Linnaean system is the method that taxonomists use to identify taxon names on the Linnaean hierarchy.

However, in 1994, some vertebrate and herpetological researchers proposed to replace the Linnaean system with a phylogenetic nomenclature system because "the current nomenclatural system's basis in Linnaean taxonomic categories promotes neither explicitness, universality nor stability with regard to the phylogenetic meanings of taxon names" (de Queiroz and Gauthier 1994, p. 27). As the fundamental language and tool of biological communication, it is considered essential that it must be unambiguous, explicit, universal, and stable, as the link between names and taxa must be a one-to-one match. The proponents of the phylogenetic system criticize the non-evolutionary character of the Linnaean system. They claim that the main failure of the Linnaean system is revealed by the "taxonomic practices of unification and division" (de Queiroz and Gauthier 1994, p. 27), which can further lead to changes of scientific names of a species or the designation of a name.

According to Kevin De Queiroz, the aim of the phylogenetic system is to reflect "phylogenetic relationships accurately and efficiently" (de Queiroz 1997, p. 127). In other words, besides the principles of the Linnaean system, such as the non-overlapping taxa and the hierarchy, the phylogenetic system limits the relationship among taxa to the monophyly (one-tribe), namely, all descendants who share the common ancestors should be grouped into one taxa, and descendants of a different ancestor would be grouped into a different taxon. When members of a group of organisms share the same ancestor, they are grouped as an evolutionary clade, i.e., a single branch on the "tree of life." In this regard, the defining feature of the phylogenetic system is the reproductive and evolutionary relationships among species.

Since 2010, the International Code of Phylogenetic Nomenclature has been created and revised to clarify the monophyletic relationships among organisms. In terms of the transition from the Linnaean to the phylogenetic nomenclature, it is fair to claim that rather than the dissimilar taxonomic principles for classification, the binomial system of nomenclature have not changed. With the advancement of species identification, delimitation, taxonomy, and nomenclature, numerous scientific names of mosses have been recorded on platforms such as ITIS (Integrated Taxonomic Information System) or GBIF (Global Biodiversity Information Facility).

Records of Indigenous nomenclature for mosses or other living organisms present a contrasting image. These names are scattered in various fragments of ethnobotanical, ethnomedical, folklores, or oral stories, and because of this, Indigenous nomenclature can be studied from different perspectives. One of the common approaches is to combine case studies with ethnobotany and ethnomedicine to investigate the nomenclature system of a particular Indigenous group, often utilizing interviews, collections, and documentation to creating databases of native plant names, as well as analyzing the underlying structure of botanical nomenclature. For example, Addi and collaborators discovered that the plant folk nomenclature of the Yi People (彝族) in Xiaoliangshan, Yunnan province, China combines both binomial and non-binomial structure to name plants (Addi et al. 2022). Besides the name structure, they also discussed the characteristics of Yi nomenclature of plants (e.g. plant characteristic, habitat, plant-use, and cultural attributes), and linked Yi nomenclature with their traditional ecological knowledge (TEK), their relations to native plants, and how conservation of Indigenous nomenclature could contribute to biodiversity conservation. Similar studies also have been conducted on the Newe in northwestern North America (Chamberlin 1909), Yanomami Indians of Brazil (Milliken and Albert 1996), Mongolians in Ejina desert area (Khasbagan and Soyolt 2008), Tibetan people (Guo et al. 2022), etc.

Another approach would proceed by focusing on a particular group of plants to investigate the Indigenous biosystematics. This approach presents a cultural, intellectual, and linguistic complexity. Indeed, studies about indigenous nomenclature of Palms of Fiji (Doyle and Fuller 1998), African rattans (Sunderland 2004), Andean potatoes (De Haan et al. 2006), and general plant taxa (Wang et al. 2019), not only have documented names, taxa, and families of indigenous botanical nomenclature, but they have also interpreted interspecies relationships.

Based on empirical studies, we examine general principles of ethnobotanical nomenclature through the lens of comparative ethnobiology. American anthropologist Brent Berlin, a pioneer and leading researcher in ethnobiology has identified widespread regularities in the classification and naming of plants and animals among peoples of traditional societies. In his groundbreaking book *Ethnobiological Classification: Principles of Categorization of Plants and Animals in Traditional Societies* (Berlin 1992), he integrated the principles of ethnobotanical classification and he also discussed the evolution of ethnobotanical categories. Following this comparative perspective, Canadian ethnobotanist Nancy Turner has investigated processes through which indigenous people of the Pacific Northwest have developed their ethnobotanical and ethnoecological knowledge systems. In her two volumes work *Ancient Pathways, Ancestral Knowledge: Ethnobotany and ecological wisdom of Indigenous people of northwestern North America,* Turner (2014) presents a comprehensive study of the ethnobotanical knowledge of northwestern North American Indians. She covers indigenous nomenclature and taxonomy, ethnomedicine, cultural institutions related to ethnobotanical knowledge and their transmission, exchange, management, and the worldviews and belief systems that ethnobotanical knowledge exhibits. Remarkably, some of the indigenous names associated with bryophytes and related plants are only marginally documented (in one of the book appendices).

Based on these publications, we have learned that Indigenous systems of nomenclature are deeply embedded in their languages, cultures, worldviews, norms, and everyday practices of subsistence. It is for these reasons that we agree with the biocultural perspective that affirms that names of organisms allude to diverse frameworks of understanding and subtly different expectations of the co-inhabiting relationships among species (Rozzi and Tauro 2023).

19.4 Epistemic Dimension of Bryophytes Nomenclature

Human-Moss interrelationships firstly are manifested at the epistemic dimension, namely, how mosses are recognized, identified, and understood. We do not intend to create a new dualism or dichotomy between Western sciences and Indigenous knowledge. Both are pluralistic, varied, and heterogeneous. However, there are some justifications for discussing Indigenous ways of knowing because various Indigenous knowledge systems share some common features in understanding interspecies relationships. As ethnobotanist Robin Wall Kimmerer once noted that,

> In indigenous way of knowing, we say that a thing cannot be understood until it is known by all four aspects of our being: mind, body, emotion, and spirit. The scientific way of knowing relies only on empirical information from the world, gathered by body and interpreted by mind. (Kimmerer 2003, p. vii)

We agree with Kimmerer's characterization of scientific knowledge. A scientific body of knowledge, including bryological scientific knowledge, must be gradually constructed through countless field studies, specimen collections of generations of biologists, comparably complete documentation and inheritance through publications, and probably molecular DNA matches nowadays, etc. In this sense, the mind understands and interprets the empirical evidence collected by the body.

We also agree with Kimmerer's statement about Indigenous ways of knowing. They are more holistic and intersubjective without the typical presence of mind/body dualism. Names tell. Kimmerer knows this firsthand, since she is a member of the Citizen Potawatomi Nation, which protects Native American spiritual beliefs, historic values, and the celebration of these unique traditions, language, and sovereignty.

Bryologists developed systematic knowledge on bryophytes' life cycle, morphology, anatomical structure, sexual strategy, ecophysiology, photosynthesis, interspecies relationships with birds, insects, mammals, habitats, and the ecological services they offer. These scientific studies reveal a vivid and vibrant moss cosmos in which these tiny plants are ecologically linked with other organisms, including humans, and with environments composed of the full range of organisms and inorganic matter through ecological interactions. Some scientific names disclose relevant information; although, most scientific names are notorious for being tedious and meaningless.

Among those scientific names whose meanings can be traced, they contain information related to morphological features which were usually observed either by naked eyes (parts of a body) or hand-lens (extension of a body). For example, *Brachythecium subplicatum* (Fig. 19.3, left), a moss which is distributed in southern South America and New Zealand, derives its generic name from its morphological characteristics, as *Brachythecium* is dissected into *Brachy-* (short, stout) and -*thereon* (little vessel, container), possibly alluding to the short and fat leaves while they harbor adequate water (Fig. 19.3, right) (Goffinet et al. 2012, p. 208). Generic names such as *Orthotrichum* [*ortho(upright) + trich (hair), Platyneuron* [*platy (flat, broad) + neuron (nerve)*], *Philonotis* [*philos (love) + notis (moisture)*] are like the generic names of *Brachythecium subplicatum,* as they all contain information of the

Fig. 19.3 Left: *Brachythecium subplicatum,* Jaeger, 1878, collected in Chile, by The New York Botanical Garden (licensed under http://creativecommons.org/licenses/by/4.0/). Right: Leaf Cluster of Rough Stalked Feather Moss *(Brachythecium rutabulum),* growing on a boulder in an urban environment (busy filling station access road), Uttoxeter, Staffordshire, December, 2017. Photographer: Barry Turner. Both photos are in the public domain

morphological features or the habits of the species. The vibrancy, energetics, and vigor of mosses revealed with the aid of a microscope or a magnifying glass, are contained in these generic names.

Names could also contain information of interspecies interactions, mainly how people interact with other species. *Hypnum skottsbergii,* a moss that inhabits tree trunks and logs with leaves that are strongly bent and curved in one direction, derived its genus name *Hypnum* from its Greek word *Hypno* (sleep), alluding to the ancient use of this genus as "filler for pillows and cushions" (Goffinet et al. 2012, p. 220). *Grimmia trichophylla*, common name as "Grimmia dry rock moss," a cosmopolitan moss that grows on bare rocks, was given its genus name *Grimmia* after John F. K. Grimm (1737–1821), a German physician and botanist. Jacob F. Ehrhart (1742–1795), who was also a German botanist and a student of Carl Linnaeus, coined the genus name *Grimmia* in memory of Grimm (Meagher 2011). The generic name reflects nothing about the morphological features or habits of the species. However, its specific name *trichophylla* combines *trichos-* (hairs) and *-phyllum* (leaf), referring to its morphological features. Giving bryophytes a generic name after a botanist was also a strategy to name them.[1] It is usually to honor a bryologist's outstanding contribution to the field.

The accumulation of scientific knowledge and the proliferation of binomial scientific names are beyond the reach of Indigenous traditional knowledge and nomenclature. Around 20,000 species of bryophytes on the globe, each has their scientific names (Goffinet and Shaw 2009). While reflecting on the dominance of scientific knowledge, one should not lose sight of the interactions between scientists and the world of organisms. In the case of our subject, it is the efforts of generations of bryologists (amateurs and professionals) that present a vibrant, silent, serene, and enduring moss world.

The association between the dominance of scientific knowledge and colonialism is beyond the scope of this paper. However, it is important to mention that losses of Indigenous languages and cultures are derived from the genocide during the colonial period (Veronelli 2015), the absence of written records of Indigenous culture/knowledge, urbanization and migration to urban areas (Moya-Santiagos and Quiroga-Curín 2022), or the replacement of Indigenous education by global biocultural homogenization (Rozzi 2018). Consequently, today we are challenged in our research efforts about Indigenous names of bryophytes and other organisms.

We can address the former challenge by examining bryophyte names given by Indigenous people across the world. These traditional cultures have observed and named mosses paying close attention to their morphological characteristics, their habitats, and cultural relationships of co-inhabitation with them.

---

[1] Nor is this a technique unique to bryology or botany. It also happens in other disciplines, such as in astronomy, where newly discovered stars or phenomena are named after distinguished astronomers, artists, and writers, etc. It is well-known that a lunar impact crater was given the name after the Danish astronomer Tycho Brahe (1546–1601).

For the Newe (Gosuites or Goshute, in English), an Indigenous group who lives in the desert region southwest of the Great Salt Lake, the species of *Bryum sp.* is named as *so'-go-ba-gwĭp* or *so'-ko-ri-bo-ûmp*. These two names refer to life habits. The first name *so-go* means earth, and this species grows on the ground. The second name includes the term *so'-ko-ri* (deer), referring to the fact that the species is eaten by deers. *Ûmp* frequently occurs in the names of plants, conveying the meaning of plants used for some purpose. The species *Bryum sp.* is also widely used on burns by the Newe (Chamberlin 1911). Mosses of the genus *Hypnum are* called *pá-oñ-gûp,* meaning moss in general. Apparently, the Newe name *pá-oñ-gûp* is used as part of the naming of other plants including vascular plants such as *Polygonum amphibrium*, *Polytrichum juniperinum,* and *Rachunculus aquatilis* (Chamberlin 1911). Some bryologists in the early twentieth century attempted to match scientific and Indigenous Newe names of mosses, but only a few species could be matched (Flowers 1957). The Newe names of plants reveal their observations and classifications of plant habits and habitat, showing their keen observation of the places that plants inhabit, such as floating upon the surface of water, growing over the rock, on mountains, submerging in water, etc.

Other nomenclatures of mosses in traditional cultures pay attention to the plants' habitats. *The Collection of Mosses* is a Chinese book compiled by Wang Xian (1721–1771), which exclusively collected literature about bryophyte-related plants. The book covers medical uses, habitats, odes, and poems. A remarkable feature of this book is that it not only classifies and names mosses according to their habitats, but it also includes verses about the mosses. In other words, all bryophyte-related plants were classified by the Chinese word, "處所" (Pinyin as Chù-Su⍰), which literally refers to the dwelling place or habitat. Correspondingly, all bryophyte-related plants were classified by various kinds of places. In this regard, all mosses observed on old tombs would be named as *Gŭ-Zh⍰ng-Tái* (古冢苔, literally old tomb moss), and mosses occupying trails would be name *Jìng-Zhōng-Tái* (徑中苔, Trail moss). The variety of dwelling places reveals both the diversity of habitats and the complexity of interspecies interactions. Another key Chinese text, completed between 1552 and 1578, is the *Compendium of Materia Medica*, or *Běncăo Gāngmù* (Pen-ts'ao Kan-mu, 本草綱目). This work illustrates traditional Chinese herbology and medicine, which describe the habits, habitats, odors, morphological features of more than one thousand plants. Among these plants, 16 types are bryophytes-related plants (Li 2022). Among them, mosses dwelling on old roofs are called *Wă-Sōng* (瓦松, Pine on the roof tile, Fig. 19.4), *Wă-Xiăn* (瓦蘚, Moss on the roof tile), or *Wū-Yóu* (屋游, Roof drift).

Mexico is one of the countries with the highest biocultural diversity (Franquesa-Soler and Azahara Mesa-Jurado 2026. There are records of indigenous names of non-vascular plants. Multi-linguistic research has documented names such as: *Con* and *Tsot-zé* (in Tzelta Indigenous language); *Tsa'nk* (Mixe); *Xibeu' Mbaxhs*, od *Yamasa* (Zapotec); *Tecata* (Spanish) (Berlin et al. 1974; Delgadillo 2000). *Mbaxhs* and *Yamasa* may also apply to liverworts and lichens as well. Indigenous people in Mexico often classify mosses according to the "substrates" or habitats where they grow (Hernández-Rodríguez and Delgadillo-Moya 2021). For example, mosses that

Fig. 19.4 *Wǎ-Sōng*(瓦松), which means "Pine on the roof tile." (Photo by Li Zhang)

grow on rocks are called *Teamoxtli* in Náhuatl, the language spoken by about 1.7 million Nahua people, and most of whom live in Central Mexico. *Teamoxtli* literally means rock-inhabiting moss, which was illustrated by the drawing in the Aztec manuscript (Fig. 19.5). Another example is offered by the Tzeltal language. Mosses are called *Con,* those living on trees are called *Con te?* (Tree moss), and those living on ground are called *Con lumilal* (Ground moss) (Hernández-Rodríguez and Delgadillo-Moya 2021).

Kimmerer (2013, p. 261) stated that "the history of the plants is inextricably tied up with the history of the people, with the forces of destruction and creation." European colonialism decimated Indigenous languages, including ecological knowledge and their names for native organisms. Given the deep links between biodiversity and cultural diversity, so the question now is, in the face of more than 20,000 scientific names for bryophytes, as well as a patchwork of Indigenous names, we ask: What vision of the future should we hold in order to conserve the vibrancy of the moss world, and to be native to the land again?

19.5 Nomenclatural Synergies and a Polyculture of Thousands Shades of Biocultural Co-inhabitancy

In *Gathering Moss*, Kimmerer (2003) describes a case where the instrumental value and intrinsic value of bryophytes were oddly and abusively twisted together. At a wealthy owner's invitation, she was asked to "create an exact replica of the flora of

Fig. 19.5 *Teamoxtli* (right in the picture): Moss with Indigenous name to imply the place where it grows. Source: Martín de la Cruz; translated into Latin by Juan Badianus, *The Libellus de Medicinalibus Indorum Herbis, an Aztec herbal manuscript from the 16th century* (1552). Public domain image

the Appalachians" to assure "mosses were included in the restoration" of a native plant garden (Kimmerer 2003, p. 125). Mosses are expected to be instrumentally relocated so that they add up to the authenticity of the garden since they possess unique intrinsic values. According to Kimmerer, "mosses have an intense bond to their **places** that few contemporary humans can understand" (Kimmerer 2003, p. 131). In addition to their dependence on their place and habitat, their time scales are very different from those of humans. They grow not on the time scale of humans, but on their own time scale. The lesson is that, as all other-than-human organisms vary on various time scales, the contemporary exploitation of many species in a commercialized society is solely based on the time scale of humans, seeking efficiency and instant availability. The time scale of other species and the bond between them and their habitats reveal vital ecological links in various ecosystems. Peat moss takes time to grow so that it can serve as a carbon sink and water reservoir. Biocultural co-inhabitancy calls ethically attending to the time scale of bryophytes, and many other-than-human beings.

Kimmerer (2003) presented an interactive study of *Tetraphis,* and she wanted to solve the conundrum of what factors made the moss choose between sexual and asexual strategies to reproduce. Experimental designs and studies that have long been impenetrable yielded unexpected results when researchers faced their studies and approached the problem from the perspective of the moss. For Kimmerer, success rests on the subjectivity of the moss, expressing their secret to a scientist on their own time scale. The difference between a scientist's "discoveries" and mosses' "expressions" reveals a divergence of human-bryophyte interaction. As a scholar dedicated to the integration of indigenous ecological knowledge with scientific knowledge, Kimmerer (2013, pp. 158–159) affirms that scientific experimentation is "not about discovery but about listening and translating the knowledge of other beings."

Most scientists possess a "discovery" mentality, and few are able to have research subjects express themselves as Kimmerer does. It is also through her insight into the expressiveness of *Tetraphis* that the synergy between the scientific and the Indigenous knowledge is revealed. In this regard, Indigenous knowledge is the fruit of "generations of data collection and validation through time builds up to well-tested theories" (Kimmerer 2013, p. 159). She articulates her envision of a polyculture of complementary knowledge to replace the intellectual monoculture of science with the metaphor of the three sisters, namely, corn as traditional ecological knowledge, bean as curious science, and squash as the ethical habitat for coexistence and mutual flourishing (Kimmerer 2013, p. 139) .

Polyculture is what we expect for the scientific and Indigenous nomenclature of bryophytes and many other organisms. Scientific (encompassing both Linnaean and phylogenetic) nomenclature of bryophytes entails classifying them, and it carries the idea that despite the complexity of molecular process and laboriousness of reconstruction, it is possible to develop a planetary-scale understanding of all life forms through a single "tree of life" (Delsuc et al. 2005; Hinchliff et al. 2015). *Blindia magellanica, Sphagnum magellanicium, Ulota magellanica,* and *Gackstroemia magellanica* are the scientific names of four species of mosses native to the sub-Antarctic Magellanic ecoregion(including Torres del Paine, Cape Horn, Tierra del Fuego Island, and the Strait of Magellan) (Goffinet et al. 2012). The specific names "magellanica" alludes to the southernmost region of the Americas, which is characterized by its mountains, fjords, glaciers, channels, lakes, and rivers (Rozzi et al. 2012). The wonder of scientific knowledge is that by inventing concepts such as geographic distribution, endemic species, cosmopolitan species, and so on, information of flora based in various places can be integrated into a relatively coherent and universal body of knowledge.

Native American philosopher Brian Burkhart (2019) introduced the contrasting terms of "delocality" and "locality," which help us to understand an innate limitation that science has in its aim for a planetary apprehension of the living world. Burkhart defines delocality as a modality of "floating free from the land," which works with universality and abstraction towards an epistemic detachment of the land. In contrast, Indigeneity is tied more literally/deeply to locality. The land, knowing, and meaning are rooted in a single voice of culture or nation (Burkhart 2019).

From these perspectives, we can understand that Indigenous names such as *pá-oñ-gûp* (water-weed moss) or *q*^{w}*zem-éyq* (wood-tree moss) do not refer to a precise geographic region, but to the geographic range in which the Indigenous people inhabited or engaged in their activities. Indigenous knowledge of native plants and their ecological relations have often amazed ethnobotanists, as expressed by Chamberlin when he refers to the Newe's knowledge about the native plants:

> Living close to nature and impelled by strict necessity, they knew the plants of their region with a thoroughness truly surprising. From root to fruit they knew the plants in form and color, texture and taste, and according to season and habitat (Chamberlin 1911, p. 337).

These names of native plants disclose the deep connection that Indigenous people have to their dwelling places. The weight that places or localities have for Indigenous people, is illustrated in Chamberlin's description of the Newe and their habitats:

> The home of the Gosiute Indians was formerly all of the generally desert territory bordering Great Salt Lake on the south and extending westward into eastern Nevada. To the passing traveler this whole region … appeared so utterly desolate ... Yet the Gosiute ... In the truer, deeper sense still his are its gray desert stretches, its rugged mountains and steep-walled canons, its scattered springs, hidden seeps, and slender streams, about which cluster a host of legends and traditions he treasures from the past and teaches to his children's children; its animals and plants, the haunts and habits and uses of which it has been his love and much of his practical education to know …(Chamberlin 1911, p. 331).

Nobody understands this desert territory, the native flora and fauna, their habits better than the Newe do. This Native American culture has inhabited the desert for millennia and has maintained this locality through their intergenerational biocultural practices in the desert. In this deep connection with the land, an entirely different understanding of being animated and alive is revealed. In Burkhart's indigenizing of environmental ethics, "being alive or being in relation... is what gives a thing sacredness" (2019, p 200). Terms such as intrinsic and instrumental values would be unnecessary since Indigenous ethics is not based on values at all. That is also what Kimmerer called "the Grammar of Animacy" which cannot be grasped by the precise language that scientists speak. For Indigenous people, it is not necessary for them to perceive the animacy of mosses through a rehydration experiment, because everything is sacred. Through an Indigenous eye, mind, heart, and spirit, "water, land, and even a day, the language a mirror for seeing the animacy of the world, the life that pulses through all things, through pines and nuthatches and mushrooms" (Kimmerer 2013, p. 55). Through mosses.

Kimmerer had a vision how Linnaeus, the founder of binomial nomenclature of modern botany and zoology, meets Nanabozho, "the Anishinaabe Original Man, our teacher, part man, part manido," an Indigenous spirit:

> They stroll along discussing the names for things. … Linnaeus lends Nanabozho his magnifying glass so he can see the tiny floral parts. Nanabozho gives Linnaeus a song so he can see their spirits. And neither of them are lonely (Kimmerer 2013, p. 55).

That is the nomenclature synergies and a polyculture of biocultural co-inhabitancy. Let science and Indigenous knowledge flourish together.

The name "Anthropocene," which geologists used to mark the geological epoch, for conveying that human beings have shaped the surface of the earth's landscape on a spatial and temporal scale that has generated planetary effects, was rejected in 2024 by the International Union of Geological Sciences. However, in the document "the Anthropocene," it states that "[the concept will] continue to be widely used not only by Earth and environmental scientists, but also by social scientists, politicians and economists, as well as by the public at large" (IUGS-ICS 2024, p. 3). Eileen Crist critically examined the term "Anthropocene," claiming that Anthropocene, with its tenaciously anthropocentric discourse and a self-serving ontology of the world, blocks the possibility and freedom to choose an alternative way of life than this current way of life congested with massive extinction, monoculture, global migrations, ecological degradation, toxic land and watershed. Her question is, "why not choose a name whose higher calling we must rise to meet," such as Thomas Berry's "Ecozoic?"(Crist 2013, p. 142).

Names are never just names. Names are arts, understandings, and appreciation. Names are possibilities. Eradication of names is a loss of possibilities, the alternative ways that we coexist with the other-than-human beings. It would be valuable to understand how peoples in various biocultural traditions have created names for their flora and fauna, and to apprehend the intellectual, worldview, practical, and ethical messages embedded in the meanings of their nomenclature. In this sense, maybe "moss-cosmos" or "bryophyte-verse" are good terms for renewing human-bryophyte relationships in local regions and planetary scale, and foster a biocultural co-inhabitation in the interspecies sense.

Acknowledgments We acknowledge the support from the Chilean National Agency for Research and Development through the Cape Horn International Center, CHIC (ANID/BASAL FB210018).

References

Addi YW, Zhang Y, Ding XY, Guo CA, Wang YH (2022) A study of the plant folk nomenclature of the Yi people in Xiaoliangshan, Yunnan Province, China, and the implications for protecting biodiversity. J Ethnobiol Ethnomed 18(1):18. https://doi.org/10.1186/s13002-022-00504-0

Berlin B, Breedlove D, Raven P (1974) Principles of tzeltal plant classification: an introduction to the botanical ethnography of a mayan-speaking people of highland chiapas. Academic, New York

Berlin B (1992) Ethnobiological classification: Principles of categorization of plants and animals in traditional societies. Princeton University Press, Princeton, NJ

Burkhart B (2019) Indigenizing philosophy through the land: a trickster methodology for decolonizing environmental ethics and indigenous futures. MSU Press

Chamberlin RV (1909) Some plant names of the Ute Indians. Am Anthropol 11(1):27–40

Chamberlin RV (1911) The ethno-botanny of the Gosuite Indians. Mem Am Anthropol Assoc 2(5): 331–405

Crist E (2013) On the poverty of our nomenclature. Environ Humanities 3(1):129–147. https://doi.org/10.1215/22011919-3611266

De Haan S, Bonierbale M, Ghislain M, Nuñez J, Trujillo G (2006) Indigenous biosystematics of Andean potatoes: folk taxonomy, descriptors and nomenclature. In: Moscone EA, Scaldaferro MA, Grabiele M, Cecchini NM, García YS, Jarret R (eds) VI international solanaceae conference: genomics meets biodiversity, vol 745, pp 89–134

De Queiroz K (1997) The Linnaean hierarchy and the evolutionization of taxonomy, with emphasis on the problem of nomenclature. Aliso: J Syst Florist Bot 15(2):125–144

De Queiroz K, Gauthier J (1994) Toward a phylogenetic system of biological nomenclature. Trends Ecol Evol 9(1):27–31

Delgadillo MC (2000) Native moss names in Mexico. Evansia 17:84–86
Delsuc F, Brinkmann H, Philippe H (2005) Phylogenomics and the reconstruction of the tree of life. Nat Rev Genet 6:361–375. https://doi.org/10.1038/nrg1603
Doyle MF, Fuller D (1998) Palms of Fiji—I. Endemic, indigenous, and naturalized species: changes in nomenclature, annotated checklist, and discussion. Harvard Papers Bot:95–100
Flowers S (1957) Ethnobryology of the Gosuite Indians of Utah. Bryologist 60(1):11–14
Franquesa-Soler M, Azahara Mesa-Jurado M (2026) Photovoice: a community-based research tool for wildlife conservation and co-existence. In: Zhu D, Rozzi R, Tsuji R, Johnson B, Caviness T, Castro-Jorquera C, Xu L, Zhang L (eds) Linking arts with biocultural conservation, restoration, and communication, Ecology and ethics, vol 8. Springer, Cham, pp 397–415. https://doi.org/10.1007/978-3-032-04830-1_25
Goffinet B, Buck WR (2013) The evolution of body form in bryophytes. Annu Plant Rev 45:51–89. https://doi.org/10.1002/9781118305881.ch2
Goffinet B, Shaw AJ (eds) (2009) Bryophyte biology, 2nd edn. Cambridge University Press, Cambridge
Goffinet B, Rozzi R, Lewis L, Buck W, Massardo F (2012) Miniature forests of Cape Horn: ecotourism with a hand lens. University of North Texas Press, Denton
Guo CA, Ding X, Hu H, Zhang Y, Yang H, Wang Y (2022) An ethnobotanical study on wild plants used by Tibetan people in Gyirong Valley, Tibet. China J Ethnobiol Ethnomed 18(1):67. https://doi.org/10.1186/s13002-022-00565-1
Hernández-Rodríguez E, Delgadillo-Moya C (2021) The ethnobotany of bryophytes in Mexico. Bot Sci 99(1):13–27. https://doi.org/10.17129/botsci.2685
Hinchliff CE, Smith SA, Allman JF, Burleigh JG, Chaudhary R et al (2015) Synthesis of phylogeny and taxonomy into a comprehensive tree of life. Proc Natl Acad Sci 112(41):12764–12769. https://doi.org/10.1073/pnas.1423041112
IUGS-ICS (2024) The Anthropocene: IUGS-ICS statement. Accessible through https://wwwiugsorg/_files/ugd/f1fc07_ebe2e2b94c35491c8efe570cd2c5a1bfpdf on April 12, 2025
Khasbagan, Soyolt (2008) Indigenous knowledge for plant species diversity: a case study of wild plants' folk names used by the Mongolians in Ejina desert area, Inner Mongolia, Pr China. J Ethnobiol Ethnomed 4:1–6. https://doi.org/10.1186/1746-4269-4-2
Kimmerer RW (2003) Gathering Moss: a natural and cultural history of mosses. Oregon State University Press, Corvallis
Kimmerer RW (2013) Braiding sweetgrass: indigenous wisdom, scientific knowledge and the teachings of plants. Milkweed editions
Lewis LB, Gottschalk-Druschke C, Saldías C, Mackenzie R, Malebrán J, Goffinet B, Rozzi R (2018) Cultivando un jardín de nombres en los bosques en miniatura del Cabo de Hornos: extensión de la conservación biocultural y la ética a seres vivos poco percibidos. Magallania 46(1):103–123. https://doi.org/10.4067/S0718-22442018000100103
Li S, translated by Rong Y (2022) Ben Cao Gang Mu (本草綱目): creeping herbs, water herbs, herbs growing on stones, mosses, cereals, vol V. University of California Press, Oakland
Meagher D (2011) An etymology of Australian bryophyte genera. 2–mosses. Muelleria 29(1):33–61
Milliken W, Albert B (1996) The use of medicinal plants by the Yanomami Indians of Brazil. Econ Bot 10–25
Moya-Santiagos P, Quiroga-Curín J (2022) An overview of indigenous peoples in Chile and their struggle to revitalise their native languages: the case of Mapudungun. Lond Rev Educ 20(1):10
Rozzi R (2010) Field environmental philosophy: regaining an understanding of the vital links between the inhabitants, their habits and the regional habitats. Dialogue Univ 1(2):85–109. https://doi.org/10.5840/du20102011/127
Rozzi R (2012) Biocultural ethics: recovering the vital links between the inhabitants, their habits, and habitats. Environ Ethics 34(1):27–50. https://doi.org/10.5840/enviroethics20123414
Rozzi R (2013) Biocultural ethics: from biocultural homogenization toward biocultural conservation. In: Linking ecology and ethics for a changing world: values, philosophy, and action, Ecology and ethics, vol 1. Springer, Dordrecht, pp 9–32. https://doi.org/10.1007/978-94-007-7470-4_2

Rozzi R (2018) Biocultural homogenization: a wicked problem in the Anthropocene. In: Rozzi R, May RH, Chapin FS, Massardo F, Gavin MC et al (eds) From biocultural homogenization to biocultural conservation, Ecology and ethics, vol 3. Springer, pp 21–48. https://doi.org/10.1007/978-3-319-99513-7_2

Rozzi R, Armesto JJ, Gutiérrez JR, Massardo F, Likens GE et al (2012) Integrating ecology and environmental ethics: Earth stewardship in the southern end of the Americas. BioScience, 62(3): 226-236. https://doi.org/10.1525/bio.2012.62.3.4

Rozzi R (2019) Taxonomic chauvinism, no more! Antidotes from Hume, Darwin, and biocultural ethics. Environ Ethics 41(3):249–282. https://doi.org/10.5840/enviroethics201941325

Rozzi R, Tauro A (2023) The multiple lenses of ecotourism with a hand lens: fundamental concepts and practices. In: Rozzi R, Tauro A, Avriel-Avni N, Wright T, May RH Jr (eds) Field environmental philosophy, Ecology and ethics, vol 5. Springer, Cham. https://doi.org/10.1007/978-3-031-23368-5_3

Rozzi R, Massardo F, Anderson C et al (2006) Ten principles for biocultural conservation at the southern tip of the Americas: the approach of the Omora Ethnobotanical Park. Ecol Soc 11(1): 43. Available via http://www.ecologyandsociety.org/vol11/iss1/art43/

Rozzi R, Arango X, Massardo F, Anderson C, Heidinger K, Moses K (2008a) Field environmental philosophy and biocultural conservation: the Omora Ethnobotanical Park educational program. Environ Ethics 30(3):325–336. https://doi.org/10.5840/enviroethics200830336

Rozzi R, Armesto J, Goffinet B, Buck W, Massardo F et al (2008b) Changing lenses to assess biodiversity: patterns of species richness in sub-Antarctic plants and implications for global conservation. Front Ecol Environ 6:131–137. https://doi.org/10.1890/070020

Rozzi R, Anderson CB, Pizarro JC, Massardo F, Medina Y et al (2010) Field environmental philosophy and biocultural conservation at the Omora Ethnobotanical Park: methodological approaches to broaden the ways of integrating the social component ("S") in long-term socio-ecological research (LTSER) sites. Rev Chil Hist Nat 83(1):27–68. https://doi.org/10.4067/S0716-078X2010000100004

Schenk G (1997) Moss gardening: including lichens, liverworts, and other miniatures. Timber Press, Portland

Sunderland TC (2004) Indigenous nomenclature, classification, and utilization of African rattans. Adv Econ Bot 15:175–213

Tishkov AA (1986) Primary succession in arctic tundra on the west coast of Spitsbergen (Svalbard). Polar Geogr 10(2):148–156

Turner N (2014) Ancient pathways, ancestral knowledge: ethnobotany and ecological wisdom of indigenous peoples of northwestern North America. McGill-Queen's Press, Montreal, Quebec and Kingston, Ontario

Turetsky MR, Mack MC, Hollingsworth TN, Harden JW (2010) The role of mosses in ecosystem succession and function in Alaska's boreal forest. Can J For Res 40(7):1237–1264

Veronelli GA (2015) The coloniality of language: race, expressivity, power, and the darker side of modernity. Wagadu: A J Trans Women's Gender Stud 13(1):5

Wang J, Seyler BC, Ticktin T, Zeng Y, Ezhu Z (2019) Indigenous botanical nomenclature used by the Yi people in Liangshan Prefecture, Sichuan Province, China. Econ Bot 73:325–340

Zhu D (2023) Bridge the channel, enhance the inclusivity: a comparison between flagship species-centered and moss-centered conservation in Chile and China. In: Rozzi R, Tauro A, Wright T, Avriel-Avni N, May RH Jr (eds) Field environmental philosophy: education for biocultural conservation, Ecology and ethics, vol 5. Springer, Dordrecht, pp 457–482. https://doi.org/10.1007/978-3-031-23368-5_28

Zuo Q, Zhang L (2026) Small but essential: how to introduce the public to easily overlooked organisms?. In: Zhu D, Rozzi R, Tsuji R, Johnson B, Caviness T, Castro-Jorquera C, Xu L, Zhang L (eds) Linking arts with biocultural conservation, restoration, and communication. Ecology and ethics, vol 8. Springer, Cham, pp 243–263. https://doi.org/10.1007/978-3-032-04830-1_15

Chapter 20
Communion

Terrance Caviness

Abstract Across cultures and time people have sought refuge within and consultation from the natural world. How we share our experiences and our individual realizations after these experiences are shared in stories. Stories connect across generations, cultures, and landscapes, and often touch on the deeply personal and the universal. In this chapter, I share a bit of how I developed my relationship with the natural world and my own experiences going to the natural world to consult on the next phase of my life. Through storytelling I hope to communicate the narrative and the emotion.

Keywords Cranberry lake · Plant encounter · Forest community · Biocultural experience

After a boat ride across the lake, a short drive down some back roads we parked along the side of the road near a lightly tread path easily mistaken for a deer path, perhaps that's how it began. The four of us, my friends and teachers, one in the same, entered the forest there to join the established trail to circumnavigate Cranberry Lake. Our destination was equally as lofty as that 50-mile trek, but more attainable in our short time. We were visiting the Big Pine, a nearly 5-foot across 150-foot-tall white pine that has watched over the lake and the Oswegatchie River for hundreds of years. The four of us, all botanists of varying years, hiked at the leisurely pace that only botanists walk, stopping every few meters at an interesting tree, a budding flower, a favorite moss, or an unknown seedling. We chatted about plants, family stories, life, and work until we reached the white pine after about an hour of walking.

Standing at the base of her you can't help but stand in awe and admiration. When she was just beginning her life, she grew amongst millions of her kind surrounded by trillium, beeches, maples, lady slippers, catberry, and pipsissewa on land stewarded

T. Caviness (✉)
Department of Environmental Biology, State University of New York College of Environmental Science and Forestry, Syracuse, NY, USA
e-mail: tcaviness@esf.edu

D. Zhu et al. (eds.), *Linking Arts with Biocultural Conservation, Restoration, and Communication*, Ecology and Ethics 8,
https://doi.org/10.1007/978-3-032-04830-1_20

Fig. 20.1 The "Big Pine." Image Credit: NatureupNorth.org. Open accessible through: https://www.natureupnorth.org/encounter/bill-hill/big-pine-trail-wanakena

since time immemorial by the Haudenosaunee and Anishinaabe peoples who call this place home. Since then, she has witnessed the removal of these peoples, watched colonizers cut her relatives for lumber and leave their remains behind igniting fires that raged across the lake over decades, and has seen a new forest grow up around her.

She is the last white pine near her. Her canopy is higher than the sea of beeches, spruces, maples and birches that surround her (Fig. 20.1).

That summer, my third as Assistant Director of Cranberry Lake Biological Station, between work and paddling on the lake I was trying to make some life decisions; actually, just one large one with a hundred smaller bundled into it. I have thought about the idea of adoption as a pathway to parenthood, but never mentally committed. In the view of my life, parenthood has been the one throughline in every iteration, at times there is a spouse (*who* shifts frequently), other times I am single. Throughout the summer I had had similar conversations with coworkers who are friends, and friends who are family. I was trying to solicit feedback and input, but I think I was just looking for an easy answer, and most of the time there isn't one; nor really should there be in such a big decision.

That day at the Big Pine I returned to ask my oldest friends, the plants, what I should do. Ever since I was a child growing up in my family's greenhouse I would talk to the plants. At my great-grandmother Netty Annutto's side, I learned how to ask the plants who were ready to be repotted, who wanted to be placed side by side in a pot; I learned from her to read the stems and the roots, the buds and the leaves. She would never know, and I wouldn't be able to articulate it until decades later, but it was by her side with the smell of warm, damp soil and the breath of thousands of plants that I knew that the plants and I loved each other. I was shy, an "old soul" I was frequently told, who found more in common with the ageratum than with children my own age. My relationship with the plants began in the greenhouse but soon led me out into the forest. It was the plants that lead me to my college, the mosses that I followed to Southern Chile, the warm-season grasses that got me a job in New York City, then back to Chile, and finally the mosses and the maples that called me back to New York and Cranberry Lake. There is something about a maple forest that immediately lets me know I am home, and the reason that I knew I would never stay away too long.

It seemed reasonable that I should consult the plants with this next phase of my life, and who better to ask about parenthood than a tree who has watched millions grow up around her. As we discussed this, my friend explained an article she read about how white pines, morphologically and physiologically, can act as grounding lighting rods, drawing in the lighting, grounding and dispersing the charge. This can protect the surrounding forest and has non-lethal implications for the tree.

As we sat at her base, some made offerings of tobacco, but for me to do such would feel insincere. When I am in the forest, I have come up with my own way of giving thanks to the plants, if I eat some berries from a bush, I will plant a few seeds from some older fruit, if I take a sample from tree or herb, I will help to spread their seeds if they have any ready for the wind. This has been my way to give back in some way to the forest that gives so much readily.

While planting some seeds, in silence I explained to her, the pine, what I was thinking and asked her what she thought, if I could do it, if I could do it *alone*. I waited for a response; I heard her canopy shaking in the wind louder and in a different tone than the broadleaf canopy below her. In my interpretation it was an affirmation, and I felt relief. If she can stand for the last two centuries keeping watch and protecting the children of other species, doing the best she can to disperse the worst of what is thrown at them, maybe I can too.

Being Catholic, or at least being raised Catholic, a distinction and conundrum not lost on those that understand the difference between the two phrases, I'm not unfamiliar with interpreting signs. My grandmother Marie would share a story of when her sister was battling breast cancer in the 80's, she and my great-grandmother would pray to St. Thérèse of Lisieux, The Little Flower. When prayers are heard by her, she said she would "let down a shower of roses from the heavens." On the day my great aunt passed, they returned home to see a trail of rose petal from the vase in the kitchen where roses had been in bloom to the door. An answer that the prayers had been heard.

When I made my confirmation in the church, I chose Thérèse as my name, not only for this story or that she is the patron saint of florists (I expanded that definition to be the patron of plant lovers), but for her message. She was declared a Doctor of the Church because of her writings and teachings. Principle among them is her "little way," doing ordinary things, or small acts, with great love. To me, at 16, this was the faith that I knew reflected to me and while I fluctuate between being Catholic and being *raised* Catholic, my faith in doing ordinary things with great love has not waivered. This informed my "thank you" to the plants, maybe my planting or dispersing their seeds doesn't do much or is futile, but I believe that the plants know my intent and the love behind it. St. Thérèse said "frequently, only silence can express my prayer" perhaps in my silent ask to the Big Pine she was also listening. It is nice to imagine The Big Pine and St. Thérèse working together to send me that message, one sending the wind and the other waving her needles.

To many people making a life decision through consultation with a saint makes perfect sense, but consulting a tree, exceptionally odd. While I can find value in both, who better to consult about life than the leafy green beings who have 470 million years of experience on this earth. Even if all this communication is wholly in my mind, maybe that is enough—my finding the affirmation I wasn't sure I was seeking.

If we can find ourselves reflected in the natural world, if the plants around us are not *objects,* but beings whose consultation is necessary for major decisions in our lives, how different life could be. When plants and other beings are friends, therapists, and mentors, their clearcutting, poisoning, and continued existence is deeply personal. The daily environmental tragedies, sanitized and abstracted through the 24-h news cycle, are reimbued with their emotional weight; that of a tragedy much closer to home.

Twenty-eight years ago, nearly a month to the day before I was born, a microburst passed through the Adirondacks and large swaths of the forest were laid down. Gouges, like scars in the landscape, could be seen from satellites. Those far away tragedies were suddenly not so far away. Neighbors checked on friends and family making sure everyone had what they needed to make it through. Fearing for their old friend, residents from Cranberry Lake and Wanakena made pilgrimages to check on the Big Pine too, to make sure she was still standing, that she was all right; she was.

I am thankful that she was and that we were able to have our conversation. I pour out my thinking; she rustles. It is simple to play the "what if" game, any series of events could have led to our *not* having that conversation. Generations of people worked together to protect her, from Indigenous land stewardship to legislation surrounding the protection of the Adirondacks to community members caring for her. The forest community we walked through may have looked different in a thousand different ways if not for these relationships. How many conversations have been lost through history due to the mistreatment of our plant relatives, and what would the world look like if they had been had? Can we survive the passing of an era when we no longer seek the counsel of the pines, mosses, and many other plants?

Part III
Biocultural Communication and Conservation Across Urban, Rural, and Remote Ecosystems

Chapter 21
Introduction to Part III: Biocultural Communication and Conservation Across Urban, Rural, and Remote Ecosystems

Danqiong Zhu, Terrance Caviness, and Ricardo Rozzi

Abstract Biocultural homogenization and the "extinction of experience" are key drivers of rapid losses of biological and cultural diversity. However, a lack of awareness about the causes and consequences of these losses prevails in global society. To transform this social alienation and contribute to conserving biocultural diversity, it is essential to develop biocultural communication that connects society with the diversity of worldviews and cultures that value biodiversity. This biocultural diversity occurs across gradients of urban, rural, and remote ecosystems. In our introduction to Part III, we briefly refer to three case studies that inspired our book and showcase forms of biocultural communication in place-based educational settings: Omora Ethnobotanical Park (Puerto Williams, Chile), the Cranberry Lake Biological Station (New York State, USA), and the H.J. Andrews Experimental Forest (Oregon, USA). We then summarize six biocultural conservation concepts and practices presented in the chapters of Part III, which draw on innovative initiatives from Asia and Latin America. (1) Flagship species, such as the Chinese medicinal plant *Cao-guo* or Chinese black cardamom (*Lanxangia tsao-ko*), inspire citizens to engage in biocultural conservation. (2) Eco-themed art exhibitions increasingly captivate the public as well as ecologists, artists, and citizens, offering

D. Zhu (✉)
Sub-Antarctic Biocultural Conservation Program, Department of Biological Sciences, University of North Texas, Denton, TX, USA

Cape Horn International Center (CHIC), Universidad de Magallanes, Puerto Williams, Chile
e-mail: Danqiong.Zhu@unt.edu

T. Caviness
Department of Environmental Biology, State University of New York College of Environmental Science and Forestry, Syracuse, NY, USA
e-mail: tcaviness@esf.edu

R. Rozzi
Sub-Antarctic Biocultural Conservation Program, Department of Philosophy and Religion and Department of Biological Sciences, University of North Texas, Denton, TX, USA

Cape Horn International Center (CHIC), Universidad de Magallanes, Puerto Williams, Chile
e-mail: Ricardo.Rozzi@unt.edu

D. Zhu et al. (eds.), *Linking Arts with Biocultural Conservation, Restoration, and Communication*, Ecology and Ethics 8,
https://doi.org/10.1007/978-3-032-04830-1_21

a novel form of contemporary biocultural communication. (3) Innovative combinations of landscape painting and poetic dance drama amalgamate traditional and contemporary art forms. (4) Integration of rational knowledge and emotions is illustrated with photovoice, a qualitative research method used in community-based participatory research that involves gathering photographs and narratives taken by community members, as opposed to outside researchers. (5) Nature-based paths for rural revitalization developed in Chinese urban-rural co-innovation regions, are especially relevant to villages with rich natural and cultural heritage, which can be valued and protected using market-oriented approaches to sustainability in ways that meet the demands of both rural and urban residents. (6) Ancestral practices, such as the Lamkaang Naga Indigenous Traditional Healing Systems of northeastern India, express forms of collective interdependence among humans, the land, medicinal plants, and the entire community of living beings with whom the habitat is shared. These six forms of biocultural communication illustrate vital interconnections between the life habits of human and other-than-human co-inhabitants in shared habitats. In this way, biocultural communication discloses just and sustainable forms of co-inhabitation among diverse cultures and species.

Keywords Arts · Biocultural homogenization · Flagship species · Rural revitalization · Urban ecology

21.1 Introduction

The conservation of global biocultural diversity faces unprecedented challenges. Biocultural homogenization has occurred and continues to occur around the globe at different scales, impacting urban, rural, and even remote ecosystems. The impact of biocultural homogenization across anthropogenic gradients has been demonstrated by numerous studies (Celis-Diez et al. 2017; Rozzi 2018). To address these challenges, novel educational methodologies emerge in the field, such as constructivist learning approach (Arık and Yılmaz 2020), mixed reality (XR) (Aguayo and Eames 2023), and field environmental philosophy (FEP) (Rozzi et al. 2023). Several case studies in our book have adopted field environmental philosophy as a methodology to foster "direct encounters" with biocultural diversity (Rozzi et al. 2008, 2023; Tauro et al. 2021). Examining the interrelationships between biocultural homogenization and the "extinction of experience," Alexandria Poole (2023) proposed that it is necessary to make adjustments to the contemporary formal education system so that "localized knowledge" and "globalized knowledge" can interact and thus foster biocultural conservation.

From a biocultural perspective, we can value the complementarity between modern "globalized knowledge" and ancestral "localized knowledge." Forms of globalized knowledge associated with Western sciences, such as biology and ecology, taught in formal educational systems, would be enriched if they also consider traditional ecological knowledge as well as knowledge inherent to the cultural identity of local communities. In this regard, liberation theologian May Jr (2023) has combined his experience in formal and non-formal education in Latin America. He has criticized Eurocentric approaches to education for strengthening the coloniality of power, shaping social structures, mentalities, and lifestyles in ways that have oppressed Indigenous and other local communities (May Jr 2023). May Jr. as well as other Latin American ecologists and philosophers, such as Alejandra Tauro (2023), Alicia Bugallo (2012) and Rene Moreno-Terrazas and collaborators (2023), have proposed that the research, educational, and action methodology of field environmental philosophy can offer a strategy for conserving the integrity of traditional ecological knowledge and global biodiversity. Chinese environmental philosopher Yubo Song elaborated on the vital relevance of integrating local culture and knowledge into biocultural conservation and restoration (Song 2026).

Field environmental philosophy incorporates the arts, as they play a significant role in biocultural conservation and restoration by engaging local communities and fostering a deep connection with biocultural diversity. In various regions of the world, artists often collaborate with members of local communities to create site-specific and place-based works that reflect an area's unique biocultural heritage (Curtis et al. 2012; Ojeda et al. 2023; Costa-Carvalho 2023). By involving community members in the creative process, the arts raise awareness of the values of local and regional biocultural diversity. At the same time, they empower residents to actively participate in the conservation and restoration of their unique biocultural heritages, landscapes, and community ties (Song 2026; Zuo and Zhang 2026).

Parts I and II of our book included case studies that illustrate concepts and practical knowledge rooted in biocultural diversity. These cases express place-based worldviews and "localized" practical knowledge, where the arts have played powerful roles in biocultural conservation and restoration. Part III of the book emphasizes that biocultural communication is also an integral component of biocultural conservation and restoration. This is demonstrated by diverse artistic formats taking place in urban, rural, and remote ecosystems that envision future forms of biocultural communication where the arts will take even more central roles.

21.2 Biocultural Communication in Three Places That Inspired This Part

In our introduction to Part III, we briefly refer to three case studies that inspired our book and showcase forms of biocultural communication in place-based educational settings: Omora Ethnobotanical Park (Puerto Williams, Chile), the Cranberry Lake Biological Station (New York State, USA), and the H.J. Andrews Experimental Forest (Oregon, USA).

Three place-based biocultural conservation initiatives in the Americas disclose how precious biocultural communication is for reconnecting global citizens with biological and cultural diversity and their interrelationships. These cases are: (1) the biocultural research/educational/conservation program of Omora Ethnobotanical

Park in the Cape Horn Biosphere Reserve, Chile, at the southern tip of South America; (2) the educational and community engagement programs at the Cranberry Lake Biological Station, College of Environmental Science and Forestry (ESF), State University of New York (SUNY), in the Adirondacks, USA; (3) the H.J. Andrews Experimental Forest Long-Term Ecological Research (LTER) site in the Western Cascades, Oregon, USA.

The Omora Ethnobotanical Park Initiative was created in 2000 to provide a "physical and conceptual space" for long-term biocultural research, education, and conservation (Rozzi et al. 2006). As a *physical space*, it is a biological reserve that protects the Robalo River watershed, the source of drinking water for the town of Puerto Williams. As a *conceptual space*, biocultural research and educational programs of Omora Park integrate sciences, philosophy, and arts. Located 3 kilometers west of the world's southernmost city, Puerto Williams, the Omora Ethnobotanical Park team has developed concepts and methodologies of great significance for biocultural conservation and communication. Its Field Environmental Philosophy workshops and courses involve educational strategies that incorporate *in-situ* communication among students, members of the Indigenous Yaghan community, other local residents, navy people, authorities, teachers, tourism guides. Participants engage in storytelling, linguistic sharing and analysis, and artistic activities involving Indigenous and scientific knowledge of plants, birds, other organisms, and ecosystems (Rozzi et al. 2008, 2014; Rozzi and Jiménez 2014).

Practically, in a conventional classroom setting for a course in ornithology, one might study the physiology of the organism, the taxonomic differences, the skeletal structure, even the ecological interactions, but this all happens *ex-situ*, with the organisms separated from their habitat and other co-inhabitants. The language used is technical and precisely communicates. Place-based knowledge may appear as the environment the organisms live in. There is also a typical dissemination of information from instructors to students. Whereas in a class at Omora Ethnobotanical Park, connection to place is prioritized, whether local or visitor, participants are invited to bring their experiences with them. Language, names of organisms, and what they can tell us about how humans interact and perceive other-than-human beings are central. Communication is plurilateral as opposed to hierarchical. Through conversation amongst all participants (instructor included), a more holistic understanding of the organisms and the place-based biocultural diversity is attained rather than a strict biological analysis.

Rozzi considers biocultural communication occurs in the field among participants of the programs through face-to-face contacts, interpersonal communication, intercultural and interspecies encounters, and it also reaches out and extends to a broader audience about the discoveries, activities, and projects through multiple media, such as publications, documentary films, newspaper, radio programs, or social media (Rozzi et al. 2006; Rozzi 2023). These messages sent through multimedia and multiple channels are both expressions and invitations. However, Rozzi emphasizes that:

> In … urban global society, communication is carried out through media communities, to the point of turning face-to-face encounters with biological and cultural diversity into an atavistic memory … Communication without face-to-face encounters and sensory experiences, however, leads to a knowledge about biological and cultural diversity that ends up not being felt. For this reason… Omora Park's field environmental philosophy programs emphasize experiences of "direct encounter" with mosses, lichens, trees, rocks, water, birds, and other living beings (humans and other-than-humans). With its emphasis on direct encounters and the immediate and holistic perception of nature, Omora Park's educational experience aims to compensate for the current excess of mediated information (Rozzi 2023, p. 169).

Through these experiences at Omora Park, students and researchers become actors and social communicators of biocultural diversity. Today, linguistic and cultural diversity is even more "invisible" than biological diversity. To counterbalance this deficit, the educational approach of field environmental philosophy involves both intercultural and interspecies encounters (Rozzi et al. 2008; Contador et al. 2023; Crego et al. 2023; Méndez-Herranz et al. 2023; Ojeda et al. 2023).

Cranberry Lake Biological Station (CLBS) is located in a remote area of the Adirondack region, New York State, surrounded by Adirondack Wild Forest and Wilderness lands and can only be accessed by boat. For over 100 years the station has served as a place to cultivate connection with land and place (SUNY ESF 2024). Primarily an educational field station, students learn with professors directly from the beings they are studying. Their keystone course, Ecological Monitoring and Biodiversity Assessment, places value on "direct encounters" with other-than-human organisms in their native habitats, and personal reflection about these interspecies encounters. Students are required to keep a nature journal about the organisms they encounter, including their broader ecological surroundings and how they are engaging with other beings. Through art and scientific curiosity students develop a connection to place over the duration of the course. In their final week of the course, they use their journal to reflect on what they have seen and learned, both in class and through their personal journals to develop research questions. These questions are often tied closely to the biological station and reflect their scientific curiosity rooted in ecological interaction rather than extracting an organism from their surroundings. Other courses offered include ethnobotany and respectful field research approaches. Both courses include the blending of Indigenous and western sciences to inform how we learn from the natural world, how we engage with the natural world, and how to develop research projects that are founded on respect for the beings of study.

In recent years there have been initiatives to expand beyond the classroom. Through service-learning components the local communities are brought into the collective learning of the students. New programs such as lecture series and scientific illustrator programs are broadening the impact of research at the station, sharing information, and learning from participants about their perceptions of place and what they have seen change ecologically over the years (SUNY ESF 2025b). New voices are invited to the station through the Indigenous Writer in Residency Program, three writers are hosted yearly, aligned with student course sessions, to provide a place for them to engage with research, find solace in place, and have the creativity to write

(SUNY ESF 2025a). The residency is welcoming with few constraints on projects or deliverables to ensure that the place itself can provide inspiration for the work of the writers. In addition, residency writers also have the opportunity to join courses and give talks to students to ensure that cross-pollination happens among every participant. These programs seek to incorporate different perspectives and worldviews into the biological station and provide space for exchange and discussion as a method of biocultural communication.

The H.J. Andrews Experimental Forest in the Oregon Cascade Range, USA, is an experimental forest in the US Forest Service system, encompassing 6400 ha of public land of old conifer forest, forest plantation, and fast-running streams. It is a Long-Term Ecological Research (LTER) site that has gone through four stages since the 1950s (Swanson 2023). Early communication at Andrews Forest occurred primarily among professionals, such as Forest Service researchers and land managers who advocated the development and exploitation of the forests. In Frederick Swanson's words,

> [The] conversion of native forests on Forest Service lands in the Pacific Northwest was based on an instrumental/utilitarian view of the value of the forests—they are valuable only for human use as wood products. The native forest was being erased... imposing a disconnect between humans and the natural world. In this era [1950s], science was conducted at the Andrews Forest by a small cadre of Forest Service researchers mainly to support utilization, while minimizing environmental impacts. Communications efforts targeted scientists and land manager audiences, but interaction with land managers was otherwise minimal. In a sense, this was "domesticated science" ... and had no direct engagement with the notion of non-utilitarian values of the forest (Swanson 2023, p. 541).

Communication with the public emerged in the 1970s, primarily by presenting professional scientists' research on the complexities and wonders of old-growth forests to the public through a variety of media. Beginning in the 2000s, because of the trend toward art-literature-science synergies in the environmental field, the Andrews Experimental Forest has also been initiating programs to promote engagement with the humanities, such as the LTER Reflections Residency Program, the Nelson's Conservation Ethics Program, and the outreach programs (Swanson 2023). Through these programs, writers, artists, scholars, philosophers, graduate students, middle-and-high school students, and other citizens are invited to the sites to conduct reflection, creative, educational, and other field activities.

A novel educational approach at Andrews Experimental Forest combines digital technology with field experiences. Since 2018, an inquiry-based curriculum is formatted on portable tablet computers (iPads) shared by groups of two to three students enabling them to have hybrid sensorial-technological place-based experiences (Markiewicz et al. 2023). Digital technology use in STEM (Science, Technology, Engineering, and Mathematics) learning environments is common, but the use of digital technology in field settings has been poorly explored (Markiewicz et al. 2023). The tablets provide a non-invasive way to archive student assessment data and facilitate the communication of their place-based experiences. However, Julie Markiewicz and collaborators (2023) admit that there is a methodological and philosophical tension with taking digital technology outdoors because it mediates field experience in ways that could prevent emotional connections to the natural world, and attentive place-based learning.

21.3 An Overview of Part III Chapters

The chapters in this part are further elaboration, exemplification, and response to approaches to biocultural communication and conservation presented in previous volumes of the *Ecology and Ethics* series and the first two parts of our book. It includes art-science initiatives across gradients of urban, rural, and remote ecosystems. Among these, some initiatives are as mature as the research and education programs described above for Omora Park, Cranberry Lake, and Andrews Experimental Forest, and other initiatives are at an exploratory stage. However, together they cover cross-cultural theoretical discussions about biocultural communication as well as practical techniques in the field. Putting these chapters together, we expect to expose readers to a thriving and multicolor landscape of biocultural communication experiences taking place contrasting socio-ecosystems.

The use of a single, culturally important, "flagship species" is useful as an entry point for people to engage on a deeper level with the natural world (Zhu 2023). This is particularly useful to connect people with remote areas that most citizens have not directly experienced. In Chap. 22, "A Biocultural Sketch of Cao-guo (*Lanxangia Tsao-ko*)," Zhang and Xu (2026) examine a Chinese medicinal plant as a flagship species. They apply the 3Hs (habitats, habits, and co-inhabitants) model of the biocultural ethic and the field environmental philosophy methodology to explore biological, ecological, cultural, aesthetic, and socio-economic values of a flagship species, which is more integrally understood as biocultural keystone species. Reflecting on the journey to visit Cao-guo in its native habitats, Zhang and Xu (2026) analyze the potential of employing watercolor paintings to communicate the biocultural significance of this species rooted in traditional medicine and food culture.

A novel form of biocultural communication is explained by Li et al. (2026) in Chap. 23: "Immersion and Interactiveness: Eco-themed Art Exhibitions and Biocultural Communication in East Asia." Li et al. (2026) illustrate how interdisciplinary collaborations take place at museums, art festivals, and exhibitions, which offer effective platforms for biocultural communication. Exhibitions and events in various locations in China, Japan, and South Korea, such as the insect exhibitions in Quantum Farm (China), Guanzhong Mangba Arts Festival (China), Our Ecology: Toward Planetary Living (Japan), Sea Art Festival (South Korea), and Gangwon International Triennale (South Korea), are examined through the lens of biocultural ethics. They discover that these eco-themed art exhibitions generate effective biocultural communication because they carry copious messages regarding multiple species entanglement, interconnectedness of all things, and the aspiration of people to reconnect with their traditions, habits, and history. Li et al. (2026) document how curators and artists of these exhibitions proactively employ immersive and interactive methods to associate with the visitors and local people, engaging their multiple senses to sustain artistic creations and narratives of biocultural traditions in East Asia. In summary, eco-themed art exhibitions increasingly captivate the public as well as ecologists, artists, and citizens, and offer a contemporary blending of traditional and technological concepts-and-practices to communicate core values of biocultural diversity.

Urban residents tend to contribute more to the status quo of global biocultural homogenization, and are largely unaware of its detrimental impacts. Chinese artist Juan Wang exerts dance to tackle this problem in metropolitan areas of the globe. In her chapter "A Mere Touch of Green: Reconnecting People and *Shanshui* through Poetic Dance Drama," Wang (2026) presents her work concerning biocultural communication through arts. In Chap. 24, she examines the connections between a contemporary poetic dance drama, *A Mere Touch of Green,* and a national treasured landscape painting, *A Panorama of Mountains and Rivers*. Wang (2026) discovers a striking similarity between the mindsets of ancient Chinese literati and modern urbanites: the attachment to *Shanshui*, the Chinese word for landscapes hosting mountains and rivers. The paradox of being absent in the *Shanshui* and hunted by this landscape urges people to reconnect to their native biocultural habitats. Through her blending of paintings and experimental dance, Wang (2026) ultimately achieves immersing herself into the native landscapes. This art experience led her to realize that green infrastructures (such as a low-maintenance, nature-based, rewilded urban park, and green space in university campus), offer places that are at hand in everyday urban life for reconnecting people with *Sanshui.* In this way, Wang (2026) offers an affordable and amiable art practice to cultivate an appreciation for the land, and to catalyze biocultural communication through dance pedagogy.

The combination of technology with field experiences is approached by Mexican ecologists Franquesa-Soler and Mesa-Jurado (2026) in "Photovoice: A community-based research and communication tool for wildlife conservation and coexistence." Photovoice is a qualitative research method used in community-based participatory research that involves gathering photographs and narratives taken by community members, as opposed to outside researchers. Franquesa-Soler and Mesa-Jurado (2026) adopt this method to integrate rational knowledge and emotions in the participation of local people in conservation. In Chap. 25, they explore how art can be used by researchers to empower residents' voices and drive wildlife conservation. This art-science method also enhances understanding about residents' perceptions of their local environment, and dialogic communication via photography and discussion. The shift of power from researchers to members of local communities opens novel paths for the identification of socio-ecological problems, and the cocreation of solutions. Based on a case study in southern Mexico, Franquesa-Soler and Mesa-Jurado (2026) highlight how photovoice served as a methodology for communication engaging students, adults, and other stakeholders in analyses of socio-ecological changes. This communication methodology enhanced students' perception of wildlife and facilitated the identification of positive and negative interactions between wildlife and the community, thereby informing community processes focused on natural resource management.

Habitat management can be critical in urban ecosystems. In "Nature-based Solutions to the Revitalization of Wuyuan County," Kongjian Yu and his collaborators explain a nature-based rural revitalization project in Xunjiansi Village, Wuyuan County, Jiangxi Province, China. Yu et al. (2026a) use the *Wangshan Life* framework, which they designed for revitalization of traditional agricultural wisdom, restoration of ecological infrastructure, conservation of cultural landscape, and reconstruction of village life habits in Xunjiansi. Aligned with biocultural ethics which values the vital links between the life habits of human and other-than-human co-inhabitants in their shared habitats (Rozzi 2012, 2013), the *Wangshan Life* framework addresses the complexities of relationships among people, other-than-human beings, and their landscapes (Yu et al. 2026a). A nature-based ideal of co-inhabitation has reached out to domestic and international audiences through *in-situ* seminars on human settlement and vernacular, cultural landscapes as well as continuous programs at the National Mayor Training Institute and related educational/research activities. In this initiative, arts play a central role in educational and field activities. By inviting artists and villagers to create artworks jointly, local biocultural heritage is conserved. Their local, place-based aesthetics, namely "Big-foot Aesthetics" values healthy ecosystems and guides prudent use of nature in conservation. Regional cultural landscape and corresponding biocultural heritages have endured generations, and the ecological wisdom revealed is worthy of reverence, protection, learning, and passing on. This is the key message for biocultural communication and conservation stated by Yu et al. (2026a).

The former concept is further elaborated through other case studies in the chapter "The *Wangshan Life* Model for The Revitalization of Rural Villages: The Xixinan Case." In Chap. 27, Yu et al. (2026b) explains their *Wangshan Life* model, which includes five interconnected components:

- poetic dwelling,
- organic food,
- free touring,
- immersive education, and
- the arts.

Yu et al. (2026b) explain how the integration of these five elements is enhanced by improving ecological infrastructure and rebuilding residential spaces as well as by incorporating ecological learning, research, and artistic practices into village branding. Their initiative has fostered a co-governance system involving local government, village residents, experts, and enterprises. Yu's experimental model has yielded remarkable results, including improved ecosystems, increased local incomes, revitalization of the rural community (with younger generations returning from large cities), and the preservation of cultural heritage. The *Wangshan Life* model offers a promising path toward urban-rural co-prosperity and biocultural conservation, which could be essayed in other regions of the world.

The immersion of local cultures and their native land is presented in a personal and reflective way by Sumshot Khular in "Indigenous *Lamkaang Naga Loh Kchet*: Collective Interdependence in Northeastern India." In Chap. 28, Khular (2026) describes her Lamkaang Naga Indigenous community, where she was born and raised. Khular combines Brian Burkhart's model for cultural resurgence and colonial resistance with Ricardo Rozzi's 3Hs model of the biocultural ethic to interpret and analyze Lamkaang Naga practices. She employs Burkhart's term "locality" as a conceptual tool to reference how being, meaning, and knowing are rooted in the land. She uses the 3Hs (Habits, co-in-Habitants, and Habitats) to highlight the vital interconnections among the life practices of humans and other-than-human beings who share ecological and cultural spaces. Khular stresses the collective interdependence and relationality that Lamkaang people maintain with their land and all living and non-living beings in their shared habitat. Traditional practices, such as healing rituals and collective labor like *Loh Kchet,* are deeply rooted in Indigenous worldviews that respect the land as a living entity. Based on personal experience grounded in ethnographic studies and other empirical research, Khular (2026) documents how Lamkaang Naga people ask permission from the land, plants, animals, and water before engaging in various activities. This is a reflection of their deep respect for the agency and sacredness of other beings, and the land as a whole.

Part III ends with a short storytelling essay, which is emblematic for biocultural communication. In "Carekeepers," Yakovleva (2026) explores the interplays between mushrooms and her story, language, and family history. Yakovleva narrates her youth with her grandparents and parents foraging in the forests around Moscow, and the family stories that accompanied these excursions. Remembering family history, cultural stories, and respectfully engaging with the natural world, Yakovleva (2026) generates a captivating type of biocultural communication that nourishes a unique biocultural ethics. Echoing a central invitation of our book *Linking Arts with Biocultural Conservation, Restoration, and Communication*, Yakovleva encourages readers to connect with their own family and cultural histories as well as communities of other-than-human co-inhabitants to achieve a sense of biocultural communication, and foster conservation of biocultural diversity.

Acknowledgments We acknowledge the support from the Chilean National Agency for Research and Development through the Cape Horn International Center, CHIC (ANID/BASAL FB210018).

References

Aguayo C, Eames C (2023) Using mixed reality (XR) immersive learning to enhance environmental education. J of Environ Educ 54(1):58–71

Arık S, Yılmaz M (2020) The effect of constructivist learning approach and active learning on environmental education: A meta-analysis study. Int Electron J of Environ Educ 10(1):44–84

Bugallo AI (2012) Filosofía ambiental y conservación biocultural: experiencias de integración en la Reserva de Biosfera Cabo de Hornos. In: Mendoza-Celina L (ed) La cuestión ambiental: temas y problemas. Proyecto Ecoepisteme, FEPAI, Buenos Aires

Celis-Diez JL, Muñoz CE, Abades S, Marquet PA, Armesto JJ (2017) Biocultural homogenization in urban settings: public knowledge of birds in city parks of Santiago. Chile Sustain 9(4):485. https://doi.org/10.3390/su9040485

Curtis DJ, Reid N, Ballard G (2012) Communicating ecology through art: what scientists think. Ecol Soc 17(2):3. https://doi.org/10.5751/ES-04670-170203

Contador TA, Rendoll-Cárcamo J, Gañan M, Ojeda J, Kennedy JH, Convey P, Rozzi R (2023) Underwater with a hand lens: ecological sciences and environmental ethics to value freshwater biodiversity. In: Rozzi R, Tauro A, Wright T, Avriel-Avni N, May RH Jr (eds) Field environmental philosophy: education for biocultural conservation, Ecology and ethics series, vol 5. Springer, Cham, pp 53–69. https://doi.org/10.1007/978-3-031-23368-5_4

Costa-Carvalho M (2023) Small, silent, and (in) significant: childhood as a minoritarian experience of education. In: Rozzi R, Tauro A, Avriel-Avni N, Wright T, May RH Jr (eds) Field environmental philosophy: education for biocultural conservation, Ecology and ethics, vol 5. Springer, Cham, pp 339–357. https://doi.org/10.1007/978-3-031-23368-5_21

Costa-Carvalho C (2023) Small, silent and (in)significant: childhood as a minoritarian experience of education. In: Rozzi R, Tauro A, Wright T, Avriel-Avni N, May RH Jr (eds) Field environmental philosophy: education for biocultural conservation. Ecology and ethics, vol 5. Springer, Dordrecht, pp 339–358. https://doi.org/10.1007/978-3-031-23368-5_21

Crego RD, Ward N, Rozzi R (2023) The eyes of the tree: applying field environmental philosophy to tackle conservation problems at long term socio-ecological research sites. In: Rozzi R, Tauro A, Wright T, Avriel-Avni N, May RH Jr (eds) Field environmental philosophy: education for biocultural conservation, Ecology and ethics series, vol 5. Springer, Cham, pp 101–111. https://doi.org/10.1007/978-3-031-23368-5_7

Franquesa-Soler M, Mesa-Jurado MA (2026) Photovoice: a community-based research tool for wildlife conservation. In: Zhu D, Rozzi R, Tsuji R, Johnson B, Caviness T, Castro-Jorquera C, Xu L, Zhang L (eds) Linking arts with biocultural conservation, restoration, and communication, Ecology and ethics, vol 8. Springer, Cham, pp 397–415. https://doi.org/10.1007/978-3-032-04830-1_25

Khular S (2026) Indigenous Lamkaang Naga Loh Kchet: collective interdependence in northeastern India. In: Zhu D, Rozzi R, Tsuji R, Johnson B, Caviness T, Castro-Jorquera C, Xu L, Zhang L (eds) Linking arts with biocultural conservation, restoration, and communication, Ecology and ethics, vol 8. Springer, Cham, pp 469–488. https://doi.org/10.1007/978-3-032-04830-1_28

Li Z, Jia M, Ma K (2026) Immersion and interactiveness: eco-themed art exhibitions and biocultural communication in East Asia. In: Zhu D, Rozzi R, Tsuji R, Johnson B, Caviness T, Castro-Jorquera C, Xu L, Zhang L (eds) Linking arts with biocultural conservation, restoration, and communication, Ecology and ethics, vol 8. Springer, Cham, pp 367–382. https://doi.org/10.1007/978-3-032-04830-1_23

Markiewicz J, Goralnik L, O'Connell K (2023) Screens on trails: digital environmental science, arts, and humanities learning for biocultural conservation. In: Rozzi R, Tauro A, Avriel-Avni N, Wright T, May RH Jr (eds) Field environmental philosophy, Ecology and ethics, vol 5. Springer, Cham, pp 555–576. https://doi.org/10.1007/978-3-031-23368-5_33

May RH Jr (2023) Liberation philosophy and biocultural education. A Latin American journey. In: Rozzi R, Tauro A, Avriel-Avni N, Wright T, May RH Jr (eds) Field environmental philosophy: education for biocultural conservation, Ecology and ethics, vol 5. Springer, Cham, pp 203–227. https://doi.org/10.1007/978-3-031-23368-5_13

Méndez-Herranz M, Marini G, Rozzi R (2023) Sub-Antarctic high Andean "gardeners:" cultivating caring relationships. In: Rozzi R, Tauro A, Avriel-Avni N, Wright T, May RH Jr (eds) Field environmental philosophy: education for biocultural conservation, Ecology and ethics series, vol 5. Springer, Cham, pp 71–86. https://doi.org/10.1007/978-3-031-23368-5_5

Ojeda J, Tauro A, Contador T, González-Calderón J, Malebran J et al (2023) "Pay attention, dive with eyes wide open": a field environmental philosophy activity to Foster reciprocity between people and nature. In: Rozzi R, Tauro A, Avriel-Avni N, Wright T, May RH Jr (eds) Field environmental philosophy: education for biocultural conservation, Ecology and ethics, vol 5. Springer, Cham, pp 87–100. https://doi.org/10.1007/978-3-031-23368-5_6

Poole AK (2023) Education as a driver of extinction of experience or conservation of biocultural heritage. In: Rozzi R, Tauro A, Avriel-Avni N, Wright T, May RH Jr (eds) Field environmental

philosophy: education for biocultural conservation, Ecology and ethics, vol 5. Springer, Cham, pp 247–262. https://doi.org/10.1007/978-3-031-23368-5_15
Rozzi R (2012) Biocultural ethics: recovering the vital links between the inhabitants, their habits, and habitats. Environ Ethics 34(1):27–50. https://doi.org/10.5840/enviroethics20123414
Rozzi R (2013) Biocultural ethics: from biocultural homogenization toward biocultural conservation. In: Rozzi R, Pickett ST, Palmer C, Armesto JJ, Callicott JB (eds) Linking ecology and ethics for a changing world: values, philosophy, and action, Ecology and ethics, vol 1. Springer, pp 9–32. https://doi.org/10.1007/978-94-007-7470-4_2
Rozzi R (2018) Biocultural homogenization: A wicked problem in the anthropocene. In: Rozzi R, May RH, Chapin FS, Massardo F, Gavin MC et al (eds) From biocultural homogenization to biocultural conservation, Ecology and ethics, vol 3. Springer, pp 21–48. https://doi.org/10.1007/978-3-319-99513-7_2
Rozzi R (2023) Introduction to part I. Field environmental philosophy: concepts and case studies. In: Rozzi R, Tauro A, Avriel-Avni N, Wright T, May RH Jr (eds) Field environmental philosophy: education for biocultural conservation, Ecology and ethics, vol 5. Springer, Cham, pp 17–25. https://doi.org/10.1007/978-3-031-23368-5_2
Rozzi R, Jiménez JE (eds) (2014) Magellanic sub-Antarctic ornithology: first decade of long-term bird studies at the Omora ethnobotanical park, Cape Horn biosphere reserve, Chile. University of North Texas Press, Denton
Rozzi R, Massardo F, Anderson CB, Heidinger K, Silander JA Jr (2006) Ten principles for biocultural conservation at the southern tip of the Americas: the approach of the Omora Ethnobotanical Park. Ecol Soc 11(1):43. https://doi.org/10.5751/ES-01709-110143
Rozzi R, Arango X, Massardo F, Anderson C, Heidinger K, Moses K (2008) Field environmental philosophy and biocultural conservation: the Omora Ethnobotanical Park educational program. Environ Ethics 30(3):325–336. https://doi.org/10.5840/enviroethics200830336
Rozzi R, Jiménez JE, Massardo F, Torres-Mura JC, Rijal R (2014) The Omora Park long-term ornithological research program: study sites and methods. In: Rozzi R, Jiménez JE (eds) Magellanic subantarctic ornithology: first decade of forest bird studies at the Omora Ethnobotanical Park, Cape Horn biosphere reserve. University of North Texas Press, Denton, pp 3–40
Rozzi R, May RH Jr, Tauro A, Avriel-Avni N (2023) Introduction to field environmental philosophy: a new methodological approach for biocultural education and conservation. In: Rozzi R, Tauro A, Avriel-Avni N, Wright T, May RH Jr (eds) Field environmental philosophy: education for biocultural conservation, Ecology and ethics, vol 5. Springer, Cham, pp 1–13. https://doi.org/10.1007/978-3-031-23368-5_1
Song Y (2026) Interspecies Co-inhabitation: A Dialogue between Chinese philosophical traditions and Biocultural Ethics. In: Zhu D, Rozzi R, Tsuji R, Johnson B, Caviness T, Castro-Jorquera C, Xu L, Zhang L (eds) Linking arts with biocultural conservation, restoration, and communication, Ecology and ethics, vol 8. Springer, Cham, pp 95– 109. https://doi.org/10.1007/978-3-032-04830-1_6
SUNY ESF (2024) Cranberry Lake biological station. ESF. https://www.esf.edu/clbs/index.php
SUNY ESF (2025a) Indigenous writer in residency. ESF. https://www.esf.edu/clbs/iwr.php
SUNY ESF (2025b) Scientific illustrator fellowship. ESF. https://www.esf.edu/clbs/opportunities.php
Swanson FJ (2023) The HJ Andrews experimental Forest long-term ecological research program, Oregon, USA: a historical biocultural perspective. In: Rozzi R, Tauro A, Avriel-Avni N, Wright T, May RH Jr (eds) Field environmental philosophy: education for biocultural conservation, Ecology and ethics, vol 5. Springer, Cham, pp 537–554. https://doi.org/10.1007/978-3-031-23368-5_32
Tauro A, Ojeda J, Caviness T, Moses KP, Moreno-Terrazas R et al (2021) Field environmental philosophy: a biocultural ethic approach to education and ecotourism for sustainability. Sustainability 13(8):4526. https://doi.org/10.3390/su13084526
Wang J (2026) A mere touch of green: reconnecting people and Shanshui through poetic dance Drama. In: Zhu D, Rozzi R, Tsuji R, Johnson B, Caviness T, Castro-Jorquera C, Xu L, Zhang L

(eds) Linking arts with biocultural conservation, restoration, and communication, Ecology and ethics, vol 8. Springer, Cham, pp 383–395. https://doi.org/10.1007/978-3-032-04830-1_24

Yakovleva T (2026) Carekeepers. In: Zhu D, Rozzi R, Tsuji R, Johnson B, Caviness T, Castro-Jorquera C, Xu L, Zhang L (eds) Linking arts with biocultural conservation, restoration, and communication, Ecology and ethics, vol 8. Springer, Cham, pp 489–492. https://doi.org/10.1007/978-3-032-04830-1_29

Yu K, Li J, Zhou Y, Xie M, Hu Y (2026a) Nature-based solutions to the revitalization of Wuyuan County. In: Zhu D, Rozzi R, Tsuji R, Johnson B, Caviness T, Castro-Jorquera C, Xu L, Zhang L (eds) Linking arts with biocultural conservation, restoration, and communication, Ecology and ethics, vol 8. Springer, Cham, pp 417–432. https://doi.org/10.1007/978-3-032-04830-1_26

Yu K, Li J, Zhou Y, Zheng C, Hu Y (2026b) The Wangshan life model for the revitalization of rural villages: the Xixinan case. In: Zhu D, Rozzi R, Tsuji R, Johnson B, Caviness T, Castro-Jorquera C, Xu L, Zhang L (eds) Linking arts with biocultural conservation, restoration, and communication, Ecology and ethics, vol 8. Springer, Cham, pp 433–468. https://doi.org/10.1007/978-3-032-04830-1_27

Zhang S, Xu L (2026) A biocultural sketch of Cao-guo (Lanxangia Tsao-ko). In: Zhu D, Rozzi R, Tsuji R, Johnson B, Caviness T, Castro-Jorquera C, Xu L, Zhang L (eds) Linking arts with biocultural conservation, restoration, and communication, Ecology and ethics, vol 8. Springer, Cham, pp 349–365. https://doi.org/10.1007/978-3-032-04830-1_22

Zhu D (2023) Bridge the channel, enhance the inclusivity: A comparison between flagship species-centered and moss-centered conservation in Chile and China. In: Rozzi R, Tauro A, Avriel-Avni N, Wright T, May Jr RH (eds) Field Environmental Philosophy. Ecology and Ethics, vol 5. Springer, Cham, pp 457-482. https://doi.org/10.1007/978-3-031-23368-5_28

Zuo Q, Zhang L (2026) Small but essential: how to introduce the public to easily overlooked organisms? In: Zhu D, Rozzi R, Tsuji R, Johnson B, Caviness T, Castro-Jorquera C, Xu L, Zhang L (eds) Linking arts with biocultural conservation, restoration, and communication, Ecology and ethics, vol 8. Springer, Cham, pp 243–263. https://doi.org/10.1007/978-3-032-04830-1_15

Chapter 22
A Biocultural and Artistic Sketch of the Chinese Black Cardamom, Cao-guo (*Lanxangia tsao-ko*)

Stephen Zhang and Li Xu

Abstract In this chapter, we select Cao-guo (草果), or Chinese black cardamom (*Lanxangia tsao-ko)* as a focal species for the purposes of biocultural conservation, restoration and communication. We employ the methodology of field environmental philosophy (FEP) and adopt the theoretical framework of biocultural ethics, especially the 3Hs, habits, habitats, and co-inhabitants, to explore the biological, ecological, cultural, aesthetic, and socio-economic values of *Cao-guo*. Despite its significance, the species has received limited attention in biocultural studies. Through an integration of a species-centered approach and FEP, one of the authors, Zhang conducted field investigation in Tengchong, Yunnan province, Southwest China. Based on the field trip, Zhang also created watercolor paintings for further biocultural communication. In this study, combining field research and artistic creation, we aim to illustrate the biocultural importance of *Cao-guo*, foster greater awareness of the responsibility of co-inhabitation in various local places on the planet, and examine the medium of watercolor paintings for biocultural communication.

Keywords *Cao-guo (Lanxangia tsao-ko)* · Field environment philosophy · Biocultural ethics · Species-centered approach · 3H theory · Co-inhabitation

S. Zhang
Department of Design, College of Visual Arts and Design, University of North Texas, Denton, TX, USA
e-mail: Stephen.Zhang@unt.edu

L. Xu (✉)
Department of Art Education, College of Visual Arts and Design, University of North Texas, Denton, TX, USA
e-mail: Li.Xu@unt.edu; xuliza@gmail.com

D. Zhu et al. (eds.), *Linking Arts with Biocultural Conservation, Restoration, and Communication*, Ecology and Ethics 8,
https://doi.org/10.1007/978-3-032-04830-1_22

22.1 Introduction

Today's world is facing a serious environmental crisis on multiple fronts. Global warming is wreaking havoc in almost every corner of the world, although each region has its own unique situation. The anthropocentrism inherent to the prevailing Western schools of deontological and utilitarian ethics have overlooked essential relationships that humans have with the natural environments they inhabit, and with other-than-human beings (Rozzi 2013). Driven not only by anthropocentrism, but also by other cultural, economic, and political drivers, biocultural homogenization (the reduction of biological and cultural diversity) has become a pervasive major socio-environmental problem worldwide (Rozzi 2012). To confront this trend that is causing severe losses of biological and cultural diversity, it is crucial to rethink the ways in which global citizens perceive and value the world, and their place in it (Thompson & Xu, 2023). A novel way to address this challenge has been proposed by Chilean—US philosopher and conservation biologist Ricardo Rozzi, in particular his "3Hs" (habits, habitats, and co-inhabitants) model of the biocultural ethic. By clarifying the concepts underlying each of the "3Hs" and exploring their interrelations, Rozzi (2012, 2013) seeks to construct a broader, more integrated framework for understanding dwelling and promoting more ethical forms of co-existence.

Another response to the acceleration of the socio-environmental crisis has come from artists and designers who understand the necessity to address these urgent issues. Today, artists and designers have become actively engaged in raising public awareness and fostering a deeper connection between human beings and the natural world (Bertling 2023). Through creative expressions, artists illuminate intricate relationships between beings and their environments. They recognize that humans are integral components of broader ecosystems. These creative practitioners explore innovative approaches to represent symbiotic relationships among all forms of life, including humans, plants, and animals. As they confront these pressing challenges, it is essential for artists and designers to investigate creative processes to engage with environmental concerns.

The 3Hs model created by Rozzi (2012), delineates interconnected concepts to orient relations of co-inhabitation. His biocultural ethics offers conceptual clarity and actional pathways to strengthen inter-cultural and inter-species relations of co-inhabitation (Rozzi 2023a). Through the 3Hs biocultural lens, we explore the interrelationships among the natural environment and various human societies. Biocultural ethics calls for understanding the dynamic and ecological interactions between humans and other-than-human species. By connecting biophysical, cultural-linguistic, and social-institutional dimensions, we gain deeper insights into the relationship between humanity and the broader ecological community, as shown by the examination of interspecies kinships between Pewen trees and the indigenous *Pewenche* and *Mapuche* people (Rozzi 2013).

In our work, we explore the application of the 3Hs model of the biocultural ethic to the Cao-guo, or Chinese black cardamom (*Lanxangia tsao-ko)* as a focal species for initiating biocultural conservation, restoration and communication. This plant

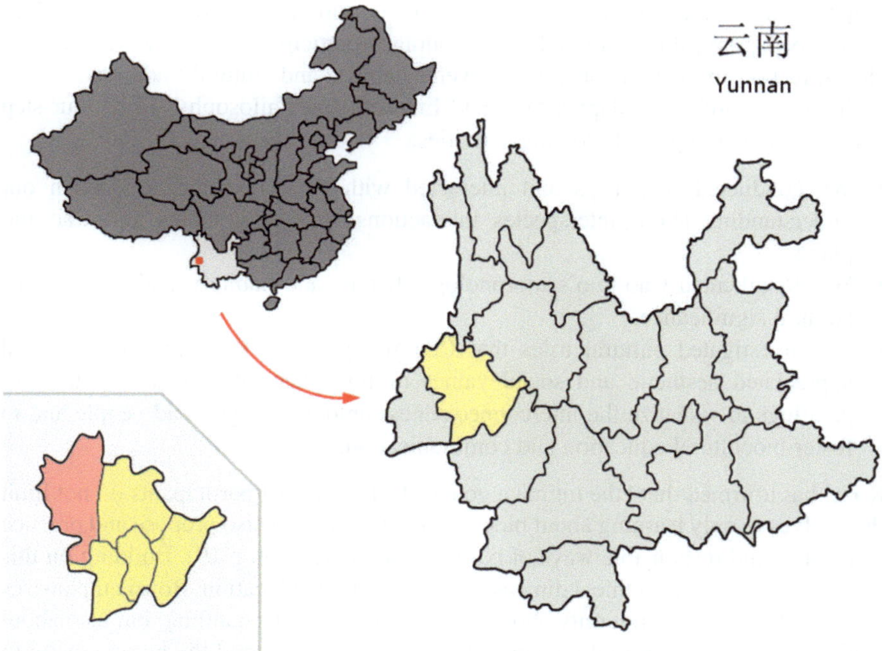

Fig. 22.1 Study Area: Tengchong. Location of Tengchong (red) in Baoshan Area (Yellow), Yunnan Province (light grey), China. (Maps from Wiki commons)

has significant economic, and cultural connections in China, particularly within the Yunnan region (Fig. 22.1).

22.2 Methodology and Study Area

Field Environment Philosophy (FEP) as the methodological approach aims to foster connections between people and biocultural diversity to enhance citizens' understanding of the values of biological diversity and cultural diversity (Rozzi et al. 2023). FEP includes a four step cycle that plays an important role, because this educational process encompasses not only research but also dialogue, communication, participation as well as inter-cultural and interspecies interactions (Rozzi et al. 2008b). In Rozzi's words,

> FEP methodology includes ethical reflections based on field experiences guided by biocultural concepts that stimulate an awareness of cohabiting and sharing habitats with diverse human beings belonging to different ethnic, socioeconomic, and linguistic groups and with myriads of other living and ecological beings (Rozzi 2023b, p.19).

Through this integrative approach, FEP connects the academic world and community practices. It also promotes active engagement with biocultural diversity. By emphasizing interconnectedness of ecosystems and human societies as well as ethical pedagogy, this methodology encourages participants to acknowledge their shared responsibilities in conserving diverse habitats and cultural traditions.

In our research, we adapted the Field Environment Philosophy (FEP) four-step cycle, implementing the following activities.

- We conducted field trips and interacted with local residents to deepen our understanding about interspecies interactions between humans and Cao-guo plants.
- We documented Cao-guo's morphology, habits, and habitats, and explored its cultural significance.
- We investigated cultural roles that Cao-guo plays in local communities and appreciated aesthetic and social values of this plant. We created watercolor paintings to visualize the interconnectedness among Cao-guo and people and to foster biocultural education and communication.

Rozzi has affirmed that "the ultimate goal of FEP … is that participants do not limit themselves to only learning about biocultural diversity, but also propose and practice respectful, and responsible ways of co-inhabitation"(2023b, p.19). Building on this idea, to underscore the ethical dimension of biocultural education, Rozzi emphasizes that co-inhabitation is not only about knowledge and understanding, but also about cultivating a sense of moral responsibility that extends beyond the human realm to the other-than-human beings. In his own words,

> The understanding of biocultural relationships of co-inhabitation entails the ethical imperative of having moral considerations and relations of care that are not limited to the human sphere but that contemplate the well-being of all beings (Rozzi 2023b, p.19).

In summary, the core purpose of Field Environmental Philosophy (FEP) is to not only learn, but also foster ethical, respectful, and practical engagement with biocultural diversity. This approach advocates living responsibly within ecological systems and caring for all forms of life as moral co-inhabitants of shared environments. This is the biocultural lens and methodological approach we adopted for our field work in Yunan, China.

Yunnan is known in China as "the kingdom of plants and animals" and the "gene bank of species", and its rich biological diversity has been globally recognized . Our study area is the county of Tengchong in Yunnan Province (Fig. 22.1). It is located in the Gaoligong Mountains range, bordering northern Myanmar, and covering an area of 5693 square kilometers. Tengchong has two distinct seasons each year: the rainy season that spans from May to October, and the dry season that lasts from November to April. Marked by its rich volcanic soil, dense forests, and a warm, humid subtropical monsoon climate with over 2, 000 higher plants. Tengchong is recognized as one of the four biodiversity hotspots in China and one of 34 worldwide (Cai et al. 2019; Myers et al. 2000;). The mountains near the Gaoligongshan National Protected Area in Tengchong, harbors a rich floristic diversity, including from small mosses to towering angiosperms. This region contains a variety of habitats such as old-growth forests, secondary forests, plantation forests, natural freshwater wetlands (barrier lakes from historical volcanic activities), and human-intervened habitats (i.e., reservoirs, ponds, agricultural lands) (Chan et al. 2019). These biophysical and ecological conditions create a fertile mosaic of "habitats" that is ideal for the growth of *Cao-guo*. This plant grows particularly well in humid soils, under the shade generated by luxuriant canopies of mixed evergreen forests.

Culturally, the region is home to several ethnic minorities,", including the Dai, Lisu, and Achang ethnic groups who have coexisted with Tengchong's biodiversity over many centuries. These ethnic groups have developed language, knowledge systems, and practices that sustain both their communities and local ecosystems. Historically, Tengchong has been the Southwest frontier since the Chinese imperial era and a fortress town with geographical significance. It also occupies an important position in transnational trade, such as the historical Southern Silk Road and the ancient Tea-Horse Road, which are all associated with it. Since the 2000s, Tengchong has been undergoing reconstruction and symbolization as a place of attraction for its biodiversity, natural beauty, and rich cultural traditions (Zhou 2013).

On one hand, the regional government is committed to revitalize the sense of place. On the other hand, the national and local governments expect that Tengchong becomes a new "bridgehead" for international trade. Consequently, the region has become more deeply involved in processes of globalization, which could have negative impacts of biocultural homogenization. We chose Tengchong as our study area for two reasons. First, the challenges and needs generated by the complexity of the above entanglement among biodiversity, cultural diversity, and international trade. Second, it is the resident home of one of the authors' parents (Stephen Zhang).

22.3 Study Species and Field Trip

The Chinese black cardamom, in Chinese known as Cao-guo, is an endemic plant in Southwest China and was listed as "near threatened" species by the International Union for Conservation of Nature (IUCN). Cao-guo thrives in "symbiosis with forest trees", fungi, and understory plants, contributing to soil stability and forest microclimates. It is also a medicinal and culinary plant in the ginger family and deeply embedded in the ecological, cultural, and economic life of Tengchong people.

As Rozzi stated, "tiny co-inhabitants are precious in themselves and are critical to the health of humans and ecosystems" (2023b, p.19). For local residents, *Cao-guo* is more than a plant. Rather, it is part of their life "habits". Residents have harvested this plant seasonally from forested hillsides using traditional methods that minimize ecological disturbance. After being harvested by local people, Cao-guo's fruits are sun-dried and used in herbal medicine to treat colds and digestive issues or as a spice in local cuisine, prized for its aromatic, peppery flavor. It can be said that this plant participates in local people's everyday life. The plants grow wild or cultivated; the dried fruits appear in the marketplace; their fruits are put in cooking pots; the ingredients connect the plant and human bodies. *Cao-guo* exists embedded in a strong interconnectedness of culinary practices, traditional medicine, and biodiversity. It embodies how local communities in Tengchong adapt to and nurture their surroundings through sustainable practices rooted in a deep awareness that reminds us about the 3Hs model about the vital links among co-inhabitants, their life habits and shared habitats. This awareness raises an ecological and ethical wisdom.

22.3.1 Species Distribution and Cultivation

Cao-guo is a perennial evergreen herb of the ginger family and the *Cardamom* genus. It thrives in tropical and subtropical regions of Yunnan, Guizhou, and Guangxi, China, and northern highland of Vietnam and Laos, where it grows at an altitude of approximately 1200–2000 meters above the sea level, with an average annual temperature of 17–19 °C, an average annual precipitation of 1000–1600 mm, and a relative humidity of 80–85% (Yang et al. 2017; Dou et al. 2023; He et al. 2023). The plant can grow beyond the above specific area and condition; however, it will not bear fruit (Yang et al. 2017). Its habitats are typically the understory valleys and slopes with sparse woods (about 50% light penetration), or in humus-rich, moist, but well-drained broadleaf evergreen forests.

Cao-guo can reach up to 3 meters in height, and the whole plant emits a pungent odor, with the underground part resembling ginger. The leaves are green, smooth in surface, slightly pointed at the tip, and oval-shaped, approximately 40 to70 centimeters long and 10–12 cm centimeters wide. The ligule is entire, apically obtuse, generally 0.8–1.2 centimeters long; flowers yellow or pale-yellow densely borne on stout inflorescence axis, with 5–30 flowers per inflorescence, pedicels 10 centimeters or longer, surrounded by scales (Fig. 22.2e). Plants usually bloom from April to June; fruits (Fig. 22.2c) are gray-brown or reddish-brown, ellipsoid, densely packed together when fresh, distinctly longitudinally furrowed and ribbed, with raised basal columns at the top, rounded, and with a pedicel or pedicel scar at the base (He et al. 2023).

Fruits have tough skin, removing skin, white or yellowish-brown septum in the middle, and the seeds are conical polyhedral, reddish-brown, and covered with grayish-white membranous, its fruiting period is generally from September to December. Dry fruits are about 2–5 centimeters long, with gray-brown to brown pericarp, longitudinal grooves and ribs, and a distinctive fragrance (Fig. 22.2f).

Cao-guo has a red fruit that grows in clusters on the ground. In the verdant environment, clusters of the fruits adorn the hillsides and the stream banks, nestled adjacently to boulders and rocks. Towering trees surround the area, though they do not directly overshadow the plants. These tall trees offered the requisite 50% shade that is essential for the plant to grow. These red fruits look strikingly beautiful against the gray-green landscape on rainy days.

Fig. 22.2 Field investigation of the Cao-guo (*Lanxangia tsao-ko*): (**a**) habitat, (**b**) community patch, (**c**) plant with fresh red fruitage, (**d**) field trip, (**e**) flowers, (**f**) dried fruits. (Photographs credit: Stephen Zhang)

Yunnan Province is the largest cultivation area of Cao-guo in China (yielding over 90% of the national production) (Yang et al. 2017; Zhang et al. 2011). Notably, altitude influences the yield of Cao-guo. When the species is transplanted from a low-latitude or low-altitude environment to a high-latitude or high-altitude area, the number of flowers per inflorescence increases and the fruit rate decreases (Yang et al. 2014).

22.3.2 *"One Root of Medicine and Food" (药食同源)*

Cao-guo holds significant importance in the daily lives of the local population, intricately woven into their cultural fabric over centuries. It subtly yet profoundly connects individuals in ways that often go unnoticed. It has been utilized in cooking since ancient times, as documented nearly 800 years ago during the Southern Song Dynasty in the culinary text "*Records from Pujiang Wu Family.*" In Yunnan Province, dried Cao-guo fruit is a household staple. The fruit releases its fragrance slowly, so it is often used in cooking methods that take a longer time, such as grilling, stewing, braising, boiling, making sauces, and marinating, among which braising is the most common. The culinary uses of the fruit are similar to those of Sichuan peppercorns and cinnamon.Its spicy taste can also stimulate the appetite. Due to its unique flavor, Cao-guo is widely embraced across China's major culinary traditions, including Sichuan, Cantonese, and Shanghai cuisines. When combined with other complementary spices, it can prolong and enhance the fragrance effect. Therefore, it is often used to blend various compound spices, such as five-spice powder and curry powder.

Cao-guo has a strong effect in removing unpleasant odors, eliminating fishy smells and enhancing flavor. It is particularly suitable for stewing beef and mutton. Excessive amounts may overpower a dish with spiciness and bitterness. With its unique bitter and spicy flavor, typically only one piece is used per dish, using both the skin and seeds. Typically, adding one or two pieces to hot pot dishes can create a rich and long-lasting flavor. It can also be added to braised dishes, but the amount for each dish should not exceed.

Overall, Cao-guo plays an important role in Chinese food culture. It is mainly used as a condiment to enhance flavor while removing the fishy taste of meat. The plant grows in the cool mountainous regions and symbolizes warmth and nourishment. Traditionally, it has been used as a seasoning as well as a medicinal plant, reflecting the idea of "One Root of Medicine and Food"(药食同源) in Chinese food culture and traditional Chinese medicine, and deeply in tune with the account of living habits in the framework of biocultural ethics.

"One Root of Medicine and Food" is a crystallization of the wisdom of traditional ecology summarized by Chinese through the ages in their long-term practice of food and medicine, reflecting the functions of food in health care and medical treatment. Ancient Chinese writings on edible herbs have documented a large number of "medicinal and food" herbs, with detailed descriptions of the principles, effects, contraindications, and usage. The secondary metabolites of herbal plants usually contain phenylpropanoids, quinones, flavonoids, tannins, terpenoids, steroids, glycosides, alkaloids and other substances, which are the main material basis for studying the efficacy of natural medicines, especially traditional Chinese medicines, today (Shan et al. 2015).

Over centuries, the Cao-guo has evolved into a trusted staple in Chinese cuisine and medicine with enduring significance. As a traditional Chinese medicinal herb, it has been widely used in Traditional Chinese Medicine (TCM) for therapeutic purposes. Research has found that a total of 10 ethnic minorities use Cao-guo to treat diseases. The earliest record of using Cao-guo can be found in "Taiping Huimin Heji Bureau Prescription" in Song Dynasty around 1078 C.E (He et al. 2023). Typically used in dried form, sometimes grounded into powder or decocted in herbal teas, Cao-guo is often combined with other herbs in formulas to treat complex diseases such as distension and pain in the epigastrium, vomiting, malaria, plague chills and fever (Cheng and Liu 2012). In the southwest regions of China (e.g., Yunnan, Sichuan), Cao-guo is commonly used in soups or medicinal broths. There are many prescriptions about Cao-guo with significant medicinal effects (He et al. 2023). Preparation methods include boiling in decoctions or incorporating the ground power into prescriptions.

Modern pharmacological research has confirmed some of these traditional uses, showing that essential oils extracted from Cao-guo possess "antibacterial, antifungal, and anti-inflammatory" properties (Yang et al. 2008), further validating its importance in both historical and contemporary medical contexts.

22.3.3 *Field Trip*

In the summer of 2023, Stephen Zhang conducted an exploratory study of biocultural diversity in Tengchong, Yunnan Province, China. An experienced local tour guide served as Zhang's guide in exploring the local ecoregion where he observed Cao-guo and its wild habitat. The expedition coincided with the monsoon season, adding a hazy poetic quality to the alpine scenery. The rain and humidity cast a soft mist over the mountains, creating a picturesque scene (Fig. 22.2a).

In its wild habitat, from a distance, Zhang and his tour guide encountered patches of Cao-guo that resembled corn fields, yet the plants reach an impressive height of 8–9 feet (Fig. 22.2b). Following the trail, they found themselves among the plants, sheltering under their leaves (Fig. 22.2c). Under a rain-tinted sky that casts a bluish-gray hue over the plants, the sight of vibrant red fruits scattered across the ground was truly captivating (Fig. 22.2d). In the verdant forest, clusters of fruits adorn the hillsides and the stream banks, nestled adjacently to boulders and rocks. Towering trees surround the area, though they do not directly shade the plants. These tall trees offered the requisite 50% shade that Cao-guo needed to grow. These red fruits look strikingly beautiful against the gray-green landscape on rainy days. Through his fieldwork in the mountainous region in Tengchong, Zhang also documented his observations of habitats where humans and Cao-guo, as well as numerous other-than-human species coexist.

Back in town, as Zhang strolled through the local farmers' market in Tengchong, his attention was drawn to vendors offering dried Cao-guo alongside other aromatic spices like star anise (*Illicium verum*) and dried red chili pepper. Through interviews and observations in a variety of settings, including fields, marketplaces, and homes, Zhang gained insight into people's perceptions of the interaction between Cao-guo and the local communities. These observations and interviews are recorded in the form of watercolor sketches.

As demonstrated in the loop of FEP's methodology, field trips are essential for face-to-face contact with other-than-human beings so that an ethical sense of co-inhabitation can be cultivated. Zhang's field trip not only offered him direct experiences of face-to-face contacts, but also boosted an epistemic and aesthetic journey to appreciate the ecological, medical, aesthetic, and cultural values of Cao-guo. The participatory approach of FEP encourages us to combine the artworks to demonstrate the richness of an ecological world centered around Cao-guo that reflects the local biodiversity in Tengchong, Yunnan. In this regard, artworks are both a representation of the uniqueness of Cao-guo and the abundance of local biocultural diversity, as well as a medium for powerful biocultural communication.

FEP does not end at the creation of artworks or completion of field activities. Our research could also be developed into a series of theoretical and practical resources for art and environmental education. For example, we can use Cao-guo as a case study to guide locals in understanding, appreciating, and recognizing the ecosystem. The species-centered approach can also be applied to develop a series of art education resources that reveal the close connection between cultural keystone species and their habitats to display the intricate interrelations between the species and the local communities within various ecoregions. Particularly, for our study, we plan to present our artworks in a variety of local and international venues, such as classrooms, local communities, libraries, and biocultural-themed international art exhibitions, to convey the message of co-inhabitant biocultural ethics to various audiences.

22.4 Art Creation for Biocultural Communication

While conducting biocultural education and communication, the first limitation is the public's lack of access to sufficient information and data about most plant species. The second is the overlooking of ecological interactions. Public participation is essential for effective ecological restoration and conservation efforts (Rozzi et al. 2006). To foster biocultural conservation and contribute to a better future for the planet, it is crucial to bridge the gap between academia and the public (Rozzi et al. 2008a, 2012). By bridging this gap, we can promote citizens' awareness about the ecological and cultural significance of plants as well as increasing access to the information and data. We urgently need to cultivate an appreciation for the diversity of life, including the flora and fauna that sustain biocultural diversity across the globe. The world is more vibrant and resilient because of the complex ecological connections between humans and more-than-human beings. However, not all of these connections are readily visible to societies or individual citizens (Rozzi 2019).

Biocultural communication is defined as an invitation and welcoming of alternative worldview, knowledge, and message to dominant Euro-centred ones to embrace biocultural conservation and restoration (Zhu and Rozzi 2026). Visual arts play a pivotal role in offering positive examples and making the abstract concept tangible through informative visualization and visual storytelling. Artists and designers can make significant contributions to biocultural conservation and communication by leveraging their strategic, conceptual, and creative skills. While the range of artistic formats and styles is vast, the essential criterion is that the work offers leap thinking, stimulates insights, and elicits emotional responses. We also refer to these insights from the visual arts when creating the following biocultural artworks.

In our exploration of the possibility of employing watercolor paintings as a tool for biocultural communication, we adopted FEP and a species-focused approach. Cao-guo growing in wilderness is both a source of artistic inspiration and the focus of this chapter's theoretical considerations. As mentioned above, one of the authors of this chapter, Stephen Zhang, an artist and designer, spent the summer of 2023 in Tengchong, Yunnan Province, China, where he conducted fieldwork that included observing the habitat, morphology, and growth of Cao-guo, and its role in the daily lives of local communities. These personal experiences have provided him with a deeper appreciation of the socio-ecological-cultural significance of this species in local communities. Biocultural framework offers a unique lens through which the links between local biodiversity and cultural diversity can be revealed. In conjunction with his fieldwork, Stephen Zhang has created three watercolor paintings that aim to visually express the imagery of Cao-guo in its ecosystem and cultural context, and to convey the calling of a biocultural ethic of co-inhabitation. Thus, these artworks (Figs. 22.3, 22.4, and 22.5) are more than illustrations, but visual representations, expression, and communication of biocultural ethics.

Chengli (Fig. 22.3), depicts a lush scene of Cao-guo coexisting with other vegetation in an alpine montane environment. The title of the painting derives from a classic work of Chinese Taoist philosophy, the book Zhuangzi. In the chapter "Knowledge Rambling in the North" of the book, it says "all things have their complete and distinctive constitutions, but they say nothing about them" (萬物有成理而不說) (*Zhuangzi*). Taoist philosophers consider that all things (including living and inorganic matters) have all the necessary and appropriate elements to become themselves under the operations of Heaven, Earth, and Nature. The ethical teaching from this completeness of things for humans is to trace, understand, and admire the operations and laws rather than to disturb the operations or violate the laws in nature. This painting vividly visualizes the completeness of Cao-guo in its biophysical environment.

Shi (Fig. 22.4) corresponds to the concept of "habits" in the biocultural ethical framework. In the classical book, Book of Change, or the *I Ching*, it says, "He [Shen-nong] caused markets to be held at midday, thus bringing together all the

Fig. 22.3 *Cheng Li* (成理, Completeness). 40 in × 40 in (101.6 cm × 101.6 cm), watercolor painting by Stephen Zhang, 2025

people, and assembling in one place all their wares. They made their exchanges and retired, every one having got what he wanted ((神農)日中為市, 致天下之民, 聚天下之貨, 交易而退, 各得其所) (*Book of Change*: Xi Ci II). In other words, the marketplace was originally a place set up to facilitate trading and exchange of materials. Exchanging materials, not direct currency, or the oldest form, "barter" (moneyless market exchanges) was considered important for reciprocity, solidarity, affection, respect, gift, equity, sustainability, women's collectives, families, communal assemblies, citizen federations, etc., as demonstrated by the Lares valley community in Peru (Argumedo and Pimbert 2010). However, the establishment of the monetary system and the operation of capital may distort the original intent of setting the marketplace. The purpose of this painting is not only to express that the

Fig. 22.4 *Shi* (市, marketplace). 40 × 40 in (101.6 cm × 101.6 cm), watercolor painting by Stephen Zhang, 2025

Cao-guo connects people in various roles, such as Cao-guo harvesters, vendors, and home cooks (usually women), but also to remind people to consider the original purpose of setting up a marketplace.

Caoguo Yin-Yang (Fig. 22.5), symbolizes the interrelationships between biological diversity and cultural diversity. The title of the painting originates from the book of *Dao De Jing* (道德經), or *Lao Tzu* (老子), in which it says, "All things leave behind them the Obscurity (out of which they have come), and go forward to embrace the Brightness (into which they have emerged), while they are harmonised by the Breath of Vacancy" (萬物負陰而抱陽, 沖氣以為和)(*Dao De Jing*: 42). Leaving obscurity and embracing brightness reveals the Yin and Yang attributes in the composition of all things, as well as their dynamism in operations, which can be manifested in the emergence and extinction, birth and death, growth and decay of

Fig. 22.5 *Caoguo Yin-Yang* (草果陰陽). 36 × 36 in (91.44 cm × 91.44 cm), Watercolor painting by Stephen Zhang, 2024–2025

living things. The continuous interaction between the Yin attributes of things (e.g., stillness, astringency, softness, and invisibility) and the Yang attributes (e.g., movement, expansion, rigidity, and manifestation) creates a complex world of dynamic equilibrium. This painting presents the form of Cao-guo in its biophysical habitat and its cultural value in human society in the form of a Yin-Yang diagram, which not only expresses the Chinese philosophical concept of the dynamic equilibrium of Yin and Yang, but also shows the correlation between the biophysical and cultural-symbolic-linguistic dimensions of this species, like Pewen seeds in the biogeochemical and cultural-linguistic circle in the Pewenche and Mapuche societies (Rozzi 2013).

22.5 Conclusion

Rozzi (2018) has pointed out that biocultural homogenization is a "wicked problem", one that continues to erode biocultural diversity across many regions of the globe. One contributing factor is the taxonomic bias in wildlife conservation, which tends to favor flagship species such as mammals, birds, and vascular plants (Bonnet et al. 2002; Rozzi 2019). Kalinkat (2017) points out, this bias can lead to the marginalization of a vast number of plant species, which remain largely unknown to the public despite their critical roles in maintaining global and regional biodiversity. To address these problems, in this chapter, we employed the methodology of "field environmental philosophy" (FEP) and "3Hs" (habitats, habits, and co-inhabitants) (Rozzi 2018) as the theoretical framework to examine the Cao-guo by using the species-centered approaches in Tengchong, Yunnan, China. Based on the research, we explore different ways to bridge the channel between academia and the public. We investigated how to explore the possibility of conducting biocultural studies at the local level for biodiversity conservation, restoration, and communication.

Through these sketches, we attempt to link biodiversity and cultural diversity through research and artistic expression. The biocultural sketches from the fieldwork became the basis for the subsequent artworks. Based on these sketches, the artist and designer, Stephen Zhang, completed a series of biocultural watercolor paintings designed to convey the ethical message of a multi-species co-habitat. Additionally, these sketches and watercolor paintings will lay the groundwork for future education and communication regarding arts, design, and the environment. We plan to facilitate biocultural communication through exhibition, publication, pedagogy, and interdisciplinary collaboration.

Acknowledgments We would like to thank Danqiong Zhu and Ricardo Rozzi for their valuable comments on draft versions of this chapter. Special thanks to Danqiong Zhu for her biocultural interpretation of the watercolor paintings.

References

Argumedo A, Pimbert M (2010) Bypassing globalization: barter markets as a new indigenous economy in Peru. Development 53(3):343–349

Bertling JG (2023) Art education for a sustainable planet: embracing Ecopedagogy in K-12 classrooms. Teachers College Press

Bonnet X, Shine R, Lourdais O (2002) Taxonomic chauvinism. Trends Ecol Evol 17(1):1–3

Book of Changes. trans. of Chinese: Legge J. Chinese text project. Accessed through: https://ctext.org/book-of-changes

Cai J, Yu WB, Zhang T, Wang H, Li DZ (2019) China's biodiversity hotspots revisited: a treasure chest for plants. PhytoKeys 130:1–24

Chan BPL, Bi Z, Duan SZ (2019) Introduction to a four-year biodiversity survey of Tengchong section of Gaoligongshan National Nature Reserve, in the footsteps of pioneering naturalists in western Yunnan. China J Threat Taxa 11(11):14391–14401

Cheng Y, Liu Z (2012) The clinical treasures of Chinese medicine through the ages series: Ciyu Xinshu. People's Military Medical Publishing House, Beijing

Dou Y, Li Y, Li M, Chen X, Zhao X (2023) The role of agroforestry in poverty alleviation: a case study from Nujiang prefecture, southwestern China. Sustainability 15(15):12090

He G, Yang S, Wang Y (2023) The potential of Amomum Tsao-ko as a traditional Chinese medicine: traditional clinical applications, phytochemistry and pharmacological properties. Arab J Chem 16(8):104936

Kalinkat G, Cabral JS, Darwall W, Ficetola GF, Fisher JL et al (2017) Flagship umbrella species needed for the conservation of overlooked aquatic biodiversity. Conservation Biology, 31(2): 481-485

Lao Tzu. Dao De Jing. trans. of Chinese: Legge J. Chinese text project. Accessed through: https://ctext.org/dao-de-jing

Myers N, Mittermeier RA, Mittermeier CG, da Fonseca GA, Kent J (2000) Biodiversity hotspots for conservation priorities. Nature 403(6772):853–858

Rozzi R (2012) Biocultural ethics: recovering the vital links between the inhabitants, their habits, and habitats. Environ Ethics 34(1):27–50. https://doi.org/10.5840/enviroethics20123414

Rozzi R (2013) Biocultural ethics: from biocultural homogenization toward biocultural conservation. In: Rozzi R, Pickett ST, Palmer C, Armesto JJ, Callicott JB (eds) Linking ecology and ethics for a changing world: values, philosophy, and action, Ecology and ethics, vol 1. Springer, Dordrecht, pp 9–32. https://doi.org/10.1007/978-94-007-7470-4_2

Rozzi R (2018) Biocultural homogenization: a wicked problem in the Anthropocene. In: Rozzi R, May RH Jr, Chapin FS III et al (eds) From biocultural homogenization to biocultural conservation, Ecology and ethics, vol 3. Springer, Dordrecht, pp 21–47. https://doi.org/10.1007/978-3-319-99513-7_2

Rozzi R (2019) Taxonomic chauvinism, no more! Antidotes from Hume, Darwin, and biocultural ethics. Environ Ethics 41(3):249–282. https://doi.org/10.5840/enviroethics201941325

Rozzi R (2023a) Inter-species and inter-cultural encounters: the biocultural education program of the Omora Ethnobotanical Park. In: Rozzi R, Tauro A, Wright T, Avriel-Avni N, May RH Jr (eds) Field environmental philosophy: education for biocultural conservation, ecology and ethics series, vol 5. Springer, Dordrecht, pp 153–174. https://doi.org/10.1007/978-3-031-23368-5_10

Rozzi R (2023b) Introduction to part I. Field environmental philosophy: concepts and case studies. In: Rozzi R, Tauro A, Avriel-Avni N, Wright T, May RH Jr (eds) Field environmental philosophy, Ecology and ethics, vol 5. Springer, Cham, pp 17–25. https://doi.org/10.1007/978-3-031-23368-5_2

Rozzi R, Massardo F, Anderson C, Heidinger K, Silander J Jr (2006) Ten principles for biocultural conservation at the southern tip of the Americas: the approach of the Omora Ethnobotanical Park. Ecol Soc 11(1):43. https://doi.org/10.5751/ES-01709-110143

Rozzi R, Arango X, Massardo F et al (2008a) Field environmental philosophy and biocultural conservation: the Omora Ethnobotanical Park educational program. Environ Ethics 30:325–336. https://doi.org/10.5840/enviroethics200830336

Rozzi R, Armesto J, Goffinet B et al (2008b) Changing lenses to assess biodiversity: patterns of species richness in sub-Antarctic plants and implications for global conservation. Front Ecol Environ 6(3):131–137. https://doi.org/10.1890/070020

Rozzi R, Armesto JJ, Gutiérrez J et al (2012) Integrating ecology and environmental ethics: earth stewardship in the southern end of the Americas. Bioscience 62(3):226–236. https://doi.org/10.1525/bio.2012.62.3.4

Rozzi R, Tauro A, Avriel-Avni N, Wright T, May RH Jr (eds) (2023) Field environmental philosophy: education for biocultural conservation, Ecology and ethics, vol 5. Springer, Cham. https://doi.org/10.1007/978-3-031-23368-5

Shan F, Huang L, Guo J, Chen M (2015) History and development of "One Root of Medicine and Food" (药食同源的历史和发展概况). Life Sciences (生命科学), 27(08): 1061–1069
Thompson M, Xu L (2023) A biocultural dialogue between Thoreau and Taoist thought: rethinking environmental ethics, nature, spirituality and place. Religion 14(7):829
Yang Y, Yan RW, Cai XQ, Zheng ZL, Zou GL (2008) Chemical composition and antimicrobial activity of the essential oil of Amomum tsao-ko. J Sci Food Agric 88(12):2111–2116
Yang YW, Liu XL, Pu CX, Qian ZG, Guan KY (2014) The influence of altitude and latitude on breeding of Amomum Tsaoko (Zingiberaceae). J Biosci Med 2(1):9–17. https://doi.org/10.4236/jbm.2014.21002
Yang ZQ, Hu YF, Nong PY, Yang RL, Zhou DL (2017) Investigation of Amomum tsaoko planting area and climate factors analysis of ecological suitability of Amomum tsaoko in Yunnan. China J Agric Resour Reg Plan 38:178–186
Zhang V, Yang SC, Wei X, Xie SC (2011) Developing status and strategies of Amomum tsao-ko plantation in Yunnan. World Sci Technol Modern Tradit Chinese Med 13(5):899–903
Zhou Y (2013) Branding Tengchong: globalization, road building, and spatial reconfigurations in Yunnan, Southwest China. In: Blumenfield T, Silverman H (eds) Cultural heritage politics in China. Springer, New York, pp 247–259
Zhu D, Rozzi R (2026) Biocultural Communication: Arts and Sciences for Conserving Biological and Cultural Diversity. In: Zhu D, Rozzi R, Tsuji R, Johnson B, Caviness T, Castro-Jorquera C, Xu L, Zhang L (eds) Linking arts with biocultural conservation, restoration, and communication. Springer, Cham, pp 243– 263. https://doi.org/10.1007/978-3-032-04830-1_15
Zhuangzi. Zhuangzi. trans. of Chinese: Legge J. Chinese text project. Accessed through: https://ctext.org/zhuangzi

Chapter 23
Immersion and Interactiveness: Eco-themed Art Exhibitions and Biocultural Communication in East Asia

Zihan Li, Meng Jia, and Kun Ma

Abstract Biocultural communication is an important component of field environmental philosophy (FEP) and an indispensable element of biocultural conservation and restoration. This study aims to explore how eco-themed art exhibitions promote biocultural communication. We focus on exhibitions that have taken place in three countries of East Asian: China, Japan, and South Korea. As a form of contemporary biocultural communication, eco-themed art exhibitions increasingly attract the attention of ecologists, artists, and citizens. In response to the prevailing socio-ecological predicament in East Asian countries, urban artists have been actively collaborating with urban and rural communities to present eco-themed art exhibitions and to raise citizens' awareness of and sensitivity to climate change, biocultural homogenization, and other complex socio-environmental problems Eco-themed art exhibitions, such as the Insect Exhibition in Quantum Farm in Hangzhou, China, the Guanzhong Mangba Arts Festival in Shaanxi, China, Our Ecology Toward Planetary Living in Japan as well as the Gangwon International Triennale and Sea Art Festival in South Korea, are characterized by immersiveness and interactivity. Drawing on the 3Hs (co-inhabitants, Habits, and Habitats) framework of biocultural ethics, in this chapter, we explore how immersive and interactive eco-themed art exhibitions facilitate biocultural communication.

Keywords East Asia; · Eco-themed art exhibition; · Biocultural ethics; · Immersion; · Interactiveness

Z. Li
School of Chinese Language and Literature, Catholic University of Korea, Bucheon, Gyeonggi Province, South Korea

M. Jia (✉)
Department of Philosophy, Xian Jiaotong University, Xi'an, China
e-mail: jiameng@stu.xjtu.edu.cn

K. Ma
Hangzhou Huacheng Inc., Hangzhou, Zhejiang, China

D. Zhu et al. (eds.), *Linking Arts with Biocultural Conservation, Restoration, and Communication*, Ecology and Ethics 8,
https://doi.org/10.1007/978-3-032-04830-1_23

23.1 Introduction

At the beginning of the twenty-first century, eco-art exhibitions of various scales were organized around the world in response to the severity of the global environmental crisis, resource depletion, food security, and climate change. Some of these exhibitions focused on urban sustainability, such as the 2010 World Expo held in Shanghai, China, with the theme "Better City, Better Life." In these exhibitions, artists and other parties advocated for the establishment of "harmonious cities," i.e., cities based on harmony between humans and natural environment, and more generally, between humans and the material world for a more resilient future. Some exhibitions focused on food and agriculture, such as BioFach in Nuremberg, Germany and Salon International de l'Agriculture in Paris, France. These exhibitions included displays of world's leading organic food and sustainable agriculture and provided platforms for producers, retailers and consumers with an interest in both organic and environmentally friendly products to network with each other. These events gave origin to a variety of eco-themed exhibitions that serve multiple functions, such as raising public awareness of ecological issues, introducing new technologies and products for sustainability and environmental conservation, educating citizens and consumers about the ecosystem, and promoting collaboration among organizations and professions.

In this chapter, we focus on eco-themed art exhibitions with more educational/communicational functions, especially those that have emerged in the 2010s in East Asia, to demonstrate the importance of biocultural conservation and restoration. These exhibitions often combine natural environments, arts, technology, and local habitats, presenting the complex interactions between people and their environment through highly interactive approaches in an immersive setting. The incorporation of environmental elements into the exhibition space helps people to recognize, experience, and have contact with other-than-human co-inhabitants in a fresh multisensory manner. To plan eco-themed art exhibitions, curators and artists have to work on the integration of their aesthetic perspectives and socio-ecological literacy. Through interactive and immersive processes, these eco-themed art exhibitions, generate innovative expressions of biocultural communication. To illustrate how complex interactions among co-inhabitants, habits, and habitats (i.e., "3Hs" model of the biocultural ethic) can be appreciated by curators, artists, and citizens, we present case studies from east Asia.

23.2 Theoretical Framework

We inhabit an era of ecological degradation and social injustice (Rozzi 2015; May 2018). While urban population is constantly growing, in addition to the problems of urban ecological security (Hodson and Marvin 2009), urban poverty (Wratten 1995), and urban environmental injustice (Wilson et al. 2008), people's access to nature is also marked by inequality. According to a quantitative study of racial/ethnic and socioeconomic disparities in urban green space accessibility, neighborhoods of African Americans or with low socioeconomic status have poor access to green spaces (Dai 2011). Similar disparities have been found in analyses of urban green space accessibility that compare groups of contrasting socio-economic status and race/ethnicity in cities in the Global South (Schetke et al. 2016). These studies prove that environmental and social issues can not be divorced from each other. A specific, pervasive socio-environmental problem, called biocultural homogenization, is driven by rural-urban migration, linguistic homogenization, and reduction of philosophy education (Rozzi 2013). To address this wicked problem, Rozzi has proposed the 3Hs framework of biocultural ethics to "advance the understanding about the interrelationships among life habits, habitats, and co-inhabitants" (Rozzi 2018).

In the late twentieth-century, Chilean philosopher and conservation biologist Ricardo Rozzi and his team further created the field environmental philosophy (FEP) as a pedagogical methodology and paradigm for formal and non-formal education to empower students learning about the value of biocultural diversity (Rozzi et al. 2008, 2023). FEP invites students and other participants to appreciate and practice various ways of intercultural and interspecies co-inhabitation. This requires them to leave the classroom and have different field experiences in natural habitats as well as urban settings. As a methodological approach, FEP integrates epistemological, ontological, ethical, ecological, and aesthetic dimensions of understanding, catalyzing field activities such as "ecotourism with a hand-lens" (Tauro et al. 2021) or "pay attention, dive with eyes wide open" (Ojeda et al. 2023).

Besides FEP, other forms of informal activities can also facilitate citizens to understand the interrelationships among the 3Hs. For example, to bridge the epistemological gap between ecological sciences and citizens' awareness of biocultural diversity, exhibitions are utilized both in flagship species conservation and moss conservation in China. This helps people to understand that pandas and mosses are humans' co-inhabitants, with their own value and life-purpose (Zhu 2023). Consequently, pandas, mosses, and other living beings should be valued as co-inhabitants who are active subjects, and not mere "natural objects" to be used by humans.

In this chapter, we propose that biocultural communication can also play an effective role to improve the understanding of citizens about the vital links among the diversity of co-inhabitants, their life habits, and shared habitats. We focus on a particular approach to biocultural communication: eco-themed art exhibitions. We define them as events, displays, or series of activities specifically designed to raise citizens' awareness, to educate, and to communicate about socio-ecological crises, resilient practices, and local biocultural diversity. In our view, eco-themed art exhibitions also function to counter two senses of exclusion of biocultural diversity as defined by Zhu (2023, p. 459), namely, "exclusion of general citizens from the plethora of ecological information" and "exclusion of many invertebrates and nonvascular plants from biological conservation." Based on our case studies, we aim to demonstrate that eco-themed art exhibitions can overcome both kinds of exclusion. Hence eco-themed art exhibitions can be effective to connect citizens with local biological and cultural diversity, leading to interactive/immersive forms of biocultural communication (Zhu et al. 2026).

23.3 Case Studies of Eco-themed Art Exhibitions in East Asia

23.3.1 Conceptual and Methodological Approach

To analyze our case studies, we integrate concepts and practices of FEP and biocultural communication with other conceptual and methodological approaches contained in eco art education. Many eco-exhibitions incorporate sciences and arts to improve collective behaviors towards biocultural conservation, restoration, and regeneration based on multidisciplinary expertise, nurturing ecological awareness and aesthetic appreciation. According to Inwood (2008), eco-art education broadens the boundaries of environmental education, which has been heavily rooted in science education. Instead, art education encourages a strong interdisciplinary approach to environmental education. Given that eco- art education is best conducted in an experiential, interactive, creative, and imagination-based manner, as this fosters a sense of empowerment and interconnectedness among participants (Blandy and Hoffman 1993), theoretically, we assume that eco-themed art exhibitions would benefit from this approach. Particularly, many practitioners of eco art education or eco artists recognize the interdependence and interconnectedness among all things. Accordingly, they commit to conveying the idea that we as living beings among plants, animals, other humans, planet Earth, and the universe are one breathing organism. This view converges with a relational ontology of biocultural perspective (Tsuji and Johnson 2026). At the same time, technology can also play a role. A human-technology-nature relationship may become a conduit or provide a sensorial pathway to establishing a reconnection to the natural realm through the use of digital technologies (Coles and Pasquier 2015). Equipped with this set of concepts we proceed now to describe and examine specific case studies in three East Asian countries: China, Japan, and South Korea.

23.3.2 Eco-themed Art Exhibition in China

Due to China's vast territory, rich biocultural heritage, and rapid urbanization in the past years, it is imperative to connect urban populations with nature and biocultural diversity. In the current environment of building an ecological civilization, eco-themed art exhibitions are also in full swing across China. We selected two art exhibitions, one in East China and the other in West China, to examine the mechanism.

The esat case is Qiantang Farm, or Hangzhou Quantum Farm (钱唐农园), the first landscape project launched in the second phase of Qianjiang New City in Hangzhou, Zhejiang Province, China. As a key project of Qian Investment to implement the utilization of cultivated soil, historical context and other resources in the development area of Qianjiang New City II, it aims to re-establish the connection between people and land, food and natural environment to create a diversified living space that provides various research, education and practical activities related to sustainable agricultural development and urban green-healthy lifestyle. Covering a total area of about 220 acres, the northern area (about 167 acres) offers platform for education, field for vegetable planting experience, pasture experience, arts studios, shared farm, pond irrigation, and other experiential/educational functions and the south area (about 54 acres) offer facilities to support an industrial park that brings together high-quality enterprises such as environment, agriculture, water conservancy, as well as light luxury style of catering service (Zheng 2019).

From 2019 to 2023, the farm had been organizing an annual insect exhibition, combining entomology with arts, traditional cultural-entomological knowledge, and digital technology to present the diversity and beauty of insects to communicate with citizens about ecological functions of insects, the intricate human-insect relationships, as well as provide citizens to feel and understand the entomological world through multi-sensory experiences such as lighting technology, sound, touch, physical interactions, etc. The annual exhibitions vary slightly in terms of curation themes. The 2020 exhibition was organized in the sequence of the life cycle of insects in four seasons, starting from summer to the rest three seasons, themed as "Chóng Shèng" (虫盛, Summer Vitality), Autumn as "Chóng Shēng" (虫声, Autumn Chirp), "Chóng Shēn" (虫深, Winter Deep), "Chóng Shēng" (虫生, Spring Reborn). One of the interesting things is that the Chinese pronunciations of the four themes are very similar, with only differences in tones or nasal sound. The thematic installation "The Eye of the Cosmos" in the Prologue Hall brings visitors into a universe composed of humans, insects, and numerous other creatures. Over the years, various themes such as Insects Pop-up Exhibition (昆虫快闪展, 2019), Reborn (虫生, 2020), Following Insects to Wilderness (随虫去野, 2021), Bean Kid's Journey to Wonderland (豆仔的奇境之旅, 2022), and Bean Kid's Cute "Bug" (豆仔的萌"虫"总动员, 2023), etc., created an immersive space for citizens to learn about insects (Fig. 23.1).

The case in West China is the Guanzhong Mangba Arts Festival (关中忙罢艺术节, Guanzhong Harvest Art Festival) in Caijiapo Village, Shijing Town, Shuangyi District, Xi'an, Shaanxi province. In 2018, Xiaochuan Wu and Yaqian Zhang from the Xi'an Academy of Fine Arts established the Guanzhong Art Cooperative and initiated this series of rural cultural practices in Caijiapo Village. "Mangba", in Chinese "忙罢," literally means "after being fully engaged" or "post-harvest," linking to the endured Chinese tradition that during the season of "wheat harvest and autumn planting" (麦收秋种), usually in the fourth month of the lunar calendar, the newly harvested wheat would be offered to ancestors and gods as a sacrifice.

As a celebration of the harvest tradition of the "Mangba", Mangba Arts Festival is designed in four main sections: the Zhongnan Theater Festival, the Wheat Field Art Exhibition, the Cooperative Art Project, and the Guanzhong Grain Crops Project. According to the initiator of the art festival, Xiaochuan Wu and others, professor and director of the Experimental Art Department of Xi'an Academy of Fine Arts, the primary goal of "Rebuilding villages through Arts" (艺术乡建, or art engagement in villages), is not to simply display a certain urbanized or academic style of art in the

Fig. 23.1 Left: The Cubic Spinning Wood Game. It featured in the 2021 exhibition shows Chinese characters and pinyin to attract children to participate in the game and activity to learn about cultural-entomology. Right: Classroom for children learning activities of the 2023 exhibition. (Left photo in public domain. Right photo by Kun Ma)

villages. Rather, based on the status quo of social development, the modernization and development of the townships, the local situation, and the appeals of the villagers, it is imperative to observe in depth and continuously study the evolving qualities of the villages in terms of culture, economy, habits, and customs (Wu et al. 2019). During the first art festival, around 10 international and domestic artists created artworks such as installations, behaviors, images, environmental reconstructions, etc., with the land as the carrier and the Zhongnan Mountains as the background. The exhibition not only aimed to enrich local cultural activities, but also to capture the profound relationship between people and their land, farming practices and culture, memories and childish delight, landscape and nostalgia poetry, etc. (Wu et al. 2019) (Fig. 23.2).

23.3.3 Eco-themed Art Exhibition in Japan

In recent years, Japan has also held quite several eco-themed art exhibitions, including "the Echigo-Tsumari Art Triennale" (ETAT, since 2000), "Setouchi Triennale" (since 2010), "Kenpoku Art" (2016), "ART for SDGs" (北九州未来創造藝術祭, 2021), *Singing Birds, Moving Mountains* (鸟鸣山河, 2024), and the most recent exhibitions, such as *Our Ecology: Toward Planetary Living* (2023–2024) and *Dance with All on Animals and Anima* (2025), etc. Curators and artists of these exhibitions have planned and created their works with a strong awareness of socio-ecological complexity, and the articulation and interpretation of artworks also convey rich messages on the coexistence between people and the environment. *Our Ecology: Toward Planetary Living* at the Mori Art Museum in Tokyo, Japan, in 2024 aims to show how contemporary art and artists are responding to the environmental crisis. The exhibition featured artworks by global artists as well as local Japanese artists, which provoked further public reflection and positioned the museum as a venue for dialogue on socio-environmental issues.

Fig. 23.2 Guanzhong Mangba Arts Festival. Upper left (**a**): God bless a well-fed people (上帝保佑吃饱了饭的人民). The artwork is the letter "m" (two gates) decorated with steamed buns, which were given to visiting villagers after the exhibition ended that day. Artwork by Mengmeng Wang and Xiaojuan Zhang, exhibited during the 1st Guanzhong Mangba Arts Festival, 2018. Upper right (**b**): The Book of Wheat (麦书). This work was created by dipping the wheat sheaves in ink and then tapping them on the canvas in 2018. The artist named this work "The Book of Wheat" to express the idea that human beings are not the only ones who write or create artwork. Artwork by Feng Yang, exhibited during the first Guanzhong Mangba Arts Festival, 2018. Lower left (**c**): Shadow theater performed by local artists. It is a folk cultural heritage, an ancient form of storytelling and entertainment. Flat cut-out figures are interposed between a light source and a semi-transparent veil, and the audience watches the action and narrative expressed by the shadows. Shadow puppets are voiced through the human voice. Lower right (**d**): Child of the Earth Sculpture on the site of Guanzhong Mangba Arts Festival. (Photographs (**a**), (**b**), and (**c**) were credited to the Mangba Arts Festival Committee. Photo (**d**) was taken by Lu Zhao)

Through diverse expressive artworks of 34 artists from Japan and around the globe, "Our Ecology" featured four sections, namely, "All is Connected," "Return to Earth," "The Great Acceleration," and "The Future is Within Us." Each section communicates a distinct theme. For example, section I was presented by contemporary artists, and the artworks particularly revealed the inseparably connected processes between the subject of ecology and economy, social life, and other aspects of human societies. Through these artworks, the myriad of visible or invisible processes became tangible, visible, and audible (Mori Art Museum 2023). In this section, several works by Chilean artist Cecilia Vicuña were exhibited. Vicuña's "Precarios" artworks such as *Tres Elementos* (2014), *Alas Solferino* (2014) (Fig. 23.3, left), and *Transistor* (2014), were small sculptures made from cast-off materials like shells,

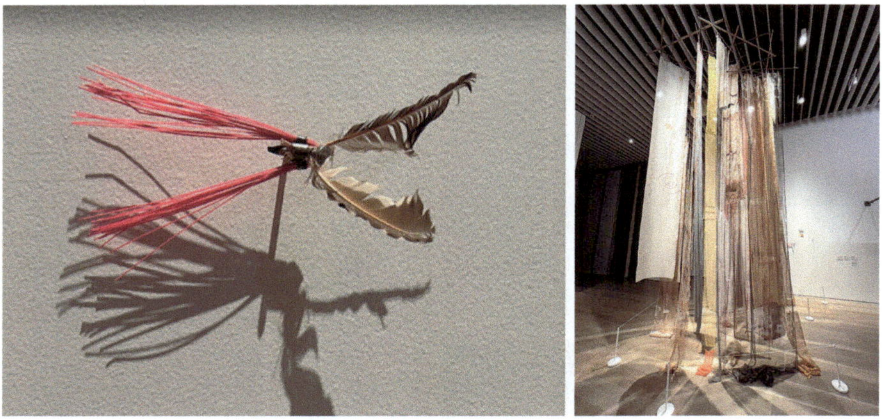

Fig. 23.3 Left: *Alas Solferino* (不安定なもの, Precarios), by Cecilia Vicuña, 2014. Right: *Quipu Girok* (キープ・ギロク), by Cecilia Vicuña, 2021. (Photographs by Meng Jia)

wool, feathers, plastic, glass, nets, and stones, etc., to reveal "the forces beyond human life which created those [spatial] poem" (Mori Art Museum 2023).

According to Chilean art historian and curator Carolina Castro-Jorquera (2026), the precarious art of Vicuña also discovered connections among all earth beings and expressed the necessary appreciation of the small and discarded things. Vicuña's other group of works, "Quipu", is also presented in "Our Ecology," most notably *Quipu Girok* (2021). *Quipu Girok* was an installed artwork of columns of painted gauze, reminiscent of pre-colonial techniques of indigenous painting in the Andes on weaving, silk, polyester, and cotton that all hang vertically from uneven bamboo sticks (Fig. 23.3, right). As a main staple of her art, *Quipus has* become the fundamental metaphor to communicate the erasure of Indigenous people and their cultures, which are also the ethical concerns of biocultural conservation and restoration.

Sweden Artist Nina Canell focuses on the physical and chemical nature of art materials, particularly on the interchanging processes of materials. In *Our Ecology*, she presented her work, *Muscle Memory (5 Tonnes)*, a transformative installation of five tonnes of seashells that cover the floor of an entire room (Fig. 23.4, left). This installation thus transformed a place of static observation into an interactive space where encounters can happen. When visitors came to *Our Ecology* walking above the shells, they were invited to participate in the collective process of crushing, grinding, sculpting the shells, making sounds, and likely, reflecting. Their visual, auditory, tactile, and even olfactory senses were engaged, enabling them to deeply perceive the messages conveyed by the artwork. The audience was no longer just passive observers or appreciators of the artwork, but instead actively influenced the content and presentation of the artwork through the act of "entering" or "participating" in completing the piece.

Japanese artist Saijo Akane (西條茜)'s ceramic artworks, *Orchard* and *Wetland,* were collected in section IV, "The Future is Within Us." This section serves to

Fig. 23.4 Left: *Muscle Memory (5 Tonnes)* by Nina Canell. Right: *Orchard* by Saijo Akane. (Photographs by Meng Jia)

present various "artist expressions that envision a future influenced by activism, indigenous wisdom, feminism, and artificial intelligence" (Mori Art Museum 2023, p. 5). Akane is known for applying ceramic as the primary material to express her thoughts on abstract sculpture, various local culture and history, personal experiences and memories, "as well as biomorphic sculptures inspired by her own body" (Mori Art Museum 2023, p. 180). The ceramic *Orchard* resembles the internal organs of living things with multiple holes into which visitors can blow to produce sounds, which generates potential examination or introspection on communications in dramatically changing societies riddled with political and ideological polarization and economic disparities through the situation of the body.

Overall, the exhibition *Our Ecology*, as the Mori Art Museum's 20th anniversary Exhibition, represents the Mori Art Museum's response to the International Council of Museums' (ICOM's) initiative to make museums accessible and inclusive places that foster diversity and sustainability. Bringing together the works of global and local Japanese artists also demonstrates the curators' commitment to the museum and arts as a venue and medium for communicating about climate, environment, energy, science, society, and cultures.

23.3.4 Eco-themed Art Exhibition in South Korea

In recent years, eco-themed art exhibitions in South Korea have emerged as important practices, incorporating a variety of themes such as climate crisis, local adaptation, social engagement, and artistic innovation. Among those practices, several eco-themed art exhibitions have received considerable public attention, notably the Sea Art Festival (바다미술제, since 1987), the Gangwon International Triennale (강원국제트리엔날레, since 2013), and Time of the Earth (대지의 시간, 2021–2022), etc. According to Gim Jungi, the chief curator of National Museum of Modern and Contemporary Art, South Korea, as "an art form that expands ecological values via the ethics, attitude, discourse, and methodology of artistic action," ecological art has become the basis for a new spiritual culture in South Korea (Jungi 2021, p. 9). Some of these exhibitions focus exclusively on local history, culture, landscape, and the connection between people and their local environment. For example, over the years, Gangwon International Triennale has exhibited artworks of nature in the18 cities and counties in Gangwon Province to seek an engagement with "various art forms and experiences," "alternative knowledge production and distribution methods," as well as foster interactions among multiple elements (Gangwon International Triennale 2024). Among these eco-art exhibitions, the Sea Art Festival and the Gangwon International Triennale are particularly important for examining the interactive mechanism of eco-art exhibitions in South Korea.

Since 1987, the Sea Art Festival has been held in an open and accommodating coastal area such as Ilgwang Beach. Located in Gijang-gun, Busan, Ilgwang Beach means "일광해수욕장" (sunshine beach) in Korean, representing a beautiful coast of sunshine and harmony, which perfectly matches the theme of the Sea Art Festival. In 2023, the Festival organized a 37-day exhibition, *Flickering Shores, Sea Imaginaries* (깜빡이는 해안, 상상하는 바다), to rethink the history of coastal villages and the origins of marine ecosystems, as well as to reimagine the relationship between human beings and the ocean. Forty-three artists from more than 20 countries submitted several works to the festival, some of which not only conveyed the artists' thoughts and concerns on marine ecosystems and human-ocean relationships but also encouraged festival visitors to interact with the ocean through artworks (Figs. 23.5 and 23.6 left).

Gangwon International Triennale was initiated in 2013 as a Pyeongchang Biennale event. Over the years, the event has been taken to a broader stage and transitioned into international triennale events. In 2021, the Gangwon International Triennale organized *Warm Revitalization* (따스한 재생) and presented artworks and programs by 139 artists from 39 countries. *Warm Revitalization* embraced themes such as technology, ecology, everyday life, and locality. Particularly, it aimed to convert idle areas in the local region into places with vitality, such as the former ammunition maintenance plant (built in 1973), Wadong branch school (originally opened in 1954 after the Korean War), Hongcheon Art Museum (built in 1956), and Hongcheon Joongang Market. According to Sung-Ho Kim, the Artistic Director of the Gangwon Triennale 2021, the exhibition halls were specifically situated in the above four sites, which represent the four dimensions of the revitalization, namely revitalization of technology (기술 재생), ecological regeneration (생태 재생), everyday rebirth (일상 재생), and regional regeneration (지역 재생) (Kim 2021). In 2024, the Gangwon Art & Culture Foundation organized *Ecological Art from Beneath* (아래로부터의 생태예술), which presented 65 artworks from 21 countries and further examined the intersections between local art and global art.

Fig. 23.5 Left: *Aquatic Garden* (수생 정원) by South Korean art studio Studio 1750 (스튜디오 1750), with filter brush, pipe, LED, sensor, 300 × 250 × 1000 cm. Considering that, through anthropogenic activities, more than 199 million tons of plastic have been released into the oceans, the artists of this installment have created an imaginary garden in which human intervention has wreaked devastation. During the Festival, visitors were invited to make a sea creature paper hat, wear the hat, walk through the installation, and perform like the species. Through this performative and interactive strategy, visitors could understand, feel, and imagine how "a new and peculiar species" has to mutate and evolve to adapt to the changed environment. Photo credit: ©Busan Biennale Organizing Committee. All rights reserved. Right: *Rumors from the Sea* (바다의 풍문) by sound artist and engineer Félix Blume. The work employs a range of materials such as bamboos, bamboo flutes, wood path, and metal structure, 150 bamboo pieces of 3 m long & 12 cm diameter placed in the sea. This work creates an orchestra of bamboo flutes played by the tides and waves, 24 h a day without ceasing, with the music changing according to the direction, rhythm, and power of the tides and waves. Visitors can listen to the oceanic music and decide on their own opening and closing, thus entering into a dialogue with the ocean. Both installations were commissioned by the Sea Art Festival 2023. (Photo credit: ©Busan Biennale Organizing Committee. All rights reserved)

Fig. 23.6 Left: *Seaweed Shrine*, created by Julia Lohmann & Kayoung Kim, with Gijang kelp, rattan, wood, workshop tools, drawings, research about local algae culture, Dimension variable. This exhibition explored the local algae culture and human-seaweed relations. Commissioned by Sea Art Festival 2023. Photo credit: ©Busan Biennale Organizing Committee. All rights reserved. Right: *The Old Garden- Blooming Life* (오래된 정원-발화하는 생명), created by South Korean artist Hoyoung Lee (이호영) and local citizens, 250 × 3000 × 340 cm, with a steel frame, recycling objects such as daily necessities, plastic bottles, plant seedlings, white strings, natural objects (stone, wood, etc.). Local citizens participated in this cubic plant pavilion by bringing objects and donating them to the installation. Artists invited visitors to rethink the cycling of life in old gardens, such as death and birth, blooming and decaying, etc. (Photo credit: open access)

23.4 Discussion

The above case studies show that a large number of eco-themed art exhibitions have emerged in China, Japan, and South Korea in recent years. Examining these exhibitions from the perspective of biocultural ethics, we find that rich biocultural messages are transmitted through interactive mechanisms. As South Korean curator Jungi claimed, "the examination of the relationship between the environment and its inhabitants, including humans, is where ecology begins," and "ecology is the pursuit of unveiling the interconnectedness between all life" (Jungi 20, p. 9). This echoes the biocultural ethic insofar it recognizes the value of the vital links among co-inhabitants, their life habits, and shared habitats (Rozzi 2013).

Specific to these case studies, some exhibitions underscored the interconnectedness of humans with other-than-human beings, particularly the interspecies relationships as emphasized in local/regional cultures. For example, the insect exhibition in Hangzhou, China, as several Chinese characters shown in Fig. 23.1: left, combines China's native cultural entomology with modern scientific entomology, so that visitors have the opportunity to be nurtured by both scientific knowledge and cultural heritage. One of the characters, 螽(zhōng), usually refers to the Chinese Green Bush Cricket (*Tettigonia chinensis*), which was a widely documented species in classical Chinese literature. This indicates that visitors can expand the scope of their knowledge by examining the rich tradition of Chinese cultural entomology.

A second example, the Seaweed Studio also offered people an opportunity to connect to their rich, local "algae or seaweed-culture." In the region of Busan in southern South Korea, since the Goryeo Dynasty (918–1392), people have cultivated a tradition of offering seaweed soup to puerperal women for them to recover from giving birth. In Gijiang County, where the studio was held, seaweed would be served to the family when a child was born. The artists were also aware that the installation of the studio acknowledged the multispecies entanglement and relations in the local community (SAF 2023). Through these exhibitions, people can reconnect with their biocultural traditions and better understand the endured tradition of appreciating the coexistence between humans and other-than-human species in their daily life.

Eco-themed art exhibitions also present through interactive ways in new temporal and spatial grids, biocultural habits that have been shaped by long-term biocultural place-based practices. In the context of global migration to urban areas and biocultural homogenization, much biocultural practice and associated experience are at risk of extinction. Through exhibitions that focus on these habits and site-based practices, people (both residents and visitors) have the opportunity to learn about these nearly lost habits. These experiences can help citizen to recover these traditions with deep ecological significance. For example, the Guanzhong Mangba Arts Festival in West China has taken the time-honored ancient traditions of celebrating the harvest and giving thanks to the celestial powers for the gift of food to the people. Artists have transformed these traditions into art festivals that offer opportunities to revitalize rural communities and reunite people. These annual festivals not only link artistic creation with agricultural activities. They also express a concept of "art of living" that distinguishes itself from museum art; i.e., art detached from its original context, artificially re-situated within institutional settings such as museums and art galleries. This type of museum art can be identified as one of the two polar concepts of Chinese aesthetics.

In the world of traditional Chinese aesthetics, this is the artistic dichotomy between yang-chun-bai-xue (阳春白雪) and xia-li-ba-ren (下里巴人). The first refers to more elegant songs during the ancient Warring States Period, later used as a metaphor for profound and uncommon literature and art. The second alludes to folk songs that were popular among people of Chu during the Warring States Period (BC 475-BC221). Later, xia-li-ba-ren was also used as a metaphor for popular and widely accepted literature and art. Within the bio-cultural perspective, however, the boundaries of these dichotomized practices become vague, complementary, and conversational.

"Our Ecology" exhibition in Japan exemplifies artistic creations that revolve around and express indigenous ways of being and values can also be exhibited in museums. This is also communicated by Vicuña's artworks, which remind visitors about the fragility and beauty of the world. The awareness of living in a world of transience should foster ethical attitudes of care. Interactive approaches such as intensive explanation and on-site communication enhanced visitors' understanding and appreciation of the artworks (Fig. 23.7), carrying the educational and aesthetic impacts to deeper levels.

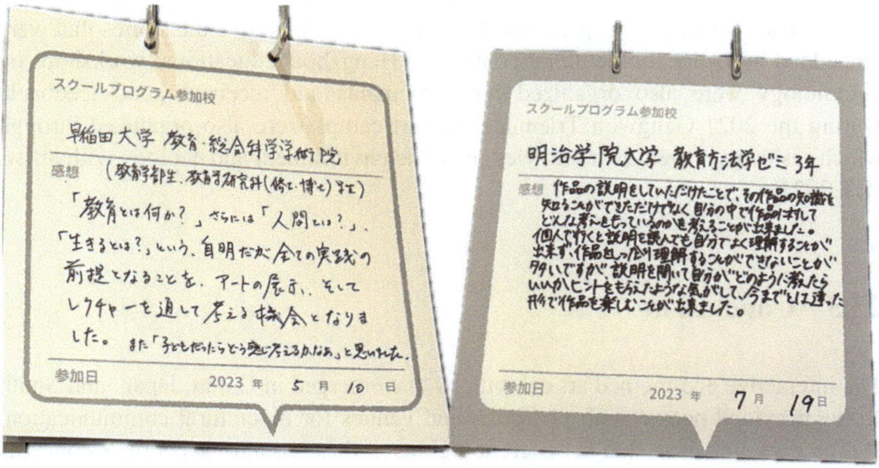

Fig. 23.7 Visitors' comments and messages on *Our Ecology*. The left message was from a first-year student in the Graduate program of the Department of Education in Waseda University, who said that the exhibition had provoked deeper questions such as "what is education?", "What is it being human?", "What is it being alive?", etc. The right message from a third-year student at Meiji Gakuin University expressed that the explanation of the artworks not only broadened the person's knowledge but also enabled the person to enjoy the artworks in a way that was different from before. (Photographs by Meng Jia)

Interactive eco-themed art exhibitions convey the curators' and artists' deep concern for the habitats they co-inhabit. Korea's 2023 Sea Art Festival clearly expressed the concerns of those involved (including curators, artists, researchers, and the public) about the pollution of the marine environment from anthropogenic activities. Irini Papadimitriou, the artistic director of the South Korea 2023 Sea Art Festival, sent a message to all participants and visitors to the Festival:

> The sea is deeply embedded in our lives and capitalist society, a vital source for our survival, but also a vast industry we exploit for food, medicines, energy, minerals, trading, travel, and so on. But increased human activity, from extensive cruise tourism, shipping, and overfishing to nuclear testing, pollution, and deep-sea mining, has been plaguing the sea, having a huge impact on marine ecosystems and habitats.
>
> Instead of viewing the sea from the coast as a divided and abstract surface for moving around commodities, *Flickering Shores, Sea Imaginaries* reminds us that we are part of this body of water. This year's Sea Art Festival aims to explore new relationships with the sea and its ecologies, enabling spaces for cooperation, collective visions, and synergies as a call to resistance and restoration. (SAF 2023, p. 5)

"Flickering shores" visualized the fragility of marine ecosystems and invited people to rethink the ways that they interact with the oceans.

In addition to exhibitions that mainly display artworks, other interactive activities such as public programs, symposiums, theatres, or workshops, were also offered during these events. The symposium "Ocean Voices" during South Korea's 2023 Sea Art Festival aimed to gather a wide range of scholars and artists to "examine the ocean as an industry exploited by capitalism and extractive practices." It also creatively empowered participants to "imagine a restorative relationship with our world's ocean through the arts" (SAF 2023). In the Guangzhong Mangba Festival in China, shadow theater was performed by local artists to narrate the stories that were embedded in the local biocultural context. In Hangzhou, educational workshops on entomology were also organized for participants and received good feedback. During the 2021 Gangwon Triennale, eco-art camps were also organized through which residents shared their knowledge on the environment and ecology with artists (GT 2021).

23.5 Conclusion

The interactive eco-themed art exhibitions that emerged in China, Japan, and South Korea provided people with platforms and venues for biocultural communication. Our case studies show that in different regions of these countries, people's concerns about the imminent loss of habitat, traditional life habits. To counteract this trend of biocultural homogenization, artists have integrated in novel ways traditional relationships between people and their environment, by transforming them into creative artistic activities. Through these interactive activities, people incorporate their biocultural traditions. They sustain biocultural narratives that are distinctive to them. As stated by the 3Hs framework of the biocultural ethic, people are embedded in certain biophysical and sociocultural environments where they shape the unique interrelations (Rozzi 2013). Now, with our own biocultural communication approach, we have shown that these interactive activities can become an inspiring and novel opportunity to creatively revitalize ancient, but dynamic, socio-environmental relationships that offer an ecologically meaningful drive and purpose for eco-themed art exhibitions.

Acknowledgement The authors sincerely thank Danqiong Zhu, Carolina Castro-Jorquera, and Ricardo Rozzi for their comments and suggestions on the structure and polishing of this chapter. Meng Jia would like to express her gratitude for receiving private funding for her research trip to Japan.

References

Blandy D, Hoffman E (1993) Toward an art education of place. Stud Art Educ 35(1):22–33

Castro-Jorquera C (2026) The precarious art of Cecilia vicuña: finding the ecological and sacred value of the small and the discarded. In: Zhu D, Rozzi R, Tsuji R, Johnson B, Caviness T, Castro-Jorquera C, Xu L, Zhang L (eds) Linking arts with biocultural conservation, restoration, and communication, Ecology and ethics, vol 8. Springer, Cham, pp 79–93. https://doi.org/10.1007/978-3-032-04830-1_5

Coles LL, Pasquier P (2015) Digital eco-art: transformative possibilities. Digit Creat 26(1):3–15

Dai D (2011) Racial/ethnic and socioeconomic disparities in urban green space accessibility: where to intervene? Landsc Urban Plan 102(4):234–244

GT (2021) Manual of Gangwon Triennale 2021 (강원국제트리엔날레2021). Commission of Gangwon Triennale

Hodson M, Marvin S (2009) Urban ecological security: a new urban paradigm? Int J Urban Reg Res 33(1):193–215

Inwood H (2008) Mapping eco-art education. Can Rev Art Educ Res Issues 35:57–73

Jungi G (2021) Ecological art as the basis for a new spiritual culture. In: Jungi G (ed) The progress and state of Korea ecological art: from nature art to ecological art. National Museum of Modern and Contemporary Art, Seoul, pp 9–10

Kim S (2021) Preface: warm revitalization (서문: 따스한 재생). In: Manual of Gangwon Triennale 2021 (강원국제트리엔날레2021). Commission of Gangwon Triennale

May RH (2018) Land grabbing and violence against environmentalists. In: Rozzi R, May RH, Chapin FS, Massardo F, Gavin MC et al (eds) From biocultural homogenization to biocultural conservation, Ecology and ethics, vol 3. Springer, Cham, pp 109–123. https://doi.org/10.1007/978-3-319-99513-7_7

Mori Art Museum (2023) Our ecology: toward planetary living. Mori Art Museum

Ojeda J, Tauro A, Contador T, González-Calderón J, Malebran J et al (2023) "Pay attention, dive with eyes wide open": a field environmental philosophy activity to foster reciprocity between people and nature. In: Rozzi R, Tauro A, Avriel-Avni N, Wright T, May RH Jr (eds) Field environmental philosophy: education for biocultural conservation. Springer, Cham, pp 87–100. https://doi.org/10.1007/978-3-031-23368-5_6

Rozzi R (2013) Biocultural ethics: from biocultural homogenization toward biocultural conservation. In: Rozzi R, Pickett ST, Palmer C, Armesto JJ, Callicott JB (eds) Linking ecology and ethics for a changing world: values, philosophy, and action, Ecology and ethics, vol 1. Springer, Dordrecht, pp 9–32. https://doi.org/10.1007/978-94-007-7470-4_2

Rozzi R (2015) Implications of the biocultural ethic for earth stewardship. In: Rozzi R, Chapin FS III, Callicott JB, Pickett STA, Power ME, Armesto JJ, May RH Jr (eds) Earth stewardship, Ecology and ethics, vol 2. Springer, Cham, pp 113–136. https://doi.org/10.1007/978-3-319-12133-8_9

Rozzi R (2018) Biocultural homogenization: a wicked problem in the Anthropocene. In: Rozzi R, May RH, Chapin FS, Massardo F, Gavin MC et al (eds) From biocultural homogenization to biocultural conservation, Ecology and ethics, vol 3. Springer, pp 21–48. https://doi.org/10.1007/978-3-319-99513-7_2

Rozzi R, Arango X, Massardo F, Anderson C, Heidinger K, Moses K (2008) Field environmental philosophy and biocultural conservation: the Omora Ethnobotanical Park educational program. Environmental Ethics 30(3):325–336. https://doi.org/10.5840/enviroethics200830336

Rozzi R, Tauro A, Wright T, Avriel-Avni N, May R (eds) (2023) Field environmental philosophy: education for biocultural conservation, Ecology and ethics, vol 5. Springer, Cham. https://doi.org/10.1007/978-3-031-23368-5

SAF (2023) Korea Sea art festival guidebook: flickering shores, sea imaginaries. Commission of 2023 Sea Art Festival

Schetke S, Qureshi S, Lautenbach S, Kabisch N (2016) What determines the use of urban green spaces in highly urbanized areas?–examples from two fast growing Asian cities. Urban For Urban Green 16:150–159

Tauro A, Ojeda J, Caviness T, Moses KP, Moreno-Terrazas R et al (2021) Field environmental philosophy: a biocultural ethic approach to education and ecotourism for sustainability. Sustainability 13(8):4526. https://doi.org/10.3390/su13084526

Tsuji R, Johnson B (2026) Biocultural ontologies, art, and conservation. In: Zhu D, Rozzi R, Tsuji R, Johnson B, Caviness T, Castro-Jorquera C, Xu L, Zhang L (eds) Linking arts with biocultural conservation, restoration, and communication, Ecology and ethics, vol 8. Springer, Cham, pp 61–77. https://doi.org/10.1007/978-3-032-04830-1_4

Wilson S, Hutson M, Mujahid M (2008) How planning and zoning contribute to inequitable development, neighborhood health, and environmental injustice. Environ Justice 1(4):211–216

Wratten E (1995) Conceptualizing urban poverty. Environ Urban 7(1):11–38

Wu XC, Zhang YQ, Zeng XZ (2019) An examination on "building villages through art" in the Guanzhong area ("艺术乡建"的关中探索). Shanghai Art Rev 04:53–55

Zheng J (2019) Hangzhou Qiantang farm grand opening: promising urbanites an idyllic dream (杭州钱唐农园 盛大开园:许都市人一个田园梦). Qianjiang Wanbao

Zhu D (2023) Bridge the channel, enhance the inclusivity: a comparison between flagship species-centered and moss-centered conservation in Chile and China. In: Rozzi R, Tauro A, Avriel-Avni N, Wright T, May RH Jr (eds) Field environmental philosophy: education for biocultural conservation. Springer, Cham, pp 457–482. https://doi.org/10.1007/978-3-031-23368-5_28

Zhu D, Caviness T, Rozzi R (2026) Introduction to part III: biocultural communication across urban, rural, and remote ecosystems. In: Zhu D, Rozzi R, Tsuji R, Johnson B, Caviness T, Castro-Jorquera C, Xu L, Zhang L (eds) Linking arts with biocultural conservation, restoration, and communication, Ecology and ethics, vol 8. Springer, Cham, pp 335–347. https://doi.org/10.1007/978-3-032-04830-1_21

Chapter 24
A Mere Touch of Green: Reconnecting People and *Shanshui* Through Poetic Dance Drama

Juan Wang

Abstract When dance meets Chinese Landscape Painting (CLP), the Chinese and global audience are offered a poetic dance drama, *A Mere Touch of Green*. With its remarkable interpretive skills, the poetic dance drama has not only been hailed as a phenomenal achievement in the field of Chinese contemporary art, but has also become a relentless effort to reconnect people with nature in the context of China's narrative of ecological civilization. Taking as a source of inspiration *A Panorama of Rivers and Mountains*, a legendary landscape painting by Wang Ximeng of the Northern Song Dynasty, the drama skillfully integrates the mood of verdant landscapes into the dance expression through carefully choreographed dance vocabulary, stage art design rich in Song Dynasty characteristics, and highly distinctive music composition, thus creating a unique audio/visual immersion. The goal of this chapter is to provide an in-depth analysis of the traditional Chinese aesthetic connotations and its environmental aesthetic effects presented in the poetic dance drama. Through exploring its poetic expression, color restoration and reinterpretation, audio-visual fusion, and narrative structure, it reveals the innovation and transcendence of *A Mere Touch of Green*. Inspired by the dance drama, the author performed an experimental dance in Xi'an Yannan Park characterized by low maintenance, resilient design, and rewilding, and integrated the visit experience into the dance pedagogy. During the experimental dance the author accomplished an innovation in narrative structure that bestowed new values and meanings to the dance.

Keywords *A Mere Touch of Green* · Poetic dance drama · *A Panorama of Rivers and Mountains* · Biocultural communication

---

J. Wang (✉)
Department of Arts, School of Humanities and Arts, Xidian University, Xi'an, China
e-mail: wj59@xidian.edu.cn

D. Zhu et al. (eds.), *Linking Arts with Biocultural Conservation, Restoration, and Communication*, Ecology and Ethics 8,
https://doi.org/10.1007/978-3-032-04830-1_24

24.1 Introduction

During the Covid-19 Pandemic, on August 18, 2021, the premiere press conference of the poetic dance drama, *A Mere Touch of Green* (只此青绿, another translation is *Only This Green*)—A Choreographed Portrait of *A Panorama of Mountains and Rivers* (千里江山图, literally translated as *A Thousand Li of Rivers and Mountains*), was launched at the People's Daily Network's Studio 1. The drama premiered at the National Center for the Performing Arts (NCPA) in Beijing from August 20th to 22nd and started its national premiere tour in September 2021. On January 31, 2022, on the Lunar New Year's Eve, excerpts from the drama were featured in the CCTV Spring Festival Gala 2022, and on April 21, *A Mere Touch of Green* was included in the White Paper" Chinese Youth in the New Era," released by the Information Office of the State Council. From August 24th to 28th, 2023, the poetic dance drama *A Mere Touch of Green* made its debut at the Venetian Theatre Macao, and on the second anniversary of its premiere, the drama officially embarked on a global tour. The drama was presented in the cultural performance "Journey along the Silk Road" at the Third "One Belt, One Road" International Cooperation Summit Forum in 2023, and also appeared in the dedicated cultural performance of "Chinese Culture Day" in Uzbekistan. By the end of 2023, *A Mere Touch of Green* has traveled to 55 cities in China and the international community, with 440 performances in theaters.

Given the significant impact of the poetic dance drama, *A Mere Touch of Green* has had on both the international and domestic cultural spheres, through this chapter I examine some of the poetic dance drama's revelations regarding biocultural communication and the general environmental aesthetics. I begin with an introduction to the meeting between the poetic dance drama and Chinese Landscape Painting, that is, the relevance of *A Mere Touch of Green* to *A Panorama of Mountains and Rivers*. I then address four aspects of the value of the drama: poetic expression, color presentation, audiovisual immersion, and the spirit of landscape. Next, I reflect on the narrative and metaphors of the dance drama in an attempt to investigate the relevance of the poetic dance drama as an artistic expression to biocultural communication. Finally, immersing myself in a more "natural" environment, explored the possibilities of experimental dance and the creativity of dancing pedagogy.

24.2 *A Mere Touch of Green*: Being and Absence

The poetic dance drama *A Mere Touch of Green* is co-produced by the Palace Museum, China Oriental Performing Arts Group, and People's Daily Online. It revolves around the classical landscape painting *A Panorama of Rivers and Mountains* and tells a story with seven sections, including "Zhan Juan" (展卷, Scroll Unfolding,), "Wen Zhuan" (问篆, Seal Tracing), "Chang Si" (唱丝, Silk Reeling), "Xun Shi" (寻石, Stone Searching), "Xi Bi" (淬墨, Brush Making), "Cui Mo"

Fig. 24.1 Complete and partial view of Wang Ximeng's *A Panorama of Rivers and Mountains* or *A Thousand Li of Rivers and Mountains* (千里江山图). The viewer can see the houses amidst the greenery and people engaging in their routines. In the partial view, in front of the house in the center is a watermill that provides power, a benefit of living by the water. Complete in 51.3 × 1191.5 cm, collected in the Palace Museum, Beijing. Open access

(习笔, Ink Grinding), and "Ru Hua" (入画, Into the Painting), through which a research fellow at the Palace Museum unfolds a beautiful scroll of landscape painting, wanders in the legendary aesthetic pleasures of traditional Chinese art, and visualizes the scene when the artist Wang Ximeng created the painting *A Panorama of Rivers and Mountains*.

The painting itself, which is housed in the Palace Museum, Beijing, is a silk painting in color by Wang Ximeng (around 1096–1119) of the Northern Song Dynasty. The painting is also the sole surviving work of the artist himself, who was 18 years old when he completed it. According to the inscription on the painting, it was completed in 1113, the third year of the Zhenghe Period in the Northern Song Dynasty. The painting is 51 centimeters wide and 11.9 meters long, depicting a magnificent view of thousands of kilometers of mountains and rivers, with rolling mountains and endless waterscape, making it one of the most representative works among the surviving Blue-Green Shanshui (青绿山水) or Blue-Green landscape paintings. In this painting, the audience can see the human settlements among the vast landscape (Fig. 24.1), such as houses, villages, bridges, ferries, temples, pagodas, pavilions, etc. In addition, various details of people's activities and their interactions with the environment, such as traveling, living in seclusion, fishing, watching waterfalls, touring, etc., can also be glimpsed in the painting. This artwork inherits and develops the essential techniques of Chinese Tang Dynasty Blue-Green landscape painting, focusing on seeking variations in the colors of blue and green, and intermingling actual landscapes with imagined scenes, making it a milestone piece in the history of Chinese landscape painting.

Just as *A Panorama of Rivers and Mountains* has been acclaimed as a national treasure in the cultural relics sector, it is also obvious that the present dance-poetry drama is also praised as a creative transformation of traditional aesthetic culture. The aesthetic experience of "immersive" viewing of landscape paintings in the Song Dynasty enabled people to unfold the landscape scrolls as if they were walking, viewing, touring, and relaxing amidst the landscape, especially for those who had long lived in towns and cities (usually referred to as the Chinese literati, educated scholar-officials, who had the sense of appreciating and enjoying landscape paintings) to wander in the landscape. The transcendental quality of landscapes was tacitly recognized by the Chinese literati. In addition to political discontent, it probably conveyed a bond with the transcendental force. In other words, the landscape as an aesthetic object will always be superior to the cultivated gardens, which are a mere simulation of the landscape, transferring the elements of the *Shanshui* into the human environment. The remote landscape, on the other hand, or *Shanshui*, is a secret realm where the gods live; it can be a place of retreat and spiritual restoration after suffering from political discontent.

It is interesting to have a cross-reference here with the Western interpretation of Chinese landscape paintings. Dissatisfied with Jay Appleton's biological-evolutionary interpretation of Western landscape paintings as devices of prospect (seeing) and refuge (avoiding being seen), Turner (2009) stepped out of the Western tradition to address Chinese landscape paintings for a fresh taste of valuing nature aesthetically. Turner considers that the artists' techniques (brushstrokes, ink-wash, etc.), poetry, and the Chinese literati essential for appreciating the unique feature of a Chinese landscape painting, particularly the Spirit Resonance, which would help Westerners to "see the features of the natural world in their multiplicity, variety, and particularity" (Turner 2009, p. 118).

Considering that Turner's ultimate aim is to address contemporary Western debate on aesthetic appreciation of nature, to be specific, Allen Carlson's scientific approach and Emily Brady's metaphysical imagination, it is reasonable for Turner to recognize that the aesthetic experience of Chinese landscape paintings eschews abstracting nature and instead appreciates nature itself. However, from a relational perspective, whether abstracting nature or appreciating nature itself is still an objectification of nature. The dilemma encountered when dealing with Chinese aesthetics using Western discourses and paradigms can be detected here, whereby appreciating landscape painting expresses the ambiguity, vacillation, and embarrassment, revealed by the tension between the existential-spiritual and dichotomized scenarios, biophysically inhabited (being) and spiritually haunted (absence). For ancient Chinese literati, biophysically inhabiting the landscape came at the cost of seclusion and unfulfilled political ambitions, whereas being absent in the landscape would be haunted by its perpetuity.

This existential-spiritual tension can also explain the stylishness of the poetic dance drama *A Mere Touch of Green* among Chinese urbanites and their international kind. China's large-scale urbanization has resulted in a dramatic outflow of population from rural to urban areas, and studies of urban public green spaces have shown that the development of public green spaces has been hampered by local

Fig. 24.2 Two stills from the dance-poetry drama *A Mere Touch of Green.* The top half features silhouettes of performers in traditional attire against a vibrant yellow-green backdrop, creating a dramatic shadow effect. The bottom half reveals the same performers in colorful costumes, illuminated under stage lights, with reflections visible on the floor. In the dance drama, the digital background of the stage is exactly the painting of *A Thousand Li of Rivers and Mountains.* (Photographs: Xufeng Wang)

governments' pursuit of maximizing revenue from the commercialization of urban land, and that there are social inequalities in public access to public green spaces (Chen and Hu 2015). The thirst of the urban population for access to nature is undeniable and can also be observed in the current national policy of revitalization of the rural areas (Yu et al. 2026). To a certain extent, the melancholy of urbanites' inaccessibility to the mountains and rivers, or their absence in the landscape, is compensated for precisely in the poetic space of the *Shanshui* served by *A Mere Touch of Green*. In other words, while once landscape paintings remedied the faint melancholy of the absence of Chinese literati from the landscape, *A Mere Touch of Green* compensates for the despondency resulting from the limited or no accessibility to public green spaces. This is the spiritual bonding that mends for the biophysical absence, a virtual recognition with the perpetual landscape out there (Fig. 24.2).

24.3 A Touch of Human-Landscape Narratives Through Costumes

As far as the visual creation of the dance drama is concerned, *A Mere Touch of Green* firstly presents the dramatic hues of *A Thousand Li of Rivers and Mountains* in terms of color, i.e. the blue-green. It is well-known that the painters of the blue-green *Shanshui* usually depict landscapes in lime green color using minerals such as stone blue and stone green as main pigments. Thus, the blue-green costumes of the female dancers, indeed, are not actually blue or green. The gradient design of the two colors contains so many shades, which makes the expression of dance movements filled with cornucopian changes. Through the stretch of the dance movements, the blue-green becomes dynamic, as if the majestic and rolling mountains, the billowing rivers, the drifting clouds, and the mist among the mountains were presented on stage.

In order to present the mountains and rivers in the landscape painting in a more stylized way, the drama director designed a series of movements for the group of female dancers dressed in blue-green, including silent waiting, moon watching, cloud falling, pondering, walking in solitude, cliffy peak, and crouching rock. According to Cao (2022), the dance movement "cliffy peak" symbolizes the steep and dangerous mountain peaks in the painting, therefore, the dancers' movements are presented as a bow step with the right foot in front, supported by the left foot, with the upper body slowly tilted backward to 90° and then back to the upright position (Fig. 24.3). The movements are sturdy and powerful, representing the

Fig. 24.3 Dance movement: Cliffy Peak. The dancer is visualizing the steep cliff through her dance movements on stage. (Photograph by Xufeng Wang)

majestic mountains, while their facial expressions are extremely calm and emotionless, designating the boundless waters of the landscape paintings. In this design of movement, the dancers' bodies are symbolized as mountains and rivers in the painting. For the audience, the most unexpected design of the dance drama may be that the static rolling mountains become dynamic due to the fluidity of the bodily movements, while flowing rivers in the painting turns out to be calm and peaceful, conveying a sense of tranquility.

The overall color scheme of the blue-green section also carries a profound meaning. Since the group of female dancers is a symbol of the landscape, they are dressed in blue-green, matching the color of the painting. Donglin Yang, the costume designer, extracts the blue-green of the clothing from the painting and designs the sleeves, skirt style, waist rapping, and the layering of the cloth based on the cascading mountains in the painting. Although the dancers in blue-green are silent or the landscape seems to be silent, when they lift their hands and raise their feet, they (the landscape) become(s) phenomenal and flowing. This is based on the traditional aesthetic value expressed by Chinese Taoist Lao Tzu, "the most astounding sound (大音, da-yin) is not audible (希声, xi-sheng); the great phenomenon (大象, da-xiang) manifested by Tao has no shape nor form (无形)" (Lao Tzu et al. 2020). According to Zhang (2022), blue-green in the painting conveys threefold meanings: as a color of the empire, it represents the idea of everlasting kingdom where thousands of years of blue-green represents boundless mountains and rivers; blue-green is linked to Taoism philosophy, cannoting longevity, healing, and avoidance of plague; the flamboyant and magnificent of the blue-green echos the Xuanhe style (宣和画风) espoused by Emperor Huizong (宋徽宗, 1082–1135). This threefold meaning was transformed by the directors of the poetic dance drama. The political and theological significance of blue-green was erased and converted to purely aesthetic values, signifying the return and regeneration of the spirit of classical Chinese landscapes (Zhang 2022).

While the blue-green colors touch the mountains and rivers in the landscape painting, other colors in the dance drama touch the artistic embodiment of traditional landscape painting. The birth of a traditional Chinese landscape painting involves a number of traditional crafts that are part of the Intangible Cultural Heritage, including seal carving, silk weaving, quarrying and grinding, brush manufacturing, and ink making. These processes are also manifested in the dance drama through various sections. Meanwhile, the colors of the dancers' costumes link to their professions. To complete a traditional Chinese landscape painting requires not only artists' talent, but also a great deal of interactions with the environment from the perspective of people in various traditional professions, thus providing artists with the materials they need to complete their artistic creation, such as the brush, ink, silk, and the seal, etc.

Five main professions are presented in the dance drama, namely seal carvers (dark floral cyan embellished with the red), silk weavers (light mugwort green), stone grinders (ochre as the base and partially stained with cobalt blue), brush makers (in dark carmine), and ink makers (basalt supplemented by lampblack gray). From the perspective of the costume designer, the five artisans' costumes are styled after

Fig. 24.4 Still of Wang Ximeng, from *A Mere Touch of Green*. Artist Wang Ximeng is dancing gracefully under soft glow of a lantern, holding a brush in his hand - the primary tool for creating Chinese landscape paintings. (Photograph by Xufeng Wang)

Song Dynasty folk artisans, whose costume colors are profoundly related to their craftsmanship. For instances, the costume color of the seal carver represents his profession in the sense that floral dark blue produces a sense of steadfastness in carving the seal and the embellished red represents the clay; the costume of the silk weaver is in light mugwort green, indicating the season of the silk weaving (Yang 2021). For the artist of the painting, Wang Ximeng, his costume color is extracted from silk, similar to cotton white or ivory, as silk has been widely used as the drawing paper for Chinese landscape paintings. This cotton white color makes the character extraordinarily elegant (Fig. 24.4).

Overall, the dance movements of the characters, the design of the colors and textures of the costumes carry profound meanings, gesturing not only to the expressive power of traditional Chinese landscape paintings, but also to the variety of traditional techniques involved in both the painting and the preparation of the indispensable painting materials. These skills exceed far beyond the artist's painting on silk, exposing the complexity of the human-environment connection in the creation of art. In a sense, if the audience does not have the knowledge of these traditional techniques, they will not properly appreciate this dance drama. But if one views this dance drama as a stranger, the viewer will indeed learn about the legacy of these traditional arts and their cultural significance, as well as how people of all ages, such as the restorer of artifacts in the drama, craftspeople, artist Wang Ximeng, and even the viewer oneself as an audience, can reconnect with such traditions. Since these crafts involve human interaction with the environment, they possess, intrinsically, biocultural attributes which are beyond the scope of this chapter.

24.4 Connect with *Shanshui* Through Experimental Dance

Echoing the theme, I raised in 1.2 about being and absence, as stated earlier, the painting *A Thousand Li of Rivers and Mountains* expresses the tension between the two. And the dance drama *A Mere Touch of Green* expresses this tension even more to the extreme. That is to say, the tension between being in the landscape and being absent in the landscape, which may rarely occur in the life world of the ancient Chinese literati, is presented in the dance drama as a tension felt by contemporary Chinese (and even global) urban citizens, a tension between yearning for *Shanshui* but lacking a meaningful connection with it. In this regard, when the poetic dance drama visualizes *Shanshui* through dance movements, *Shanshui*/landscape is both absent and present in the theater. People (the audience) are thus both immersed in/absent from the *Shanshui*/landscape. In such a paradoxical situation, what are people, the performers and the audience, connecting with? To answer this question, the author, as a dance artist and educator, and also as an urbanite, decided to perform an experimental dance in a more "natural" setting to better address this question.

The stage, or this more "natural" setting, is Xi'an Yannan Park. The 121-hectare low-maintenance park is located at Xi'an, Shaanxi province, China, designed by Turenscape with multiple aims, such as to protect an ancient archeological site which contains 25 ancient tombs and other archeological relics dating from the Han Dynasty (206 BC-220 AD) back to the Neolithic Yangshao Culture, create a recreational area for local communities of the megacity Xi'an with a population of over ten million, design a resilient "green sponge" for stormwater management and rewilded urban natural space (Yu 2023), etc. These advanced design concepts, such as the coexistence of urban natural spaces and cultural heritage, rewilding, sponge, and resilient design, follow the principles of nature-based solutions (Yu et al. 2026). For instance, as revealed in Fig. 24.5 (right), the concrete drainage channels were transformed into terraces and low-maintenance vegetation that filter nutrient-rich drainage water. It is in this nature-based sense that I would like to call Yannan Park a more "natural" place or environment.

Fig. 24.5 Xi'an Yannan Park in different seasons. In the left photograph, Yan Nan Park serves as a gathering place during autumn and winter. Citizens congregate in the covered trail areas in Yan Nan Park, practicing the erhu (a traditional two-stringed Chinese bowed instrument) and Qin Qiang (a genre of folk Chinese opera specific to Shaanxi Province (as Qin region)). City buildings are visible in the distance. The right photograph shows the trails, water bodies, and other scenic areas during the transition between spring and summer, presenting a verdant environment. (Photos by Turenscape)

Fig. 24.6 Dance in *Shanshui.* The author wanted to taste and experience what it would be like to dance in a natural environment while dressed in the costumes which were designed for *A Mere Touch of Green*. The colors of the dance costumes and of the water match each other here, creating a sense of immersion and serenity. The photo was taken at Yannan Park, Xi'an, Shaanxi Province. (Photo credit: Juan Wang)

In a more "natural" environment, parks can bring out quite unique feelings in different seasons. One cannot enter a low-maintenance park in winter yet expect to see a lush, green, and vibrant landscape. The withered flora in Yannan Park in winter evokes a strong sense of vicissitude of seasons and life, something that high maintenance parks cannot offer people. For a dance artist, it seems that on a bleak winter day, wearing a thin dance costume and sensing the chillness, the body was genuinely immersed in the "natural" environment and glimpsing the realm of *Shanshui*. It is a realm that cannot be easily glimpsed by an urbanite. Historically, most hermits (in the companion of *Shanshui*) were known for their austere living yet noble characters. During this experimental dance performed by an urbanite, the slowly raised arms were my tribute to the land (Fig. 24.6); the waving long sleeves were my greetings to the water (Fig. 24.7). At the moment, I was present in the *Shanshui*/landscape, connected to it, and it also shaped my experimental dance and dance pedagogy.

After completing the experimental dance, and then reflecting on the dance-poetry drama, I recognized some of the implications of connecting with nature. The narrative of the poetic dance drama, *A Mere Touch of Green,* represents a connection with the spirit of *Shanshui* and the cultural heritage of the aesthetics of landscape. The landscape becomes a magnificent context, or rather, the dance drama takes the landscape as a carrier to actually narrate the inheritance and re-creation of a

Fig. 24.7 Dance with long sleeves. Long sleeves are a major costume feature of classical Chinese dance, which usually show the elegance, flowing and changing of long sleeves through the movement of the arms, featuring the spirituality of water. Long sleeves can drape, flutter, and tumble, thus being an extension of inner sentiments to express complex feelings. The photo was taken at Yannan Park, Xi'an, Shaanxi Province. (Photo credit: Juan Wang)

humanistic tradition that cannot be separated from an abstract yet poetic spirit of *Shanshui*. To this point, based on former discussion, it is clear that humanistic traditions in ancient China cannot be understood in separation with the natural environment. During the experimental dance, the uniqueness of the ambient setting has presented Yannan Park and landscape no longer just a backdrop, a place, or a carrier, but an integral element of a novel narrative structure. In other words, the dance artist would have to appreciate the design philosophy of the Yannan Park before she could bring new meanings and values to the experimental dance and renovate pedagogy. From a biocultural perspective, human beings and more-than-human co-inhabitants have on all occasions dwelled in their specific environments and shaped their unique ethos because of the interrelations among co-inhabitants and the environment (Rozzi 2012).

The experimental dance has therefore empowered an artistic transformation and pedagogy with a *Shanshui*/landscape dimension. The location of dance instruction will not be limited to the classroom or the stage; the outdoor space can also become the venue for teaching (Fig. 24.8). Further, the tension between absence amidst the landscape and yearning for it is released through dancing in the square. The more important message is that dance artists and learners can sense the environment

Fig. 24.8 Outdoor dance lessons. Photos were taken in a small plaza (left) and outdoor space in woods (right) on Xidian University campus. Bringing students to the outdoor in the chilly spring air allows them to develop a deeper appreciation for their surroundings while learning dance. With everything still dormant, the campus may burst into greenery in just a week or two. In other words, two weeks from now, our bodies' interactions (such as visual, auditory, or tactile) with the environment will be quite different. (Photo credit: Juan Wang)

physically and express their gratitude for the place they dwell in through dance. One cannot say that he or she is connected to the natural environment, or that a true sense of the spirit of the *Shanshui*/landscape has been revealed, until one can express gratitude to the place one settles in.

24.5 Conclusion

The focus of biocultural ethics on the profound connectedness between co-inhabitants and their habitats has proved to be a powerful framework for understanding and exploring the interaction of contemporary traditional cultures with the contemporary socio-environmental issues. Local biocultural communication will inevitably involve cultural accounts and interpretations of the nature-human relationship in the local or regional context. Reflection on the association between poetic dance drama and landscape painting in the *Shanshui/landscape* spirit has prompted experimental dance and pedagogical creativity for novel biocultural communication. Overall, the sustainability and well-being of global and local communities will depend on how we modify our habitats (which is clearly irreversible), bestow meanings, and provide younger generations with avenues for shaping their ecologically and ethically embedded connections to the places they inhabit. Although the effects of experimental dance and creative dancing on socio-environmental complexes need to be further analyzed, it is evident that the initial results are remarkable: not only in the fact that experimental dance has become an innovative way of moulding the relationship between the dance artist and the surrounding environment, but also in the fact that the intervention of biocultural communication has shaped the narrative structure of the dance.

Acknowledgement The author would like to express her gratitude to Danqiong Zhu for translating the original Chinese manuscript into English. She also appreciates Danqiong Zhu, Terrance Caviness, and Ricardo Rozzi for taking the time to comment on this chapter.

References

Cao X (2022) Imagery creation and real generation in the dance Drama a mere touch of green (舞剧《只此青绿》的意象营造与意境生成). Oriental Art (东方艺术) 03:53–59

Chen WY, Hu FZY (2015) Producing nature for the public: land-based urbanization and provision of public green spaces in China. Appl Geogr 58:32–40

Lao Tzu, Poon KS, Poon KK (2020) An English translation and the correct interpretation of Laozi's Tao Te Ching. The SenSeis, Toronto

Rozzi R (2012) Biocultural ethics: recovering the vital links between the inhabitants, their habits, and habitats. Environ Ethics 34(1):27–50. https://doi.org/10.5840/enviroethics20123414

Turner M (2009) Classical Chinese landscape painting and the aesthetic appreciation of nature. J Aesthetic Educ 43(1):106–121

Yang D (2021) Contemporary interpretation of song dynasty aesthetics: a talk on the costume making of the poetic dance drama "A Mere Touch of Green" (宋代美学的当代解读——舞蹈诗剧《只此青绿》服装创作谈)[EB/OL]. Accessed through https://mp.weixin.qq.com/s/ZjyY7AZuQggvVtpyEwCgkA

Yu K (2023) Urban balcony embracing Rewilded nature —Xi'an Yannan Park phase II, China. Landsc Archit Front 11(06):88–91. https://doi.org/10.15302/J-LAF-1-040033

Yu K, Li J, Zhou Y, Xie M, Hu Y (2026) Nature-based solutions to the revitalization of Wuyuan County. In: Zhu D, Rozzi R, Tsuji R, Johnson B, Caviness T, Castro-Jorquera C, Xu L, Zhang L (eds) Linking arts with biocultural conservation, restoration, and communication, Ecology and ethics, vol 8. Springer, Cham, pp 417–432. https://doi.org/10.1007/978-3-032-04830-1_26

Zhang X (2022) *Only This Green*: the return and regeneration of the spirit of classical landscape(只此青绿":古典山水精神的返场与再生). Arts Critic 05:55–64. https://doi.org/10.16364/j.cnki.cn11-4907/j.2022.05.002

Chapter 25
Photovoice: A Community-Based Research and Communication Tool for Wildlife Conservation and Coexistence

Montserrat Franquesa-Soler and M. Azahara Mesa-Jurado

Abstract Socio-ecological systems offer a framework to guide the social transformations required for sustainability. However, disconnects within these systems have been identified, particularly in the relationship between scientific knowledge and the emotional or cultural connections people have with wildlife. This gap hinders the development of effective and inclusive conservation strategies. Our research highlights the potential of arts-based approaches to integrate rational knowledge and emotions to address this disconnection. In this chapter, we describe and assess experiences with primate conservation education in schools from Southern Mexico. These experiences incorporate artistic expressions to explore, collect, analyze, communicate, and highlight issues related to socio-ecological processes and biocultural conservation. We use photovoice's theoretical and practical basis to explore and generate solutions for wildlife conservation issues. In this methodology participants of all ages take pictures of various scenarios or situations related to a specific topic to structure a dialogue that can diagnose environmental problems in a given context and then advance socio-environmental action. As the community listens and reacts to the participants' perspectives, this process can guide them toward finding joint solutions that benefit both wildlife and human communities, fostering coexistence. Although this technique has been widely applied in various disciplines,—primarily sociocultural and health research issues—we use it here in a novel context: wildlife conservation. We have found that photovoice is an effective participatory and educational technique because it: (1) served as a connection tool linking students, adults, and other stakeholders from the community, including the municipality, artisans, and farmers; (2) helped examine social, environmental, and socio-ecological change; (3) enhanced students' perception of

M. Franquesa-Soler
Instituto de Neuroetología, Universidad Veracruzana, Xalapa, Veracruz, Mexico

Miku Conservación, AC, Xalapa, Veracruz, Mexico

M. A. Mesa-Jurado (✉)
El Colegio de la Frontera Sur, Unidad Villahermosa, Villahermosa, Tabasco, Mexico
e-mail: mmesa@ecosur.mx

D. Zhu et al. (eds.), *Linking Arts with Biocultural Conservation, Restoration, and Communication*, Ecology and Ethics 8,
https://doi.org/10.1007/978-3-032-04830-1_25

wildlife and facilitated the identification of positive and negative interactions between wildlife and the community; and (4) provided input into community processes focused on natural resource management. Although this chapter focuses on a local case study in a Mexican community, its methodology could be used at other locations and support biocultural conservation at broader scales by identifying new pathways to foster respectful coexistence in biocultural diversity.

Keywords Participatory photography · Wildlife conservation · Community conservation · Mexico

25.1 Introduction

Many current and traditional approaches to biodiversity conservation have emphasized the role of humans as a threat to the environment. From this approach, research has focused mainly on studying how extractive and transformative human activities modify ecosystems (Jaureguiberry et al. 2022). Although the high pressure of humans on the environment is undeniable, this perspective underlies a philosophical view that portrays humans as external and separate from nature and considers our interactions with the natural environment as primarily attempts to dominate it (Eldredge 1995). Within this perception of separation, it is essential to recognize that not all human groups have been treated equally, and power imbalances often emerge (Rozzi 2012, 2015).

As a result of colonialism's erosion to biodiversity conservation, numerous local Indigenous communities have been displaced from their land and prohibited from engaging in or pursuing traditional methods of livelihood, such as harvesting wild resources or hunting (Guha 1989; Guha and Martínez-Alier 1997; Bray and Velazquez 2009). This was done under the motto of 'protecting' wildlife and maintaining the ideal of pristine, people-free, environments; in contrast, hunting by colonial and visiting European elites was allowed (Fabricius et al. 2004). Therefore, conservation models characterized by top-down approaches, often called "fortress" models, were prevalent throughout the nineteeth and twentieth centuries (Archer et al. 2022).

There is a notable shift towards a more holistic approach in analysis, moving from narrowly focused perspectives to broader and deeper ones (Berkes 2004). For instance, researchers are transitioning towards embracing the paradigm of Socio-ecological Systems (Bennett et al. 2017), which involves redefining ecosystem services to include relational and cultural values (Pascual et al. 2023). This shift advocates for truly participatory processes that enhance social justice. While considerable progress has been made, a complete transition has yet to occur in conservation science. Criticism has arisen concerning the effectiveness of these approaches in changing power dynamics and supporting community-led conservation efforts that genuinely benefit local populations (Shackleton et al. 2023; Petriello et al. 2024).

In seeking truly participatory processes, several biases embedded in mainstream conservation and public discourse have been identified as barriers to more effective and inclusive efforts (Escobar 2008; Bennet et al. 2017; Shackleton et al. 2023). One such bias is **logocentrism**, which privileges verbal or written language over other forms of communication. This emphasis can exclude individuals and communities who communicate through alternative modes, such as visual art, storytelling, or performance—forms that are equally valid for expressing environmental knowledge and values (Finlay 2008; Bowers 2002). Another common bias is **Eurocentrism**, which tends to prioritize European epistemologies and values, often dismissing or marginalizing the diverse environmental worldviews of Indigenous peoples and local communities (Escobar 2008; Smith 1999 Nadasdy 2005). This can result in power imbalances and the systemic exclusion of locally grounded ecological knowledge. Additionally, **anthropocentrism**—the centering of human needs and benefits—frequently underpins conservation efforts, thereby undervaluing the intrinsic worth of non-human life and limiting the scope of ethical engagement with nature (Kopnina et al. 2018; Crist 2012).

Another added bias that still predominates in conservation biology is 'Positivism', a philosophical approach highlighting the importance of empirical evidence and scientific methods in understanding the world (Bustamante et al. 2010). This approach allowed significant advancements in conservation science, but it only values quantitative methods, undervaluing qualitative or anecdotal evidence, particularly from Indigenous knowledge systems. Positivism also seeks to simplify complex phenomena into discrete variables that can be measured and analyzed, which is important to understanding ecosystem processes and recognizing patterns. However, it may overlook the interconnection and complexity of socio-ecological systems without being able to address conservation issues completely. This approach aspires to objectivity and impartiality in advancing knowledge, but researchers' values, interests, and power dynamics underlie research questions. This 'apparent' objectivity is a subjectivity of the dominant narrative. And because it is primarily Western scientists' paradigms that continue to dominate policy and decision-making spaces, other systems of knowledge, particularly Indigenous and local epistemologies, remain at risk of marginalization, ultimately weakening the legitimacy, inclusivity, and effectiveness of conservation efforts.

Addressing the aforementioned biases requires a more inclusive and holistic approach to conservation that values biodiversity, cultural diversity, and social equity. This may involve integrating Indigenous and local knowledge, promoting diverse voices and perspectives in conservation decision-making, advocating for other methods of communication besides verbal, and acknowledging the interconnectedness of human and ecological well-being. To overcome these limitations, 'biocultural approaches' to conservation offer a distinct path for conservation efforts (Rozzi et al. 2006, Gavin et al. 2015, Bridgewater and Rotherham 2019). These approaches are based on learning from past studies on the diversity of both biological and cultural aspects, understanding how humans interact with their environment (socio-ecological systems theory), and considering various conservation models that prioritize the involvement and well-being of local communities (Berkes 2004; Bennett et al. 2017; Archer et al. 2022). Biocultural conservation involves preserving the natural and cultural aspects of dynamic and interconnected socio-ecological systems. This approach invites us to rethink historical biological conservation efforts in Mexico.

25.1.1 From 'Biological Conservation-Based' on Positivist Science to 'Biocultural Community-Based conservation' in Mexico

Mexico is often celebrated for its rich biocultural diversity, encompassing its biological and cultural heritage. It is one of the world's megadiverse countries, meaning it harbors a significant portion of Earth's species and an equally outstanding linguistic diversity (Sarukhán 2008; UNEP-WCMC 2014; Eberhard et al. 2022). Its diverse ecosystems range from deserts to tropical rainforests, mountains, and coastal areas. The country is home to a vast array of plant and animal species, many of which are endemic (Sarukhán 2008). Coevolving with this diversity (Maffi 2007), Mexico has many Indigenous cultures that have coevolved harmoniously with their environments.

These cultures possess deep knowledge of local ecosystems, biodiversity, and sustainable practices. Indigenous peoples in Mexico have accumulated centuries of traditional ecological knowledge (TEK). This knowledge encompasses a wide range of topics, including medicinal plants, sustainable agriculture, water management, and biodiversity conservation. TEK is often passed down, mostly by elders, through oral traditions and is integral to Indigenous communities' cultural identity and resilience. Recognizing the importance of its biocultural heritage, Mexico has implemented various conservation initiatives to protect its biodiversity and support Indigenous communities (Tran et al. 2020; Méndez-López et al. 2014). For instance, Martin et al. (2011) found that the abundance and diversity of Indigenous and Community Conserved Areas in Oaxaca, such as the 126 community conservation sites, covered 375,000 ha, illustrating the successful integration of cultural, ecological, and historical elements in biocultural conservation efforts.

Overall, Mexico's biocultural diversity represents a unique and invaluable heritage that is integral to the country's identity and sustainability. Few projects have been based on biocultural approaches (Martin et al. 2011; Hernández-Rodríguez and López-Santiago 2022), and most still fall under traditional conservation approaches (Guibrunet et al. 2021). This perpetuates the above-mentioned biases, disallowing progress towards a more inclusive and fair conservation. If not appropriately addressed, it may foster biocultural homogenization, a significant but often overlooked global driver of biological and cultural diversity loss (Rozzi 2018). Biocultural homogenization occurs when local knowledge systems, practices, and biodiversity are replaced or diluted by globalized, often Western-centric, approaches. This process leads to a loss of both unique biological species and the traditional cultural knowledge that has sustained them for generations. As cultures and ecosystems become more standardized, both social and environmental injustices arise, disproportionately affecting Indigenous and local communities that rely on these ecosystems for their livelihood and cultural identity. As Rozzi (2013) notes, the impact of biocultural homogenization is profound, with consequences not just for biodiversity conservation but for social equity, cultural survival, and environmental sustainability. While biodiversity conservation remains crucial, preserving ecological diversity — or biological diversity—holds even greater significance. In addition, biodiversity conservation must be understood within the context of safeguarding the biocultural systems that sustain it, emphasizing the importance of both biological and cultural resilience. While biodiversity conservation is important, preserving *ecological* diversity—or biological diversity—holds even greater significance.

In Mexico, conservation biology education has been dominated by the "positivism" perspective and focuses on the cognitive domain (Franquesa-Soler et al. 2023). At the individual level, human-nature connectedness comprises three interrelated dimensions: cognitive (knowledge and beliefs about nature), affective (feelings and emotions toward nature), and experiential (actions and experiences in nature) (Zylstra et al. 2014). While it is crucial to understand these three dimensions of human-nature connection better, many studies focus primarily on cognitive connections with nature (Restall and Conrad 2015), often using quantitative approaches. Arts-based methods have the potential to engage not only the cognitive but also the affective and experiential dimensions of the human-nature connection. These methods transcend language dependence and reveal emotional and experiential nuances of the connection that are often difficult to capture through traditional cognitive approaches. Since most art forms involve a nonverbal component, their use in research can evoke tacit knowledge, making arts-based methods particularly suited to exploring and enhancing emotional and experiential connections with nature (Muhr 2020).

Although biocultural approaches to conservation are increasingly better defined and the importance of addressing them is highlighted, studying them raises epistemological and methodological challenges. Given the conceptual critique, which is crucial to address, it is also important to consider "how to facilitate this change," i.e., the methodological aspect. In this chapter, we aim to introduce readers to emerging artistic methods and approaches for biocultural communication, as well as new and innovative ways of articulating, transmitting, and communicating biocultural diversity that has been previously overlooked (Li et al. 2026; Song 2026; Tauro and Rozzi 2025; Tsuji and Johnson 2026; Tsuji et al. 2026; Zhang and Xu 2026). We offer insights from the field regarding participatory visual methods and arts (e.g., photovoice). Additionally, we discuss various challenges encountered during the research process and propose photovoice as a potential methodology in response to biocultural homogenization.

Arts-based methods play a crucial role in addressing emotional aspects overlooked by traditional approaches, thereby overcoming biases in understanding human-nature connectedness and fostering more inclusive sustainability efforts (Castro-Jorquera 2026; Chiarini 2026; Hu and Yang 2026). Engaging with emotions and diverse forms of expression offers a pathway to deepen individuals' connection to nature beyond cognitive dimensions, enriching our understanding of sustainability challenges and solutions (Muhr 2020). In addition, in the last 10 years, arts-based approaches helped migrate to a biocultural approach through a Participatory Action Research (PAR) epistemological perspective, which also helps to overcoming environmental and social injustices. PAR approach holds relevance in amplifying the voices of vulnerable groups often excluded from decision-making processes, such as young generations (Brydon-Miller et al. 2020). Research suggests that PAR engenders new insights into children's ecological experience, by facilitating meaningful and transformative interactions between humans and nature while fostering efficacy both in education and conservation endeavors (Eastwood et al. 2023).

25.1.2 Arts-Based Approaches for Biocultural Conservation Focused on Children/Young Generations: Insights from the Conservation Field in Mexico

The integration of art, ecology, aesthetics, and ethics in biocultural conservation represents a comprehensive, interdisciplinary strategy that recognizes the symbiotic relationship between human cultures and nature. This approach leverages the power of art to foster connections between people and nature, promote conservation awareness, and engage diverse communities in conservation efforts. It also encompasses diverse forms of expression, including visual arts, performance, music, dance, storytelling, and traditional crafts (Heras et al. 2016). By embracing a broad and cross-cultural interpretation of art, conservation initiatives can engage diverse audiences and cultural perspectives, fostering inclusivity and nurturing empathy towards natural environments.

Art in biocultural conservation goes beyond aesthetics to incorporate ecological principles and ethical considerations. Through artistic expressions, such as environmental installations, eco-art, or multimedia presentations, artists can convey messages about biodiversity conservation, ecosystem services, and sustainable living practices, inspiring individuals to take action to protect natural resources. Integrating art into biocultural conservation contributes to conservation goals, enriches the artistic process, and empowers artists as agents of change. Artists who engage with environmental themes can deepen their understanding of ecological processes, expand their creative repertoire, and amplify their impact by fostering connections between art and activism. Art-based approaches can catalyze the participation of multiple actors, including students, teachers, community members, and various stakeholders, in wildlife conservation initiatives. By providing opportunities for creative expression and collaborative action, arts can provide a creative and valuable pedagogical approach particularly useful for young generations. Integrating art, ecology, aesthetics, and ethics in biocultural conservation with young adults can foster a deeper understanding of the interconnectedness between cultural practices, ecological systems, and ethical considerations. It also promotes environmental stewardship by encouraging a sense of responsibility towards the environment and cultural heritage, incorporating lasting conservation and sustainability values. These exchanges of knowledge are happening through social and interdisciplinary learning. Artistic expression can inspire and motivate young people to become advocates for biocultural conservation, sparking passion and dedication for preserving biodiversity and cultural diversity.

25.2 Photovoice as an Arts-Based Community Biocultural Conservation Tool

Incorporating the photovoice method into biocultural conservation endeavors offers a creative and valuable pedagogical approach to reconnecting citizens with wildlife and biocultural diversity. Photovoice is a participatory research method that empowers individuals to document their experiences and perspectives through photography, facilitating dialogue, reflection, and community action (Wang and Burris 1997). In biocultural conservation, photovoice can help citizens acknowledge, value, and respect biological and cultural diversity that may be overlooked or marginalized in their daily lives due to global biocultural homogenization.

Photovoice enables citizens to participate actively in conservation efforts by providing a platform to express their unique perspectives, experiences, and connections to wildlife and biocultural diversity (Derr and Simons 2020). Through photography, individuals can capture the beauty, complexity, and significance of local ecosystems and cultural practices, fostering a sense of ownership and agency in conservation decision-making processes.

Many people have become disconnected from nature and biodiversity in an increasingly urbanized and technologically driven world (Muhr 2020). Hence, photovoice can serve as a tool for reconnecting citizens with wildlife and natural environments by encouraging them to explore and appreciate the surrounding biodiversity. In addition, being an arts-based approach can help overcome the cognitive domain and enhance the emotional and experiential domains as well. Globalization and cultural homogenization have led to the erosion of traditional knowledge systems and the loss of cultural diversity worldwide (Rozzi 2018). Photovoice can help citizens recognize and celebrate their communities' rich biological and cultural diversity tapestry. By documenting and sharing images of local flora, fauna, landscapes, and cultural practices, participants can raise awareness about the value of diversity and the importance of preserving traditional knowledge for future generations.

Photovoice encourages dialogue, reflection, and collective action around conservation issues by providing a visual platform for sharing stories, experiences, and aspirations. Through group discussions, exhibitions, and community forums, participants can engage in meaningful conversations about the connections between biodiversity conservation, cultural heritage, and sustainable development, leading to collaborative efforts to address local environmental challenges.

The photovoice methodology offers a powerful and participatory approach to biocultural conservation education and advocacy, enabling citizens to reclaim their role as stewards of biodiversity and cultural heritage. By harnessing the power of photography and storytelling, photovoice can foster greater awareness, appreciation, and stewardship of wildlife and biocultural diversity, contributing to more sustainable and resilient communities.

25.2.1 Case Study in Southern México: Methodological Insights from Primate/Wildlife Conservation Education

The methodology of photovoice that was employed in primate conservation education in schools in Southern Mexico serves as a valuable case study for understanding the effectiveness of art-based approaches in biocultural conservation and communication.

25.2.1.1 Study Context: A Broader Project

Our case study was part of the follow-up stage of a broader project focused on Primate Conservation Education in 12 primary schools of Southern Mexico (Franquesa-Soler and Serio-Silva 2017; Franquesa-Soler et al. 2018, 2020). It started as MFS' PhD (2013–2018) to design arts-based educational strategies for the conservation of the black howler monkey (*Alouatta pigra*) (Franquesa-Soler et al. 2018). The study sites were selected based on the geographic distribution of the black howler monkeys (i.e., states of Tabasco, Chiapas, Campeche, Yucatán, and Quintana Roo, Fig. 25.1). The project was developed in four stages.

- The first stage was a diagnosis of both children's learning preferences, perceptions, and knowledge about black howler monkeys (Franquesa-Soler and Serio-Silva 2017; Franquesa-Soler et al. 2018).
- The second stage included the design of several arts-based education strategies based on the information from the first stage. We applied to the same group of students from the twelve participating schools (Franquesa-Soler et al. 2020).
- The third stage consisted of a post-evaluation using the same diagnosis tools applied during the initial stage. We aimed to evaluate the effectiveness of arts-based education strategies used, as well as children's satisfaction concerning the activities and our performance as facilitators (Franquesa-Soler et al. 2020). Arts played a predominant role, from methods (Franquesa-Soler and Serio-Silva 2017) to the educational content (cognitive and emotional).
- The fourth stage (i.e., follow-up) maintained the same arts-based approach: participatory mural paintings (Franquesa-Soler et al. 2023).

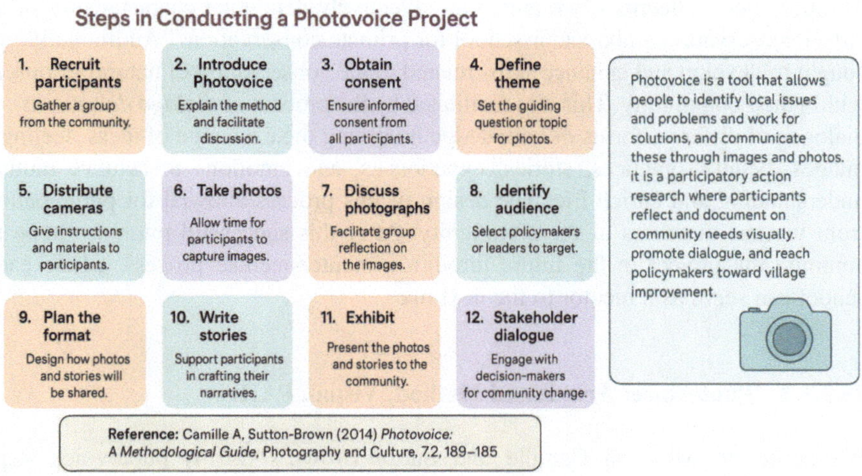

Fig. 25.1 Suggested sequence of steps to follow during the photovoice activity, adjusted based on Sutton-Brown (2014)

In the 5 years since the project was created, the facilitators and authors of the project went through various personal and professional processes. The doctoral process that initiated the project was completed, the *Miku Conservación* NGO was created, focused on primate conservation hand in hand with community well-being, and external funding was obtained from UNESCO with a focus on the Sustainable Development Goals. In this sense, there was reflection on how to approach the project in a more comprehensive, sensitive, inclusive manner, considering the local agenda.

For the aforementioned reasons, we decided to create a new project, while continuing to follow up on the previous work, called "*Escuelas Ya'ax*," with the aim of establishing a network of conservation ambassador schools (with the howler monkey as the flagship/representative species of the region). This would involve not only schools that are aware of the issues affecting their natural environments but also actively engaged in efforts to reverse or mitigate the negative impacts associated with the socio-ecosystem.

Due to project constraints a pilot of this program was established with a school located in the State of Yucatán, in the municipality of Valladolid, specifically in the community of San Francisco Tinum. Our goal was to promote increased awareness of conservation culture within the school and encourage reflection on several questions, such as:

- What does conservation mean to me?
- What natural resources in my environment deserve preservation?
- What actions can my school take?
- What actions can I, as an individual/student/teacher, undertake?

Through these reflections, we aimed to collectively define the characteristics of an "ideal conservation ambassador school for primate conservation." Additionally, we sought to develop and enhance skills related to the conservation of natural resources within their community. This was facilitated by the process of *diálogo de saberes*—a dialogue of different forms of knowledge involving the exchange of ideas, feelings, images, beliefs, practices, stories, experiences, and emotions to achieve mutual understanding and enrich life. The design of this process allowed for participation from various members of the community. After this successful pilot, we hope to continue the project in the future through a mentor-mentee process, where each school can serve as a mentor to the next one.

25.2.1.2 Photovoice: Arts-Based Method, Visual, PAR

We based on work on Camille and Sutton-Brown's (2014) photovoice steps (Fig. 25.1). However, we modified some parts of the process to adapt it to a local context and for the participation of children (Fig. 25.2).

Fig. 25.2 Group discussion. Photographs are discussed in small or large groups, and effective facilitation is based on the principles of Paulo Freire (2000). (Photograph by Lucía Jorge Sales)

The photovoice process in our case study took the following phases:

(a) Introduction to photovoice: Provide an overview of what photovoice is and its significance as a participatory research method.
(b) Questions for participants and their families: Encourage participants to reflect on their community, daily life, and any environmental concerns they may have. Pose questions that elicit personal and community experiences.
(c) Familiarization with cameras: Ensure participants are comfortable using cameras. Offer basic training on camera operation and techniques for effective photography.
(d) Equipment and research objectives: Clarify the purpose of the project, emphasizing the exploration of natural resources or environmental issues. Provide an understanding of the goals and objectives.
(e) Photography in the community: Guide participants in capturing images related to the identified concerns in their community. Encourage a diverse range of photographs to represent various perspectives.
(f) Selection of photographs together: Facilitate a collaborative process for participants to choose the most significant photographs and discuss the reasons behind their choices.
(g) Printing selected photographs: Assist in printing the chosen photographs for further analysis and presentation.
(h) Joint analysis session: Conduct a group session to analyze the selected photographs collectively. Encourage participants to express their thoughts and interpretations.
(i) Sharing *Sentipensamientos* (feelings and thoughts in Spanish): Create a platform for participants to share their emotions, thoughts, and stories associated with the photographs.
(j) Extending involvement to school and community: Discuss the possibility of expanding the photovoice project to involve the school and broader community. Explore ways to engage a wider audience and promote awareness of environmental concerns (Fig. 25.2).

25.3 Main Findings and Discussion

25.3.1 Environmental Issues/Challenge Identified as a Community

Through community analysis conducted via the photovoice activity, five main environmental issues were identified:

1. Deforestation: The clearing of forests, often for agriculture or logging, leading to the loss of biodiversity and disruption of ecosystems.

2. Decrease in corn and other crop yields: Reduction in the harvest of maize and other crops, possibly linked to factors such as changing environmental conditions or agricultural practices.
3. Water scarcity: Limited availability of water, which can have adverse effects on agriculture, ecosystems, and communities.
4. Inorganic waste: Accumulation of non-biodegradable waste materials, contributing to environmental pollution and degradation.
5. Mistreatment of domestic and wild animals: Abuse or ill-treatment of both domesticated and wild animals, posing ethical concerns and potentially disrupting ecosystems.

Addressing these issues requires a comprehensive, community-driven approach that integrates both practical and sustainable solutions tailored to the local context. To combat deforestation, the community must implement strategies that promote reforestation, forest management, and the use of agroforestry techniques that allow for sustainable agricultural practices without further degrading natural habitats. Regarding the decrease in crop yields, introducing diversified farming practices, including the promotion of climate-resilient crops and soil conservation techniques, can help mitigate the impact of environmental changes and improve food security. Water scarcity can be tackled by adopting water-saving technologies and implementing a rainwater harvesting system that ensures a more reliable and sustainable water supply for both agriculture and domestic use. To address the growing issue of inorganic waste, community-led initiatives, such as waste separation, recycling programs, and the reduction of single-use plastics, are essential. Additionally, promoting the ethical treatment of domestic and wild animals is crucial for restoring balance to local ecosystems and ensuring the well-being of both humans and animals.

These actions should be seen as interconnected solutions that promote not only environmental sustainability but also social cohesion, creating a resilient community equipped to face ongoing environmental challenges. Through collective effort and engagement, these issues can be addressed in ways that prioritize both the health of the environment and the well-being of the community.

25.3.2 Potential Solutions Arising from the Community

The community's journey towards identifying these sustainable solutions began with a participatory process that engaged residents, educators, and environmental experts. Through a series of workshops, discussions, and visual documentation (e.g., photographs), the community collectively reflected on its environmental challenges and priorities. The process involved identifying key issues such as waste management, water conservation, and the need for educational initiatives to foster empathy and environmental stewardship. These actions were not presented as a ranked list but as a series of interrelated solutions that emerged through consensus. While there were varying perspectives, the final set of proposals represents a unified effort to address the most pressing concerns. The solutions are practical and grounded in the community's daily experiences, offering a path towards positive, sustainable change. Five key practical solutions were identified:

1. Create a community nursery/school garden.
2. Diversify crops and promote meliponiculture (beekeeping with stingless bees).
3. Change water habits and implement a rainwater harvesting system.
4. Negotiate municipal garbage collection and treatment: Purchase plates/cutlery at school; Place containers in the community and negotiate municipal garbage collection and treatment.
5. Theater to work on empathy/Collaborate with teachers to work it in the classroom: Use theater to promote empathy/Have teachers incorporate it into the classroom. These actions propose practical and sustainable solutions to address environmental challenges and encourage positive change in the community.

As a suggestion from teachers, we created a challenge calendar, to ensure that nobody would be left behind (Fig. 25.3). These community-driven solutions demonstrate the value of amplifying local voices and co-creating pathways for collective ownership and leadership in conservation efforts, positioning local schools as key stakeholders and drivers of change.

25.3.3 Ethical Considerations

We identified six ethical considerations in research that are crucial to ensure that participants are treated with respect, and that the data collected is managed responsibly.

First, it is necessary to obtain official permissions for presentations or interventions, which vary depending on the level of access and the nature of the project.

Second, anonymity of the data is essential to protect participants' identities, so informed consent must be obtained, clearly explaining how their data will be used.

Third, clear guidelines should be established regarding the use of images, considering who will appear in them, where they will be taken, and their intended reach.

Fourth, when addressing sensitive topics, it is important to approach them with care, respecting participants' boundaries and avoiding imposing personal opinions or judgments.

Fifth, actively listening and not imposing external views is crucial for ensuring that the community's voice is heard authentically.

Sixth, adopting a biocultural approach involves recognizing the relevance of local cultures and knowledge, ensuring that the project's agenda aligns with the needs and priorities of the community rather than imposing external frameworks.

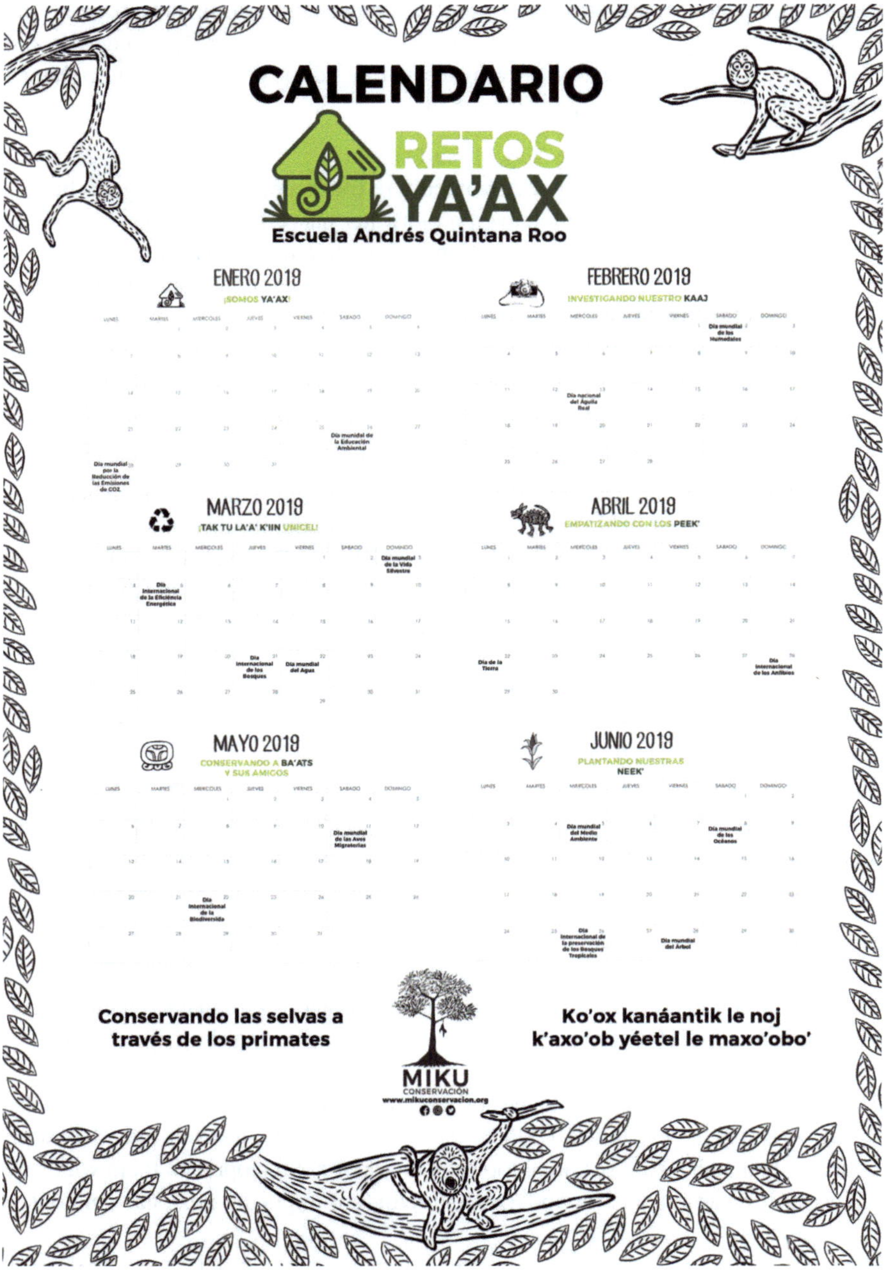

Fig. 25.3 Example of challenge calendar developed to encourage participation and keep participants engaged. (Figure was prepared by the author)

25.3.4 Major Difficulties Faced and Major Lessons Learned

+	−
Empowerment	Time strains
Allows individuals to showcase the perspective of their surroundings	"Abstract or complex ideas" difficult to capturate through a camera
Enables thinking differently about oneself, others, and the environment	A lot of flexibility and patience!
Redistribuite community governance	Analysis of certain topics can create negative feelings
Community and stakeholder engagement	Not everyone has the same level of experience
Creates or strengthens a sense of belonging	It can be expensive, but (when possible) a cellphone camera works perfectly.

Finally, while the process was successful, several methodological and logistical challenges emerged. Arts-based methods, such as photovoice, present challenges throughout the research process, including data collection, interpretation, and representation (Muhr 2020). Incorporating visual and artistic elements adds layers of complexity to these stages, requiring careful consideration and integration alongside traditional quantitative and qualitative methods. In the case of photovoice, where participants use photography to express their perspectives and experiences, challenges arise in ensuring the accuracy and reliability of the data collected, as well as in interpreting the visual narratives captured. Additionally, we found it difficult to present the findings in a way that conveyed the authenticity and intentions of the participants while also meeting scholarly standards of rigor, which can be demanding.

25.4 Concluding Remarks

This project highlights how participatory, arts-based methods like photovoice can foster meaningful connections between communities and conservation efforts. By merging scientific perspectives with emotional and cultural knowledge, the process empowered participants to recognize environmental challenges and co-create solutions grounded in their own realities. Ultimately, these experiences show that when communities are given the tools and the voice to lead, conservation becomes not just more inclusive and fair, but more effective and sustainable. The following five key insights summarize the main outcomes of our project, and illustrate the transformative potential of participatory conservation approaches:

(a) *Community engagement and empowerment*: The photovoice method proved invaluable in engaging community members and empowering them to identify and voice environmental concerns. By giving them a platform to express their perspectives and concerns, the community became active participants in the conservation process.
(b) *Biocultural approach*: The project's adoption of a biocultural approach facilitated a deeper understanding of the interconnectedness between culture, environment, and well-being. Recognizing and valuing traditional knowledge systems alongside scientific insights enriched the conservation dialogue and fostered more holistic solutions.
(c) *Co-identified environmental issues*: The project identified several pressing environmental issues, including deforestation, decreased crop yields, water scarcity, inorganic waste accumulation, and mistreatment of animals. These findings underscore the urgent need for action to address these challenges and safeguard both the environment and community well-being.
(d) *Community-led solutions*: The community proposed practical and sustainable solutions to address the identified environmental issues, such as establishing community nurseries/school gardens, promoting meliponiculture, implementing rainwater harvesting systems, negotiating municipal waste management, and utilizing theater for empathy-building. These solutions reflect the community's commitment to environmental stewardship and their capacity for positive change.
(e) *Call to action*: The project concludes with a call to action for stakeholders at local, regional, and national levels to support and collaborate with the community in implementing the proposed solutions.

Acknowledgments Our deepest gratitude goes to the schools and families of the Mexican communities who generously shared their time, knowledge, and creativity throughout this process. We thank Lucía Jorge Sales for her heartfelt collaboration in the design and implementation of the activities, and for bringing her passion and care into every step. We are especially grateful to the editors of this book for their kind invitation to contribute and for their insightful suggestions and enriching conversations along the way. MF-S thanks SECIHTI for a postdoctoral fellowship. We also thank *La Vallée des Singes* and *Le Conservatoire pour la Protection des Primates* for their support in funding the implementation of this project.

References

Archer LJ, Müller HS, Jones LP, Ma H, Gleave RA et al (2022) Towards fairer conservation: perspectives and ideas from early-career researchers. People Nat 4(3):612–626

Bennett NJ, Roth R, Klain SC, Chan K, Christie P et al (2017) Conservation social science: understanding and integrating human dimensions to improve conservation. Biol Conserv 205: 93–108

Berkes F (2004) Rethinking community-based conservation. Conserv Biol 18:621–630. https://doi.org/10.1111/j.1523-1739.2004.00077.x

Bowers CA (2002) Toward an Eco-justice Pedagogy. Environmental Education Research, 8(1): 21–34

Bray DB, Velazquez A (2009) From displacement-based conservation to place-based conservation. Conserv Soc 7(1):11–14

Bridgewater P, Rotherham ID (2019) A critical perspective on the concept of biocultural diversity and its emerging role in nature and heritage conservation. People Nat 1(3):291–304

Brydon-Miller M, Kral M, Ortiz Aragón A (2020) Participatory action research: International perspectives and practices. International Review of Qualitative Research, 13(2):103–111

Bustamante TD, Shadler E, White H, Pope B (2010) Positivism to post-positivism in conservation: a call for a philosophical paradigm shift. The George Wright Forum 27(1):31–46

Castro-Jorquera C (2026) The precarious art of Cecilia vicuña: finding the ecological and sacred value of the small and the discarded. In: Zhu D, Rozzi R, Tsuji R, Johnson B, Caviness T, Castro-Jorquera C, Xu L, Zhang L (eds) Linking arts with biocultural conservation, restoration, and communication, Ecology and ethics, vol 8. Springer, Cham, pp 79–93. https://doi.org/10.1007/978-3-032-04830-1_5

Chiarini VS (2026) How the invisible becomes visible: situating scientific illustration in biocultural communication. In: Zhu D, Rozzi R, Tsuji R, Johnson B, Caviness T, Castro-Jorquera C, Xu L, Zhang L (eds) Linking arts with biocultural conservation, restoration, and communication, Ecology and ethics, vol 8. Springer, Cham, pp 149–169. https://doi.org/10.1007/978-3-032-04830-1_9

Crist E (2012) Abundant Earth and Population. In P. Cafaro and E Crist (Eds.), Life on the Brink: Environmentalists Confront Overpopulation (pp. 141–153). Athens, GA: University of Georgia Press

Derr V, Simons J (2020) A review of photovoice applications in environment, sustainability, and conservation contexts: is the method maintaining its emancipatory intents? Environ Educ Res 26(3):359–380

Eastwood A, Juárez-Bourke A, Herrett S, Hague A (2023) Connecting young people with greenspaces: the case for participatory video. People Nat 5(2):357–367

Eberhard DM, Simons GF, Fennig CD (2022) Ethnologue: languages of the world. 25, SIL International, Dallas. Online version: http://www.ethnologue.com

Eldredge N (1995) Dominion. New York: Henry Holt

Escobar A (2008) Territories of Difference: Place, Movements, Life, Redes. Duke University Press

Fabricius C, Koch E, Magome H, Turner S (2004) Rights, resources and rural development: community-based natural resource manage-ment in Southern Africa. Earthscan, London

Franquesa-Soler M, Serio-Silva JC (2017) Through the eyes of children: drawings as an evaluation tool for children's understanding about endangered Mexican primates. Am J Primatol 79(12): e22723

Finlay L (2008) Reflexivity: A Practical Guide for Researchers in Health and Social Sciences. Wiley-Blackwell

Franquesa-Soler M, Barraza L, Serio-Silva JC (2018) Children's learning preferences for the development of conservation education programs in Mexican communities. J Educ Res 112(1):28–37

Franquesa-Soler M, Jorge-Sales L, Aristizabal JF, Moreno-Casasola P, Serio-Silva JC (2020) Evidence-based conservation education in Mexican communities: connecting arts and science. PLoS One 15(2):e0228382

Franquesa-Soler M, Jorge Sales L, Silva-Silva Rivera E (2023) Participatory action research for primate conservation: a critical analysis of a nonformal education program in southern Mexico. Am J Primatol 85(5):e23450

Freire P (2000) Pedagogy of the oppressed. The Continuum International Publishing Group Inc, New York

Gavin MC, McCarter J, Mead A, Berkes F, Stepp JR, Peterson D, Tang R (2015) Defining biocultural approaches to conservation. Trends Ecol Evol 30(3):140–145

Guha R (1989) Radical American environmentalism and wilderness preservation: a third-world critique. Environ Ethics 11:71–83

Guha R, Martínez-Alier J (1997) Varieties of environmentalism: essays North and South. Earthscan Publications, London

Guibrunet L, Gerritsen PRW, Sierra-Huelsz JA, Flores-Díaz AC, García-Frapolli E et al (2021) Beyond participation: how to achieve the recognition of local communities' value-systems in conservation? Some insights from Mexico. People Nat 3(3):528–541

Heras M, Tabara JD, Meza A (2016) Performing biospheric futures with younger generations: a case in the MAB Reserve of La Sepultura. Mexico Ecol Soc 21(2):14. https://doi.org/10.5751/ES-08317-210214

Hernández-Rodríguez E, López-Santiago J (2022) Uses and traditional knowledge of Dendropogonella rufescens (Bryophyta: Cryphaeaceae) in a Zapotec community of southeastern Mexico. Bot Sci 100(1):153–168

Hu J, Yang P (2026) A biocultural encounter with the four gentlemen: art education in Chinese ecological and aesthetic traditions. In: Zhu D, Rozzi R, Tsuji R, Johnson B, Caviness T, Castro-Jorquera C, Xu L, Zhang L (eds) Linking arts with biocultural conservation, restoration, and communication, Ecology and ethics, vol 8. Springer, Cham, pp 111–129. https://doi.org/10.1007/978-3-032-04830-1_7

Jaureguiberry P, Titeux N, Wiemers M, Bowler DE, Coscieme L et al (2022) The direct drivers of recent global anthropogenic biodiversity loss. Sci Adv 8(45):eabm9982

Kopnina H, Washington H, Taylor B et al (2018) Anthropocentrism: More than Just a Misunderstood Problem. J Agric Environ Ethics 31:109–127

Li Z, Jia M, Ma K (2026) Immersion and interactiveness: eco-themed art exhibitions and biocultural communication in East Asia. In: Zhu D, Rozzi R, Tsuji R, Johnson B, Caviness T, Castro-Jorquera C, Xu L, Zhang L (eds) Linking arts with biocultural conservation, restoration, and communication, Ecology and ethics, vol 8. Springer, Cham, pp 367–382. https://doi.org/10.1007/978-3-032-04830-1_23

Maffi L (2007) Biocultural diversity and sustainability. The SAGE Handbook of Environment and Society, 267–278

Martin GJ, Camacho Benavides CI, del Campo García CA, Anta Fonseca S, Chapela Mendoza F, Ortíz MAG (2011) Indigenous and community conserved areas in Oaxaca, Mexico. Manage Environ Qual Int J 22(2):250–266

Méndez-López ME, García-Frapolli E, Pritchard DJ, González MCS, Ruiz-Mallén I, Porter-Bolland L, Reyes-Garcia V (2014) Local participation in biodiversity conservation initiatives: a comparative analysis of different models in south East Mexico. J Environ Manag 145:321–329

Muhr MM (2020) Beyond words–the potential of arts-based research on human-nature connectedness. Ecosyst People 16(1):249–257

Nadasdy P (2005) The Anti-Politics of TEK: The Institutionalization of Traditional Ecological Knowledge. Anthropologica, 47(2):215–232

Pascual U, Balvanera P, Anderson CB, Chaplin-Kramer R, Christie M et al (2023) Diverse values of nature for sustainability. Nature 620(7975):813–823. https://doi.org/10.1038/s41586-023-06406-9

Petriello MA, Redmore L, Sène AL, Katju D, Barraclough L, Boyd S, Madge C, Papadopoulos A, Yalamala RS (2024) The scope of empowerment for conservation and communities. Conserv Biol e14249. https://doi.org/10.1111/cobi.14249

Restall B, Conrad E (2015) A literature review of connectedness to nature and its potential for environmental management. J Environ Manage 159:264–278. https://doi.org/10.1016/j.jenvman.2015.05.022

Rozzi R (2012) Biocultural ethics: recovering the vital links between the inhabitants, their habits, and habitats. Environ Ethics 34(1):27–50. https://doi.org/10.5840/enviroethics20123414
Rozzi R (2013) Biocultural ethics: from biocultural homogenization toward biocultural conservation. In: Rozzi R, Palmer C, Armesto JJ, Callicott JB, Pickett S (eds) Linking ecology and ethics for a changing world: values, philosophy, and action, Ecology and ethics, vol 1. Springer, Dordrecht, pp 9–32. https://doi.org/10.1007/978-94-007-7470-4_2
Rozzi R (2015) Implications of the biocultural ethic for earth stewardship. In: Rozzi R, Chapin FS III, Callicott JB, Pickett STA, Power ME, Armesto JJ, May RH Jr (eds) Earth stewardship: linking ecology and ethics in theory and practice, Ecology and ethics, vol 2. Springer, Cham, pp 113–136. https://doi.org/10.1007/978-3-319-12133-8_9
Rozzi R (2018) Biocultural homogenization: a wicked problem in the Anthropocene. In: Rozzi R, May RH, Chapin FS, Massardo F, Gavin MC et al (eds) From biocultural homogenization to biocultural conservation, Ecology and ethics, vol 3. Springer, pp 21–48. https://doi.org/10.1007/978-3-319-99513-7_2
Rozzi R, Massardo F, Anderson C, Heidinger K, Silander J Jr (2006) Ten principles for biocultural conservation at the southern tip of the Americas: the approach of the Omora Ethnobotanical Park. Ecol Soc 11(1):43. https://doi.org/10.5751/ES-01709-110143
Sarukhán J (2008) Conocimiento Actual de la Biodiversidad. Vol. 1 of Capital Natural de México. Comisión Nacional para el Conocimiento y Uso de la Biodiversidad, Mexico City
Shackleton RT, Walters G, Bluwstein J, Djoudi H, Fritz L et al (2023) Navigating power in conservation. Conserv Sci Pract 5(3):e12877
Smith LT (1999) Decolonizing Methodologies: Research and Indigenous Peoples. London and New York: Zed Books
Song Y (2026) Interspecies co-inhabitation: a dialogue between Chinese philosophical traditions and biocultural ethics. In: Zhu D, Rozzi R, Tsuji R, Johnson B, Caviness T, Castro-Jorquera C, Xu L, Zhang L (eds) Linking arts with biocultural conservation, restoration, and communication, Ecology and ethics, vol 8. Springer, Cham, pp 95–109. https://doi.org/10.1007/978-3-032-04830-1_6
Sutton-Brown CA (2014) Photovoice: A methodological guide. Photogr Cult 7(2):169–185
Tauro A, Rozzi R (2025) Biocultural ethics and earth stewardship: a novel integration to revitalize multiple values of nature. Ecol Soc 30. https://doi.org/10.5751/ES-16362-300335
Tran TC, Ban NC, Bhattacharyya J (2020) A review of successes, challenges, and lessons from indigenous protected and conserved areas. Biol Conserv 241:108271
Tsuji R, Johnson B (2026) Biocultural ontologies, art, and conservation. In: Zhu D, Rozzi R, Tsuji R, Johnson B, Caviness T, Castro-Jorquera C, Xu L, Zhang L (eds) Linking arts with biocultural conservation, restoration, and communication, Ecology and ethics, vol 8. Springer, Cham, pp 61–77. https://doi.org/10.1007/978-3-032-04830-1_4
Tsuji R, Castro C, Johnson B, Rozzi R (2026) Introduction to part I: theoretical inquiries linking arts with biocultural CRC. In: Zhu D, Rozzi R, Tsuji R, Johnson B, Caviness T, Castro-Jorquera C, Xu L, Zhang L (eds) Linking arts with biocultural conservation, restoration, and communication, Ecology and ethics, vol 8. Springer, Cham, pp 25–33. https://doi.org/10.1007/978-3-032-04830-1_2
UNEP-WCMC [UN Environment Programme World Conservation Monitoring Centre] (2014) Biodiversity A-Z website, Cambridge, UK. Accessible via https://www.biodiversitya-z.org/
Wang C, Burris MA (1997) Photovoice: concept, methodology, and use for participatory needs assessment. Health Educ Behav 24(3):369–387
Zhang S, Xu L (2026) A biocultural sketch of the Chinese black cardamom, Cao-guo (*Lanxangia tsao-ko*). In: Zhu D, Rozzi R, Tsuji R, Johnson B, Caviness T, Castro-Jorquera C, Xu L, Zhang L (eds) Linking arts with biocultural conservation, restoration, and communication, Ecology and ethics, vol 8. Springer, Cham, pp 349–365. https://doi.org/10.1007/978-3-032-04830-1_22
Zylstra MJ, Knight AT, Esler KJ, Le Grange LL (2014) Connectedness as a core conservation concern: an interdisciplinary review of theory and a call for practice. Sci Rev 2(1–2):119–143. https://doi.org/10.1007/s40362-014-0021-3

Chapter 26
Nature-Based Solutions of Rural Revitalization: The Xunjiansi Village of Wuyuan County

Kongjian Yu, Jianing Li, Yan Zhou, Mengyun Xie, and Yifan Hu

Abstract This paper proposes a nature-based path for rural revitalization: for villages with rich ecological and cultural endowments, by protecting natural and cultural heritage, and utilizing market-oriented approaches to sustainably realize the value of ecosystem and cultural services, it is possible to meet the demand of urban residents for high-quality living and traveling experiences, while promoting the flow of urban capital, talent, and knowledge into rural areas, thus advancing rural revitalization. Xunjiansi Village in Wuyuan, Jiangxi, is a typical example of this model. Faced with challenges in ecology, economy, community, and culture, the team, while improving the village's ecological infrastructure, also drew on traditional wisdom in architecture and agriculture. They implemented a "five-in-one" development model combining poetic dwelling, ecological agriculture, comprehensive local tourism-based development, learning in practice and the development of cultural and creative arts, successfully creating the Wangshan Life brand and establishing multiple links between the village's natural environment, culture, industry, and art. The team also facilitated the establishment of a stakeholder

Electronic supplementary material: The online version of this article (https://doi.org/10.1007/978-3-032-04830-1_26) contains supplementary material, which is available to authorized users.

K. Yu · Y. Hu
College of Architecture and Landscape, Peking University, Beijing, China
e-mail: kjyu@urban.pku.edu.cn

J. Li
College of Urban and Environmental Sciences, Peking University, Beijing, China
e-mail: 2201112237@pku.edu.cn

Y. Zhou
Systemic Restoration Department, Land Consolidation and Rehabilitation Center, Ministry of Natural Resources, Beijing, China

M. Xie (✉)
School of Urban Design, Wuhan University, Wuhan, China
e-mail: 00033987@whu.edu.cn

D. Zhu et al. (eds.), *Linking Arts with Biocultural Conservation, Restoration, and Communication*, Ecology and Ethics 8,
https://doi.org/10.1007/978-3-032-04830-1_26

cooperation network in the village, ensuring the sustainable development of rural revitalization through a shared governance model. This practice has yielded remarkable results in the ecological, economic, community, and cultural aspects. By realizing the value of ecosystem and cultural services, it has fostered the flow of elements and cultural interaction between urban and rural areas. The ideas and measures of this approach have the potential to be replicated and promoted in other regions.

Keywords Rural revitalization · Ecosystem service · Traditional village · Common prosperity · Xunjiansi Village · China

26.1 Introduction

Today socio-ecological problems are rapidly increasing; climate change, loss of biocultural diversity, resource depletion, frequent and devastating large-scale natural disasters, food security, and rural depletion, are all on the rise. A variety of strategies have emerged to address these problems, such as Long-term Socio-ecological Research and Management (Redman et al. 2004; Anderson et al. 2008), Biocultural Conservation (Rozzi 2013), Earth Stewardship (Chapin III et al. 2011), and Nature-based Solutions (Seddon et al. 2020). These strategies show a distinct mindset from reliance on technological and industrial strategies alone. Among them, Nature-based solutions (NbS) are defined by IUCN as “actions to protect, sustainably manage and restore natural or modified ecosystems that address societal challenges effectively and adaptively, simultaneously providing human well-being and biodiversity benefits” (Cohen-Shacham et al. 2016). Based on our practical experience regarding NbS in China, we believe that “natural forces, instead of the engineering approach symbolized in industrial civilization,” may make substantial contributions to mitigating climate change, regulating floods, conserving/restoring biocultural diversity, upgrading urban resilience, and revitalizing the countryside (Yu 2017; Yu 2020). In this chapter, we present our NbS practices in Xunjiansi Village, Wuyuan as a case study in the context of biocultural communication (Zhu and Rozzi 2026; Zhu et al. 2026).

Wuyuan County is located in the northeast of Jiangxi Province, at the junction of Anhui and Zhejiang province (Fig. 26.1). It is described by local residents in the format of word-of-mouth transmission as “85% mountains, 10% fields, and 5% waterways and manors.” It belongs to one of six counties shared by one governance of the ancient Huizhou with a wide distribution of traditional villages and Ming and Qing Dynasty ancient buildings. Xunjiansi (巡检司, Patrol and inspection Division Village), as one of the ancient Huizhou villages with excellent natural resources and a profound culture of farming and education, is currently one of the five natural villages under the administration of Yantian Village of Fuchun Town. Influenced by the limited resources and frequent local disasters in the basin, people in Xunjiansi

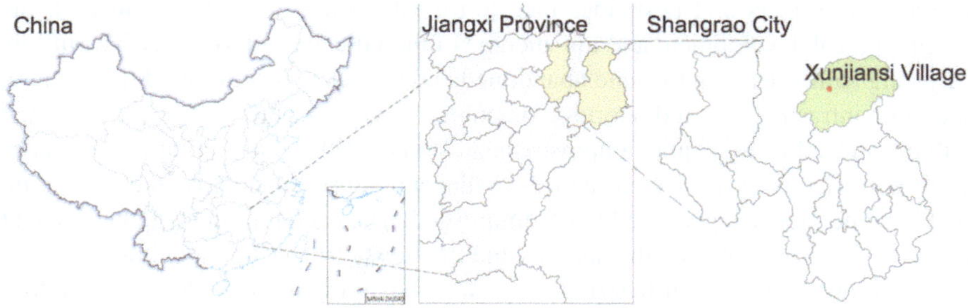

Fig. 26.1 Location of Xunjiansi Village. (Map was prepared by Jianing Li)

have developed ecological moderation (Yu 1992). Ecological moderation fosters sustainable use of the overall agricultural production and environment, including land conservation by building villages in accordance with the topography without destroying the landscape patterns, vegetation conservation by protecting Feng Shui forests to conserve water and regulate the climate, and water conservation by maintaining water inlets/outlets, digging ponds to reserve water, and constructing buildings to retain water. In this way, ecological moderation exemplifies nature-based sustainable development in the agricultural era.

Since the 1980s, with accelerated industrialization and urbanization, the enduring mechanism and traditional agro-ecological landscape of self-sufficiency and self-stability of Xunjiansi village have been destroyed, leading to a massive population outflow and the predicament of gradual depletion of natural resources. In 2015, Yu Kongjian of Peking University led the team of *Wangshan Life* and Turenscape designers to experiment in the Xunjiansi village with the framework of "conservation of roots and heritage, catalyzing through intervention, co-existence of the old and the new, and seeking common prosperity with people" (Yu 2017). *Wangshan Life* team designs and practices a lifestyle experiment where people can view mountains, rivers and have nostalgic memories through which the path towards realizing the countryside's high-quality development, good quality of life, and the claim of the central government that "clean water and green mountains are mountains of silver and gold" are explored.

It promotes the revitalization of the countryside and the construction of a beautiful China through a green mode of production and lifestyle, as well as fulfilling the people's aspiration for a better life in the era of ecological civilization. Now, the five-in-one *Wangshan Life* initiatives, namely, poetic dwelling, organic food, free touring, immersive education, and arts, are taking root and blooming in the Xunjiansi Village. While creating a better life for urbanites, these initiatives have led to the revitalization of villages, along with better protection and efficient management of natural ecosystems. Depleted villages and communities are now being revitalized. The case of Xunjiansi is a successful practice of nature-based solutions for rural revitalization in China.

Before the initiation of our project, the problems in the Xunjiansi village were considerable. First, the concrete channelization of rivers and the excessive use of chemical fertilizers and pesticides had destroyed natural ecosystems. Second, the stagnant local development and low income of the villagers led to a massive outflow of young adults, leaving the land abandoned and houses to deteriorate. Furthermore, the poor infrastructure and shortage of public services in the village rendered the culture and education quite impoverished. Worse still, local heritage landscapes were not effectively protected, and the villagers' sense of identity and belonging had become increasingly fragile and diminished. These conditions together created negative feedback between the habitats and life habits triggering biocultural homogenization processes (sensu Rozzi 2018), which have implied local losses of biological and cultural diversity as well as social and ecological degradation (Yu 2021; Li 2022).

Based on the above-mentioned problems, the goals of our project are to conduct the *Wangshan Life* Experiment, promote rural revitalization through fully exploring the advantages of the ecological, historical, and cultural heritages, and to prioritize the conservation of ecological functions and the natural environment.

26.2 The Implementation of Nature-Based Solutions

The implementation of the project is based on the systematic organization of the natural and cultural heritage. We have undertaken a holistic planning and design of the ecological, production and living spaces, as well as initiated the reinvention of the rural governance system. Throughout the process, the five-in-one framework of the *Wangshan Life* has been actualized. As mentioned above, the five-in-one framework includes: (i) poetic dwelling, (ii) organic food, (iii) free touring, (iv) immersive education, and (v) arts. The core of these concepts involves the establishment of restored habitats and life habits in the vein of biocultural conservation (Rozzi 2013), which lead to a new urban-rural relationship and provisioning strategy based on ecosystem services, whereby the survival, health, and wellbeing are founded on the safety, health and fecundity of the ecosystems. This will also lead people to re-conceptualize the value of rural ecosystems, realize harmonious coexistence between humans and nature, and promote urban-rural co-prosperity (Fig. 26.2).

26.2.1 Improving Ecological Infrastructure

Building a spongy land and improving the ecological infrastructure are the ecological foundation for practicing *Wangshan Life*. In the case of Xunjiansi, the following specific measures are implemented. The first is to protect and rehabilitate the ancient water heritage with the basic structure of pei ponds- low weirs—and canals (Fig. 26.3). These micro-hydrological systems not only protect natural water courses and patterns but also enable water conservation, groundwater recharge, rainwater

Fig. 26.2 Satellite map of the layout of Xunjiansi. (Map was prepared by Jianing Li)

Fig. 26.3 Comparison of before (2016) and after (2018) restoration of the old weir. (Photography by *Wangshan Life*)

purification, and flood control while meeting the basic needs of both domestic and agricultural water consumption.

The second measure was to prevent the canalization and hardening works of the river channels promptly. The multiple functions of protecting natural river basins are to maintain the habitats of aquatic organisms, to purify pollution from surface sources, and to replenish groundwater (Forman 1995; Yu et al. 2005, p. 75, 149).

The third measure was the preservation and restoration of water in/outlets and woods around them. At the water in/outlets, where the village's water system joins, the creation of higher terrain on one side and the cultivation of woods would serve the functions such as soil and water conservation, windbreaks and sand fixation, as well as the regulation of the microclimate (Yu 1992; Cheng et al. 2009).

Fig. 26.4 Xunjiansi villagers are plucking tea leaves grown free of fertilizers and pesticides in 2020. (Photograph by *Wangshan Life*)

26.2.2 Reviving Traditional Agricultural Wisdom

In the course of the project, we have cultivated the Xunjiansi Optimal Agricultural Base to revive the wisdom of traditional agriculture, through fallow, crop rotation, returning straws to the field, and other measures to resume the fertility and vitality of the soil and the land. No chemical fertilizers and pesticides are used in the growing process to ensure the green quality and safety of the crops (Fig. 26.4). To give full advantage to the comprehensive functions of agricultural space, open bamboo poultry farms and honeybee farms are being established in the meantime to protect and restore healthy farmland ecosystems, in accordance with the principle of the mutual reinforcement between crop-farming and animal husbandry (Li 2022; Zhang et al. 2024; He et al. 2024).

26.2.3 Rebuilding the Poetic Dwelling Living Space

In order to rebuild the living space of poetic dwelling, we adopt the method of "insertion and collage" (Yu 2017) in the overall planning of the village construction. Using this method we transform the old residence base, insert the urban residents who have returned to their hometowns, and plan the new residential and service facilities. In terms of architectural renovation, we draw on the wisdom of the games between lighting and heat insulation, water storage and dehumidification in

Fig. 26.5 Comparison of the restored old house: before (2016) and after (2021). (Photograph by *Wangshan Life*)

Fig. 26.6 Before (2016) and after (2021): the renovation of *Wangshan Life* Center and Huixiu Bridge. (Photograph by *Wangshan Life*)

Fig. 26.7 Before (2016) and after (2021): the construction of the rain-wind corridor of Huixiu Bridge of Xunjiansi Village. (Photograph by *Wangshan Life*)

traditional Huizhou architecture, retaining and repairing the patio, opening up the wall, and applying floor-to-ceiling windows to ventilate and lighten the building to protect the existing wood carvings while simultaneously maintaining its dehumidification function (Figs. 26.5, 26.6, and 26.7).

To promote stormwater infiltration and to integrate with rapid regional transportation, we have also preserved and restored the slabstone paving of the Hui Rao ancient road. Meanwhile, open channels were constructed along the village road to connect with the river to solve drainage problems (Fig. 26.8). And an aeration tank

Fig. 26.8 Before (2016) and After (2021) Images of Street Improvement. (Photograph by *Wangshan Life*)

was designed at the outlet to reoxygenate the water through the collision of water flowing into the tiles.

26.2.4 Branding Ecological Learning and Research

In terms of free touring, we have designed and planned a three-kilometer heritage corridor and interpretation system, linking the tangible and intangible cultural heritage of Xunjiansi village and three surrounding ancient villages. In terms of research and learning practice, the National Mayor Training Institute has been launched in Xunjiansi (Fig. 26.9). With the continuation of the Ideal Settlement and Vernacular Landscape Studies (Fig. 26.10), this place has also become an in-situ education site for rural revitalization.

Regarding arts, we have curated China's first *Wangshan Life* Field Stage (Fig. 26.11), and national inaugural 5G Countryside Arts and Living Museum (Fig. 26.12), which have received more than 50 million followers, and served as China's first online live broadcasting base for rural revitalization.

26.2.5 Village Co-governance

The *Wangshan Life* Team guides and establishes a cooperative management mechanism composed of the government, experts, villagers, and enterprises to jointly discuss and review planning, supervise the protection of cultural heritage, and improve rural landscape, which ameliorated rural hygiene and social order. The team offers expertise to the government and village collectives, attracts social capital investment, and creates job opportunities for villagers. Correspondingly, the government and village collectives provide policy support, coordinate villagers' interests, and involve villagers in common building and communal activities, thereby forming a model of co-governance in the Xunjiansi Village (Fig. 26.13).

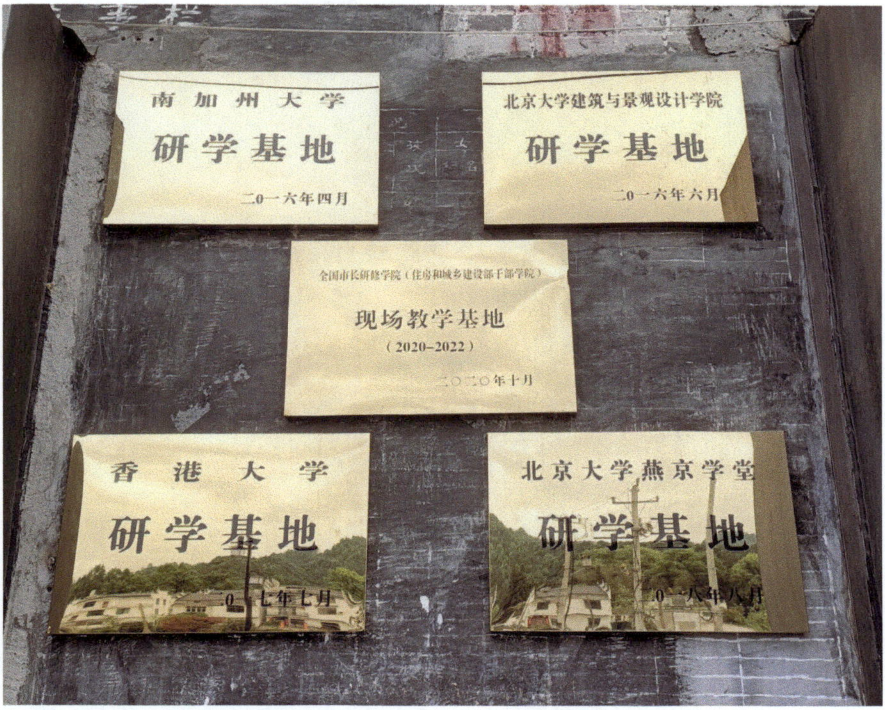

Fig. 26.9 On-site teaching center in Xunjiansi, Wuyuan, 2020. The center harbors National Mayor Training Institute, the University of Hong Kong Research and Learning Base, University of Southern California Research and Learning Base, etc. (Photograph by *Wangshan Life*)

26.3 Result

26.3.1 Dramatically Improved Ecosystem Regulation Services

Through the above endeavors, the natural ecosystem of Xunjiansi has been improved and sponges-based ecological infrastructures have been structured. In the context of overall hydro-ecology, hydro-environment, and aquatic resources these eco-infrastructures materialize the reconciliation between the structure and function of the hydro-ecological system, thus solving the water problem in an integrated, systematic, and sustainable manner. The building design makes optimal use of natural light, water, and wind to maximize comfort while minimizing energy consumption. Additionally, conservation and repair of the slow-paced system encourages a green approach to recreation. Once again restoration of the habitats fosters restoration of nature-linked life habits.

Fig. 26.10 Kongjian Yu was conducting an *in-situ* seminar on ideal human settlements and vernacular landscapes in 2021. (Photograph by *Wangshan Life*)

Fig. 26.11 *Wangshan Life* Field Stage (2021). (Photograph by *Wangshan Life*)

Fig. 26.12 Artistic Living Studio in Xunjiansi: Watercolor Pottery Art (2022). (Photograph by *Wangshan Life*)

Fig. 26.13 Villagers build and share the public space of the bridge area (2021). (Photograph by *Wangshan Life*)

26.3.2 Driving Local Development and Raising Farmers' Incomes

The approach of Xunjiansi has brought forth a sustainable model of rural revitalization that does not rely solely on State investment. This is specifically reflected in the following.

- First, the income of the population has increased. Based on our survey, the average annual income of the Xunjiansi residents in 2021 has improved by 71% compared to 2015. The number of residential buildings repaired and remodeled by residents on their own initiative has greatly increased, from only 10 in 2015 to 60 in 2021, accounting for 55% of the total number of households.
- Second, agricultural products have completed the transformation towards high quality and competitive prices. For example, compared with the former local Zhangding black tea, which until 2015 was grown with chemical fertilizers and pesticides, being sold at about CNY¥ 100 per catty (amounts to half Kilogram), the organic, healthy, and traceable Wangshan Zhangding black tea is recognized in the market at CNY¥ 4000 per catty.
- Third, the tourism industry is thriving. For example, until 2015 only a few painters visited Xunjiansi to draw sketches while after 2015, numerous villagers turned their homes into characteristic Bed and Breakfast (B&Bs) lodges, with occupancy rates reaching over 80% during the flowering season of canola and sunflower, and harvest season of rice.

26.3.3 Catalyzed Rural Vitality

Young adults chose to return to their hometowns to work, and compared with 2015, the resident population of the Xunjiansi Village in 2020 increased by 54% and the household population by 24%. The restoration of the village landscape with rich

Fig. 26.14 Father and daughter participate in a rural art class at Xunjiansi Studio (2020). (Photograph by *Wangshan Life*)

local characteristics and traditions involved several interrelated outcomes: (i) the thousand-year-old Hui-rao ancient road was protected, (ii) the 300-year-old corridor bridge was reversibly restored, (iii) the infrastructure and public service facilities were improved, and (iv) the village road (which was plagued by rain and flooding) was transformed into a gorgeous street with gurgling brook. In addition, through attending the study program and joining the public welfare arts and creativity class (Fig. 26.14), the villagers and children learned about ecological and cultural knowledge and opened up their drawing and design skills. Ultimately, every villager became a potential tour guide.

26.3.4 Cultural Heritage Preserved

With the systematic protection of tangible and intangible cultural heritage under the heritage corridor system, local residents have gradually realized the value of local natural systems and cultural landscapes and enhanced their sense of belonging to and identification with the land under their feet. The villagers have come to realize that clean rivers and lush mountains are mountains of gold and silver. They also have come to understand and accept a low-carbon, sustainable *Bigfoot Aesthetics* (大脚美学) based on healthy natural ecosystems and recognize that prudent use of nature in conservation is the long-term solution to revitalizing the villages (Yu 2017).

26.4 Concluding Remarks

We end this chapter by stating three concluding remarks.

1. **Nature-based rural development leads to rural revitalization while providing urbanites with access to high-quality ecosystem services, representing an effective way to reduce urban-rural disparities and achieve urban-rural co-prosperity.**

Megacity malaise has generated aspirations for ecosystem services and a better life among urban residents, while the rich ecosystem services of the countryside have not been effectively honored. This gap motivated *Wangshan Life* team to shift towards a nature-based approach to rural development in the Xunjiansi village, where urban-rural relationships are reconstructed, and the revitalized countryside becomes a site of green values of ecosystem services. By restoring ecosystems and promoting green production and lifestyles, the Xunjiansi case exemplifies a sustainable and replicable model of rural revitalization via establishing links between ecosystem services and human well-being.

Through the market-oriented approach to ecological compensation and transfer payments, this model allows urbanites to fully understand the value of rural ecosystem services and enjoy them while bringing knowledge and wealth to the rural area. Thus, it revitalizes the depleted rural economy and turns "clean rivers and lush mountains" into "mountains of gold and silver," thereby enabling sustainable development that eliminates poverty without reverting to poverty. The model and concepts embodied in this case have facilitated the localization of nature-based solutions (NbS) in China, which provides experience and lessons for rural governance and village revitalization in other regions.

2. **Enrich NbS from the dimension of traditional Chinese ecological wisdom.**

The solutions in maintaining the natural roots and heritage of the village, protecting, and restoring its ecosystem, and reconstructing a resilient landscape, as mentioned above, have been inspired by the traditional ecological wisdom of Huizhou. A series of sponge landscapes in the village are not only the wisdom of the ancestors to adapt to droughts and floods, but also the embodiment of the harmonious coexistence of geo-social and neighborly relations in Wuyuan, a sear on the land of the experience and wisdom at the cost of lives for thousands of years (Yu 2016).

Although this ecological wisdom of prudent use of nature has been gradually forgotten or even discarded in the vast countryside of China with the development of industrial civilization, this wisdom derived from survival experience can be adapted to monsoon climates, and some of the methods are applicable to other parts of the world as well. Programs that translate traditional ecological wisdom into scientific practice will have opportunities to be disseminated globally. Moreover, these cases in China complement and are complemented by initiatives of biocultural conservation (May 2015; Rozzi 2015; Kono 2018; Hedge and James 2018) and biocultural restoration (Klaver 2018; Toyoda 2018) that are being conducted in other regions of the world.

3. **The *Wangshan Life* Model Provides a Chinese Paradigm for Global Rural Development.**

Rural dilapidation is a global problem. Many countries have explored the model of rural reconstruction, such as Rural Development Plan of England (Bruton 1983), Germany's "Equal Living Conditions" (Germany Federal Press Office 2017), US's urban-rural symbiosis and eco-village model (Mao and Lin 2022), and Japan's One Village One Product movement (Li 2005; Feng 2021). In contemporary China, the strategy of rural revitalization is being comprehensively implemented, and the Chinese countryside is facing an even more complex road of construction. The *Wangshan Life* model developed by Yu Kongjian and his team is a model of rural urbanism in the era of urbanization and post-industrialization, particularly in the post-epidemic era. In the global rural construction process, as a market-based development model, nature-based solutions need to leverage the willingness of the affluent to pay for high-quality ecological services and cultural assets. By recognizing the economic values of green water and nostalgia, jobs are created and a sustainable development of the economy, society and culture is achieved without damaging or depleting natural and cultural assets. Consequently, it can provide empirical evidence, methods, and inspiration for biocultural conservation initiatives in other regions of the world (e.g., Rozzi et al. 2013, 2015, 2018, 2023; Mallik 2022; Valera 2023). Importantly, the *Wangshan Life* is a dynamic one, worthy of constant research and promotion (Fig. 26.15).

Fig. 26.15 Artists and villagers envisioning a better future together (2020). (Photograph by *Wangshan Life*)

References

Anderson CB, Likens GE, Rozzi R, Gutiérrez JR, Armesto JJ, Poole A (2008) Integrating science and society through long-term socio-ecological research. Environ Ethics 30(3):295–312. https://doi.org/10.5840/enviroethics200830334

Bruton MJ (1983) Local plans, local planning and development plan schemes in England, 1974-1982. Town Plan Rev 54(1):4–23

Chapin FS III, Power ME, Pickett ST, Freitag A, Reynolds JA et al (2011) Earth stewardship: science for action to sustain the human-earth system. Ecosphere 2(8):1–20

Cheng J, He F, Liu Y (2009) Progress of the research on geomantic forests of Lingnan villages (岭南村落风水林研究进展). Chinese Landsc Architect 25(11):93–96

Cohen-Shacham E, Walters G, Janzen C, Maginnis S (2016) Nature-based solutions to address global societal challenges, vol 97. IUCN, Gland, pp 2016–2036. https://doi.org/10.2305/IUCN.CH.2016.13.en

Feng C (2021) The promotion mechanism of the "one village one product" movement in Japan and the independence of rural society (日本"一村一品"运动的推动机制与农村社会自主性). World Agric 10:62–69

Forman RT (1995) Land mosaics: the ecology of landscapes and regions. Cambridge University Press, New York

Germany Federal Press Office (2017) Government Report on Wellbeing in Germany. Accessed through https://www.gut-leben-in-deutschland.de/downloads/Government-Report-on-Wellbeing-in-Germany.pdf

He K, Liu Y, Zheng J (2024) The development of integrated crop-livestock system under the carbon peaking and carbon neutrality goals ("双碳"目标下的种养结合循环农业发展). J Zhongnan Univ Econ Law 06:28–39

Hedge P, James G (2018) Challenging biocultural homogenization: experiences of the Chipko and Appiko movements in India. In: Rozzi R, May RH Jr, Chapin FS III, Massardo F, Gavin M, Klaver I, Pauchard A, Nuñez MA, Simberloff D (eds) From biocultural homogenization to biocultural conservation, Ecology and ethics, vol 3. Springer, Dordrecht, pp 427–442. https://doi.org/10.1007/978-3-319-99513-7_27

Klaver IJ (2018) Trans-scapes transitions in transit. In: Scapp R, Seitz B (eds) Philosophy, Travel, and Place: Being in Transit. Palgrave Macmillan, Cham, pp 189-207. https://doi.org/10.1007/978-3-319-98225-0_12

Kono T (2018) The garden as a representation of nature: a space to overcome biocultural homogenization? In: Rozzi R et al (eds) From biocultural homogenization to biocultural conservation, Ecology and ethics, vol 3. Springer, Cham, pp 459–475. https://doi.org/10.1007/978-3-319-99513-7_29

Li Q (2005) Japan's "one village one product" movement and its implications (日本的"一村一品"运动及其启示). World Agric 1:32–35

Li J (2022) The study of vernacular landscape interpretation in Huizhou—a case of Yantian Village, Wuyuan County (徽州乡土景观的解释学研究——以婺源县严田村为例). Peking University

Mallik B (2022) Beauty, Wonder, and Sustainability: Reconnecting Humans to Nature for "Durable Harmony". In: Mallik B, Legends in Gandhian Social Activism: Mira Behn and Sarala Behn. Ecology and Ethics. vol 4, Springer, Cham. https://doi.org/10.1007/978-3-030-95431-4_9

Mao R, Lin X (2022) Promoting integrated development of rural and urban areas amid the rural revitalization strategy: implications and lessons from main developed countries (在乡村振兴中促进城乡融合发展——来自主要发达国家的经验启示). Int Econ Rev 1:155–173+8

May RH (2015) Andean Llamas and Earth stewardship. In: Rozzi R et al (eds) Earth stewardship, Ecology and ethics, vol 2. Springer, Cham, pp 77–86. https://doi.org/10.1007/978-3-319-12133-8_7

Redman CL, Grove JM, Kuby LH (2004) Integrating social science into the long-term ecological research (LTER) network: social dimensions of ecological change and ecological dimensions of social change. Ecosystems 7:161–171

Rozzi R (2013) Biocultural ethics: from biocultural homogenization toward biocultural conservation. In: Rozzi R, Pickett S, Palmer C, Armesto J, Callicott J (eds) Linking ecology and ethics for a changing world, Ecology and ethics, vol 1. Springer, Dordrecht, pp 9–32. https://doi.org/10.1007/978-94-007-7470-4_2

Rozzi R (2015) Earth stewardship and the biocultural ethic: Latin American perspectives. In: Rozzi R et al (eds) Earth stewardship, Ecology and ethics, vol 2. Springer, Cham, pp 87–112. https://doi.org/10.1007/978-3-319-12133-8_8

Rozzi R (2018) Biocultural homogenization: a wicked problem in the Anthropocene. In: Rozzi R et al (eds) From biocultural homogenization to biocultural conservation, Ecology and ethics, vol 3. Springer, Cham, pp 21–48. https://doi.org/10.1007/978-3-319-99513-7_2

Rozzi R, Tauro A, Avriel-Avni N, Wright T, May Jr RH (eds) (2023) Field Environmental Philosophy. Ecology and Ethics, vol 5. Springer, Cham. https://doi.org/10.1007/978-3-031-23368-5

Seddon N, Chausson A, Berry P, Girardin CA, Smith A, Turner B (2020) Understanding the value and limits of nature-based solutions to climate change and other global challenges. Philos Trans R Soc B 375(1794):20190120

Toyoda M (2018) Revitalizing Local Commons: A Democratic Approach to Collective Management. In: Rozzi R, May Jr RH, Chapin III FS, Massardo F, Gavin MC et al (eds) From Biocultural Homogenization to Biocultural Conservation. Ecology and Ethics, vol 3. Springer, Cham. https://doi.org/10.1007/978-3-319-99513-7_28

Valera L (ed) (2023) Pantheism and Ecology. Ecology and ethics, vol 6, Springer, Cham. https://doi.org/10.1007/978-3-031-40040-7

Yu K (1992) Basin experience of Chinese agriculture and the ecological prudent landscape(盆地经验与中国农业文化的生态节制景观). J Beijing For Univ 4:37–44

Yu K, Li D, Liu H (2005) "Anti-planning" approach ("反规划"途径). China Architecture and Building Press, Beijing

Yu K (2016) Ecology-based water management: "Sponge City" and "sponge land" (论生态治水: 海绵城市"与"海绵国土"). Frontiers 21:6–18

Yu K (2017) New Ruralism movement in China and its impacts on protection and revitalization of heritage villages: Xixinan experiment in Huizhou District, Anhui Province ("新上山下乡运动"与遗产村落保护及复兴——徽州西溪南村实践). Bull Chin Acad Sci 32(07):696–710

Yu K (2020) The concept between two civilizations: on nature-based solutions. Landsc Architect Front 8(3):7–9

Yu K (2021) Human landscapes in big history. Landsc Architect Front 9(02):4–7

Zhang J, Wei S, Liu X et al (2024) Research progress and prospect of integrated rice-fish coculture (稻渔综合种养研究进展与展望). J South China Agric Univ 45(06):812–824+808

Zhu D, Rozzi R (2026) Introduction to the book: biocultural communication: arts and sciences for conserving biological and cultural diversity. In: Zhu D, Rozzi R, Tsuji R, Johnson B, Caviness T, Castro-Jorquera C, Xu L, Zhang L (eds) Linking arts with biocultural conservation, restoration, and communication, Ecology and ethics, vol 8. Springer, Cham, pp 1–22. https://doi.org/10.1007/978-3-032-04830-1_1

Zhu D, Caviness T, Rozzi R (2026) Introduction to part III: biocultural communication across urban, rural, and remote ecosystems. In: Zhu D, Rozzi R, Tsuji R, Johnson B, Caviness T, Castro-Jorquera C, Xu L, Zhang L (eds) Linking arts with biocultural conservation, restoration, and communication, Ecology and ethics, vol 8. Springer, Cham, pp 335–347. https://doi.org/10.1007/978-3-032-04830-1_21

Chapter 27
The *Wangshan Life* Model for the Revitalization of Rural Villages: The Xixinan Case

Kongjian Yu, Jianing Li, Changhui Zheng, and Hong Yun

Abstract The aim of "Linking Arts with Biocultural Conservation, Restoration, and Communication" acquires new relevance in the context of China's effort to construct ecological civilization, promote rural revitalization, and achieve common prosperity. In this vein, we propose in this chapter a rural revitalization model for traditional villages based on natural and cultural assets under a novel urban-rural relationship in China: *Wangshan Life* (望山生活). This model, rooted in reflection on the modern urban-rural relationship, aims to prioritize the consideration of ecology and conservation of natural/cultural heritage. Meanwhile, its theoretical basis involved an examination of the conventional urban-rural relationship through discourses such as ecosystem services, human well-being, and demand theory. By proper planning, designing, and management of rural villages, *Wangshan Life* established renewed urban-rural linkage, bringing about multiple benefits, including: (1) providing high-quality cultural and ecological services for urban and rural areas, meeting the aspirations of urban residents for a desirable rural lifestyle; (2) transforming green mountains, clean rivers, and nostalgia into economic benefits, promoting common prosperity of urban and rural residents; and (3) revitalizing rural areas. In this chapter, we summarize and present the experiences of *Wangshan Life* related to the nine-year rural revitalization practices in Xixinan Village in Huizhou, Anhui Province, China. The substance and methodology of rural revitalization based on the

K. Yu
College of Architecture and Landscape, Peking University, Beijing, China
e-mail: kjyu@urban.pku.edu.cn

J. Li
College of Urban and Environmental Sciences, Peking University, Beijing, China
e-mail: 2201112237@pku.edu.cn

C. Zheng
College of Humanities and Development Studies, China Agricultural University, Beijing, China

H. Yun (✉)
School of Design, South China University of Technology, Guangzhou, Guangdong, China
e-mail: yunhong90@scut.edu.cn

D. Zhu et al. (eds.), *Linking Arts with Biocultural Conservation, Restoration, and Communication*, Ecology and Ethics 8,
https://doi.org/10.1007/978-3-032-04830-1_27

renewed, post-modern, and vibrant urban-rural relationship is defined, offering a market-oriented and sustainable revitalization model for traditional villages.

Keywords Rural revitalization; · Renewed urban-rural relationship; · Ecosystem service; · Landscape services; · Traditional village; · Common prosperity; · Xixinan Village

27.1 Introduction

Traditional villages, commonly known as "古村落" (gǔ-cūn-luò) in Chinese, are natural villages and rural areas formed by larger settlements or multiple groups of settlements rich in tangible and intangible cultural heritage with historical value, cultural representativeness, and overall inheritance (Yu 2017; Zhou 2013). Traditional villages include, but are not limited to villages listed on the *List of Chinese Traditional Villages* (《中国传统村落名录》, since 2012-) and *Famous Chinese Historical and Cultural Villages* (《中国历史文化名村》, since 2003-) published by the Ministry of Housing and Urban-Rural Development, China. Rather, traditional villages are a category of villages that exist broadly in China. However, during the decade of 2000–2010, China has lost 140,000 villages in the administrative sense and 900,000 villages in the geographical or natural sense, and the rate of village extinction is much faster than their rates of conservation or preservation (Feng 2017). An important reason for this extinction is that, under the long-standing urban-centric and urban-prioritized view of development, villages are regarded as the hinterland of urbanization and the provision areas of agricultural products. As a result, the development momentum of villages was weakened, and the villages were deprived of development rights. Over decades of China's dramatic urbanization, the multifunctionality of the rural territorial systems and the interactions between urban and rural territorial systems have been gradually attenuated (Liu et al. 2021). Other reasons for the extinction of villages include problems such as the expanded urban-rural gap, soil and water pollution, community hollowing out, population aging, and heritage destruction in traditional villages. As a result, the sustainable development of villages is facing a serious crisis.

Generally speaking, in the highly urbanized stage, the capacity of cities to absorb new labor would decline, and the development of agricultural technology would also make the supply of agricultural products exceed the demands, which in turn would significantly reduce the value of human labor and agricultural products transported from rural to urban areas (Gutman 2007). Based on this insight, China has called for changing the one-directional flow from the rural to the urban areas, aiming at establishing a new urban-rural relationship based on what is more valuable to each other, narrowing the gap between urban and rural area, alleviating ecological degradation, and revitalizing the decaying villages.

In September 2018, the State Council of China issued the Strategic Plan for Rural Revitalization (2018–2022), which states that "ecological quality is the greatest strength of rural development" (State Council of China 2018). The ecosystem or landscape services provided by a good rural ecological environment and cultural landscape should become an important link between urban and rural areas under ecological civilization. For some traditional villages, with both of their outstanding ecological substratum and cultural heritages, they have great potential to provide urban and rural residents with multiple well-beings including both ecological services and cultural services. This also becomes a groundbreaking point for the revitalization of traditional villages.

In recent years, with the rapid growth of traditional village tourism, a new urban-rural relationship is emerging, and some villages have been able to fight against poverty and become affluent. However, some new problems have emerged. For example, the relationship between multiple entities in the urban-rural relationship is not balanced. Some local governments and external enterprises have inappropriately exploited the long-term development potential of traditional villages in order to achieve short-term economic growth, concealing the crisis of deterioration behind the ostensible prosperity. How to build a bilateral and interactive approach based on a renewed urban-rural relationship and sustainably bring into play the ecological and cultural advantages of traditional villages is the key for traditional villages to achieve new-era goals, such as ecological livability, comprehensive revitalization in ecological, social, cultural, economic and human capital terms, and achieve common prosperity.

In this chapter we examine the new urban-rural relationship from the perspective of ecosystem services (Millennium Ecosystem Assessment 2005), human well-being, and the demand theory, define the essence of traditional village revitalization, explore the methodology of rural revitalization, and summarize the strategies and results of the nine-year revitalization and regeneration experimental practices in Xixinan Village, Huizhou, Anhui Province, China to explore a market-oriented approach to common prosperity in rural areas that does not rely on state input.

27.2 The Keystone of Rural Revitalization—Renewed Urban-Rural Relationship

The space scale, population size, and social resources in the countryside make it difficult to develop momentum comparable to that of cities and towns. Hence, the revitalization of the countryside must not be driven from within the traditional villages, but should be placed in the overall urban-rural relationship, analyzing the urban-rural issues and elements, and seeking a mutually reinforcing way of development.

27.2.1 The Drawbacks of Modern Urban-Rural Relations Within Industrial Societies

Since China's economic reform and opening up, under the dominant dualistic relationship between urban and rural areas, the flow of elements between urban and rural areas has become relatively homogenous and one-directional. While cities receive the labor and agricultural products provided by the villages, they often fall into a situation like "beggar-thy-neighbor," transferring pollution and other environmental hazards to the countryside. Such a formal and structured urban-rural relationship has resulted in a stark contrast between prosperous cities and decaying villages in China, with the countryside facing a number of ecological, community, and cultural crises.

The ecological crisis is mainly manifested by the fact that the conventional urban-rural relationship in industrial societies has not given importance to the protection of the environment (Gutman 2007). In order to increase crop yields and reduce planting costs, most villagers and large-scale planters have abused the land with chemical fertilizers and pesticides, resulting in non-point source pollution. According to the results of the first National Pollution Source Census, the annual emissions of chemical oxygen demand, total phosphorus, and total nitrogen from agricultural sources respectively accounted for 43.7%, 57.2% and 67.3% of the total national emissions (Ministry of Environmental Protection, China 2010). In terms of utilization rate, the utilization rate of chemical fertilizers and pesticides for the three major food crops of rice, corn, and wheat in China was 40.2% and 40.6%, respectively, in 2021 (International Cooperation Department of the Ministry of Agriculture and Rural Affairs, China 2021), which is at least 10% lower than that of developed countries. The high intensity and low utilization efficiency of pesticides and fertilizers have introduced the pollutants to rural aquatic systems and soils. In addition to this, urban discharge of domestic and industrial pollution into rural areas further aggravates the deterioration of rural aquatic and territorial environment.

The community crisis is mainly characterized by the one-way export of labor from villages to cities, resulting in a very serious loss of robust rural laborers. In 2020, the rural population decreased by 24.4% compared to 2010, and rural employment decreased by 15.8% from 2016 to 2020. In terms of rural life quality, by the end of 2016, 26.1% of village domestic waste and 82.6% of village domestic sewage in rural China had not received any treatment, 46.5% of villages had not completed toilet improvement, and 52.3% of farm households did not have access to treated natural water (NBS 2017; Zhang and Fu 2020).

The cultural crisis is evidenced by the erosion of the integrity of cultural heritage in most villages, the loss of the authenticity of the heritage, and the homogenization of heritage assets in individual regions that otherwise have a rich diversity. These conditions pose many difficulties for the systematic inventorying of cultural heritage in traditional villages. The deeper reason is that the local government and villagers do not treasure the value of their heritage. These cause a range of problems, such as the absence of actors in rural cultural revitalization, the lack of management mechanisms, and the ineffective protection of rural cultural heritage (Sun et al. 2019; Zhang 2018).

From the viewpoint of cities, while cities are continuously absorbing rural labor and products, there are problems such as increased environmental pressure and excessive productivity. According to the statistics, the proportion of China's urban population increases from 49.68% in 2010 to 63.89% in 2020 (National Bureau of Statistics 2021). Rapid urbanization, while improving housing quality, employment, educational and cultural vitality in cities (Radford and James 2013), has also led to a series of growing "metropolitan diseases" such as biodiversity loss, urban poverty, land constraints, inadequate water supply, and rising food prices (Gibb and Huchuli 2002; Gebre and Gebremedhin 2019). In the face of these above-mentioned multiple socio-ecological crises, it is urgent to renew the rural-urban relationship.

27.2.2 The Construction of Renewed Urban-Rural Relationship Within Ecological Civilization

Reflecting on the conventional rural-urban relationship in industrial societies, many scholars realized that building multi-level and multi-dimensional rural-urban linkages will play an important role in rural development, poverty alleviation, and rural transformation (Akkoyunlu 2015; Mayer et al. 2016).

As early as the late nineteenth century, the British urbanist Howard proposed the concept of "garden cities," which combines the strength of urban and rural areas but hedges the drawbacks of both, in terms of social, economic and spatial layout (Howard 2013). Since the twentieth century, an important dimension of urban-rural linkages has been added, namely ecosystem services (Gebre and Gebremedhin 2019), and in the view of some landscape planners, protecting and improving ecosystem functions and ecosystem services has become an important way to mitigate problems associated with urbanization (Radford and James 2013). However, in the conventional urban-rural relationship in the industrialized societies, the important role of urban-rural linkages and rural ecosystems for urban development was often overlooked (Gebre and Gebremedhin 2019). In recent years, some studies have begun to focus on the value of rural ecosystem services and construct a new urban-rural relationship with ecosystem services as a link (Yu 2017, 2020).

Under the new relationship, cities are the beneficiaries of the ecological services provided by the rural areas, and should be aware of the obligation to pay for the ecological services of the rural areas. A key to the long-term maintenance of this relationship is obviously the establishment of a market-based, pluralistic ecological compensation mechanism under this discourse of "willingness to pay." In contrast to European and Northern American countries, China's ecological compensation mechanism not only stresses contractual compensation oriented to the "trade" of spatial development rights and ecosystem services, but also possesses the attributes of public policy, which is used to reconcile the contradictions between survival and development, and between environmental resources and ecological well-being (Wang et al. 2021). Ecological compensation and payment transfer from cities to villages is not only an important way to revitalize the villages, but also plays an influential role in protecting the socio-ecosystem of the villages, securing the high-quality supply of cultural services in the villages, and promoting the sustainable development of the villages (Börner et al. 2017), which can reconcile a well-off living with ecological dwelling.

Within the discourse of urban-rural relations in China, common prosperity between urban and rural areas is an important component of national common prosperity (Huang 2021), and narrowing the urban-rural gap will play a decisive role in achieving common prosperity in the future at the national level (Li et al. 2021). In 2020, the urban-rural income ratio in China is 2.56 (National Bureau of Statistics 2021), while the ratio in developed countries such as the United Kingdom and the United States is roughly between 1 and 1.3 (Li et al. 2021). In addition, the disparity between urban and rural communities is also evidenced in the infrastructure, public service, employment, human capital, education quality, medical care, social security, etc. (Jiang 2022). In Li's view (Li 2021), under a renewed urban-rural relationship, rural ecosystem services will (1) attract urban residents to experience nature and culture in the rural villages to promote economic integration, (2) also promote rural residents' attention to urban growth and urban residents' awareness of rural development to improve social inclusion, and (3) promote urban-rural cultural exchange and reduce cultural isolation.

From the perspective of culture, traditional villages often feature cultural continuity (Hu et al. 2014). In China, as a legacy of thousands of years of agricultural civilization, traditional villages are not only the experiential fruits of the farming ancestors in the course of their struggle against various natural disasters and rivals through adaptation, experimentation, and constant efforts, but also record and manifest the wisdom of our ancestors in choosing residential areas, creating farming land, cultivation, irrigation, etc. (Yu 2006). This is reflected in the landscape patterns, street layouts, traditional architecture, and hydrological heritage of traditional villages, etc. In addition, traditional villages retain unique folk customs, local regulations, craftsmanship, clan culture, and spiritual heritage, making them a gathering place for valuable cultural heritage. In contrast to urban culture, these village cultures constitute the foundation of the grassroots beliefs of the Chinese nation and the roots of traditional culture (Lu et al. 2016), which are important for maintaining the spiritual home of the common Chinese, building the overall cultural identity of China, and achieving sustainable development on the national scale. Therefore, the renewed urban-rural relationship encompasses multiple dimensions such as ecological, social, economic, and cultural dimensions, aiming to foster urban-rural co-prosperity through a more diversified and two-way flow of urban and rural elements.

27.3 Approaches of Traditional Village Revitalization Based on the New Urban-Rural Relationship

Based on the new urban-rural relationship described above, the revitalization of traditional villages should be implemented at least in a "three-dimensional strategy"—with equal emphasis on ecological, social and cultural approaches. The ecological approach builds the biophysical basis for ecosystem services in the rural area. The social approach establishes a high level of urban-rural elements flow, and forms a condition of common management of the rural areas by multiple stakeholders. Cultural approach interprets the cultural lineage behind cultural landscapes and vernacular heritage to boost public awareness, appreciation, comprehension, and preservation. In the following, all three approaches are explained in detail.

27.3.1 Ecological Approach: Nature-Based Solutions

Nature-based Solutions (hereafter NbS) is defined as the following,

... actions to protect, sustainably manage, and restore natural and modified ecosystems that address societal challenges effectively and adaptively, simultaneously benefiting people and nature. ... Nature-based Solutions are underpinned by benefits that flow from healthy ecosystems. They target major challenges like climate change, disaster risk reduction, food and water security, biodiversity loss and human health, and are critical to sustainable economic development. (IUCN 2021).

Under the new urban-rural relationship, the ecological approach and NbS for the revitalization of traditional villages is based on the relationships between ecosystem services, human well-being, and the theory of demand (Millennium Ecosystem Assessment 2005), with ecosystem services linking urban and rural areas. Using nature-based design approaches such as forest restoration, water bodies (surface and underground) cleansing, and farmland remediation, the countryside can provide human-managed ecosystem service flows vital to human survival and livelihood for urban and rural socio-ecological systems. These multiple services included support service (such as nutrient cycling), provision (such as natural resources, agricultural products, and ecological space), regulating service (such as climate regulation, regional flood remediation and water cleansing, and soil conservation), and cultural service (such as recreation, education, and aesthetic and spiritual activities) (Millennium Ecosystem Assessment 2005; Gutman 2007; Peng and Dai 2015; Zhang and Fu 2020).

The renewed urban-rural relationship is no longer a one-directional flow, where the traditional villages in rural areas give unilaterally, but rather a bidirectional flow where urban visitors are required to pay for the ecosystem services provided by the rural areas through multiple means while enjoying the services. The population, wealth, knowledge, and vitality that urbanites pour into the villages lead to the revitalization of villages and common prosperity (Fig. 27.1).

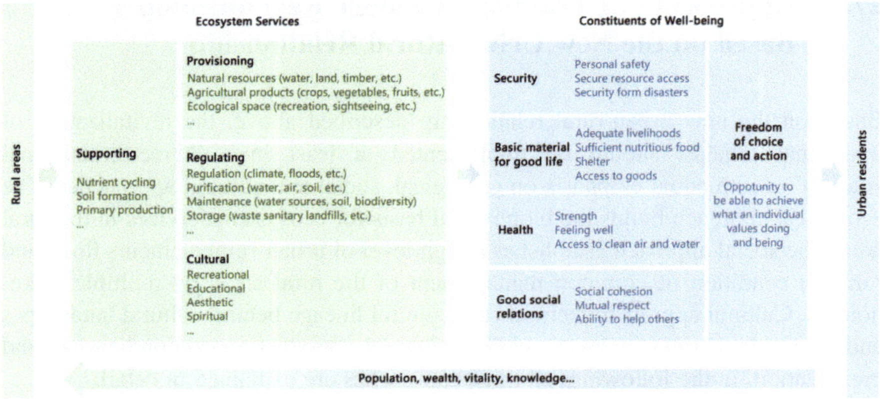

Fig. 27.1 Ecosystem service links for a new urban-rural relationship. This flow chart illustrates the relationship between Ecosystem Services and the Constituents of Well-being. On the left are the four types of Ecosystem Services: Provisioning (such as natural resources), Supporting (such as nutrient cycling), Regulating (such as climate regulation), and Cultural (such as recreational values). On the right are the Constituents of Well-being: Security (such as personal safety), Basic material for a good life (such as adequate livelihoods), Health (such as strength), Good social relations (such as social cohesion), and Freedom of choice and action (opportunity to do what people value). Arrows show a two-way flow: ecosystem services move from rural to urban areas. In turn, feedback—such as changes in population, wealth, vitality, and knowledge influenced by these services—returns to rural areas. This demonstrates the mutual interdependence of ecosystems and human well-being.

27.3.2 Social Approach—Collaboration of Urban and Rural Multi-stakeholders

In terms of the collaboration model of urban-rural multi-stakeholders, the current revitalization models of traditional villages mainly include four types: government-led, collective-led, enterprise-led, and shared governance (He 2009; Yin and Li 2018).

(i) The *government-led model* is conducive to the government's role in regulating and guiding policies, but often faces problems such as insufficient utilization of social capital, weak participation of villagers, and inefficiency of financial funds usage (Wu and Liu 2017).

(ii) The *collective-led model* is advantageous in respecting villagers' demands and safeguarding their interests, but it usually involves difficulties such as sustaining operating funds, homogeneous industrial structure, and strong reliance on the village collective leader (Zou and Li 2007; Wang 2018).

(iii) The *enterprise-led model* takes commercial value as the fundamental pursuit, which is criticized as often ignoring the interests of villagers and lacks credibility.

(iv) The *shared governance model* aims to overcome the inadequacy of sole-stakeholder revitalization of traditional villages. It is based on the co-management and sharing model of urban-rural multi-stakeholder collaboration has become the direction of rural revitalization, and now in China there are two main mechanism, namely, joint cooperation (Li 2013) and sharing management.

Joint cooperation is exemplified by Tangmo village in She County, Anhui Province. *Sharing management* is illustrated by Xixinan Village described in this chapter. Under the Xixinan model, the multi-stakeholders both discipline each other and collaborate, which is favorable to unifying the power and responsibility of village conservation and development, and encouraging common prosperity. Therefore, to ensure the sustainable supply of rural ecological and cultural services, a collaborative network of urban and rural multi-stakeholders needs to be constructed as one of the ways for urban residents to pay for these services in the long term. Multi-stakeholders will jointly weigh short-term and long-term benefits, natural and human well-being, to achieve sustainable urban and rural development.

27.3.3 Cultural Approach: Landscape Interpretation

Under the renewed urban-rural relationship, multiple urban and rural stakeholders will be deeply involved in rural development. In the process of China's conventional rural development in recent decades, tourists, government, landscape architects, developers, and even villagers themselves often have inadequate knowledge on the evolution, connotation, and value of rural cultural heritage, thus undermining the cultural bonds that link urban and rural areas. Therefore, it is imperative to systematically excavate and sort out the cultural connotation of the traditional village, including not only the tangible and intangible cultural heritage that survives in history, but also the "underfoot culture" or "value the ordinary," i.e., the culture of ordinary people (Yu 2001). The accumulated farming wisdom and experience of farmers, the traditional social customs, and the conservation and management of mountains, forests, water resources, farmland, and buildings by village committees and villagers all belong to this category. These types of traditional village cultures will become valuable heritage for future generations, so it is an obligation to cherish them, organize, and record them in a timely manner.

Very frequently, the cultural connotation of the rural villages is not visually revealed, but concealed under the figurative landscape, which needs to be "disclosed" and conveyed to the public through researchers. This approach of revealing and communicating is called "landscape interpretation." As an educational activity to reveal the meaning of landscape, it originates from heritage interpretation, aiming at the comprehensive interpretation of ecological, economic, cultural, and social information behind the landscape through first-hand information and a multidisciplinary approach (Tilden 1957; Kreisel and Reeh 2019). On the one hand, it can provide a decision-making basis for those responsible for developing and managing rural spaces and resources. On the other, it enables visitors to understand factual information, context, and the connection among various kinds of information in the process of visiting or touring, so as to link their own experience with the information obtained, achieve the purpose of understanding through interpretation, appreciation through understanding, and conserving through appreciation (Kreisel and Reeh 2019).

27.4 Concrete Initiatives for Traditional Village Revitalization

Based on the above-mentioned approaches, in this chapter we propose the following concrete initiatives for the revitalization of traditional villages based on the renewed urban-rural relationship.

27.4.1 Intensive Interpretation of the Heritage Assets: Uncovering the Values of Culture and Ecosystem Services

Drawing on the method of landscape interpretation, we examined comprehensively and thoroughly the "cultural-ecological" system of traditional villages, trace their evolution, and unveil their cultural and ecological values. This step requires the participation of scholars, who can obtain first-hand information through historical document analysis, satellite map analysis, on-site research, field interviews, and other sources of information. The cultural study needs to focus on the historical facts, tangible heritage, intangible heritage, social patterns, and the status of multiple stakeholders. By focusing on the local ecological structure and traditional ecological wisdom, ecological research links the two together to avoid unintentional destruction in the process of conservation and utilization.

27.4.2 Building a Medium for Value Transformation: Reap Economic Benefits of Cultural and Ecosystem Services

Based on the intensive interpretation of the heritage assets, specific interventions are required to build a medium for value transformation, so that the ecological and cultural services of villages can be converted into economic benefits that can be valued and paid voluntarily by urbanites (Fig. 27.2). These intervention measures should not only comprehensively consider the problems and demands of both

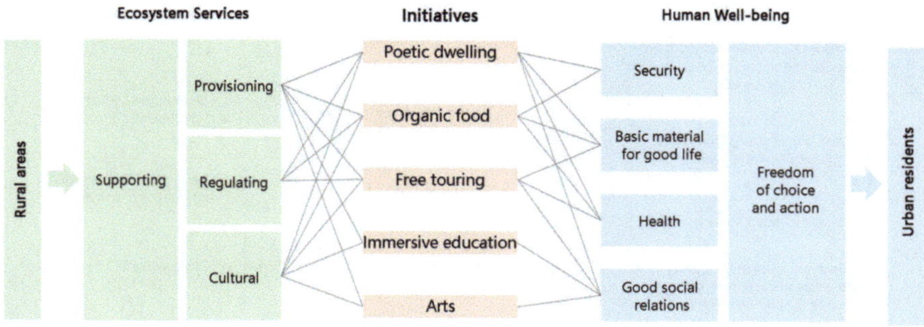

Fig. 27.2 *Five-in-one* initiatives transform ecosystem services into human well-being. The flow chart illustrates the relationship between ecosystem services of rural areas, initiatives, and human well-being of urban residents. The chart shows how supporting, regulating, and cultural ecosystem services from rural areas connect to initiatives like poetic dwelling, organic food, free touring, immersive education, and arts. These initiatives are linked to aspects of human well-being for urban residents, including security, basic material for a good life, health, good social relations, and freedom of choice and action. Arrows and lines indicate the flow and connections between these elements.

villages and cities to renew urban-rural resource linkages, but also ensure that these measures are non-depleting or irreversible. In this way, the integrity of rural ecosystems and the long-term stability of ecosystem services can be secured, thereby fulfilling the principle that "lucid waters and lush mountains are invaluable assets" (metaphorically as "clean water and green mountains are mountains of silver and gold", in Chinese "绿水青山就是金山银山").

To combine the above requirements, we propose in this chapter the *five-in-one* initiatives, namely, poetic dwelling, organic food, free touring, immersive education, and arts (Figs. 27.2 and 27.3). The *five-in-one* initiatives are currently in a dynamic stage of development with continuous adjustment and optimization in practice.

(i) **Poetic dwelling**: Village dwelling used to be considered as an underdeveloped stage lagging behind urban inhabitation, but in fact village dwelling can produce a different but equally high-quality living experience from urban living experience. Using the language of economics, village settlement is a competitive residential product with differentiation. Fully excavating and magnifying the vernacular characteristics of the villages, transforming the deteriorated villages houses into high-quality Bed and Breakfast (B&B) hotels serves multiple functions such as preserving and utilizing traditional architecture, boosting villagers' employment, and raising tourists' sense of belonging and identity.

(ii) **Organic food**: Under the current situation that Chinese population's food needs are satisfied, the villages should focus on producing agricultural and sideline products with high, medium, and low gradients of needs, especially to provide products that meet high quality urban demands. Drawing on traditional agricultural wisdom and adopting organic production methods to produce

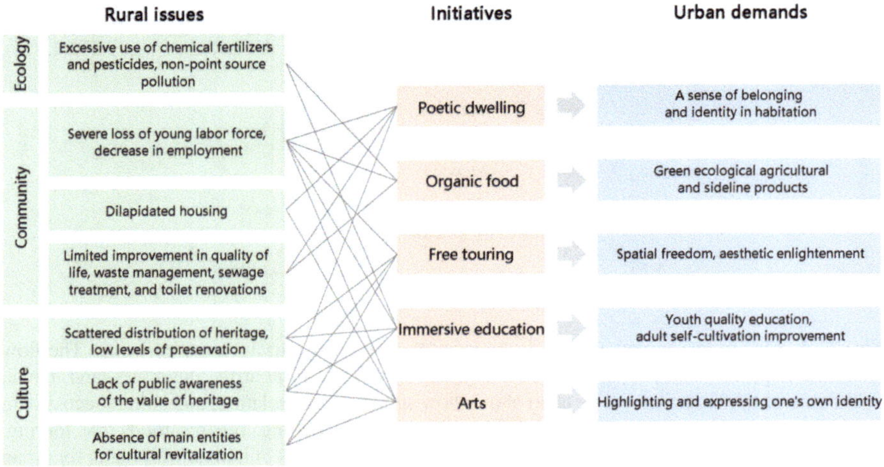

Fig. 27.3 *Five-in-one* initiatives link rural problems with urban needs. Flow chart illustrates connections between rural issues, initiatives, and urban demands. On the left, rural issues are categorized under ecology, community, and culture, including excessive use of fertilizers, loss of labor force, lack of cultural revitalization, and so on. The middle column lists initiatives like poetic dwelling, organic food, immersive education, and so on. The right column shows urban demands such as a sense of belonging, green products, youth education, and so on. Arrows and lines connect related issues, initiatives, and demands, highlighting interdependencies.

agricultural by-products can alleviate non-point source pollution caused by chemical fertilizer and pesticide misuse, and thus is worthy of being advocated and promoted by the whole society. The ecological value of these products should be converted into prices to establish farmers' concept of ecological cultivation through simple economic rationality.

(iii) **Free touring**: To raise the additional value of the village products, it is necessary to transform the tourism industry chain. The depth of it is determined by the diversity of tourism models and the breadth of the hinterland space. Free touring becomes an effective means of expanding the spatial and temporal breadth of rural tourism through regional collaboration. Combining the heritage corridor with landscape interpretation, free touring offers opportunities for people to realize ecological and cultural services of aesthetic enlightenment.

(iv) **Immersive education**: Immersive learning can maximize the value of various rural resources and provide a variety of experiential products through a combination of various resource options, increasing the additional value of rural tourism. Immersive learning in traditional villages on cultural heritage and ecological wisdom are fundamental both for the quality education of adolescents and for adults to gain intellectual pursuit for their own cultivation and growth.

(v) **Arts**: Cultural and creative arts are not only a vehicle to pass on a culture, but also a re-creation of rural resources. Featuring both greater degrees of flexibility and considerable intellectual and knowledge requirements, recreation represents a viable means of attracting city residents to work in the rural villages. The village should offer jobs that involve high knowledge levels to attract talented urban residents, who can further energize and revitalize the community with innovative development.

27.4.3 Building a Framework of "Shared Governance": Fostering the Equilibrium and Stability of the Renewed Urban-Rural Relationship

In the model of traditional village revitalization based on the new urban-rural relationship, it requires that multiple stakeholders undertake different responsibilities. Although villagers should be treated as the main actors in villages, tourism and arrival of urban residents can bring new possibilities for villages, such as reclaiming abandoned farmland, managing intangible culture, and growing featured Bed and Breakfasts (hereafter B&Bs), which create employment and generate income for villagers. The prospect of profitable markets for ecosystem services will also attract relevant enterprises to join, creating markets, and employing the knowledge of market development to increase social recognition and willingness for urban residents to pay for sustainable ecosystem services (Gebre and Gebremedhin 2019). In this process, the relevant enterprises need to put social and environmental responsibility at the core, and the government should intervene to prevent excessive domination by the enterprises at the expense of the villagers. Professionals such as landscape and heritage experts, scholars, and designers should work interdisciplinarily for a more scientific and sustainable revitalization.

A scientific and sound multi-stakeholder collaborative network is indispensable to promoting the flow of population, wealth, and knowledge between urban and rural areas, and enabling inclusive, transparent, empowering, and effective governance processes. Building a collaborative network can safeguard the realization of rural revitalization, common prosperity, and the long-term maintenance of the new urban-rural relationship.

27.5 The Revitalization Practice of the Village of Xixinan in Huizhou

Since 2014, the authors have been practicing the above theory of rural revitalization in Xixinan Village, Huizhou on the platform of *Wangshan Life*. Based on the concept of "conservation of roots and heritage, catalyzing through intervention, co-existence of the old and the new, and seeking common prosperity with people," *Wangshan Life* has attained preliminary success.

27.5.1 Overview of Xixinan Village

Xixinan Village is located in Xixinan Town, Huizhou District, Huangshan County, Anhui Province (Fig. 27.4), with a history of more than 1200 years and a village area of 7.33 square kilometers. Situated in the hilly plain area, the Fengle River, the largest tributary of the upper reaches of Xin'an River, passes through the village, forming a unique landscape of woodlands dominated by Chinese wingnut (*Pterocarya stenoptera C.DC.*, 1862) on the river bank. Known as "the wealthiest town in Sheyi," Xixinan Town has nurtured a profound clan culture and *Hui* (徽) merchant culture. Xixinan Village has preserved many ancient residential buildings,

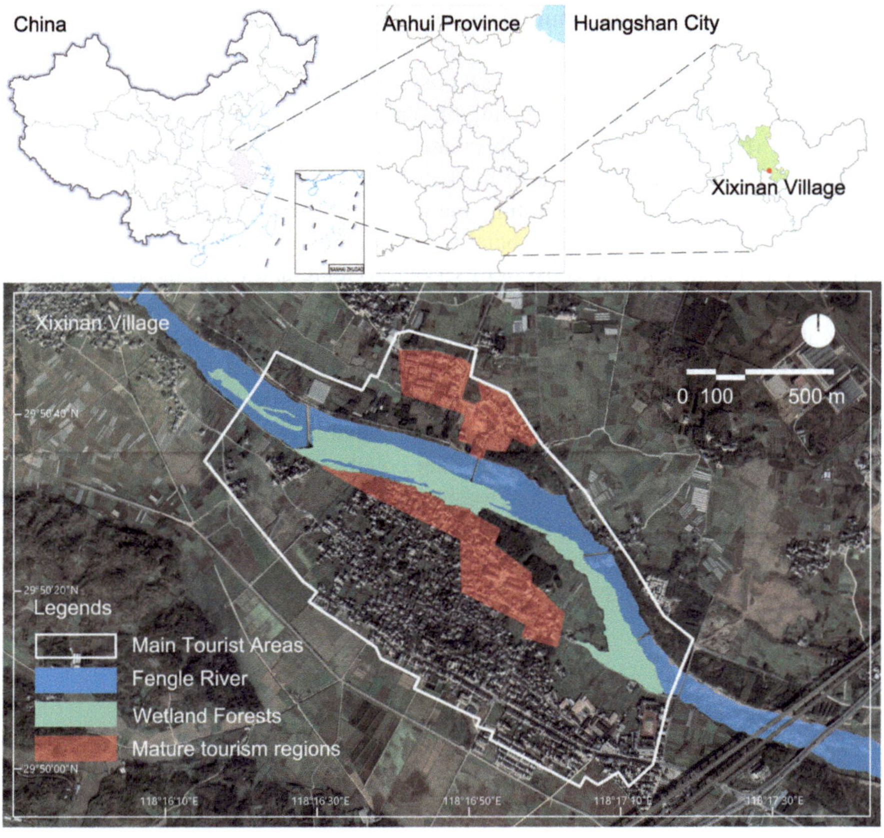

Fig. 27.4 Location of Xixinan Village planning area. The ap shows the location of Xixinan Village in Huangshan City, Anhui Province, China. The top section highlights the village's position within China, Anhui Province and Huangshan City. The bottom section is a detailed map of Xixinan Village, featuring the Fengle River, wetland forests, main tourist areas, and mature tourism regions. A legend indicates these features with color coding: blue for the river, green for forests, and red for tourism regions. Coordinates and a scale bar are included for reference.(Map was prepared by Jianing Li)

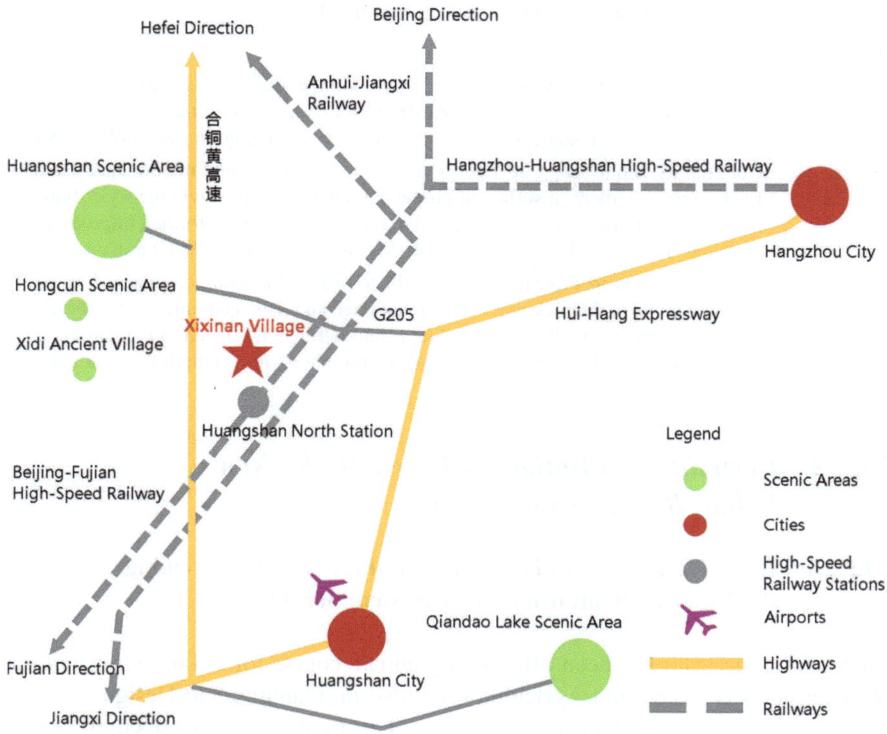

Fig. 27.5 Schematic diagram of traffic location in Xixinan Village. The diagram illustrates transportation routes and scenic areas around Xixinan Village. Key locations include Huangshan City, Hangzhou City, and Huangshan North Station. Highways, such as Hui-Hang Expressway and railways (including the Hangzhou-Huangshan High-Speed Railway), are marked. Scenic areas like Huangshan, Hongcun and Xidi Ancient Village are highlighted. (Map was prepared by Jianing Li)

ancient garden relics, and civil water heritage, and the village was listed as the third batch of Chinese traditional villages in 2014.

Today, Xixinan Village is 4 kilometers away from Huangshan North High-speed Railway Station. It only takes 1.5 hours for Hangzhou tourists and 2.5 hours for Shanghai tourists to arrive at the village. Around the village are other tourist destinations such as Huangshan Scenic Area, Chengkan Ancient Village, Hongcun Scenic Area, and Xidi Ancient Village within an hour drive (Fig. 27.5). Thus, Xixinan Village is also favored by experts, scholars, and urban residents because of its superior transportation location, high-quality natural and cultural heritage.

However, in 2014, because the high-speed railway had not yet launched, Xixinan Village was isolated and faced serious problems of economic depression, community decay, population outflow, and inappropriate protection and utilization of cultural heritage (Table 27.1). In order to avoid mindless destruction and overexploitation, the local government introduced the *Wangshan Life* team to manage the conservation and utilization of Xixinan's core resources.

Table 27.1 Analysis of the problems in Xixinan Village in 2014

Categories	Problems
Economic	60% of the labor force was employed in the village, mainly in agricultural production with low income; Slow development of tourism industry, the quantity and quality of tourism resources do not have strengths in the region, with only 2 B&Bs.
Social	Community "hollowing out" due to serious loss of young and middle-aged labor; Lack of management; insufficient public participation; hybrid village style; traditional residential buildings were repaired, renovated, and transformed improperly in the New *Hui* camp (新徽派) versus Modifying *Hui* camp (改徽派) movement.
Cultural	Villagers lack proper awareness and skills to protect the local landscape. For example, most of the rock ports were abandoned and damaged; uneven conservation of historical relics, with the cultural units and heritage not listed in the national protection list are basically unprotected; intangible heritage inheritors were not appreciated.

27.5.2 Revitalization Initiatives Based on the New Urban-Rural Relationship

27.5.2.1 Intensive Interpretation of the Heritage Assets: Revealing the Values of Cultural and Ecosystem Services

Since 2016, the team has been offering a graduate course for Peking University at Xixinan, namely, Rural and Agricultural Landscape Planning and Design. Besides that, several graduate students have proposed their master theses or doctoral dissertation to systematically study the natural and cultural landscape, history and the status quo, and tangible/intangible heritage of Xixinan and Huizhou region. Three doctoral dissertations and two master theses have been completed and successfully defended by these graduate students (Shi 2020; Yun 2021; Zheng 2021; Zhang 2021; Hu 2022).

Tangible Heritage

Combining the contents of the Republic of China's (1911–1949) *Fengnan Local Chronicles* (《丰南志》) and local interviews, the current status and evolution process of the Xixinan water conservancy system, gardens, street systems, residential buildings, public buildings, and agricultural systems were sorted out (Table 27.2).

Intangible Heritage

Xixinan Village, as historically the wealthiest village in Sheyi, possesses intangible heritage including not only *Hui* ink, *Hui* opera, and *Hui* carving, but also unique collection and inscription culture. During the Ming (1368–1644) and Qing (1636–1911) dynasties, Xixinan was the gathering place of China's collecting

Table 27.2 Historical and evolutionary features of tangible heritage in Xixinan Village

Tangible heritage	Historical forms	Status quo	Evolutionary characteristics
Water conservancy system	① Three weirs (Lei, Long, Tiao) pass through villages with integrate river-weir-channel-pond water conservancy system; ② river in/outlet landscape; ③ Chinese wingnut (*Pterocarya stenoptera C.DC.*) woods was a productive landscape.	① Old weirs were preserved intact, fewer canals and ponds remain; ② river in/outlet landscape disappeared; ③ Chinese wingnut (*Pterocarya stenoptera C.DC.*) woods became a recreation place.	The overall patterns still exist with reduced numbers of canals, ponds, and river in/outlet landscapes; productive landscapes are transformed into recreational places.
Gardens	① 41 classical gardens; ②eight scenes of Xixinan	The Remaining few classical gardens are the Orchard, Wild Trail Garden, and Snow Garden.	Large numbers of classical gardens went extinct.
Streets and alleys	① Streets and alleys were laid out like a fishbone with the Middle Street as the main street; ② Traditional buildings were mostly built along the street; ③ Street's height to width ratio ranged from 2.0 to 4.0.	① Streets and alleys were laid out like a fishbone with the Fengnan Street as the main street; ② buildings along streets are hybrid in styles and have lost their local characteristics and become homogeneous; ③ street's height to width ratio ranged from 1.0 to 2.0.	Street and alley structures remain, with varied architectural styles and overall widening of streets.
Residential buildings	① Residential buildings were centered on the ancestral hall and distributed in groups along the street and water conservancy system; ② built with high walls, deep countyards, patios, and Ma Tau Wall (马头墙, wall roofs decorated with carved horse head) ; ③ built with bricks, wood, and decorated with *Hui* carving of three materials (brick, wood, and stone).	① Old and new residential buildings were staggered, and new buildings are spreading along public roads; ② mostly adopt open style, and some with Ma Tau Wall; ③ mostly built with reinforced concrete structures and modern decoration.	The overall layout is more scattered, and the *Hui* characteristics of new residential buildings have greatly declined.
Public buildings	① During 1911–1949, there were around 20 ancestral halls, 10 folk temples, 16 buddhist and Taoist temples, and	① The only ancient public building that survived is parts of the Simu ancestral hall; ② Most of public buildings were in	The clan landscapes have drastically declined, and the new public buildings are lacking in *Hui* features.

(continued)

Table 27.2 (continued)

Tangible heritage	Historical forms	Status quo	Evolutionary characteristics
	Confucian schools, pavilions, etc. ; ② displayed typical *Hui* style of architecture.	New *Hui* Style or modern style, with reinforced concrete structures and modern decoration.	
Agricultural system	Farming was in accordance with celestial phenomena and solar terms, employing measures such as crop rotation and fallow to maintain soil fertility.	The extensive use of pesticides and chemical fertilizers leads to the nutrient disproportion of the soil and the decrease of fertility and organic matter.	Agricultural cultivation is increasingly dependent on chemicals, and traditional agricultural wisdom is gradually getting lost.

community and literati. Wu Tingzang, owner of Yu Qing Zhai (余清斋), collected a large number of masterpieces of Chinese calligraphy and paintings, such as the late Ming and early Qing dynasty painter Jianjiang's *Place for Immortality and Alchemy in the Clouds* (《云根丹室图轴》), *Fengxi Landscape Paintings* (《丰溪山水册》), and Shi Tao's(1642–1701) *Eight Scenes of Xixinan* (《溪南八景》). The *Fengnan Local Chronicles* contains records of traditional geomancy and divination, water conservancy construction, village transformation, and agricultural practices, all of which are the crystallization of the wisdom of Xixinan villagers' lives and important intangible heritage as well.

Social Patterns

The social patterns of Xixinan at each phase of development was the result of the games of multi-stakeholders, which significantly affected the landscape evolution and economic development of the village. The *Wangshan* team categorized the stakeholders at Xixinan into four groups: governmental, local, external, and social stakeholders. The team also systematically sorted out the status of each stakeholder at various historical phases (Table 27.3).

27.5.2.2 Designed and Implemented the *Wangshan Life* Model of *Five-in-One* Initiatives of Ecological and Cultural Services

To solve the economic, social and cultural problems of Xixinan Village in a comprehensive manner, the *Wangshan life* team has implemented *five-in-one* initiatives, including poetic inhabitation, eco-friendly agriculture, whole-area tourism, research-based learning for intellectual growth, and cultural and creative arts mentioned above.

Table 27.3 The Status of the multi-stakeholders at various historical phases in Xixinan

Category of stakeholders	1121–1912	1912–1987	1987–now
Governmental	Imperial power does not manage counties, the main role of supervision and coordination was played by local authorities such as the Huizhou Prefecture and the county government of She County.	During the Huizhou Special Zone period (1961.4–1971.3), villages were directly governed by the central government, which showed a single hierarchy of "county—township—village—production brigade—production team—villagers."	Currently it is governed by the three-level government structure of "Huangshan County—Huizhou District—Xixinan Township Government."
Local	The gentry self-governance system was dominated by the powers of multiple clans.	Clan power declined, and the main local stakeholders gradually evolved from "big clan—small clan—core family" (Tang 1991) to the production-centered hierarchy mentioned above.	It is mainly composed of core families, with village collectives and village committees as the organizational form.
External	It was mainly composed of extra-regional gentry, peasants and tenant farmers (Liu 2015). During the mid- and late-Qing dynasties, a large number of squatters and displaced people flowed into Huizhou and created rivalries.	From 1954 to 1958, external immigrants moved in under the government organization, such as those for the construction of Xin'an River Reservoir; from 1968 to 1972, external population were generated because of the "Down to the Countryside Movement" (知青下乡).	Cultural tourism developers, investors, new immigrants from the "100 villages and 1000 buildings" project, tourists, etc.
Social	During the Ming and Qing dynasties, Huizhou gentry formed the social stakeholder Wenhui (文会, an organization of literati) (Bian 1993; Yang 2011).	Few.	Experts and scholars, university students (e.g. Peking University team), NGOs, etc.

Poetic Dwelling

The poetic inhabitation firstly lies in preserving the traditional historical patterns of the streets, laying out streets as longitude and along the river dikes, alleys as latitudes, and houses built on both sides of the streets and alleys. The *Wangshan*

Fig. 27.6 Before(2015)-and-after(2020) comparison of Diaoxueyuan(Snow) Hotel. The left photo shows a traditional, abandoned building with whitewashed walls and a tiled roof, surrounded by trees and paved stone paths. The right photo shows the same building, but a modern, renovated building with large windows, adjacent to a lush garden with trees and a path. The garden includes outdoor seating and lush foliage.(Photographs by *Wangshan Life*)

Fig. 27.7 Before(2015)-and-after(2020) comparison of Hetianli (Lotus Pond) Hotel. The left image shows an old, narrow courtyard with weathered walls and a view of a two-story building with balconies and windows. The right image depicts a modern, glass-fronted building surrounded by lush green trees, with visible interior lighting and furniture, creating a vibrant and inviting atmosphere. (Photographs by *Wangshan Life*)

Life team also assisted the village to apply for the government's creative town, and seek national funds to preserve the historical streets and alleys, repair damaged areas, restore and dredge the water system along the streets, and improve the municipal infrastructure. Meanwhile, the team protected the structure and spatial characteristics of the heritage buildings, followed the concept of reversible restoration and "minimal intervention." It also incorporated functions such as residence, conference, research, and education to bring vitality to the traditional residential buildings (Fig. 27.6, 27.7 and 27.8).

Organic Food

Located in the largest basin in She County, Xixinan is an unparalleled major food-producing area in the entire Huizhou region, with well-developed planting and farming industries. On the one hand, *Wangshan Life* selected local high-quality

Fig. 27.8 Before(2015)-and-after(2020) images of Wangshan Commune. Two contrasting photos of a multi-story building. The left image shows an old building with a red sign featuring Chinese characters, surrounded by greenery and a concrete courtyard. The right image displays a renovated version of the same building with a modern facade, featuring wooden paneling and a clear blue sky in the background. (Photographs by *Wangshan Life*)

Fig. 27.9 Various series of eco-friendly agricultural products. A variety of packaged food products are displayed on a wooden table. The items include jars of sauces, bottles of oil, and several bags of dried goods, all labeled with Chinese text. Bowls containing different ingredients are placed in front of the packages. (Photograph by *Wangshan Life*)

farmers, optimized the supervision system, and fully leveraged ecological wisdom such as fallow and crop rotation to maintain a healthy agricultural ecosystem. To produce healthy, organic, and high value-added agricultural products, no chemical fertilizers and pesticides are used during crop cultivation. On the other hand, for Xixinan's long-established beekeeping industry, the team has ameliorated its honey products in accordance with local conditions (Fig. 27.9).

Fig. 27.10 *Wangshan Mao Feng Tea* Event: Mr. Xie Yiping performed tea stirring techniques on site. Mr. Xie Yiping is carefully handling a pile of fresh green tea leaves over a large container. The scene conveys a sense of craftsmanship and attention to detail in tea preparation. (Photograph by *Wangshan Life*)

Beside the production phase, *Wangshan Life* also actively expanded the marketing channels of agricultural products, by-products, and traditional crafts. For example, during the Dragon Boat Festival, the team held a stall festival to promote marketing local agricultural products and folk crafts through both online and offline forms of shopping. Another example is that the team invited Mr. Xie Yiping, an inheritor of intangible heritage, to perform the traditional tea stirring techniques through the launch of *Wangshan Mao Feng* (Fig. 27.10). While promoting the development of the Huizhou region through various channels, these activities and events promoted the conservation of the local intangible heritage.

Free Touring

By combining **free touring** with landscape interpretation, Xixinan planning has established the Fengle River water heritage corridor to protect Chinese wingnut wetland, village water landscape, and water conservancy heritage. Using the water conservancy heritage as the vein, the landscape, historical buildings, and intangible cultural exhibition sites are linked, and immersed projects including local cuisine, arts and design exhibitions, creative arts and industries are incorporated (Fig. 27.11).

Fig. 27.11 Comprehensive node map of the whole-area tourism in Xixinan Village. Map illustrates a geographical area with labeled landmarks and features. (Map was prepared by Jianing Li)

In terms of interpretation mechanism, the *Wangshan Life* team combined historical analysis, geographical analysis, and villager interviews to present the current situation and evolution of the natural and cultural landscape of Xixinan through "storytelling" and "landscape interpretation," thus promoting tourists' appreciation of the cultural heritage and awareness of conservation. The team also invited villagers to experience **free touring** to raise the local community's sense of place, of local identity, and of dignity.

Immersive Education

Relying on the alumni resources of Harvard University and Peking University, as well as the rich project experience of the Turenscape, the team has developed a variety of courses based on the rich natural and cultural heritage of Xixinan. Courses cover disciplines such as landscape design, human geography, environmental engineering, ecology, history, sociology, etc. These courses are offered to domestic and international students, who gain training with high quality, advanced level, based on local and global -perspectives.

Based on the different groups of audiences, the courses and training are launched under three main sections: Turenscape Academy for professionals, *Wangshan* Junior for youth education, and Insight Academy for corporate training and study tours. These programs bring urban participants to the countryside, resulting in the bidirectional flow of urban and rural populations and the construction of new cultural distribution centers in the countryside (Fig. 27.12).

Fig. 27.12 Schedule of main “research-based learning in practice” activities of *Wangshan Life* at Xixinan Village. Timeline diagram illustrates events from 2016 to 2020 related to design and cultural workshops.

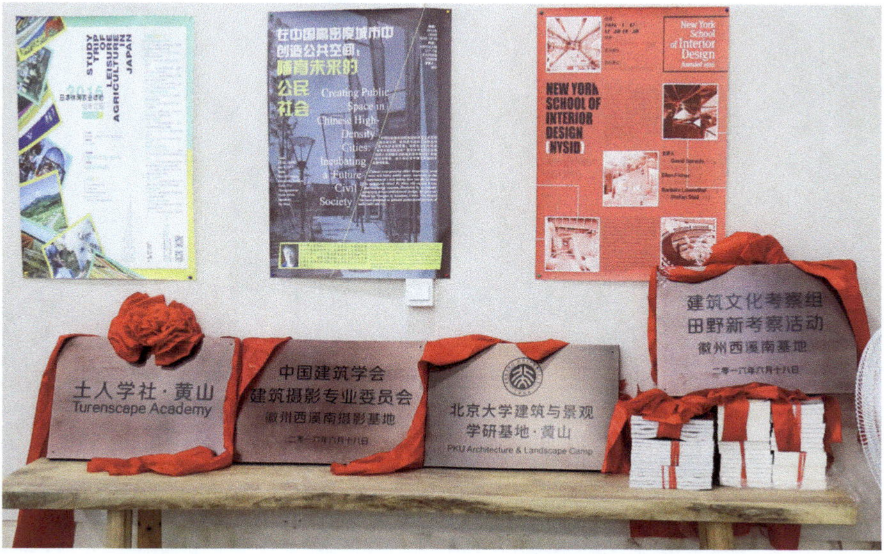

Fig. 27.13 "Research-based learning & practice" Center in Xixinan Village. The photo displays four plaques. The plaques feature text in Chinese and/or English, including Turenscape Academy, Architectural Culture Investigation Team: New Field Investigation Activities, Huizhou Xixinan Base, PKU Architecture and Landscape Camp, and Architectural Culture Investigation Team: New Field Investigation Activities, Huizhou Xixinan Base. Above, three posters are mounted on the wall. The posters contain text and images related to architecture and design of common space. (Photograph by *Wangshan Life*)

Over years of practice, Xixinan Learning Base now contains National Base for Mayors' Training, College of Architecture and Landscape Architecture Teaching Practice Base, Peking University, University of South California Internship Base, Yanjing Academy Internship Base, and Huizhou Xixinan Photography Base of the Architectural Photography Professional Committee, and Chinese Society of Architecture (Fig. 27.13).

Arts

Wangshan Life has launched many artistic and creative activities to carry on the cultural heritage and revitalize the village. The team cooperated with the "Time and Space Friendship Club" to gather more than 50 collectors in Xixinan to share their stories and experiences in collecting. In collaboration with Lawi Art Studio, the team hosted the Xixinan Super Cello Music Festival (Fig. 27.14) to revitalize the thousand-year-old village with music.

Meanwhile, it is worth mentioning that the team has also creatively renovated the traditional Master-Apprentice system, forming a bilateral mentor-apprentice mechanism of interactions between urban artists and rural craftsmen. On the one hand, a

Fig. 27.14 A thrilling performance of European cello artists in Xixinan Village. A group of musicians playing cellos outdoors under in Xixinan Village. A conductor stands in the center, energetically leading the orchestra. The scene is vibrant with natural light filtering through the leaves. (Photograph by *Wangshan Life*)

pottery studio was installed, and well-known pottery artists were invited to instruct locals in pottery skills (Fig. 27.15), and trained the village's pottery artists. On the other hand, local craftsmen were invited to teach urban artists the skills of Huizhou bamboo weaving to explore new application scenarios of traditional knowledge.

27.5.2.3 Co-building, Communion, and Co-governance of Villages

In Xixinan Village, the *Wangshan Life* team has steered a multi-stakeholder collaboration mechanism consisting of government, village collectives, villagers, markets, media groups, artist groups, etc. (Fig. 27.16).

At the institutional level, Huangshan County's innovative institutional reformation—the "Hundred Villages and Thousand Buildings" project has contributed to the establishment of a harmonious relationship among the *Wangshan Life* team, the local government, village collectives, and villagers. With the support and assistance of the government, the village collective, and villagers, the *Wangshan Life* team was able to apply for the purchase or lease of ancient residential buildings. Beginning from the overall planning of Xixinan Village and the restoration of ancient buildings, the experiment of *Wangshan Life* was carried out. During the operation, the local government was responsible for organizing and coordinating

Fig. 27.15 Chinese and foreign visitors experienced ceramic art in Liu Ying pottery studio. A group of people sitting around a table in a pottery studio, engaging in clay modeling. Various pottery tools and clay pieces are scattered on the table. The participants appear focused and joyful, enjoying the creative activity together. (Photograph by *Wangshan Life*)

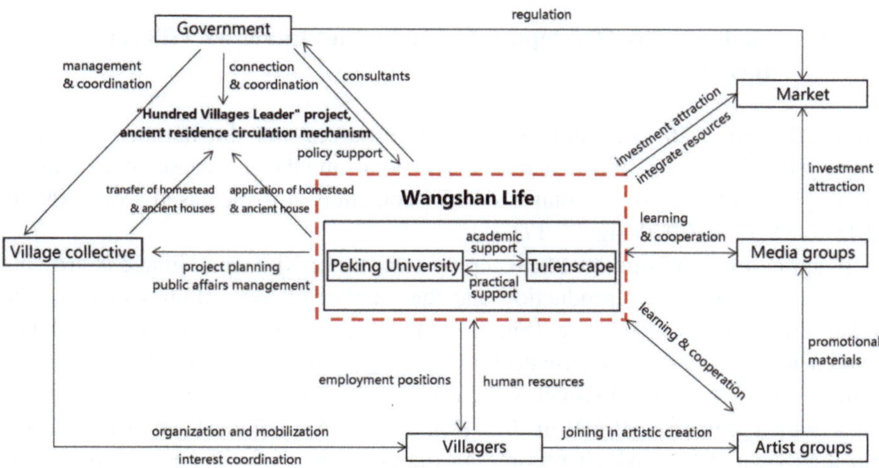

Fig. 27.16 The network of multi-stakeholders to co-build, share and co-govern Xixinan Village. Flow chart illustrating the Wangshan Life model, showing connections between various stakeholders. The government provides management, coordination, and policy support to the Hundred Villages Leader project. The village collective handles project planning and public affairs management, with homestead transfers. Peking University and Turenscape offer academic and practical support. Villagers are involved in organization, mobilization, and artistic creation. Media and artist groups engage in learning, cooperation, and promotional activities. The market focuses on investment attraction and resource integration.

the revitalization, and the village collective was entrusted with the import of *in situ* human resources. With its academic expertise and professionalism, Wangshan Life was committed to introducing external resources and overseeing the preservation and utilization of the local heritage. The *Wangshan life* team and the local government actively collaborated with international academic, artist, and media groups, not only expanding the visibility of Xixinan Village, but also revitalizing the village through arts, designing, and cultural interventions.

In terms of relationships with villagers, the various industries of *Wangshan Life* provide employment for villagers, while selling green agricultural products on their behalf. Some villagers have also converted their self-constructed houses into agritourism to increase their income. Among the benign relationship of co-building, communion, and co-governance of the rural villages, multi-stakeholders discuss and review the village planning, supervise the conservation of cultural heritage, improve health and social order of rural villages, and become the cornerstone of Xixinan revitalization.

27.5.3 Revitalization Results

The *Wangshan Life* team has initialized a sustainable development model in Xixinan Village that is completely independent of state input.

27.5.3.1 High-Quality Development of Industries Increases Villagers' Income

From 2015 to 2020, the number of labor resources and the number of people employed in Xixianan Village increased in general, and the proportion of the number of people employed to the total resident population also increased from 60% in 2015 to 81.6% in 2020(Fig. 27.17).

In agriculture, the signature beekeeping industry in Xixianan Village is growing rapidly. In 2020, honey production and the number of bee colonies increased to about three times the 2015 level (Fig. 27.18). Agricultural production has also taken a greener turn, with pesticide use decreasing from 24 tons in 2015 to 18 tons in 2019.

In terms of tourism, Xixinan Village had only 2 rudimentary guest houses in 2014, which grew to 51 B&Bs in 2020 (Fig. 27.18). Among them, 26 local villagers run their own B&Bs, which become an important part of villagers' income. The number of commercial tenants in Xixianan Village also has grown from 62 in 2014 to 227 in 2020 (Fig. 27.19). According to incomplete statistics, the average number of visitors to Xixinan Village exceeds 250,000 from 2019 to 2021.

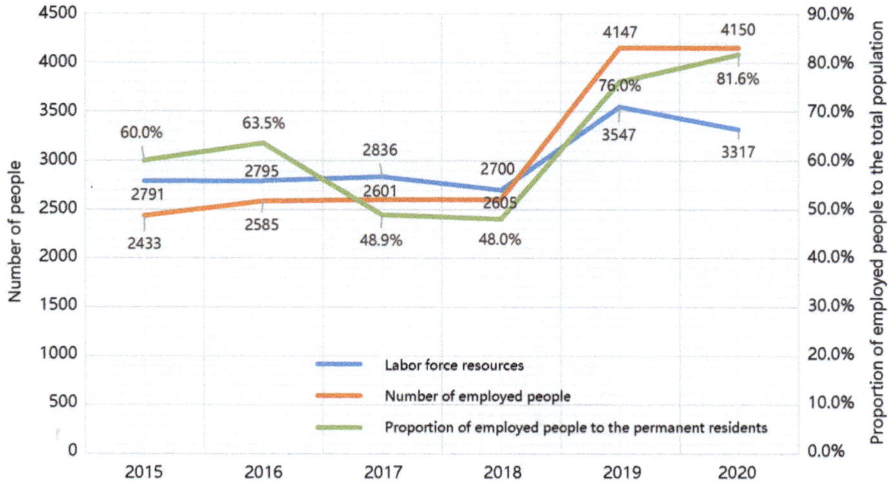

Fig. 27.17 Changes in the number of labor resources, the number of employees, and the number of employees in the proportion of the resident population in Xixinan Village. Line chart showing labor force trends from 2015 to 2020. The blue line represents labor force resources, the orange line shows the number of employed people, and the green line indicates the proportion of employed people to permanent residents. Key data points include a rise in employment numbers and proportions in 2019 and 2020, The left y-axis shows the number of people, while the right y-axis shows the proportion percentage.

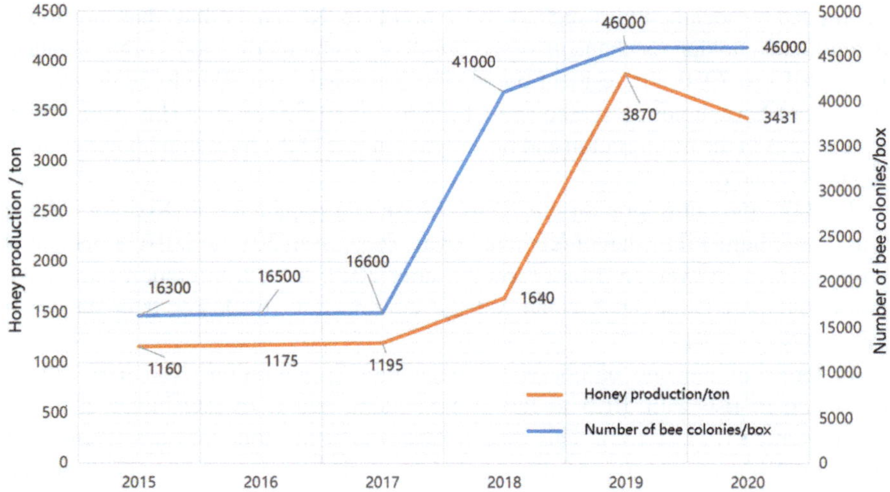

Fig. 27.18 Changes in honey production and number of bee colonies in the beekeeping industry in Xixinan Village. Line chart showing honey production in tons and the number of bee colonies per box from 2015 to 2020. The blue line represents bee colonies, starting at 16,300 in 2015, rising sharply to 46,000 by 2019, and stabilizing in 2020. The orange line represents honey production, starting at 1,160 tons in 2015, increasing to 3,870 tons in 2019, and slightly decreasing to 3,431 tons in 2020.

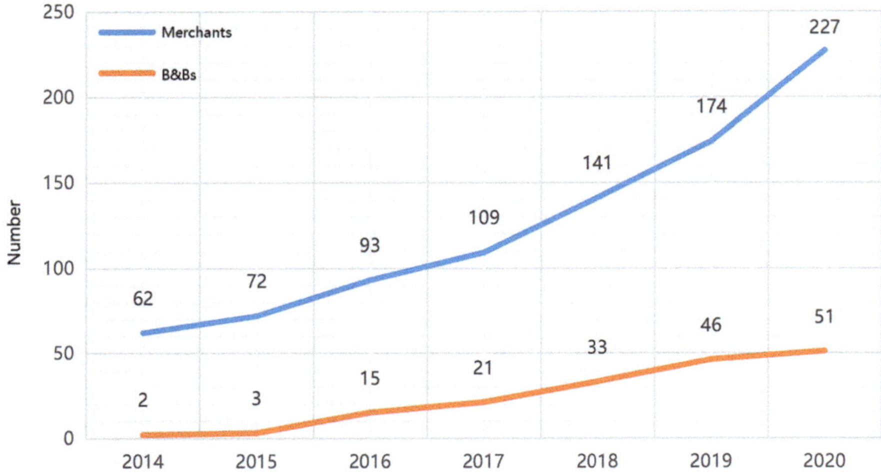

Fig. 27.19 Changes in the number of merchants and B&Bs in Xixinan Village. Line chart showing the growth in numbers of Merchants and B&Bs from 2014 to 2020. The blue line represents Merchants, increasing from 62 in 2014 to 227 in 2020. The orange line represents B&Bs, rising from 2 in 2014 to 51 in 2020.

27.5.3.2 With Co-building, Communion, and Co-governance, Community Revitalized

Facilitated by social entities such as *Wangshan* Life, Xixinan Village has developed a multi-stakeholder mechanism of co-building, communion, and co-governance. From 2015 to 2020, the resident population increased from 4056 to 5088, and the number of households increased from 1435 to 1494 (Fig. 27.20). This is related to the changes in the in-migration and out-migration rate of the household population in Xixinan Village.

In 2009, the out-migration rate of the household population in Xixinan Village was 0.49% higher than that of Xixinan Town. Then, from 2011 to 2016, it fluctuated around the town average. More recently, after 2017, the out-migration rate of the household population in Xixinan Village continued to be lower than the town average (Fig. 27.21).

In terms of the rate of household population in-migration, Xixianan Village continued to rise with 2015 as the turning point, and is overall higher than the town average (Fig. 27.22). This data reflects the rising tourism attractiveness of Xixinan Village, the returning population, and the growing vitality of the community.

With the growth of agriculture and tourism, villagers' income has increased significantly. More and more villagers are voluntarily restoring and renovating their houses, and external renters and B&B operators are also preserving and repairing traditional buildings, which has improved the appearance and quality of Xixinan residential buildings. In addition, the government, the village collective, and

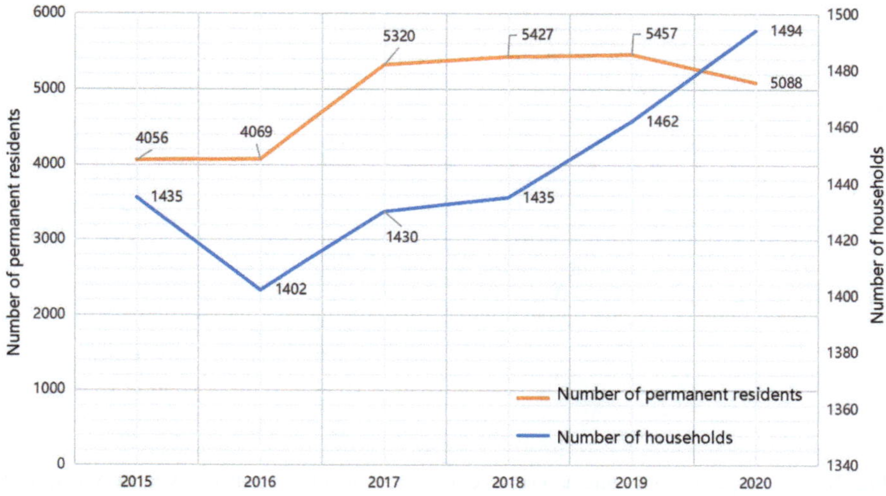

Fig. 27.20 Changes in the number of permanent residents and households in Xixinan Village. Line chart shows trends from 2015 to 2020. The orange line represents the number of permanent residents, increasing from 4056 in 2015 to 5088 in 2020, with a peak at 5457 in 2019. The blue line represents the number of households, starting at 1435 in 2015, dipping to 1402 in 2016, and rising to 1494 in 2020.

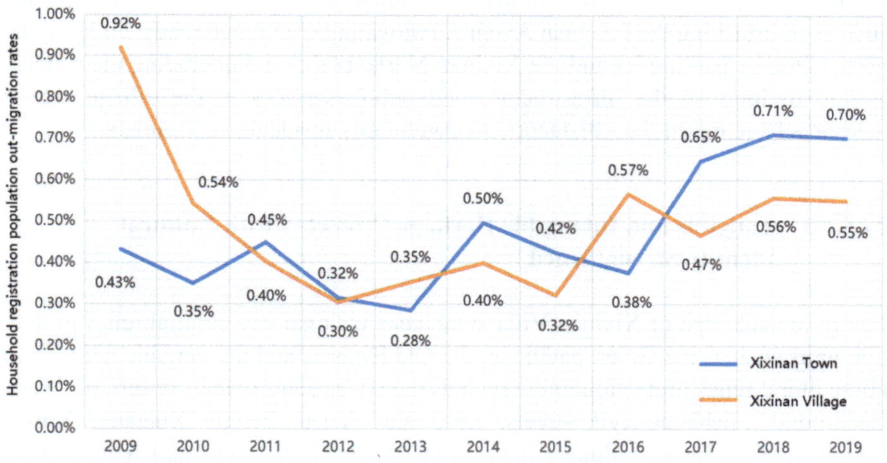

Fig. 27.21 Comparison of the out-migration rate of registered population between Xixinan Village and Xixinan Town. Line chart showing household registration population out-migration rates from 2009 to 2019 for Xixian Town and Xixian Village. The blue line represents Xixian Town, starting at 0.43% in 2009 and peaking at 0.71% in 2018. The orange line represents Xixian Village, starting at 0.92% in 2009 and decreasing to 0.55% in 2019.

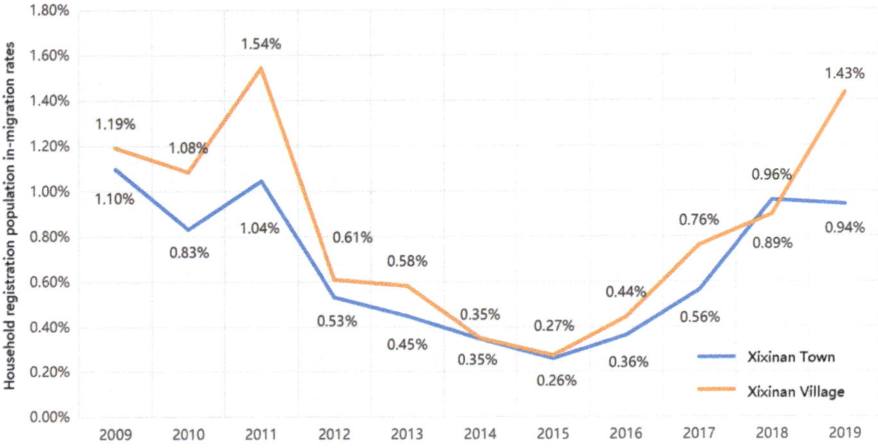

Fig. 27.22 Comparison of in-migration rates of registered population between Xixinan Village and Xixinan Town. Line chart shows household registration population in-migration rates from 2009 to 2019 for Xixinan Town and Xixinan Village. Xixinan Town is depicted with a blue line, and Xixinan Village with an orange line. Both lines show a general decline from 2009 to 2015, followed by an increase until 2019. Key data points are labeled, with peaks in 2010 and 2018 for Xixinan Village and a notable rise in 2018 for Xixinan Town.

the *Wangshan Life* team have completed a series of transportation improvements, such as refurbishing the Fengnan Avenue, renovating the village roads, building the north entrance parking lot and the Xixinan Marketplace, and upgrading the signage system to improve the infrastructure and public services in the village. These measures have raised the satisfaction of community residents and tourists.

27.5.3.3 Tangible and Intangible Heritage Protected with Cultural Identity Strengthened

The rural landscape of Xixinan Village includes national key cultural relics protection units such as the Green pavilions, the Old Houses, and the tangible heritage of non-cultural relics protection units, such as the water conservancy system of "weir-canal-pond," fishbone-style streets, Diao Xue Yuan, Fengle Xuetang, Xibian Wenyuan, etc. These buildings have been protected, restored, and reutilized. A number of intangible cultural heritages, such as Mao Feng tea stirring technique, pottery, and bamboo weaving, have also been inherited and developed. In 2019, in order to study how multiple stakeholders identify with Xixinan Village, as well as the degree of emotional attachment and degree of satisfaction, the research team used questionnaires and interviews to evaluate the sense of place of each stakeholder in Xixinan Village (Zheng 2021). The results show that the overall sense of place of each stakeholder was good, resulting in stronger place attachment and place identity (Fig. 27.23).

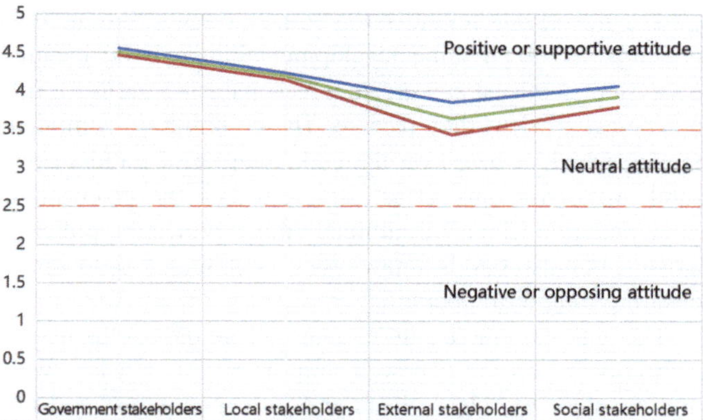

	Government stakeholders	Local stakeholders	External stakeholders	Social stakeholders
Degree of emotional attachment	4.55	4.23	3.85	4.06
Degree of satisfaction	4.46	4.14	3.43	3.79
Degree of sense of place	4.5	4.19	3.64	3.92

Fig. 27.23 Quantitative evaluation of the sense of place of the multi-stakeholders in Xixinan Village. Line chart shows attitudes of different stakeholder groups: government, local, external, and social. The y-axis represents attitude levels from 1 to 5, with 3 marked as neutral. Three lines represent degrees of emotional attachment (blue), satisfaction (red), and sense of place (green). Government stakeholders show the highest positive attitude, while external stakeholders have the lowest.

27.6 Conclusion

In this chapter, we have examined the importance of China's village revitalization based on a renewed urban-rural relationship, suggesting that multiple stakeholders of village revitalization should consider the relationship between ecosystem services and human well-being, and weigh the urban-rural relationship from the perspective of demand theory. In this regard, village revitalization in China is a form of rural urbanism in the post-urbanization and post-industrialization era, especially in the Post-Epidemic Era.

In rural revitalization, the bidirectional flow of urban and rural elements becomes an essential feature of the renewed urban-rural relationship. On the one hand, the villages provide high-quality ecosystem services and products to the cities, satisfying the aspirations of urban residents for organic products and green lifestyles. On the other, the city provides the capital, talents, and knowledge needed for the development of the villages. Overall, the urban and the rural are deeply bonded to constitute a win-win collaboration. This model is a market-based development approach through phrased ecological compensation and transfer payments, which can effectively alleviate the over-reliance of rural development on a single government entity.

In terms of revitalization methodology, we propose *five-in-one* initiatives and framework. Intensive interpretation of heritage assets aims to fully excavate the ecological and cultural values of traditional villages, the *status quo* of cultures, and its evolutionary marks, to gain a deeper understanding of the stories behind the rural landscape.

The *five-in-one* initiatives integrate ecology, productivity, and livability of the countryside into the same developmental logic, build a value conversion medium for ecosystem services, and create a reasonable scenario for urban residents to pay for ecological services they receive. The co-building, communion, and co-governance model, which is based on the collaboration of multi-stakeholders from urban and rural areas, not only effectively remedies the problem of over-reliance on the government in China's traditional village protection policies and supervision mechanisms, but also builds a new order combining top-down and bottom-up strategies, ensuring long-term operation.

The ten-year practice in Xixinan village proves the operability of village revitalization based on this renewed rural-urban relationship, which can realize the economic values of "clean rivers and lush mountains" and of nostalgia. The authors hope that the revitalization practices in Xixinan can offer reference for rural governance and sustainable development of traditional villages in other parts of China or the global societies. They also hope to examine and optimize the theoretical framework and specific measures through more extensive socio-ecological practices.

Acknowledgments We gratefully acknowledge financial support from the General Project of the Humanities and Social Sciences Research Foundation of the Ministry of Education of China (Grant No. 24YJCZH404).

References

Akkoyunlu S (2015) The Potential of Rural–Urban Linkages for Sustainable Development and Trade. Int J Sustain Dev World Pol, 4(2): 20–40

Bian L (1993) The exploration of disputes and the issue of litigious customs in Huizhou since the mid-Ming dynasty (明中叶以来徽州争讼和民俗健讼问题探论). Ming Stud J 00:75–84

Börner J, Mendoza A, Vosti SA (2017) Ecosystem services, agriculture, and rural poverty in the Eastern Brazilian Amazon: interrelationships and policy prescriptions. Ecol Econ 64:356–373

Department of International Cooperation, Ministry of Agriculture and Rural Affairs of the PRC (2021) China's Food System Sustainable development path report[EB/OL].[2021-09-23]. http://www.moa.gov.cn/xw/gjjl/202109/t20210923_6377473.htm

Feng J (2017) The development of tourism in traditional villages should be based on the premise of cultural heritage preservation (传统村落开展旅游要以文化遗产保护为前提). China Art News 2017-03-03(001)

Gebre T, Gebremedhin B (2019) The mutual benefits of promoting rural-urban interdependence through linked ecosystem services. Glob Ecol Conserv 20:e00707

Gibb H, Huchuli DF (2002) Habitat fragmentation in an urban environment: large and small fragments support arthropod assemblages. Biol Conserv 106(1):91–100

Gutman P (2007) Ecosystem services: foundations for a new rural-urban compact. Ecol Econ 62(3–4):383–387

He Z (2009) Study on the impact of tourism development in historic villages and towns in China (旅游开发对我国历史文化村镇的影响研究). Tsinghua University, Beijing

Howard E (2013) Garden cities of to-morrow. Routledge, London

Hu Y (2022) Protection and reuse of Beitang from the perspective of cost-benefit analysis——taking the Songtang Village of Huizhou as an example (成本效益视角下的陂塘保护与再利用——以徽州竦塘村为例). Peking University

Hu Y, Chen S, Cao W et al (2014) The concept and cultural connotation of traditional villages (传统村落的概念和文化内涵). Urban Dev Stud 21(01):10–13
Huang C (2021) On the internal logic and theoretical issues of rural revitalization and common prosperity (论乡村振兴与共同富裕的内在逻辑及理论议题). J Nanjing Agric Univ (Soc Sci) 21(06):1–9
International Cooperation Department of the Ministry of Agriculture and Rural Affairs, China (2021) Report on the sustainable development pathway of China's Food System (中华人民共和国农业农村部国际合作司.中国粮食系统可持续发展路径报告). [EB/OL] [2021-09-23]. http://www.moa.gov.cn/xw/gjjl/202109/t20210923_6377473.htm
IUCN (2021) A user-friendly framework for the verification, design and scaling up of NbS. IUCN, Gland
Jiang C (2022) New Pattern of Development, Common Prosperity, and Rural Industrial Revitalization (新发展格局、共同富裕与乡村产业振兴). J Nanjing Agric Univ (Soc Sci), 22(01): 1-11+22
Kreisel W, Reeh T (2019) Tourism research and landscape interpretation. Universitätsverlag Göttingen, Gotinga
Li J (2013) A study on the business model of private capital involvement in heritage protection and tourism development of ancient villages and towns (民营资本介入古村镇遗产保护与旅游开发的商业模式研究). East China Normal University
Li Z (2021) A strategic study on achieving common prosperity in China (中国走向共同富裕的战略研究). Chinese Rural Econ 10:2–23
Li S, Chen J, Teng Y (2021) Rural revitalization on the path of common prosperity: problems, challenges and suggestions (共同富裕路上的乡村振兴:问题、挑战与建议). J Lanzhou Univ (Soc Sci) 49(03):37–46
Liu B (2015) Re-searching Huizhou: folk memory and folk writing of Huizhou Village in the transition period (重寻徽州—转型期徽州乡村的民间记忆与民间书写). Guangxi Normal University Press, pp 4–6
Liu Y, Long H, Li Y (2021) Human geography research based on the new thinking of global rural-urban relationship (全球乡城关系新认知与人文地理学研究). Acta Geograph Sin 76(12): 2869–2884
Lu Y, Li Y, Song P (2016) Safeguarding and inheriting Chinese vernacular culture in course of "beautiful villages project" (乡土文化在"美丽乡村"建设中的保护与传承). J Northwest A&F Univ (Soc Sci Ed) 16(03):69–74
Mayer H, Habersetzer A, Meili R (2016) Rural-urban linkages and sustainable regional development: the role of entrepreneurs in linking peripheries and centers. Sustainability 8(8):745
Millennium Ecosystem Assessment (2005) Ecosystems and human well-being: synthesis. Island Press, Washington, DC
Ministry of Environmental Protection of the PRC, National Bureau of Statistics of the PRC, Ministry of Agriculture of the PRC (2010) Bulletin of the first national Census of pollution source [EB/OL]. [2010-02-10]. http://g.mnr.gov.cn/201701/t20170123_1428261.html
National Bureau of Statistics of the PRC (2017) Main Data Bulletin of the Third National Agricultural Census (No.1) [EB/OL]. [2017-12-14]. http://www.stats.gov.cn/tjsj/tjgb/nypcgb/qgnypcgb/201712/t20171214_1562740.html
National Bureau of Statistics of the PRC (2021) China statistical yearbook: 2021. China Statistic Press, Beijing
Peng W, Dai J (2015) The ecological relationship between cities and rural areas in ecological civilization construction (生态文明建设中的城乡生态关系探析). Ecol Econ 31(08):173–177
Radford KG, James P (2013) Changes in the value of ecosystem services along a rural-urban gradient: a case study of greater Manchester. UK Landsc Urban Plann 109(1):117–127
Shi S (2020) Sustainable mechanism of water cultural landscape in Huizhou from Ming and Qing dynasty to the Republic of China (明清民国时期徽州水文化景观的可持续机制). Peking University, Beijing

Sun X, Jia L, Lu W (2019) Culture development dilemma and path choice of rural revitalization (乡村振兴的文化发展困境及路径选择). J Shandong Univ (Philos Soc Sci) 05:135–144
Tang L (1991) The family and clan structure of Huizhou in the Ming and Qing dynasties (明清徽州的家庭与宗族结构). Hist Res 01:147–159
The CPC Central Committee and The State Council (2018) The CPC Central Committee and The State Council issued the Strategic Plan for Rural Revitalization (2018–2022) [EB/OL]. [2018-09-26]. http://www.gov.cn/zhengce/2018-09/26/content_5325534.htm
Tilden F (1957) Interpreting our heritage. University of North Carolina Press, Chapel Hill
Wang C (2018) The influence and enlightenment of the development model of collectivized rural tourism on the strategy of rural revitalization (集体化乡村旅游发展模式对乡村振兴战略的影响与启示). Shandong Soc Sci 05:34–42
Wang Z, Zhang J, Zhou Z et al (2021) The value target of ecological compensation: international experience and its enlightenment to China (生态补偿的价值目标:国际经验及对中国的启示). Chinese. J Environ Manag 13(02):27–32+19
Wu L, Liu J (2017) Common problems and subject positioning under the government-led beautiful village construction mode (政府主导型美丽乡村建设模式的共性问题与主体定位). Ecol Econ 33(10):226–230
Yang L (2011) The literature association of Huizhou in Ming and Qing dynasty (明清时期徽州文会研究). Anhui Normal University, pp 18–23
Yin Y, Li X (2018) Internal response and path selection of rural tourism under the background of rural vitalization strategy (乡村振兴战略背景下乡村旅游的发展逻辑与路径选择). J Chinese Acad Gov 05:182–186+193
Yu K (2001) The culture that has been ignored, the beauty of weeds——Qijiang Park in Zhongshan City (足下的文化与野草之美——中山岐江公园设计). New Arichiture 05:17–20
Yu K (2006) The art of survival: position landscape architecture (生存的艺术:定位当代景观设计学). Architect J 10:39–43
Yu K (2017) New Ruralism movement in China and its impacts on protection and revitalization of heritage villages: Xixinan experiment in Huizhou District, Anhui Province ("新上山下乡运动"与传统村落保护及复兴——徽州西溪南村实践). Bull Chin Acad Sci 32(07):696–710
Yu K (2020) Designing and building a beautiful city with the vision of ecological civilization (生态文明理念下美丽城市的规划设计与建设). Frontiers 04:18–36
Yun H (2021) Study on the continuity of the cultural landscape of tourism oriented Huizhou heritage villages (旅游导向型徽州遗产村落的文化景观连续性研究). Peking University
Zhang J (2018) Maintenance and reformation of protection methods of cultural heritage under rural revitalization strategy: an interpretation of rural development system in the past four decades (乡村振兴下乡村文化遗产保护的"固旧"与"维新"——兼论改革开放40年乡村发展制度解读). Planner 34(10):26–31
Zhang H (2021) Simulation of agricultural non-point source pollution at slope scale using a physical model (基于物理模型试验的坡面尺度农业面源污染模拟). Peking University
Zhang C, Fu R (2020) Goal setting and realization path of rural ecological revitalization based on ecosystem services (基于生态系统服务的乡村生态振兴目标设定与实现路径). Rural Econ 12:42–48
Zheng C (2021) Landscape evolution of rural settlements from the perspective of multiple stakeholders—a case study of Xixinan Village in Huizhou (多利益主体视角下的乡土聚落景观演变研究—以徽州西溪南村为例). Peking University
Zhou Q (2013) The current situation, problems and countermeasures of the protection of traditional villages in China (我国传统村落保护的现状问题与对策思考). China Constr News 2013-01-29(003)
Zou T, Li F (2007) Research on the development model of ancient village heritage tourism led by the community: a case study of Cuandixia ancient village in Mentougou, Beijing (社区主导的古村落遗产旅游发展模式研究——以北京市门头沟爨底下古村为例). J Beijing Int Stud Univ (北京第二外国语学院学报) 5:78–86

Chapter 28
Indigenous *Lamkaang Naga Loh Kchet*: Collective Interdependence in Northeastern India

Sumshot Khular

Abstract The current socio-environmental crisis entails severe losses of biological and cultural diversity. Hence, it represents not only an ecological crisis but also a profound cultural and ethical crisis. To solve it, individuals and societies need to assume an ethical responsibility toward nature, local communities, and their cultures. I address this ethical imperative by analyzing a case study in my home region and by examining conceptual frameworks for valuing and respecting biological and cultural diversity in ways that contribute to biocultural conservation based on just and sustainable ways of co-inhabiting. First, regarding the case study, I explain place-based practices of my *Lamkaang* Naga Indigenous community in the state of Manipur in northeastern India. I focus on our *Lamkaang Naga* Indigenous Traditional Healing Systems and *Loh Kchet,* a type of Collective Interdependence (CI), as rooted in the *Lamkaang Ksen Paam* (land) where I was born and raised. This healing system utilizes *Chukpaa* and *Kuulriing* leaves, and Lamkaang people maintain a kinship relationship with the plants used for traditional medicine. Seeking permission before harvesting is a vital practice that reflects respect for the agency of plants, and within the Lamkaang community, *Loh Kchet,* a system of collective interdependence, ensures that no one is left behind. I interpret and analyze these practices through both (1) my own lived experiences and (2) ethnographic studies I have conducted in recent years. Second, regarding the conceptual frameworks, I focus on the works by Native American philosopher Brian Burkhart and Chilean environmental philosopher and ecologist Ricardo Rozzi. Burkhart's decolonizing framework emphasizes cultural resurgence and resistance to colonialism, particularly through his concepts of locality, being-from-the-land, and knowing-from-the-land. I use his term "locality"to highlight how being, meaning, and knowledge are deeply rooted in the land. Rozzi's biocultural ethics relies on his "3Hs" model, which examines the interconnected life Habits of human and other-than-human co-in-Habitants within a shared Habitat. I use his biocultural ethic framework to emphasize the inseparable links between humans and the ecosystems they

S. Khular (✉)
Department of Philosophy and Religion, University of North Texas, Denton, TX, USA
e-mail: Sumshot.Khular@unt.edu

D. Zhu et al. (eds.), *Linking Arts with Biocultural Conservation, Restoration, and Communication*, Ecology and Ethics 8,
https://doi.org/10.1007/978-3-032-04830-1_28

co-inhabit. Rozzi's "3Hs" model and Burkhart's concept of locality are consistent with place-based practices of Lamkaang Naga Indigenous people. The comparative ecological and ethical approach used in this case study could be applicable in other communities to identify sustainable and just forms of inter-cultural and inter-species co-inhabitation.

Keywords Locality · Co-inhabitants · Traditional healing · Decolonial · Biocultural ethics

28.1 Introduction

Human relationships with nature have captured a rapidly growing academic interest since the end of the twentieth century (Gadgil 1987; Maffi 2001, 2005; Rozzi 2001; Rozzi et al. 2010; Restall and Conrad 2015; Ives et al. 2017). This research has identified vital connections between biological and cultural diversity and its implications for human wellbeing, health and sustainability (Pretty et al. 2009; Rapport and Maffi 2010; Rozzi 2010, 2013; Tam 2013; Zylstra et al. 2014; Restall and Conrad 2015; Ives et al. 2018; Hatty et al. 2020). The need to better understand and value forms of traditional ecological knowledge is more pressing than ever for at least three reasons.

First, because much work on biocultural approaches has been done in the Global South and is little known to the English-speaking world (Díaz-Reviriego et al. 2024). For example, in India and Latin America, biocultural perspectives have inspired environmental and social justice movements as part of political struggles to protect Indigenous territorial rights and sovereignty over natural resources (Singh et al. 2010, 2021; Merçon et al. 2019). Indian and Latin American academics have also extensively contributed to the conceptualizations of biocultural diversity and the implementation of biocultural approaches, which has led to the publication of widely influential work on phytodiversity (Chakravarty et al. 2012), biocultural memory (Bhatt and Nazrul Islam 2008), biocultural heritage (Argumedo 2008), and biocultural ethics (Rozzi 2012, 2013).

Second, a modern Eurocentric separation of nature from human culture has been identified as a serious problem in the conservation of both biological and cultural diversity (Rotherham 2014; Zhu 2023; Andreozzi 2025). The process has been described as cultural severance, which leads to dramatic declines in ecological richness (Rotherham 2013). In contrast, conservation of biocultural diversity brings socio-environmental justice (Shiva 1992; Rozzi 2010), sustainability (Pungetti 2013; Elands et al. 2019; Otamendi-Urroz et al. 2025), and wellbeing of human and other-than-human co-inhabitants (Rozzi 2003, 2013; Hegde and James 2019; Prescott et al. 2022).

Third, cultural and linguistic diversity of Indigenous peoples faces increasing threats, linked to climate change, land-use changes, generational gaps, global cultural processes and narrow economic policies are imposed from sectors of society with political power (Nettle and Romaine 2000; Amano et al. 2014; Rozzi 2015a).

Their knowledge and language are also threatened by climate-driven shifts in species distribution (Pecl et al. 2017). Fortunately, the use of traditional ecological knowledge is being promoted by organizations such as the International Union for Conservation of Nature, and the urgent need to integrate Indigenous knowledge into biodiversity assessments and management is widely recognized (Johnson 2012; Tom et al. 2019; McElwee et al. 2020).

Motivated by the former three reasons, in this chap. I examine place-based practices of my *Lamkaang* Naga Indigenous community in the state of Manipur in northeastern India. I focus on our *Lamkaang Naga* Indigenous Traditional Healing Systems and *Loh Kchet,* a type of Collective Interdependence (CI), as rooted in the *Lamkaang Ksen Paam* (land) where I was born and raised. This healing system utilizes *Chukpaa* and *Kuulriing* leaves, and Lamkaang people maintain a kinship relationship with the plants used for traditional medicine. Seeking permission before harvesting is a vital practice that reflects respect for the agency of plants, and within the Lamkaang community, *Loh Kchet*, a system of collective interdependence, ensures that no one is left behind. I interpret and analyze these practices through both (1) my own lived experiences and (2) ethnographic studies I have conducted in recent years.

To develop my analyses, I adopt two main conceptual frameworks, forged by Native American philosopher Brian Burkhart and Chilean environmental philosopher and ecologist Ricardo Rozzi. First, Burkhart's decolonizing framework emphasizes cultural resurgence and resistance to colonialism, particularly through his concepts of locality, being-from-the-land, and knowing-from-the-land. I use his term "locality" to highlight how being, meaning, and knowledge are deeply rooted in the land. Second, Rozzi's biocultural ethics relies on his "3Hs" model, which examines the interconnected life Habits of human and other-than-human co-in-Habitants within a shared Habitat. I use his biocultural ethic framework to emphasize the inseparable links between humans and the ecosystems they co-inhabit. I examine how Rozzi's "3Hs" model and Burkhart's concept of locality are consistent, or not, with place-based practices of Lamkaang Naga Indigenous people.

28.2 Lamkaang Naga Indigenous People in Northeastern India

The eight northeastern states of India—Arunachal Pradesh, Assam, Manipur, Meghalaya, Mizoram, Nagaland, Sikkim, and Tripura—cover 8% of India's land and share a total of over 5400 km of borders with five neighboring countries: Bangladesh, Bhutan, China, Myanmar, and Nepal. The Indigenous peoples of the Northeastern (NE) region of India have diverse histories, cultures, languages, religions, social organizations, and economies, with most tribes speaking Tibeto-Burman languages (Haokip 2011). The region is also known for its rich and biodiverse ecosystem (Chakravarty et al. 2012). In Meghalaya, 80% of the forest

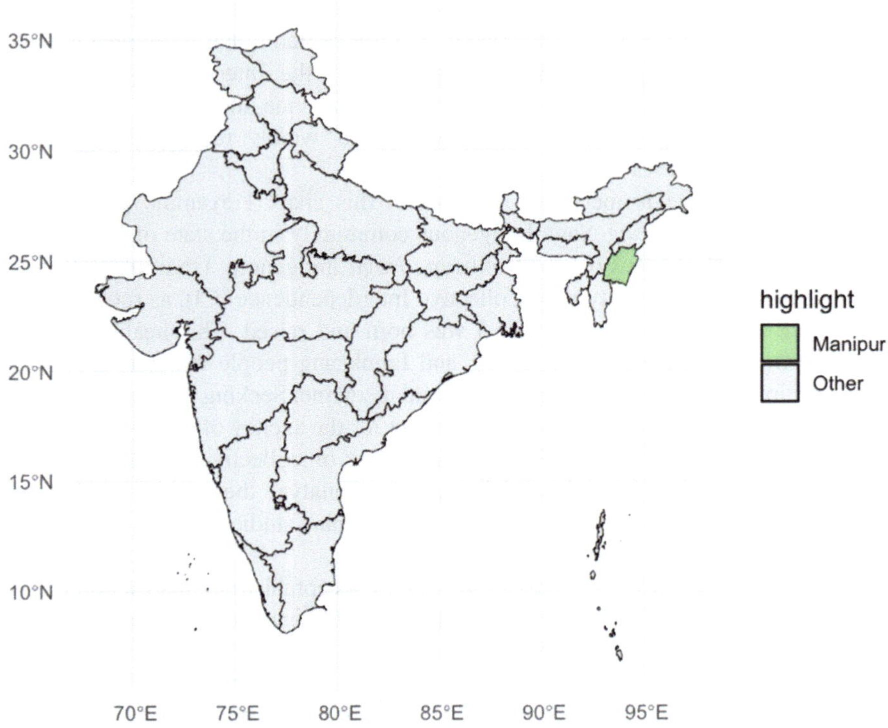

Fig. 28.1 Map of India showing the location of Manipur state (green)

land is owned by local communities and managed by Indigenous traditional institutions (Mir et al. 2022).

In the context of the Naga Indigenous peoples in the NE region and Myanmar, land is collectively owned by the community. While certain portions of land belong to clans, homestead land can be owned individually. Unlike other tribal groups, the Nagas follow a unique tradition where land cannot be sold. According to Horam (1975), approximately 88% of the land in Nagaland is collectively owned by Naga Indigenous peoples, while only 12% is individually owned (Karlsson 2011, Shimray 2007).

Manipur is bordered by Nagaland to the north, Mizoram to the south, and Assam to the west. To the east, it shares an approximately 400 km-long border with Myanmar, connecting with the Sagaing Region and Chin State (Fig. 28.1). As a strategic gateway to Southeast Asia, Manipur has gained increasing significance in the Indian government's policies, particularly under the Look East Policy, now known as the Act East Policy (Baruah 2004; Haokip 2015).

The Lamkaang Naga Indigenous people reside in the Chandel, Tengnoupal, and Senapati districts of Manipur. A small colony of Lamkaang people also lives in Dimapur, Nagaland, while a few can be found in Tamu Township, Myanmar (Sankhil 2015). Lamkaangs are one of the 16 Naga tribes of Manipur and are also

Fig. 28.2 Challong villagers Loh Kchet for Rice/Paddy cultivation. This process involves (1) seed sowing, (2) paddy growing, (3) identifying when the paddy is ready for harvest, (4) (4) collecting the cut paddy, (5) threshing the collected paddy, and (6) cleaning the chaffs after threshing is over. (Photographs 1 and 3 were taken on April 25, 2021, by Nini Khular. Photograph 2 was taken on August 17, 2022, by Sumshot. Photographs 4, 5, and 6 were taken November 3, 2023, by Sankhil Bungsong)

known as *Ksen Mii*, meaning "red people." These two names are used interchangeably (Sankhil, 2015, Khular 2018). In the traditional Lamkaang Naga way of life, there is a deep-rooted belief that they come from the land and belong to it. They do not see themselves as landowners; rather, they believe that the land owns them. As stewards (Rozzi et al. 2012), they are entrusted with the responsibility of diligently caring for the land, which they believe was given to them by *Parvaan Kruung* (God of the Heavens). Lamkaangs knowledge on how to control the soil run off, leaves and other plants for fertilizing the soil, keeping the land/soil balance are all practice-based knowledge that are transmitted from the elders to the younger generations. Such knowledge stems from everyday interactions with their local environment (Fig. 28.2).

For the Naga Indigenous peoples, land is equivalent with their identity. It connects them to their history, art, artifacts, culture, and language (Longchar 2000). As Yangkhohao Vashum states, *"The land owns the people, not the people owning the land. It is the land that gives the people their identity"* (Vashum 1998). Land is also the source of folktales, stories, knowledge, and all relationships with the land and other beings in Indigenous cosmology (Vashum 1998). Furthermore, Wati A. Longchar argues that for tribal and Indigenous peoples, land is the source of life—it is alive. The Nagas do not perceive the land as an inanimate entity or an object; rather, they believe that land has life, power, and agency in its own right (Longchar, 1999, 2000).

Both Longchar and Vashum emphasize that the Nagas regard land as a "mother" and see themselves as her children. This mother-child relationship with the land is deeply significant and shared across Indigenous peoples and nations (Vashum 1998; Longchar 2000; Varah 2013). Among the Chang Naga tribe, the word "*Manyü*" means Mother Sky, while "*Kanyü*" means Mother Earth (Pongen and Bosman 2023). This reflects the deep reverence the Nagas hold for Mother Earth and their inextricable connection to the land (Burkhart 2019; Vashum 2005). For Lamkaangs, as with most Indigenous peoples, land is not merely a physical space—it is the heart of their identity, their longing, and their sense of belonging.

28.3 Challenges Affecting Indigenous Peoples Today

The ways of life of the Lamkaang Indigenous people are increasingly threatened by large-scale developmental projects initiated by the Indian government. One such project is the construction of the six-lane Trans-Asian Highway (Ziipao 2018), which passes through Lamkaang villages, disrupting the once-peaceful equilibrium of their environment. This disruption has significantly affected their relationship with the land, water, and non-human co-inhabitants. The Look/Act East Policy (Baruah 2004, Das 2010) aims to enhance trade and commerce between India and East Asian countries. However, despite various proposed projects under this policy, railway infrastructure has yet to reach Lamkaang Naga Indigenous lands. According to Haokip (2015), the Look/Act East Policy prioritizes counterinsurgency strategies over developmental agendas.

India's political landscape plays a crucial role in these challenges. Dolly Kikon, a Naga anthropologist, argues that the Northeast region holds significant cartographic importance, where physical power dominates political discourse on citizenship and peoplehood (Kikon 2005). As a result, Indigenous people's citizenship and land rights (Shimray 2001) are frequently under threat, like the proposed Trans-Asian Railroad (Yhome 2015a, b) represents yet another large-scale developmental project that threatens not only human communities but also the ecosystems of Lamkaang lands and the entire Northeastern region. A notable example of displacement caused by such projects is the construction of the Chadong Mapithel Dam in Manipur (Yumnam and Koijam 2014).

Since 2015, this dam has completely submerged the Chadong-Ramrei villages, displacing Tangkhul Naga communities along with their paddy fields. Moreover, over 40 surrounding villages have also suffered the loss of their agricultural lands due to submergence (MDAVO). This is just one among many dam projects planned across Northeast India, particularly in Manipur. Despite strong Indigenous resistance, movements, and protests throughout the region, displacement, evictions, and land dispossession continue under the guise of development. Lamkaangs now face imminent land-grabbing for infrastructure projects such as highways and railways. This neocolonial process of land dispossession prioritizes global development policies over the interests and survival of local communities (Rozzi 2018).

Similarly, the Northeast region has faced massive onslaughts. Yet, hope persists as Indigenous communities continue to defend their land and habitats. In many places, such as Arunachal Pradesh, Assam, Meghalaya, Nagaland, Mizoram, and Manipur, certain areas remain intact and free from corporate exploitation. Indigenous peoples, including the Khasis, Nagas, and other tribal groups, have steadfastly protected their habitats from outside intrusion. This resilience is rooted in their deep relation with the land and all other-than-human beings, making such protection possible. The epistemic, ontological, and cosmological aspects of Indigenous peoples' lives, along with their physical and cultural dimensions, are deeply tied to the land and cannot be separated (Whyte 2018; Karlsson 2022; Vashum 2005).

For this reason, the Khasi Indigenous tribes of Meghalaya have continued to maintain sacred groves to this day (Karlsson 2011). If such places received stronger governmental support for preservation, it could help prevent the extinction of many exotic species native to the Northeast region. Building on this, Indigenous sovereignty as pedagogy—encompassing *territory, body, mind, home, and community*—can serve as a powerful approach to biocultural conservation and restoration (Simpson 2013, p. 67). The *Nishnaabewin* grounded normativity, which is based on the land, emphasizes reviving intellectual practices, reclaiming traditional ecological knowledge, and integrating praxis as theory. This approach, central to the *Radical Resurgence Project* (Simpson 2013, p. 173), can inspire actions across global habitats toward biocultural conservation, reinforcing practices that Indigenous peoples have long upheld.

28.4 The Lamkaang Naga Indigenous Traditional Healing Practices

In the Lamkaang traditional healing system, listening to the land and its co-inhabitants is of utmost importance. In *Ardii Kthung*, the *Thiimpuu/Thiimpii* (priest/priestess) performs a blessing ritual to seek permission from the land before engaging in activities such as farming, house construction, road building, or any other land use. Today, after the Lamkaangs embraced Christianity, pastors continue this practice by praying for guidance from the land. Lamkaang women play a vital role in preserving and promoting traditional knowledge and culture. They are also the keepers of seeds. Regarding the use of medicinal plants, leaves such as Chukpaa Ksen (*Adhatoda vasica*), herbs like Rihiing (*Blanda Benth.*), and roots such as *Kihiing* (English common name as ginger), Kihiing Kvom (English common name as black ginger), and turmeric (*Zingiberaceae*) are widely cultivated. Additionally, barks or stems from specific plants, such as the banana (*Musa paradisiaca*), are commonly grown by families and used for various medicinal purposes. Women play a crucial role in cultivating these plants, growing them in their yards, fields, and farms. Some medicinal plants also thrive among large trees, along riverbanks, or in swampy areas, further highlighting the deep connection between Lamkaang women and their land.

Lamkaangs use "Toom Siingkhaa" (bile of a bear), which remains an important ingredient in traditional medicine to this day. For most ailments, Lamkaangs use "Toom Siingkhaa" by taking a small grain-sized portion, mixing it with a spoonful of water, and having the person swallow it. They are warned not to let the drink touch their teeth, as it is believed to harm them. "Toom Siingkhaa" is extremely difficult to find, as it is not sold in the market, and not everyone can hunt a bear to collect its bile. To emphasize the challenge of finding "Toom Siingkhaa," consider the testimony of a hunter about his relationship with the bear. In an interview with Welngam Kochaai of Challong village, he shared that he encountered the bear ten times, but each time, the bear managed to escape—even on the tenth attempt, when the bear was injured but still managed to flee. According to him, "If the bear wanted to be killed, it would act so tame that one could kill it without trouble. But it is only when the bear allows itself to be killed that one can do so. If the bear wanted to be killed, it would come closer, not run away." This story reflects the deep relationship between humans and animals. It exemplifies how human institutions, intertwined with food systems, adapt to social and environmental changes.

Traditional knowledge plays an important role in preserving and healing the land, with specific practices aimed at ensuring the land is given rest (Judges 3:8) or protected from external intrusion. Lamkaangs identify local plants by their names and understand the cultural uses of these plants, relying on Indigenous knowledge. For example, in Lamkaang, Kuulring leaves are used for healing (Maffi 1998). The traditional healing practices of the Lamkaang Naga people are closely linked to the preservation and promotion of their language(Matsuura 2008; Moore et al.2002; Grant 2012; Pretty et al. 2009). Elders carry the oral histories that linguistically encode environmental knowledge, such as traditional biodiversity conservation methods (e.g., bulhok lam (small drain), terracing, manure, canal building), ethno-botanical knowledge, and the ritual regulation of forests and sacred trees. Such practices must be passed down to the younger generation to ensure the continuity of this invaluable knowledge.

28.5 Herbs in Lamkaang Naga Traditional Healing Systems

Herbal plants, as co-inhabitants of humans, are a vital component of many Indigenous traditional medicines around the world (Pan et al. 2014). In the Lamkaang tradition, Chukpaa Ksen and Kuulring leaves are among the most widely used herbs in traditional healing systems. I will present these two species of leaves as a case study.

28.5.1 Case 1: Chukpaa Ksen (Adhatoda Vasica)

Chukpaa Ksen, known in English as Malabar nut or adulsa, is a shrub native to Asia (Fig. 28.3). It is widely spread in the plains of India and ranges of Himalayan at an altitude of 1300 m, also distributed in Punjab, Bengal, Nepal, Assam, and Sri Lanka (Shoaib 2021). Chukpaa Ksen is commonly used by the Lamkaang people, and it is one of the key plant species in Chinese traditional medicine (Shamsuddin et al. 2021). The preparation of its leaves for medicinal uses includes four key steps:

(a) Boiling the leaves and consuming them to treat cold and cough.
(b) Crushing the leaves to extract their juice, which is then mixed with honey and ginger to treat the common cold.

Fig. 28.3 Chukpaa Ksen (*Adhatoda vasica*). The Chukpaa Ksen shrub is used daily by the Lamkaangs, not only as a medicinal plant but also in various forms in their daily life. It is also a favorite ingredient in Lamkaang cuisine. The flowers of Chukpaa Ksen are used in chutney, a salad-like dish in the Lamkaang traditional style, while the leaves are often fried with peas to make a delicious salad. (Photograph by Nini Khular)

(c) Boiling the leaves in hot water and using it as a bath to reduce fever.
(d) Boiling the leaves to bathe children suffering from skin rashes and scabies.

In addition to its medicinal uses, Lamkaang people plant Chukpaa Ksen shrubs around their houses as fencing, ensuring that the plant continues to be a part of their lives and knowledge. This practice helps preserve traditional herbal medicinal knowledge.

The knowledge of using Chukpaa Ksen as both food and medicine has been passed down through generations via oral traditions and practices. Even though modern medicines may help with certain ailments, the use of Chukpaa Ksen as food, medicine, and a bath ingredient remains an important part of Lamkaang culture. This knowledge is not limited to one individual but is shared collectively among the Lamkaang people. Such collective knowledge ensures the preservation and protection of Chukpaa Ksen and other plants, fostering an environment where their growth and care are prioritized. This culturally situated, community-based practice enables the younger generation to continue the relationship with the land and plants, further strengthening the bond between people and their shared habitat. This practice can be extended to many other species in our environment, contributing to biocultural flourishing in the shared landscape we inhabit with our co-inhabitants.

28.5.2 *Case 2: Kuulring (*Solanum spirale*) Leaves*

The *Kuulring* plant (*Solanum spirale*), known in English as Spiral nightshade, is a shrub that has leaves (Fig. 28.4), which are widely used in traditional healing rituals in the Lamkaang culture. In Lamkaang healing practices, knowledge about the land, plants, water, and their interconnection with humans and other-than-human beings holds immense importance. The famous folktales of "Arthluk leh Raangchel" and "Soombeel leh Huumpiipaa" illustrate how healing and bringing Raangchel back to life are closely linked to the use of Kuulring leaves. As part of the ritual, Kuulring leaves are dipped in clean water and sprinkled on the sick person as the first step in the healing process. As mentioned earlier, the *Thiimpii/Thiimpuu* (priest/priestess) must first seek permission from the Kuulring plant before plucking its branches. Only after permission is granted can the branches with leaves be carefully harvested for use in the healing rituals.

Kuulring plants or shrubs typically grow in moist areas. These practices are deeply embedded in both the cultural and environmental context, reflecting the relationality between the Lamkaang people and their habitat, where they coexist with their co-inhabitants. The practice of seeking permission and acknowledging the agency of water, plants, and non-human beings emphasizes a mutually beneficial relationship. This ongoing habit ensures the preservation of the symbiotic connection they share with the environment, preventing harm and promoting sustainability.

Fig. 28.4 Close up of leaves of Kuulring (*Solanum spirale*). The species is known in English as Spiral nightshade. (Photograph by Rex Khullar)

28.6 A Biocultural Analysis of Loh Kchet

According to Rozzi (2012, 2013), the 3Hs model of the biocultural ethic is a framework that values the vital links between the life habits of distinct humans and non-human co-inhabitants that share a common habitat. This framework provides a multi-dimensional lens through which to perceive and value biological and cultural diversity. Rozzi uses a philosophical approach to examine and appreciate the diverse forms of knowledge—spanning biophysical, linguistic-cultural, and socio-political dimensions. Therefore, to uphold these processes, Rozzi's 3Hs framework proposes three important aspects of co-inhabitants.

(a) Co-inhabitants are subjects not objects
(b) Co-inhabitants, co-constitute their identities and their well-being by dwelling with other human and other -than-human beings,
(c) Co-inhabitants share habitats that they co-share through co-inhabitation relationships. They establish ecological relationships of complementarity and reciprocity that takes place through the exchanges of matter and energy(Klaver 2013). Therefore, the care and conservation of habits is the condition of possibility for the existence and well-being of the diverse co-inhabitants (Rozzi 2018).

Below, I introduce specific co-inhabitants that interact in the Lamkaang Naga practice of *Loh Kchet.*

28.6.1 Co-inhabitants in the Loh Kchet

The Lamkaang Naga Indigenous practice of Collective Interdependence through the performance of *Loh Kchet* is a conscious effort to ensure that everyone and every family participates in the agreed-upon tasks. *Loh Kchet* fosters collective participation, ensuring that no one is left behind in agricultural work, housebuilding, or any other essential tasks needed for the community's benefit—such as building roads, creating waterways, or cleaning communal spaces like the *diirum* pond, stream, or playground.

For these purposes to be fulfilled, it is essential that the habits of the collective continue, the habitat of the collective is protected, and that all co-inhabitants live in cordial reciprocity, in harmony with each other within their shared ecosystem. The practices of *Loh Kchet* align with the 3Hs (*Habits, Habitat, co-inhabitants*) model proposed by Rozzi (2018) in crucial aspects that can be cultivated.

According to Lamkaang worldview and Rozzi's biocultural ethics, human as well as other-than-human co-inhabitants are subjects, active agents in relationships of reciprocity. For example, Lamkaang's cosmogony acknowledges reciprocity and kinship relationships between humans and animals, even via intermarriages. In the *Thampthlum* story, Thampthlum marries a snake; in the *Penpen Yur* story, a woman marries a tiger; and in the *Dingkakuk* story, a man marries a bird. These stories illustrate that Lamkaangs have maintained close relationships with many of the co-inhabitants in their land.

These co-inhabitants—land, water, animals, birds, and many other species—are integral to their identity. They co-constitute the well-being of Lamkaang people within the shared habitat, which is inhabited by both human and non-human beings in ways that align with a biocultural ethics (*sensu* Rozzi 2018, p 39). The shared habitats are co-structured through reciprocal relationships (Yazzie and Baldy 2018). According to environmental philosopher Irene Klaver, co-inhabitants, along with their habitat and habits, must interact through exchanges of matter, energy, and processes in a nonlinear way of thinking. This interaction requires multidimensional thinking for the diverse and dynamic socio-ecosystems (Klaver 2013, 2018; Rozzi 2018). Despite the challenges posed by climate change, the Lamkaangs continue to maintain their interactions with non-human beings in the Eastern Lamkaang *Kleer Loon* areas, where they uphold reciprocal relationships. The land, water, animals, birds, and many other species are thriving together with these co-inhabitants as good relatives (Yazzie and Baldy 2018). In *Loh Kchet*, the community contributes to collective interdependence by:

(a) binding the community together in close-knit relationships,
(b) uplifting widows, widowers, and orphans to be on equal footing with other community members,
(c) addressing poverty within the community,
(d) allowing access to communal grain storage for those in need,
(e) showing kindness to other co-inhabitants by following strict rules, such as not fishing or hunting during mating seasons, not over-harvesting plants, and using water only for necessary purposes without overusing it.

This care for non-human beings allows them to thrive in the face of global climate change. The responsibility of care, as good relatives/kin, is essential for maintaining a symbiotic relationship where all co-inhabitants can flourish together.

28.6.2 Loh Kchet*: Collective Interdependent Habit*

The Lamkaang Naga Indigenous practice of *Loh Kchet* embodies collective action that serves both the community and the individual. These reciprocal activities, ranging from agriculture, fishing, and hunting to housebuilding, promote collective interdependence. At the heart of *Loh Kchet* lies the deep bond between the community and their land, as well as their relationship with other-than-human beings. This practice is rooted in the spirit of oneness, where people are deeply connected to the land and believe that they come from it (Burkhart 2019; Whyte 2018). The Lamkaang practice of *Loh Kchet* is deeply spiritual, emphasizing reciprocity, respect, and responsibility. This responsibility stems from the respect Lamkaangs have for one another and their habitat. It is through *Loh Kchet* that a widow or orphan can survive without experiencing poverty, as all members of the community are considered equal and part of a larger Lamkaang family. Relationality extends beyond humans to their co-inhabitants, other-than-human beings in the environment. In Lamkaang creation stories, origin stories, and folktales, humans maintain cordial relationships with non-human entities, such as snakes, tigers, lions, and birds.

These stories depict a shared village habitat, where humans live harmoniously with various animals and birds in the environment. Such relationality allows for deep knowledge of their co-inhabitants, not just humans, but all organisms living on the land. Speaking to and listening to the land remains a vital practice. *Loh Kchet*'s collective interdependence closely mirrors the concept of collective continuance. As Whyte (2017, p 355) suggests, collective continuance refers to how community members maintain their cultural integrity, health, economic viability, and political order for the future. While Lamkaangs may not be aware of Indigenous peoples of North America, their practices share similar life ways. The continuity of knowledge and skills is passed from one generation to the next through experiential learning, oral traditions, songs, and stories. Until the twentieth century, Lamkaang language was purely oral, and written documentation of the people, history, and language was minimal. These cultural practices are rooted in the oral traditions handed down over generations. Indigenous knowledge is broadly defined as the knowledge accumulated over generations in a specific environment (Ryser 2011).

The Lamkaang people's biocultural and ecosystem knowledge encompasses local technologies, innovations, practices, skills, and beliefs that unite the community in conserving forest resources and preserving cultural values. For instance, *Pkhuui Loon* in Challong village is protected as a sacred forest/mountain due to the *Trlei Kdet* story in Lamkaang *Paomin*. Similarly, *Vaangou Paa Loon* in Charangching Khuibung village is considered sacred due to the belief that an Indigenous deity resides there. Sacred sites like these play an essential role in linking people to their territories within the Indigenous culture. Indigenous traditional ecological knowledge and cultural values have developed over time through direct contact with and a strong relationship with the land and local environment (Whyte 2018; Ives et al. 2018; Williams 2023). Therefore, Indigenous knowledge is not limited to practices, teachings, skills, and innovations. As WIPO (2012)[1] acknowledges, Indigenous traditional knowledge systems are intrinsically tied to traditional lifestyles, biodiversity, and natural resources.

The last time I participated in *Loh Kchet* was in 2017, during rice transplantation in a wet paddy field. About 30 people worked together from morning until late evening, completing the transplantation process. This experience instills self-confidence, as you realize you are not alone; the community is collectively with you and behind you, giving you a sense of fearlessness. It fosters resilience in individuals and as a collective, removing feelings of deprivation and building confidence. In *Loh Kchet* as collective interdependence, the labor contributed can either be voluntary or reciprocal. The idea is that you gather a group of people to help you, and in return, you support them when they need assistance, whether the next

[1] World Intellectual Property Organization, https://www.wipo.int/tk/en/tk/

day or at a later stage. "Today for you, tomorrow for me." *Loh Kchet* nurtures a spirit of self-sufficiency both in individuals and as a collective, with a strong foundation of interdependence.

My recent participation in *Loh Kchet* allowed me to experience a new level of meaning, providing an opportunity for more intimate discussions on environmental injustices, such as the impact of land and resource exploitation and modern development projects that displace Indigenous peoples from their ancestral lands. Lastly, I observed a shift: in some areas of work, the spirit of full interdependence is being replaced by a bond based on monetary exchange, as people increasingly seek to buy labor instead of engaging in collective efforts.

28.7 Concluding Remarks

In this chap. I have discussed the Lamkaang Naga traditional healing systems and *Loh Kchet*: Land-based healing systems and *Loh Kchet* practices are deeply intertwined with the land, and are embodied in Lamkaang people's everyday life. This collective interdependence practice is the source of their origin and identity of Lamkaang people—their habitat, and their life habits. It expresses a deep relationship of reciprocity and kinship with the land (Whyte 2018, Burkhart 2019, Yazzie and Baldy 2018, Estes 2019:67) and its co-inhabitants (Rozzi 2013).

Before anything can be done on the land, the Lamkaangs seek the blessing and permission from the land, where a land blessing ritual is performed by a *Thiimpii/ Thiimpuu* (Priestess/Priest), a ritual known as *Ardii Kthung*. This process has many elements that are involved. Most importantly, listening to the land. The Lamkaang Naga ritual of listening to the *Paam Kruung* is of utmost importance as this can mean a life-or-death situation for villagers, who may suffer dire consequences if the listening goes wrong, or the communication is not successful. Listening through dreams and the interpretation of these ritual signs requires a special connection to the land. The *Thiimpii/ Thiimpuu* must be *Plung-Kchiir-Plung-Ksheeng* ("pure hearted") and *Khut-Ksen-Khut-Ktobor Kma* ("clean hands without blood," i.e., one who does not murder). Only then will the land speak to the *Thiimpii/Thiimpuu.*

Loh Kchet's collective interdependence involves deep connections between human institutions, such as the Lamkaang *Noopaa* ways of life, rituals, ceremonies as well as the habitats and co-inhabitants—other-than-human beings. This collective interdependence contributes to the continuity of these practices, grounded in the trust and resilience it fosters within the community. As demonstrated through the practice of *Loh Kchet*, this Indigenous paradigm highlights collective interdependence within an expansive network of relationships. These include the self, fellow humans and neighbors, ancestors, future generations, their habitat, co-inhabitants, and the cosmos.

Lamkaang traditional healing systems and Loh Kchet reveal dimensions of Indigenous thought and practices that are often overlooked, namely Indigenous views on self-cultivation and spiritual transformation. Lamkaang practices those of

Mishkos Kenomagwen Indigenous people in North America, who also ask permission to plants to show respect for their personhood (Kimmerer 2013). Another Indigenous group in North America, the Ojibwa people also have practices of consent, recognition, and reciprocity (Simpson 2013). Accordingly, patterns of intimate relationships of reciprocity with the land can be found among Lamkaang Naga as well as with other Indigenous people across the world (Yazzie and Baldy 2018; Rozzi 2015b; Ojeda et al. 2022, 2023; Hu and Yang 2026; Song 2026; Tsuji and Johnson 2026; Tuck and Yang 2012; Zhang and Xu 2026).

Loh Kchet creates a shared capacity that plays a vital role in socio-ecological systems, involving humans and non-human beings such as animals, plants, spiritual entities, and the landscape, including climatic regions. Such interconnectedness enhances Indigenous peoples' resilience and adaptation to climate change. Furthermore, it offers a mechanism for adapting to environmental changes caused by development pressures and economic policies that impact both local and global communities. In this global context, I offer the comparative ecological and ethical approach used to examine Lamkaang Naga traditional healing systems, as a biocultural approach that could be used to better understand and value traditional ecological knowledge systems of other indigenous communities. This might be a critical step to identify existing sustainable and just forms of inter-cultural and inter-species co-inhabitation.

Acknowledgement The author thanks Ricardo Rozzi, Terrance Caviness, and Danqiong Zhu for their comments on the manuscript.

References

Amano T, Sandel B, Eager H, Bulteau E, Svenning JC, Dalsgaard B et al (2014) Global distribution and drivers of language extinction risk. Proc R Soc B Biol Sci 281(1793):20141574

Andreozzi M (2025) Reimagining ethics. Non-anthropocentric perspectives on morality, Ecology and ethics, vol 7. Springer, Cham

Argumedo A (2008) The Potato Park, Peru: conserving agrobiodiversity in an Andean indigenous biocultural heritage area. Protect Landsc Agrobiodivers Values 1:45–58

Baruah S (2004) Between South and Southeast Asia: Northeast India and the look east policy, CENISEAS papers number 4. Centre for Northeast India, South and Southeast Asian Studies, Guwahati

Bhatt MS, Nazrul Islam AKM (2008) Biocultural diversity in India: prospects and challenges. ICFAI J Environ Econ 6(3):46

Burkhart B (2019) Indigenizing philosophy through the land: a trickster methodology for decolonizing environmental ethics and indigenous futures. Michigan State University Press, East Lansing

Chakravarty S, Suresh CP, Anju P, Shukla G (2012) North-East India, the geographical gateway of India's phytodiversity. Indian Forester 138:702–709

Das SK (2010) India's look east policy: imagining a new geography of India's northeast. India Q 66(4):343–358. https://doi.org/10.1177/097492841006600040

Díaz-Reviriego I, Hanspach J, Torralba M, Ortiz-Przychodzka S, Frias CB et al (2024) Appraising biocultural approaches to sustainability in the scientific literature in Spanish. Ambio 53(4): 499–516
Elands BH, Vierikko K, Andersson E, Fischer LK, Goncalves P et al (2019) Biocultural diversity: a novel concept to assess human-nature interrelations, nature conservation and stewardship in cities. Urban For Urban Green 40:29–34
Estes N (2019) Our history is the future: standing rock versus the Dakota access pipeline, and the long tradition of indigenous resistance. Verso Books, London/New York
Gadgil M (1987) Diversity: cultural and biological. Trends Ecol Evol 2(12):369–373
Grant C (2012) Analogies and links between cultural and biological diversity. J Cult Herit Manag Sustain Dev 2(2):153–163
Haokip P (2011) The languages of Manipur: a case study of the Kuki-Chin languages. Linguist Tibeto-Burman Area 34(1):85–118
Haokip T (2015) India's look east policy: prospects and challenges for Northeast India. Stud Indian Polit 3(2):1–14
Hatty MA, Smith LDG, Goodwin D, Mavondo FT (2020) The CN-12: a brief, multidimensional connection with nature instrument. Front Psychol 11(1566):1–14. https://doi.org/10.3389/fpsyg.2020.01566
Hegde P, James G (2019) Challenging biocultural homogenization: experiences of the Chipko and Appiko. In: From biocultural homogenization to biocultural conservation, vol 3. Springer, Cham, p 427
Horam M (1975) Naga polity. B.R. Publishing Co., New Delhi
Hu J, Yang P (2026) A biocultural encounter with the four gentlemen: art education in Chinese ecological and aesthetic traditions. In: Zhu D, Rozzi R, Tsuji R, Johnson B, Caviness T, Castro-Jorquera C, Xu L, Zhang L (eds) Linking arts with biocultural conservation, restoration, and communication, Ecology and ethics, vol 8. Springer, Cham, pp 111–129. https://doi.org/10.1007/978-3-032-04830-1_7
Ives CD, Abson DJ, von Wehrden H, Dorninger H, Klaniecki K, Fischer J (2018) Reconnecting with nature for sustainability. Sustain Sci 13(5):1–9. https://doi.org/10.1007/s11625-018-0542-9
Ives CD, Giusti M, Fischer J, Abson DJ, Klaniecki K et al (2017) Human–nature connection: a multidisciplinary review. Current opinion in environmental sustainability, 26:106-113
Johnson S (ed) (2012) Indigenous Knowledge. The White Horse Press, Cambridge. http://www.environmentandsociety.org/node/3641
Karlsson BG (2011) Unruly Hills: a political ecology of India's northeast (NED-new edition, 1). Berghahn Books. http://www.jstor.org/stable/j.ctt9qcmsq
Karlsson BG (2022) Destroying one's own home: resource frontiers and indigenous. Governance in Northeast India. Contemp South Asia 30:298–300
Khular S (2018) The status of Lamkaang Naga women, the place of women in Naga Society, Naga Women's Union (NWU) International Work Group for Indigenous Affairs (IWGIA) Henry Martyn Institute (HMI), p 80
Kikon D (2005) Engaging Naga nationalism: can democracy function in militarised societies? Econ Polit Wkly 40(26):2833–2837
Kimmerer RW (2013) Braiding sweet grass. Indigenous wisdom, scientific knowledge, and the teachings of the plants, Milkweed edn, Minnesota
Klaver IJ (2013) Environment imagination situation. In: Rozzi R, Pickett STA, Palmer C et al (eds) Linking ecology and ethics for a changing world: values, philosophy, and action, Ecology, and ethics, vol 1. Springer, Dordrecht, pp 85–105. https://doi.org/10.1007/978-94-007-7470-4_7
Klaver IJ (2018) Reclaiming rivers from homogenization: meandering and riverspheres. In: Rozzi R, May RH Jr, Chapin FS III, Massardo F, Gavin M, Klaver I, Pauchard A, Núñez MA, Simberloff D (eds) From biocultural homogenization to biocultural conservation, Ecology, and ethics, vol 3. Springer, Dordrecht, pp 49–69. https://doi.org/10.1007/978-3-319-99513-7_3

Longchar WA (1999) Dancing with the land: significance of land for doing tribal theology. In: Longchar AW, Davis LE (eds) Doing theology with tribal resources: context and perspective. Tribal Study Centre, Jorhat, pp 16–28

Longchar W (2000) The tribal religious traditions in Northeast India: an introduction. Eastern Theological College, Jorhat

Maffi L (1998) Language: a resource for nature, nature, and resource. The UNESCO J Environ Nat Resour 34(4):12–21

Maffi L (ed) (2001) On biocultural diversity. Linking language, knowledge, and the environment. Smithsonian Institution Press, Washington, DC

Maffi L (2005) Linguistic, cultural, and biological diversity. Annu Rev Anthropol 34:599–617

Matsuura K (2008) The international year of languages 01. Director General of UNESCO. Message for international year of languages

McElwee P, Fernández-Llamazares Á, Aumeeruddy-Thomas Y, Babai D, Bates P, Galvin K et al (2020) Working with indigenous and local knowledge (ILK) in large-scale ecological assessments: reviewing the experience of the IPBES global assessment. J Appl Ecol 57(9):1666–1676

Merçon J, Vetter S, Tengö M, Cocks M, Balvanera P, Rosell JA, Ayala-Orozco B (2019) From local landscapes to international policy: contributions of the biocultural paradigm to global sustainability. Glob Sustain 2(e7):1–11. https://doi.org/10.1017/sus.2019.4

Mir AH, Sarma K, Upadhaya K (2022) Assessing the effectiveness of community managed forests for plant diversity conservation in Meghalaya. Northeast India Plant Divers 44(3):243–254

Moore JL, Manne L, Brooks T, Burgess ND, Davies R, Rahbek C et al (2002) The distribution of cultural and biological diversity in Africa. Proc R Soc London Ser B Biol Sci 269(1501): 1645–1653

Nettle D, Romaine S (2000) Vanishing voices: the extinction of the world's languages. Oxford University Press, Oxford

Ojeda J, Salomon AK, Rowe JK, Ban NC (2022) Reciprocal contributions between people and nature: a conceptual intervention. Bioscience 72(10):952–962

Ojeda J, Tauro A, Contador T, González-Calderón J, Malebran J et al (2023) "Pay attention, dive with eyes wide open": a field environmental philosophy activity to foster reciprocity between people and nature. In: Rozzi R, Tauro A, Avriel-Avni N, Wright T, May RH Jr (eds) Field environmental philosophy: education for biocultural conservation, Ecology and ethics, vol 5. Springer, Cham, pp 87–100. https://doi.org/10.1007/978-3-031-23368-5_6

Otamendi-Urroz I, Quintas-Soriano C, Hanspach J, Requena-Mullor JM, Lagies AS, Castro AJ (2025) Exploring biocultural diversity: a systematic analysis and refined classification to inform decisions on conservation and sustainability. Ambio, pp 1–17

Pan SY, Litscher G, Gao SH, Zhou SF, Yu ZL et al (2014) Historical perspective of traditional indigenous medical practices: the current renaissance and conservation of herbal resources. Evid Based Complement Alternat Med 1:525340

Pecl GT, Araújo MB, Bell JD, Blanchard J, Bonebrake TC et al (2017) Biodiversity redistribution under climate change: impacts on ecosystems and human Well-being. Science 355:eaai9214 10.1126/science.aai9214

Prescott SL, Logan AC, Bristow J, Rozzi R, Moodie R et al (2022) Exiting the anthropocene: achieving personal and planetary health in the 21st century. Allergy 77(12):3498–3512. https://doi.org/10.1111/all.15419

Pretty J, Adams B, Berkes F, De Athayde SF, Dudley N et al (2009) The intersections of biological diversity and cultural diversity: towards integration. Conserv Soc 7(2):100–112

Pungetti G (2013) Biocultural diversity for sustainable ecological, cultural and sacred landscapes: the biocultural landscape approach. In: Fu B, Jones K (eds) Landscape ecology for sustainable environment and culture. Springer, Dordrecht, pp 55–76. https://doi.org/10.1007/978-94-007-6530-6_4

Rapport D, Maffi L (2010) The dual erosion of biological and cultural diversity: implications for the health of ecocultural systems. In: Nature and culture. Routledge, pp 103–120

Restall B, Conrad E (2015) A literature review of connectedness to nature and its potential for environmental management. J Environ Manag 159:264–278. https://doi.org/10.1016/j.jenvman.2015.05.022

Rotherham ID (ed) (2013) Cultural severance and the environment: the ending of traditional and customary practice on commons and landscapes managed in common. Springer, Dordrecht

Rotherham ID (2014) Eco-history: a short history of conservation and biodiversity. The White Horse Press, Cambridge

Rozzi R (2001) Éticas ambientales latinoamericanas: raíces y ramas. In: Primack R, Rozzi R, Feinsinger P et al (eds) Fundamentos de Conservación Biológica: perspectivas Latinoamericanas. Fondo de Cultura Económica, México, pp 311–362

Rozzi R (2003) Biodiversity and social wellbeing in South America. In: Encyclopedia of Life Support Systems (EOLSS). UNESCO-EOLSS. http://www.eolss.net, http://greenplanet.eolss.net/EolssLogn/mss/C14/E1-37/E1-37-04/E1-37-04-10/E1-37-04-10-TXT.aspx

Rozzi R (2010) Field environmental philosophy: regaining an understanding of the inextricable links between the regional habitats, the inhabitants and their habits. Dialogue Univ 21(11–12): 85–109. https://doi.org/10.5840/du201020111/127

Rozzi R (2012) Biocultural ethics: the vital links between the inhabitants, their habits and regional habitats. Environ Ethics 34(1):27–50. https://doi.org/10.5840/enviroethics20123414

Rozzi R (2013) Biocultural ethics: from biocultural homogenization toward biocultural conservation. In: Rozzi R, Pickett STA, Palmer C, Armesto JJ, Callicott JB (eds) Linking ecology and ethics for a changing world: values, philosophy, and action, Ecology, and ethics, vol 1. Springer, Dordrecht, pp 9–32. https://doi.org/10.1007/978-94-007-7470-4_2

Rozzi R (2015a) Implications of the biocultural ethic for earth stewardship. In: Rozzi R, Chapin FS III, Callicott JB, Pickett STA, Power ME, Armesto JJ, May RH Jr (eds) Earth stewardship: linking ecology and ethics in theory and practice, Ecology and ethics, vol 2. Springer, Cham, pp 113–136. https://doi.org/10.1007/978-3-319-12133-8_9

Rozzi R (2015b) Earth stewardship and the biocultural ethic: Latin American perspectives. In: Rozzi R, Chapin FS, Callicott JB, Pickett STA, Power ME, Armesto JJ, May RH Jr (eds) Earth stewardship: linking ecology and ethics in theory and practice, Ecology and ethics, vol 2. Springer, Cham, pp 87–112. https://doi.org/10.1007/978-3-319-12133-8_8

Rozzi R (2018) Biocultural homogenization: a wicked problem in the Anthropocene. In: From biocultural homogenization to biocultural conservation, Ecology, and ethics, vol 3. Springer, Cham, pp 21–48. https://doi.org/10.1007/978-3-319-99513-7_2

Rozzi R, Massardo F, Anderson C, McGehee S, Clark G, Egli G, Ramilo E, Calderón U, Calderón C, Aillapan L, Zárraga C (2010) Multi-ethnic bird guide of the sub-Antarctic forests of South America. UNT Press, Denton

Rozzi R, Armesto JJ, Gutiérrez J, Massardo F, Likens G, Anderson CB, Poole A, Moses K, Hargrove G, Mansilla A, Kennedy JH, Willson M, Jax K, Jones C, Callicott JB, Kalin MT (2012) Integrating ecology and environmental ethics: earth stewardship in the southern end of the Americas. Bioscience 62(3):226–236. https://doi.org/10.1525/bio.2012.62.3.4

Ryser RC (2011) Indigenous peoples and traditional knowledge, Berkshire Encyclopedia of sustainability 5/10, Ecosystem management and sustainability. Center for World Indigenous Studies. Berkshire Publishing Company, Great Barrington

Sankhil S (2015) A short history of the Lamkang tribe. City Printers, Chumukedima, Dimapur, Nagaland,India

Shamsuddin T, Alam MS, Junaid M, Akter R, Hosen S, Ferdousy S, Mouri NJ (2021) *Adhatoda vasica* (Nees.): a review on its botany, traditional uses, phytochemistry, pharmacological activities and toxicity. Mini Rev Med Chem 21(14):1925–1964

Shimray UA (2001) Ethnicity and socio-political assertion: the Manipur experience. Econ Polit Wkly 36:3674–3677. https://doi.org/10.2307/4411161. https://www.epw.in/journal/2001/39/commentary/ethnicity-and-socio-political-assertion.html

Shimray UA (2007) Ecology and economic system: a case study of the Naga community. Regency Publication, New Delhi

Shiva V (1992) Women's indigenous knowledge and biodiversity conservation. India Int Centre Q 19(1/2):205–214

Shoaib A (2021) A systematic ethnobotanical review of Adhatoda vasica (L.). Nees Cell Mol Biol 67(4):248–263

Simpson LB (2013) As we have always done, Indigenous freedom through radical resistance. Wilfrid Laurier University Press, Waterloo

Singh RK, Pretty J, Pilgrim S (2010) Traditional knowledge and biocultural diversity: learning from tribal communities for sustainable development in Northeast India. J Environ Plan Manag 53(4): 511–533

Singh RK, Singh A, Ksherchokpa L, Rallen O, Taniang B et al (2021) Grassroots approaches for sustaining biocultural diversity and livelihood security: insights from Indian eastern Himalaya. Environ Manag 68:17–37

Song Y (2026) Interspecies co-inhabitation: a dialogue between Chinese philosophical traditions and biocultural ethics. In: Zhu D, Rozzi R, Tsuji R, Johnson B, Caviness T, Castro-Jorquera C, Xu L, Zhang L (eds) Linking arts with biocultural conservation, restoration, and communication, Ecology and ethics, vol 8. Springer, Cham, pp 95–109. https://doi.org/10.1007/978-3-032-04830-1_6

Tam KP (2013) Concepts and measures related to connection to nature: similarities and differences. J Environ Psychol 34:64–78. https://doi.org/10.1016/j.jenvp.2013.01.004

Tom MN, Sumida-Huaman E, McCarty TL (2019) Indigenous knowledges as vital contributions to sustainability. Int Rev Educ 65:1–18

Tsuji R, Johnson B (2026) Biocultural ontologies, art, and conservation. In: Zhu D, Rozzi R, Tsuji R, Johnson B, Caviness T, Castro-Jorquera C, Xu L, Zhang L (eds) Linking arts with biocultural conservation, restoration, and communication, Ecology and ethics, vol 8. Springer, Cham, pp 61–77. https://doi.org/10.1007/978-3-032-04830-1_4

Tuck E, Yang KW (2012) Decolonization is not a metaphor, in decolonization: indigeneity. Educ Soc 1(1):1–40

Varah F (2013) Situating the humans relationship with nature in the Tangkhul Naga's. Lifeworld J Human Ecol 41(3):247–254

Vashum Y (1998) Theology of land: a Naga Perspect. In: Longchar AW, Vashum Y (eds) The tribal worldview and ecology. Tribal Study Centre, Jorhat, p 71

Vashum R (2005) Nagas' right to self-determination: an anthropological-historical perspective. Mittal Publications, New Delhi

Whyte KP (2017) Food sovereignty, justice and indigenous peoples: an essay on settler colonialism and collective continuance. In: Barnhill A, Doggett T, Egan A (eds) Oxford handbook on food ethics. Oxford University Press, Oxford, pp 345–366

Whyte KP (2018) Settler colonialism, ecology, and environmental injustice. Environ Soc 9(1): 125–144

Williams L (2023) From indigenous philosophy in environmental education to indigenous planetary futures: what would it take? Aust J Environ Educ 39(3):320–335. https://doi.org/10.1017/aee.2023.23

World Intellectual Property Organization (WIPO)(2012) World Intellectual Property Indicators. Accessed through: https://www.wipo.int/edocs/pubdocs/en/intproperty/941/wipo_pub_941_2012.pdf

Yazzie M, Baldy CR (2018) Introduction: indigenous peoples and the politics of water. Decolonization: indigeneity. Educ Soc 7(1):1–18

Yhome K (2015a) From 'look East'to 'act east': what it means for India's Northeast. Eastern Q 11(1–2):18–29

Yhome K (2015b) The Burma roads: India's search for connectivity through Myanmar. Asian Surv 55(6):1217–1240. https://doi.org/10.1525/as.2015.55.6.1217

Yumnam J, Koijam P (2014) People assessment of Mapithel. Centre for Research and Advocacy, Manipur, Imphal

Zhang S, Xu L (2026) A biocultural sketch of Cao-guo (Lanxangia Tsao-ko). In: Zhu D, Rozzi R, Tsuji R, Johnson B, Caviness T, Castro-Jorquera C, Xu L, Zhang L (eds) Linking arts with biocultural conservation, restoration, and communication, Ecology and ethics, vol 8. Springer, Cham, pp 349–365. https://doi.org/10.1007/978-3-032-04830-1_22

Zhu D (2023) Bridge the channel, enhance the inclusivity: a comparison between flagship species-centered and moss-centered conservation in Chile and China. In: Rozzi R, Tauro A, Wright T, Avriel-Avni N, May RH Jr (eds) Field environmental philosophy: education for biocultural conservation, Ecology and ethics, vol 5. Springer, Dordrecht, pp 457–482. https://doi.org/10.1007/978-3-031-23368-5_28

Ziipao RR (2018) Look/act east policy, roads, and market infrastructure in north-east India. Strat Anal 42(5):476–489. https://doi.org/10.1080/09700161.2018.1523082

Zylstra MJ, Knight AT, Esler KJ, Le Grange LLL (2014) Connectedness as a core conservation concern: an interdisciplinary review of theory and a call for practice. Springer Sci Rev 2:119–143. https://doi.org/10.1007/s40362-014-0021-3

Chapter 29
Carekeepers

Tusha Yakovleva

Abstract This chapter explores intergenerational land-based knowledge and cultural continuity through personal reflections on ethnobotanical practices in post-Soviet Russia and as an emigrant living on Turtle Island. Drawing on childhood experiences of gathering wild food with family, it highlights how tending to plants, storytelling, and ecological reverence transmit biocultural values, strengthening reciprocal relationship to place.

Keywords Mushroom · Care · Reciprocity · Remembering

Born in spring in Moscow, I met my first mushrooms and berries the following summer, strapped to my mom's back, taking in the soft acidic air of dark coniferous forests, the humpbacked glacial erratics wrapped in lichens, the smoky pastels of heather. We unpacked a car trunk full of baskets (birch bark and willow) and usually returned them at sunset filled with fragrant forest gifts.

My parents and four grandparents helped me learn. Before I could walk, they placed mushrooms into my palms and I squeezed them until the pores united into a rubbery mass, until the gills papered the ground like fish scales. They lowered me onto the thick damp down of sphagnum and pointed to luminous lingonberries hovering just above the mosses and pearlescent, yeasty bilberries suspended above my small body. They taught me the smells of dinner mushrooms and those that were meant to remain in the forest, for the forest. They taught me the Russian language which in turn taught me too, giving my memory strength through story. Some names came with a map to their arboreal relations: *podberezoviki*—under birch mushroom, *podosinoviki*—under aspen mushroom, *mohovik*—by the mosses mushroom; other names came with a recipe: *siroezhki*—eat raw mushroom, *broosnika*—ripe when red berry. They introduced me to the carekeepers: Leshy—keeper of the trees;

T. Yakovleva (✉)
Center for Native Peoples and Environment, The State University of New York College of Environmental Science and Forestry, Syracuse, NY, USA
e-mail: teyakovl@esf.edu

D. Zhu et al. (eds.), *Linking Arts with Biocultural Conservation, Restoration, and Communication*, Ecology and Ethics 8,
https://doi.org/10.1007/978-3-032-04830-1_29

sphagnum steward—Vodyanoy; Baba Yaga—ecotone queen, and other biocultural keystone creatures, the scary and wise ones, who warrant respect for the land and know all the wild foods and medicines.

We lived in the city but like many Russian families, sought places where wild edible plants and mushrooms grow. Harvesting was a joyful time of togetherness for three generations of my small family as well as time to be together in intimate reverence of plants and animals in our midst. We gathered in forests that had endured heavy disturbances; in our meanders we encountered post-Soviet industrial relics, sudden new suburbs, scarred over trenches left from past wars. The days spent in soil were accompanied by endless stories: we remembered our ancestors' time with plants and mushrooms, retold Slavic folktales of foraging and gardening, sang and composed poems dedicated to the inhabitants of northern boreal forests, seen and unseen.

One grandpa relished the hunt more than any other aspect of foraging. He was in a lifelong competition against himself. I remember him filling an entire bathtub with *gruzdi* mushrooms (to lacto-ferment them) and the palpability of his contentment washing over the entire house, albeit for a moment. The other grandpa was an adventurous eater. Our relatives made fun of him for pouring over field guides for days before triumphantly emerging from the lowland forest behind his house with a basketful of species he was to try for the first time. He lived through the 900-day blockade of Leningrad as a child, then lived through the rest of life with an incurable sense of food insecurity. His spouse, my grandma, worriedly blamed him for gathering *poganki* (the "bad" mushrooms), until he eventually wore her questions down with scientific evidence and she would fry up the unknown mushrooms with ample salt and butter for him, for him only. She was in it for the cooking, I think. Other grandma liked to see everyone getting along and chances of that were highest in the forest. She too grew up in war and was sent into Siberian forests along with other children to gather and fill trains with mushrooms that then whisked them back to the hungry cities, her part in ameliorating the national wartime food shortage. My parents? I am not sure what about wild mushrooms, berries, greens drew them in. They became adults in the sunrise of capitalism, nicknamed their baby (me)-Perestroika, and were already exhausted from overwork. I think they liked the slowness and the good air.

The two branches of my family were in radical opposition: one thrived under the Soviet regime, the other ushered in democracy. One grandfather described himself as a devout atheist, the other committed to Orthodox Christianity. Both branches, however, considered themselves tremendously Russian. They told the story of our family as a Russian family. But, over the years, anecdotes and relics cohered into a different origin story: the cosmopolitan Moscow family was a thinly plastered facade over Ukrainian Jews, Poles, Armenians, Georgians and other peoples who left their homelands to become Russian. My relatives, like so many families from that part of the world and at that time, lived in an internal displacement of sorts within the place they considered their homeland. The paths into the past had grown over and the future taking shape in front of them didn't look at all like they imagined.

I think my relatives needed a common root, to pull on something strong to feel grounded in a culture, in family, in place. The strongest one they found was in holding onto a few traditional land care practices (foraging, planting, pickling…) and protecting the stories that went along with those practices. An essential part of preserving such practices, of course, was in passing them on to the next generation and, in my family, I happened to be the sole recipient of these ways of knowing.

Only after many years of living far from my relations and homeland have I realized that one true protagonist of my family's biocultural pathways is hunger. My grandparents lived through a great food shortage and so, taught my parents about the land. My parents grew up with an influx of "western" abundance, excitedly witnessing the opening of the country's first McDonald's, and also with great uncertainty about the future, having witnessed the tumbling of the Soviet empire. So, my parents taught me. Our intergenerational hunger spurred renewed respect for old knowledge, renewed motivation for protecting ecosystems and biodiverse lands, reverence for wild food in one's surroundings, and an urge to pass on cultural knowledge as biocultural inheritance.

In my life, hunger continues to wield its power. As an emigrant on Turtle Island, engaging with plants with whom I share an ecosystem is the best way I know to feel safe belonging. As someone who has been accompanied by the ecological and social threat of climate change through adult life, I've learned that tending to edible and medicinal plants, getting to know the weeds, gardening, are powerful ways to help others remember their relationship with land and, concurrently, ensure that good food and medicine continues to be available in my communities, human and otherwise.

My relatives and ancestors found ways to impart lessons of land care, teaching me in equal parts through practice and reverence for story. They remembered Baba Yaga, Leshy, Vodyanoy and other Slavic ecoregion guardians who care for land by guiding people toward respect and reciprocity. They were the first to teach me about honorable engagement with land, by giving voice to these biocultural creatures, repeating their stories while we picked berries. Stories of beings who are sometimes angry, sometimes content, sometimes helpful, on occasion dangerous; who ask for certain treatment in exchange for maintaining peace and balance; who remind that knowing one's right place within a given community of diverse species isn't always obvious, but upholding one's responsibilities to the earth is; who are neither deities nor gods and who reveal that the world is inspirited. These stories have stayed with me through all the forests, fields, swamps I've had the honor of walking into. It's my responsibility to carry these lessons forward no matter where I make home.

There is much Euro-Asian boreal forest heritage strewn across Turtle Island and so much hunger for finding cultural continuity. Slavic ecological folktales, rooted in a reverence for Mother Earth, can light the way for those who share ancestral paths to piece together some of the stories that have long sustained our families and the lands they call home, and to find a bit of home in one another. There are lessons about respectful engagements, about mutual care in biocultural stories that apply in any ecosystem. They are cultural reminders about how to interact with gratitude and reciprocity, about the aliveness of all, about caring harvests embedded within traditional seasonal feasts, songs, the stories we tell our children, the superstitions we uphold, the medicines we honor. Remembering is carekeeping too.

Index

D. Zhu et al. (eds.), *Linking Arts with Biocultural Conservation, Restoration, and Communication*, Ecology and Ethics 8,
https://doi.org/10.1007/978-3-032-04830-1

MIX
Papier aus verantwortungsvollen Quellen
Paper from responsible sources
FSC® C105338

If you have any concerns about our products,
you can contact us on
ProductSafety@springernature.com

In case Publisher is established outside the EU,
the EU authorized representative is:
Springer Nature Customer Service Center GmbH
Europaplatz 3, 69115 Heidelberg, Germany

Printed by Libri Plureos GmbH
in Hamburg, Germany